ANNALS OF
THE NEW YORK ACADEMY
OF SCIENCES

Volume 1011

EDITORIAL STAFF

Director, Publishing and New Media
SARAH GREENE

Managing Editor
JUSTINE CULLINAN

Associate Editor
MARION L. GARRY

The New York Academy of Sciences
2 East 63rd Street
New York, New York 10021

THE NEW YORK ACADEMY OF SCIENCES
(Founded in 1817)

BOARD OF GOVERNORS, September 2003 – September 2004

TORSTEN N. WIESEL, *Chairman of the Board*
GERALD D. FISCHBACH, *Vice Chairman*
JOHN T. MORGAN, *Treasurer*
ELLIS RUBINSTEIN, *Chief Executive Officer* [ex officio]

Honorary Life Governors
WILLIAM T. GOLDEN   JOSHUA LEDERBERG

Governors

KAREN E. BURKE	PETER B. CORR	R. BRIAN FERGUSON
RONALD L. GRAHAM	MARNIE IMHOFF	WENDY EVANS JOSEPH
JACQUELINE LEO	RODERT W. LUCKY	PAUL MARKS
BRUCE McEWEN	RONAY MENSCHEL	JOHN F. NIBLACK
SANDRA PANEM	PETER RINGROSE	DAVID D. SABATINI
	DEBORAH WILEY	

VICTORIA BJORKLUND, *Counsel* [ex officio]   LARRY R. SMITH, *Secretary* [ex officio]

MITOCHONDRIAL PATHOGENESIS
FROM GENES AND APOPTOSIS TO AGING AND DISEASE

ANNALS OF THE NEW YORK ACADEMY OF SCIENCES
Volume 1011

MITOCHONDRIAL PATHOGENESIS FROM GENES AND APOPTOSIS TO AGING AND DISEASE

Edited by Hong Kyu Lee, Salvatore DiMauro, Masashi Tanaka, and Yau-Huei Wei

The New York Academy of Sciences
New York, New York
2004

Copyright © 2004 by the New York Academy of Sciences. All rights reserved. Under the provisions of the United States Copyright Act of 1976, individual readers of the Annals are permitted to make fair use of the material in them for teaching or research. Permission is granted to quote from the Annals provided that the customary acknowledgment is made of the source. Material in the Annals may be republished only by permission of the Academy. Address inquiries to the Permissions Department (editorial@nyas.org) at the New York Academy of Sciences.

Copying fees: *For each copy of an article made beyond the free copying permitted under Section 107 or 108 of the 1976 Copyright Act, a fee should be paid through the Copyright Clearance Center, Inc., 222 Rosewood Drive, Danvers, MA 01923 (www.copyright.com).*

♾ *The paper used in this publication meets the minimum requirements of the American National Standard for Information Sciences—Permanence of Paper for Printed Library Materials, ANSI Z39.48-1984.*

Library of Congress Cataloging-in-Publication Data

Asian Society for Mitochondrial Research and Medicine. Scientific Meeting (1st : 2003 : Seoul, Korea)
  Mitochondrial pathogenesis : from genes and apoptosis to aging and disease / edited by Hong Kyu Lee ... [et al.].
    p. ; cm. — (Annals of the New York Academy of Sciences ; v. 1011)
  The result of the First Scientific Meeting of the Asian Society for Mitochondrial Research and Medicine, held on Feb. 5–6, 2003 in Seoul, Korea.
  Includes bibliographical references and index.
  ISBN 1-57331-490-0 (cloth : alk. paper) — ISBN 1-57331-491-9 (pbk. : alk. paper) 1. Mitochondrial pathology—Congresses.
  [DNLM: 1. Mitochondrial Diseases—physiopathology—Congresses. 2. Cell Death— genetics—Congresses. WD 200.5.M6 A832m 2004] I. Lee, Hong Kyu. II. Title. III. Series.
  Q11.N5 vol. 1011
  [RB147.5]
  500 s—dc22
  [616/ .0

2004006924

GYAT / PCP
Printed in the United States of America
ISBN 1-57331-490-0 (cloth)
ISBN 1-57331-491-9 (paper)
ISSN 0077-8923

ANNALS OF THE NEW YORK ACADEMY OF SCIENCES

Volume 1011
April 2004

MITOCHONDRIAL PATHOGENESIS FROM GENES AND APOPTOSIS TO AGING AND DISEASE

Editors
HONG KYU LEE, SALVATORE DIMAURO, MASASHI TANAKA, AND YAU-HUEI WEI

This volume is the result of The First Scientific Meeting of the Asian Society for Mitochondrial Research and Medicine, held February 5–6, 2003, in Seoul, Korea.

CONTENTS

Preface. *By* HONG KYU LEE.	ix
Overview. *By* HONG KYU LEE	1

Part I. Human Evolutionary History and Mitochondrial Diseases

Mitochondrial Genome Single Nucleotide Polymorphisms and Their Phenotypes in the Japanese. *By* MASASHI TANAKA, TAKESHI TAKEYASU, NORIYUKI FUKU, GUO LI-JUN, AND MIYUKI KURATA	7
Mitochondrial DNA Variation in the Aboriginal Populations of the Altai-Baikal Region: Implications for the Genetic History of North Asia and America. *By* ILIA A. ZAKHAROV, MIROSLAVA V. DERENKO, BORIS A. MALIARCHUK, IRINA K. DAMBUEVA, CHODURAA M. DORZHU, AND SERGEY Y. RYCHKOV	21

Part II. Mitochondrial Function and Diseases

Mitochondrial ALDH2 Deficiency as an Oxidative Stress. *By* SHIGEO OHTA, IKUROH OHSAWA, KOUZIN KAMINO, FUJIKO ANDO, AND HIROSHI SHIMOKATA	36
Mitochondrial Reactive Oxygen Species Generation and Calcium Increase Induced by Visible Light in Astrocytes. *By* MEI-JIE JOU, SHUO-BIN JOU, MEI-JIN GUO, HONG-YUEH WU, AND TSUNG-I PENG	45

Radical Metabolism Is Partner to Energy Metabolism in Mitochondria. *By* JIAN-XING XU 57

Part III. Mitochondrial Transcription Factor

Mitochondrial Nucleoid and Transcription Factor A. *By* TOMOTAKE KANKI, HIROSHI NAKAYAMA, NARIE SASAKI, KOJI TAKIO, TANFIS ISTIAQ ALAM, NAOTAKA HAMASAKI, AND DONGCHON KANG 61

Regulation of Mitochondrial Transcription Factor A Expression by High Glucose. *By* YON SIK CHOI, KI-UP LEE, AND YOUNGMI KIM PAK 69

Regulation and Role of the Mitochondrial Transcription Factor in the Diabetic Rat Heart. *By* YOSHIHIKO NISHIO, AKIO KANAZAWA, YOSHIO NAGAI, HIDETOSHI INAGAKI, AND ATSUNORI KASHIWAGI 78

Part IV. Mechanism of Mitochondrial Damage

The Mitochondrial Production of Reactive Oxygen Species in Relation to Aging and Pathology. *By* MARIA LUISA GENOVA, MILENA MERLO PICH, ANDREA BERNACCHIA, CRISTINA BIANCHI, ANNALISA BIONDI, CARLA BOVINA, ANNA IDA FALASCA, GABRIELLA FORMIGGINI, GIOVANNA PARENTI CASTELLI, AND GIORGIO LENAZ 86

Biological Significance of the Defense Mechanisms against Oxidative Damage in Nucleic Acids Caused by Reactive Oxygen Species: From Mitochondria to Nuclei. *By* YUSAKU NAKABEPPU, DAISUKE TSUCHIMOTO, AKIMASA ICHINOE, MIZUKI OHNO, YASUHITO IDE, SEIKI HIRANO, DAISUKE YOSHIMURA, YOHEI TOMINAGA, MASATO FURUICHI, AND KUNIHIKO SAKUMI 101

Mitochondrial Swelling and Generation of Reactive Oxygen Species Induced by Photoirradiation Are Heterogeneously Distributed. *By* TSUNG-I PENG AND MEI-JIE JOU 112

Mitochondrial Dysfunction via Disruption of Complex II Activity during Iron Chelation–Induced Senescence-like Growth Arrest of Chang Cells. *By* YOUNG-SIL YOON, HYESEONG CHO, JAE-HO LEE, AND GYESOON YOON 123

Part V. Oxidative Stress and Apoptosis

Mitochondrial DNA Mutation and Depletion Increase the Susceptibility of Human Cells to Apoptosis. *By* CHUN-YI LIU, CHENG-FENG LEE, CHIUNG-HUI HONG, AND YAU-HUEI WEI 133

Resistance of ρ^0 Cells against Apoptosis. *By* MYUNG-SHIK LEE, JA-YOUNG KIM, AND SUN YOUNG PARK 146

Mitochondrial DNA 4,977-bp Deletion in Paired Oral Cancer and Precancerous Lesions Revealed by Laser Microdissection and Real-Time Quantitative PCR. *By* DAR-BIN SHIEH, WEN-PIN CHOU, YAU-HUEI WEI, TONG-YIU WONG, AND YING-TAI JIN 154

Role of Oxidative Stress in Pancreatic β-Cell Dysfunction. *By* YOSHITAKA KAJIMOTO AND HIDEAKI KANETO 168

Initiation of Apoptotic Signal by the Peroxidation of Cardiolipin of Mitochondria. *By* YASUHITO NAKAGAWA 177

Part VI. Mitochondria and Diabetes

Diabetes Mellitus with Mitochondrial Gene Mutations in Japan.
By SUSUMU SUZUKI... 185

Accumulation of Somatic Mutation in Mitochondrial DNA and Atherosclerosis in Diabetic Patients. *By* TAKASHI NOMIYAMA, YASUSHI TANAKA, LIANSHAN PIAO, NOBUTAKA HATTORI, HIROSHI UCHINO, HIROTAKA WATADA, RYUZO KAWAMORI, AND SHIGEO OHTA........ 193

Changes of Mitochondrial DNA Content in the Male Offspring of Protein-Malnourished Rats. *By* HYEONG KYU PARK, CHENG JI JIN, YOUNG MIN CHO, DO JOON PARK, CHAN SOO SHIN, KYONG SOO PARK, SEONG YEON KIM, BO YOUN CHO, AND HONG KYU LEE.................... 205

Part VII. Diagnosis and Treatment of Mitochondrial Diseases

Mitochondrial Encephalomyopathies: Diagnostic Approach. *By* SALVATORE DIMAURO, STACEY TAY, AND MICHELANGELO MANCUSO............ 217

Mitochondrial Encephalomyopathies: Therapeutic Approach. *By* SALVATORE DIMAURO, MICHELANGELO MANCUSO, AND ALI NAINI.............. 232

Comprehensive Molecular Diagnosis of Mitochondrial Disorders: Qualitative and Quantitative Approach. *By* LEE-JUN C. WONG................... 246

Molecular Pathogenetic Mechanism of Maternally Inherited Deafness.
By MIN-XIN GUAN... 259

Genetic and Functional Analysis of Mitochondrial DNA–Encoded Complex I Genes. *By* YIDONG BAI, PEIQING HU, JEONG SOON PARK, JIAN-HONG DENG, XIUFENG SONG, ANNE CHOMYN, TAKAO YAGI, AND GIUSEPPE ATTARDI... 272

Genome-Wide Analysis of Signal Transducers and Regulators of Mitochondrial Dysfunction in *Saccharomyces cerevisiae*. *By* KESHAV K. SINGH, ANNE KARIN RASMUSSEN, AND LENE JUEL RASMUSSEN............. 284

Part VIII. Short Communications

Enhanced Detection of Deleterious Mutations by TTGE Analysis of Mother and Child's DNA Side by Side. *By* HAEYOUNG KWON, DUAN JUN TAN, REN-KUI BAI, AND LEE-JUN C. WONG............................. 299

Quantitative PCR Analysis of Mitochondrial DNA Content in Patients with Mitochondrial Disease. *By* REN-KUI BAI, CHERNG-LIH PERNG, CHANG-HUNG HSU, AND LEE-JUN C. WONG........................ 304

Somatic Mitochondrial DNA Mutations in Oral Cancer of Betel Quid Chewers. *By* DUAN-JUN TAN, JULIA CHANG, WOAN-LING CHEN, LESLEY J. AGRESS, KUN-TU YEH, BAOTYAN WANG, AND LEE-JUN C. WONG.. 310

Association of the Mitochondrial DNA 16189 T to C Variant with Lacunar Cerebral Infarction: Evidence from a Hospital-Based Case-Control Study. *By* CHIA-WEI LIOU, TSU-KUNG LIN, FENG-MEI HUANG, TZU-LING CHEN, CHENG-FENG LEE, YAO-CHUNG CHUANG, TENG-YEOW TAN, KU-CHOU CHANG, AND YAU-HUEI WEI............ 317

Mechanisms of Cell Death Induced by Cadmium and Arsenic. *By* SHIRO JIMI, MASANOBU UCHIYAMA, AYA TAKAKI, JYUNJI SUZUMIYA, AND SYUJI HARA 325

Cadmium-Induced Nephropathy in Rats Is Mediated by Expression of Senescence-Associated Beta-Galactosidase and Accumulation of Mitochondrial DNA Deletion. *By* AYA TAKAKI, SHIRO JIMI, MASARU SEGAWA, AND HIROSHI IWASAKI 332

Investigation of Common Mitochondrial Point Mutations in Korea. *By* SEON-JOO KWON, SUNG-SUP PARK, JONG-MIN KIM, TAE-BEOM AHN, SEUNG HYUN KIM, JUHAN KIM, SUNG-HYUN LEE, CHOONG-KUN HA, MOO-YOUNG AHN, AND BEOM S. JEON 339

Leber's Hereditary Optic Neuropathy: The Spectrum of Mitochondrial DNA Mutations in Iranian Patients. *By* M. HOUSHMAND, F. SHARIFPANAH, A. TABASI, M.-H. SANATI, M. VAKILIAN, SH. LAVASANI, AND S. JOUGHEHDOUST 345

Index of Contributors 351

Financial assistance was received from:
- **BUKWANG PHARMACEUTICALS**
- **DALIM PHARMACEUTICALS**
- **MITOCON INC.**

The New York Academy of Sciences believes it has a responsibility to provide an open forum for discussion of scientific questions. The positions taken by the participants in the reported conferences are their own and not necessarily those of the Academy. The Academy has no intent to influence legislation by providing such forums.

Preface

HONG KYU LEE

Department of Internal Medicine, Seoul National University College of Medicine, 28 Yongon-dong Chongno-gu, Seoul, 110-744, Korea

Identification of the mitochondrion has been a tremendous achievement of outstanding scientists who have tried to understand the essence of life, life energy. We now know the mitochondrion is the intracellular powerhouse that generates ATP, a chemical energy source, through a chain of reactions called oxidative phosphorylation. The realm of mitochondrial research has been subsequently steadily broadened, and the mitochondrion is currently known to play major roles in many disease processes—neuromuscular disorders, Parkinson's disease, Alzheimer's disease, diabetes mellitus, senescence, programmed cell death, and carcinogenesis, to name a few. Furthermore, mitochondria tell us the history of the evolution of life and our migration out of Africa. To understand and implement the vast knowledge gained by the study of mitochondria related to humanity, many scientific societies have been founded worldwide.

In this background, the Asian Society for Mitochondrial Research and Medicine (ASMRM) was founded, and the first scientific meeting was held on February 5–6, 2003 in Seoul, Korea. Leading scientists from Asia and around the world gathered to discuss recent progress in the field of mitochondriology. The sessions covered a wide range of complementary research areas—human evolutionary history and mitochondrial diseases, control of mitochondrial function, mitochondrial transcription factor, mechanism of mitochondrial damage, oxidative stress and apoptosis, diagnosis and treatment of mitochondrial diseases.

To establish a new scientific society is a huge task. This new society could not have been born without the belief that this kind of organization is needed. The enthusiastic leadership of Professor Wei from Taiwan, Professors Tanaka and Ohta from Japan, and Professor Xu from China should be commended in this regard. However, without people who responded to the call and worked hard with such great energy and enthusiasm, ASMRM would not have been established. The success of this meeting is reflected by this extraordinary collection of scientific papers, including numerous short communications. I would like to deeply thank the contributors to the ASMRM meeting and the authors of these resulting proceedings. Professors DiMauro, Tanaka, and Wei devoted much effort in making the final volume an excellent one. As there are relatively few books or monographs on mitochondria in health and disease, we hoped the publication of the proceedings of this meeting would also help to fill a gap in the literature and give some momentum in advancing the science of mitochondriology. I would like to express my special gratitude to the Publishing and New Media department of the New York Academy of Sciences for seeing these proceedings through the press.

I look forward to the second scientific meeting of ASMRM and to much progress in the field of mitochondrial research and medicine.

Overview

HONG KYU LEE

Department of Internal Medicine, Seoul National University College of Medicine, 28 Yongon-dong Chongno-gu, Seoul, 110-744, Korea

Identification of the mitochondrion has been a tremendous achievement of accomplished scientists who tried to understand the essence of life, life energy. We now know that the mitochondrion is the intracellular powerhouse that generates ATP, the chemical energy source, through a chain of reactions called oxidative phosphorylation. The realm of mitochondrial research has been steadily expanded since, and now the mitochondrion is known to play major roles in many important health problems of the twenty-first century: degenerative neuromuscular disorders, including Alzheimer's disease; metabolic disorders such as diabetes mellitus; senescence; and carcinogenesis. Furthermore, mitochondria tell us the history of the evolution of life and our past, including migrations out of Africa. This new awareness has sparked many new and expanded areas of mitochondrial research of great interest to a wide variety of scientists.

In this volume of the *Annals of the New York Academy of Sciences*, a wide range of complementary research areas is presented: human evolutionary history and mitochondrial diseases, mitochondrial function and diseases, the mechanism of mitochondrial damage and apoptosis, mitochondria and diabetes, diagnosis and treatment of mitochondrial diseases, and many short papers. The purpose of this introductory review is to provide an overview of the various aspects of biologic events related to mitochondria.

HUMAN EVOLUTIONARY HISTORY AND MITOCHONDRIAL DISEASES

More than a billion years ago, aerobic bacteria colonized primordial eukaryotic cells that lacked the ability to use oxygen. A symbiotic relationship evolved over time, accompanied by the loss of redundant genes and the transfer of genes from the prokaryote to the eukaryotic nucleus.[1,2] In this historical background, mitochondria reside in the cytoplasm of animal cells and have their own DNA (mtDNA), which is partially independent of the nuclear DNA (nDNA). The human mtDNA is a 16,569-bp-long, double-stranded, circular molecule containing 37 genes. The mtDNAs have a high mutation rate, 5–10 times faster than nDNA. As a result, the human popula-

---

Address for correspondence: Hong Kyu Lee, Department of Internal Medicine, Seoul National University College of Medicine, 28 Yongon-dong Chongno-gu, Seoul, 110-744, Korea. Voice: +82-2-760-2266; fax: +82-2-765-7966.

hkleemd@plaza.snu.ac.kr

tion currently harbors a high level of population-specific mtDNA polymorphisms. Analysis of the population-specific mtDNA polymorphism has permitted the reconstruction of human prehistory.[3] In this volume, Zakharov and colleagues show the mtDNA variations in the aboriginal populations of the Altai-Baikal region, which have implications for the genetic history of North Asia and America. As a consequence of the long history of human journey and environmental adaptation, various mtDNA lineages may be now qualitatively different. Analysis of this mtDNA variation may ultimately reveal our evolutionary past that has shaped us. The analysis could also help us understand the pathophysiology of mtDNA diseases, including modern bioenergetic disorders such as obesity, diabetes, and senescence. Considering this aspect, we should learn much about mtDNA variations among different people and their functional significance. The contribution from Tanaka and colleagues on the mitochondrial haplogroups and single nucleotide polymorphisms (SNPs) among the Japanese provides a firm ground to ask questions. Enormous amounts of data on mtSNPs are currently available on the Gifu International Institute of Biotechnology's Human Mitochondrial Genome Polymorphism database website (http://www.giib.or.jp/mtsnp/index_e.html).

MITOCHONDRIAL FUNCTION AND DISEASES

The mitochondrion is the efficient powerhouse of the cell generating ATP via the citric acid cycle, and a chemiosmotic process converts oxidation energy into ATP. Many metabolic pathways are exclusive to mitochondria: urea cycle, fatty acid oxidation, biosynthesis of heme and ubiquinol (coenzyme Q).[4] It is thus not surprising to see disease phenotypes associated with mitochondrial abnormalities. On the other hand, if mitochondria are protected, people will become healthier. Mitochondrial aldehyde dehydrogenase 2 (ALDH2) is located in the mitochondrial matrix and plays a major role in metabolizing acetaldehyde produced from ethanol. Ohta and colleagues show that mitochondrial ALDH2 is protective against oxidative stress and involved in the pathogenesis of Alzheimer's disease. Mechanisms of mitochondrial damage are diverse. Calcium uptake by mitochondria is a physiologically important mechanism for buffering or modulating cytosolic calcium.[5] Under certain conditions they may become overwhelmed: glutamate-mediated excitotoxicity in neurons is caused by excessive Ca^{2+} uptake, followed by the triggering of apoptosis.[6,7] This process is also linked to Alzheimer's disease. To support this notion, Jou and colleagues showed that visible light irradiation increases the level of mitochondrial calcium, along with the generation of reactive oxygen species (ROS), leading to photoirradiation–induced toxicity and apoptosis. ROS are continually generated as by-products of aerobic metabolism in animal cells and cause oxidative stress. Xu proposes a cytochrome c–mediated electron-leak pathway and suggests that cytochrome c may act as an antioxidative factor to regulate ROS level of mitochondria.

MITOCHONDRIAL TRANSCRIPTION FACTOR

Both qualitative and quantitative changes in mtDNA are implicated in the pathogenesis of diabetes and insulin-resistance syndrome.[8–10] Mitochondrial transcription factor A (Tfam) is a key regulator involved in mtDNA transcription and replication.[11] Maintenance of mtDNA integrity is essential for normal function of the respiratory chain responsible for the aerobic ATP production. Tfam was purified and cloned as a transcription factor for mtDNA. Tfam indeed enhances mtDNA transcription in a promoter-specific fashion in the presence of mitochondrial RNA polymerase and transcription factor B.[12] Because replication of mammalian mtDNA is to be initiated following the transcription,[13] Tfam is also thought to be crucial for replication of mtDNA. Consistent with this notion, targeted disruption of the mouse Tfam gene is an embryonic lethal mutation causing mtDNA depletion.[14] Tfam is a DNA-binding protein with no sequence specificity, although it shows a relatively higher affinity to the light- and heavy-strand promoters, LSP and HSP, respectively.

In this volume, many interesting papers are presented on the role of Tfam. Kang and colleagues show that human mtDNA is tightly associated and wrapped with Tfam, which functions as a main constitutive factor of an mtDNA/protein complex as well as a transcription factor. Pak and associates suggest that high glucose–induced Tfam transcription might be mediated by NRF-1. Nishio and colleagues show that mitochondrial dysfunction in the heart of diabetic rat is mediated by reduction in binding activity of Tfam. These novel areas should be explored further.

MECHANISM OF MITOCHONDRIAL DAMAGE

Nucleoside-analogue reverse transcriptase inhibitors (NRTIs) are major drugs in anti-HIV treatment. The clinical manifestations of NRTIs toxicity, such as myopathy, cardiotoxicity, diabetes, peripheral neuropathy, and others are quite similar to those of oxidative phosphorylation disorders.[15] These toxicities stem from the ability of NRTIs to inhibit the DNA polymerase γ, the only DNA polymerase involved in mtDNA replication.[15] Various noxious stimuli may interfere with mitochondrial function.

In this section, Lenaz and colleagues implicate the role of ROS produced by mitochondria in relation to aging and pathology. Nakabeppu and colleagues propose the importance of biological defense mechanisms against oxidative damage in nucleic acids caused by ROS. Peng and colleagues demonstrate with beautiful imaging technique that photoirradiation induces heterogeneous intracellular ROS formation and mitochondrial permeability transition pore (MPTP) opening in single intact cells, indicating the existence of heterogeneous intra- or extramitochondrial regulatory microdomains among individual mitochondrion in regulating mitochondrial ROS formation and subsequent opening of the MPTP. Since iron is an important factor in mitochondrial ATP synthesis, iron deprivation might result in mitochondrial dysfunction. Yoon and colleagues show that iron chelation inhibits complex II activity of mitochondria.

OXIDATIVE STRESS AND APOPTOSIS

Since the first reports connecting mitochondria with apoptotic cell death, researchers wondered whether the life-supporting function of mitochondria is somehow linked to their suicidal activity. Studies on the function of mitochondria in apoptotic cell death have revealed a versatility and complexity of these organelles previously unsuspected. Apoptosis is indeed an important biological process involved in embryonic development and in a number of physiological and pathological processes of the human and animals. In the past few years, mitochondria have been established to play a critical role in the early events leading to apoptotic cell death.[16] Recent studies have clearly demonstrated that mitochondria act as a major switch and arbitrator for the initiation of apoptosis in mammalian cells. The switch is believed to involve the release of several proteins that have been shown to regulate apoptosis via interaction with the MPTP. Factors influencing the opening or closing of the MPTP may be either pro-apoptotic or anti-apoptotic. Furthermore, the apoptotic pathway is amplified by the release of apoptogenic proteins from the mitochondrial intermembrane space, including cytochrome *c*, apoptosis-inducing factor (AIF, a flavoprotein), and latent forms of specialized proteases called pro-caspases. Other such proteins, including Smac/DIABLO, Endo G, and Htra2/Omi, were identified later.[16] This process can be triggered by an excessive mitochondrial uptake of Ca^{2+}, exposure to high levels of ROS, and a decline in the cellular ATP level.

In this volume, Wei and colleagues demonstrate that mutation and depletion of mtDNA increase the susceptibility of human cells to apoptosis triggered by exogenous stimuli, such as UV irradiation or staurosporine. Based on their data, Wei and colleagues propose an integrated scheme to accommodate the general mechanisms underlying the observations that human cells harboring mutated mtDNA are more susceptible to apoptosis. On the other hand, Lee and colleagues show very interesting results. They show that human hepatoma cell line (Sk-Hep1) lacking mtDNA is resistant to apoptosis, which suggests that apoptosis of Sk-Hep1 cells is dependent on intact mitochondrial function. From this observation, authors speculate that aged cells or tumor cells harboring frequent mutations or deletions of mtDNA might acquire the ability to evade apoptosis or tumor surveillance imposed by TRAIL or other death effectors *in vivo*, accounting for the selection advantage of cancer cells and frequent development of cancer in aged individuals. Indeed, Shieh and colleagues show that accumulation and subsequent cytoplasmic segregation of the mutant mtDNA (4977-bp deletion) during cell division may play an important role in the pathogenesis of oral cancer.

Apoptosis is also important in pancreatic β-cell dysfunction in diabetic milieu where oxidative stress is increased. Kajimoto and Kaneto suggest that antioxidant treatment can protect β-cells by decreasing apoptosis and inhibiting JNK pathway.

MITOCHONDRIA AND DIABETES

The mitochondrial genome has been implicated in the pathogenesis of diabetes mellitus by several mechanisms: point mutation in coding region;[17] common polymorphism in regulatory region;[18] and reduced amount of mtDNA content.[8,10] Since the mitochondrial function is crucial in insulin secretion from pancreatic β-cells, mitochondrial diabetes, a rare form of diabetes mellitus, is regarded as the consequence of pancreatic β-cell dysfunction.[19] However, after Shulman and colleagues[20] demonstrated that age-associated decline in mitochondrial function contributes to insulin resistance in the elderly, leading to type 2 diabetes, the most common form of diabetes mellitus, mitochondrial dysfunction is regarded as a precondition that disposes to diabetes by decreasing insulin secretory capacity and insulin sensitivity as well. Mitochondrial research has now become mainstream in diabetes research and extends to obesity, dyslipidemia, and hypertension, inasmuch as insulin resistance is a major pathophysiologic abnormality of the state. In this meeting, Suzuki and colleagues present the clinical characteristics of mtDNA 3243A>G point mutation in the Japanese. Interestingly, Nomiyama and colleagues showed that the mtDNA 3243 A>G substitution can be the result of somatic mutation and accumulation in diabetic milieu. Another important problem related to insulin resistance, thrifty phenotype hypothesis,[21] the increased susceptibility to type 2 diabetes in the adults who were born with small body weight (indicating intrauterine malnutrition), was considered by Park and colleagues. They showed that protein malnutrition *in utero* may cause type 2 diabetes by decreasing mtDNA content and impairing β-cell development. This study implies that decreased mtDNA contents may link fetal malnutrition (thriftiness) and diabetes in adult life.

DIAGNOSIS AND TREATMENT OF MITOCHONDRIAL DISEASES

Mitochondrial diseases have extremely heterogeneous clinical manifestations owing to the ubiquitous nature of mitochondria and the dual genetic control.[2] In this regard, mitochondrial disorders can be either multisystemic or confined to single tissue. Disease may be transmitted either by Mendelian or maternal inheritance. Therefore, to make an accurate diagnosis is often very difficult for clinicians. Treatment for mitochondrial diseases is still inadequate. DiMauro provides us with an invaluable guide into the diagnosis and treatment of mitochondrial disorders. For comprehensive molecular diagnosis of mitochondrial disorders, Wong suggests the approaches we should take—both qualitative and quantitative. However, details are really complex; Guan extensively reviews the molecular pathogenic mechanism of maternally inherited deafness; Bai and colleagues present genetic and functional analysis of mtDNA encoded complex I gene; and from the viewpoint of nucleus-mitochondrial interaction, Singh and colleagues introduce a genome-wide analytical method.

Aside from the classic mitochondrial disorders, there is a growing body of evidence that subtle mitochondrial dysfunction plays an important role in chronic metabolic disorders (e.g., type 2 diabetes) and neurodegenerative disorders. We should also focus on this relatively new field of mitochondrial disorders to detect a mild derangement in mitochondrial function and to improve mitochondrial function.

In this volume, we realize how versatile and complex mitochondria are in regard to life and death of cells and humans. Many challenges remain, and exciting insights will continue to stimulate all scientists and doctors working in the field of mitochondriology.

ACKNOWLEDGMENT

This work and the First Scientific Meeting of the Asian Society for Mitchondrial Research and Medicine were partially supported by a grant from the Korea Health 21 R&D Project, Ministry of Health & Welfare, Republic of Korea (02-PJ1-PG1-CH04-001).

REFERENCES

1. GRAY, M.W., G. BURGER & B.F. LANG. 1999. Mitochondrial evolution. Science **283**: 1476–1481.
2. DIMAURO, S. & E.A. SCHON. 2003. Mitochondrial respiratory-chain diseases. N. Engl. J. Med. **348**: 2656–2668.
3. WALLACE, D.C., M.D. BROWN & M.T. LOTT. 1999. Mitochondrial DNA variation in human evolution and disease. Gene **238**: 211–230.
4. SCHEFFLER, I.E. 2001. Mitochondria make a come back. Adv. Drug Deliv. Rev. **49**: 3–26.
5. HANSFORD, R.G. 1994. Physiological role of mitochondrial Ca^{2+} transport. J. Bioenerg. Biomembr. **26**: 495–508.
6. NICHOLLS, D.G., S.L. BUDD, R.F. CASTILHO & M.W. WARD. 1999. Glutamate excitotoxicity and neuronal energy metabolism. Ann. N.Y. Acad. Sci. **893**: 1–12.
7. REYNOLDS, I.J. 1999. Mitochondrial membrane potential and the permeability transition in excitotoxicity. Ann. N.Y. Acad. Sci. **893**: 33–41.
8. LEE, H.K., J.H. SONG, C.S. SHIN, et al. 1998. Decreased mitochondrial DNA content in peripheral blood precedes the development of non-insulin-dependent diabetes mellitus. Diabetes Res. Clin. Pract. **42**: 161–167.
9. PARK, K.S., K.U. LEE, J.H. SONG, et al. 2001. Peripheral blood mitochondrial DNA content is inversely correlated with insulin secretion during hyperglycemic clamp studies in healthy young men. Diabetes Res. Clin. Pract. **52**: 97–102.
10. SONG, J., J.Y. OH, Y.A. SUNG, et al. 2001. Peripheral blood mitochondrial DNA content is related to insulin sensitivity in offspring of type 2 diabetic patients. Diabetes Care **24**: 865–869.
11. CLAYTON, D.A. 2000. Transcription and replication of mitochondrial DNA. Hum. Reprod. **15**: 11–17.
12. FALKENBERG, M., M. GASPARI, A. RANTANEN, et al. 2002. Mitochondrial transcription factors B1 and B2 activate transcription of human mtDNA. Nat. Genet. **31**: 289–294.
13. SHADEL, G.S. & D.A. CLAYTON. 1997. Mitochondrial DNA maintenance in vertebrates. Annu. Rev. Biochem. **66**: 409–435.
14. LARSSON, N.G., J. WANG, H. WILHELMSSON, et al. 1998. Mitochondrial transcription factor A is necessary for mtDNA maintenance and embryogenesis in mice. Nat. Genet. **18**: 231–236.
15. LEWIS, W. & M.C. DALAKAS. 1995. Mitochondrial toxicity of antiviral drugs. Nat. Med. **1**: 417–422.
16. NEWMEYER, D.D. & S. FERGUSON-MILLER. 2003. Mitochondria: releasing power for life and unleashing the machineries of death. Cell **112**: 481–490.
17. KADOWAKI, T., H. KADOWAKI, Y. MORI, et al. 1994. A subtype of diabetes mellitus associated with a mutation of mitochondrial DNA. N. Engl. J. Med. **330**: 962–968.
18. POULTON, J., J. LUAN, V. MACAULAY, et al. 2002. Type 2 diabetes is associated with a common mitochondrial variant: evidence from a population-based case-control study. Hum. Mol. Genet. **11**: 1581–1583.
19. MAECHLER, P. & C.B. WOLLHEIM. 2001. Mitochondrial function in normal and diabetic beta-cells. Nature **414**: 807–812.
20. PETERSEN, K.F., D. BEFROY, S. DUFOUR, et al. 2003. Mitochondrial dysfunction in the elderly: possible role in insulin resistance. Science **300**: 1140–1142.
21. HALES, C.N. & D.J. BARKER. 1992. Type 2 (non-insulin-dependent) diabetes mellitus: the thrifty phenotype hypothesis. Diabetologia **35**: 595–601.

Mitochondrial Genome Single Nucleotide Polymorphisms and Their Phenotypes in the Japanese

MASASHI TANAKA, TAKESHI TAKEYASU, NORIYUKI FUKU, GUO LI-JUN, AND MIYUKI KURATA

Department of Gene Therapy, Gifu International Institute of Biotechnology, 1-1 Naka-Fudogaoka, Kakamigahara, Gifu 504-0838, Japan

ABSTRACT: Polymorphisms in the human mitochondrial genome have been used for the elucidation of phylogenetic relationships among various ethnic groups. Because analysis by mitochondrial genetics has detected pathogenic mutations causing mitochondrial encephalomyopathy or cardiomyopathy, most of the mitochondrial single nucleotide polymorphisms (mtSNPs) found in control subjects have been regarded as merely normal variants. However, we cannot exclude the possibility that the mitochondrial functional differences among individuals are ascribable at least in part to the mtSNPs of each individual. Human lifespan in ancient history was much shorter than that at the present time. Therefore, it is reasonable to speculate that certain mtSNPs that predispose one toward susceptibility to adult- or elderly-onset diseases, such as Parkinson's disease and Alzheimer's disease, have never been a target for natural selection in the past. Similarly, thrifty mtSNPs that had been advantageous for survival under severe famine or cold climate conditions might turn out to be related to satiation-related diseases, such as diabetes mellitus and obesity. To examine these hypotheses, we have constructed a mtSNP database by sequencing the entire mitochondrial genomes of 672 subjects: 96 in each of seven groups (i.e., centenarians, young obese or non-obese subjects, diabetic patients with or without major vascular involvement, patients with Parkinson's disease, and those with Alzheimer's disease).

KEYWORDS: mitochondrial DNA; longevity; Parkinson's disease; obesity

INTRODUCTION

Mitochondria are intracellular organelles that are responsible for ATP production via their oxidative phosphorylation system. The mitochondrial respiratory chain is the major source of reactive oxygen species within the cell, and several percent of the oxygen utilized in the body are converted into reactive oxygen species. It is widely accepted that mitochondria play an important role in the aging processes of both

---

Address for correspondence: Masashi Tanaka, Department of Gene Therapy, Gifu International Institute of Biotechnology, Naka-Fudogaoka, Kakamigahara, Gifu 504-0838, Japan. Voice: +81-583-71-4646; fax: +81-583-71-4412
mtanaka@giib.or.jp

cells and individuals. We previously identified a mitochondrial genotype, 5178C→A (ND2, Leu237Met), representing haplogroup D, to be associated with longevity in Japanese centenarians.[1] Our proposal that certain mitochondrial polymorphisms are associated with longevity is further supported by observations that haplogroups J and U are overrepresented in European centenarians.[2] Based on these findings, we have hypothesized that other haplogroups are associated with age-related neurodegeneration in Parkinson's disease or Alzheimer's disease. We also postulated that common metabolic disorders, such as obesity and type-2 diabetes mellitus, are attributable at least in part to mitochondrial polymorphisms. To examine these hypotheses, we have started comprehensive sequence analysis of the entire mitochondrial genome of centenarians, young obese or non-obese adults, patients with Parkinson's disease or Alzheimer's disease, and diabetic patients with or without angiopathy, using 96 individuals for each of these groups

FUNCTIONAL SIGNIFICANCE OF MITOCHONDRIAL GENOME POLYMORPHISMS

The size of the human nuclear genome is approximately 3 billion base pairs (bp), and the maternal and paternal genomes amount to 6 billion bp. The mitochondrial genome is a closed circular DNA molecule of 16,569 bp and is transmitted only from maternal lineage. The vast majority of the nuclear genome is occupied with non-coding regions, as only 1.5% of the nuclear genome encodes for proteins. Thus coding regions comprising only 90 million bp are functionally important. In contrast, the non-coding region of human mitochondrial genome is only 1,228 bp in total, and most (15,392 bp) of the mitochondrial genome contains 13 protein-coding genes as well as 22 tRNA and 2 rRNA genes necessary for intraorganellar protein synthesis. Assuming the number of mitochondrial genomes to be 2,000 copies per cell, a total of 30 billion bp (15,000 bp × 2,000 copies) of the mitochondrial genome are functionally expressed in the cell. In addition, the evolutionary rate of mitochondrial genome is reported to be 5–10 times higher than that of nuclear genome. Therefore, the polymorphisms in the mitochondrial genome contribute more extensively to the functional differences among individuals than those in the nuclear genome.

CHARACTERISTICS OF MITOCHONDRIAL GENOME POLYMORPHISMS

Additional important aspects of mitochondrial genome polymorphisms are haploidy and homoplasmy, both of which are related to maternal transmission. In the case of the nuclear genome, which has a biallelic nature, the deleterious effect of a polymorphism in one allele can be concealed by the normal function of the other allele. In contrast, the mitochondrial genome is exclusively maternally transmitted, and therefore should be considered as monoallelic. The mitochondrial genome is massively amplified during oogenesis; e.g., the number of mitochondrial genomes in an embryonic germinal cell is estimated to be only 100 copies per cell, whereas a mature oocyte harbors 200,000 copies of mitochondrial genome. Because of this bottleneck effect, even a new mutation that occurred in the germ-line can be propagated to become a major population in the second generation, and eventually results in homoplasmy in the third or fourth generation. Although strongly deleterious mutations causing mitochondrial dysfunctions are selected out from the human population, slightly deleterious polymorphisms would be accumulated in the population. Thus, the mitochondrial polymorphisms carried by an individual are a summation of both the haplogroup-specific mtSNPs, which are historical traits transmitted from the ancient maternal lineage, and the individual- or family-specific mutations, which had been acquired in her or his recent maternal ancestors. Because it is reported that 26–32% of naturally occurring amino acid replacements result in functional alterations,[3] and because approximately 30% of mtSNPs in the protein-coding genes are non-synonymous mutations, mitochondrial genome polymorphisms are an important contributor to differences among individuals in terms of bioenergetic efficiency, metabolic rates, oxygen consumption, and production of reactive oxygen species.

MITOCHONDRIAL SINGLE NUCLEOTIDE POLYMORPHISM DATABASE

Our institute, Gifu International Institute of Biotechnology (GiiB), has opened the GiiB-JST mtSNP (mitochondrial single nucleotide polymorphism) database in collaboration with Japan Science Technology Agency (JST) that has been operational since April 2003. This database (http://www.giib.or.jp/mtsnp/index_e.html) provides information related to the functional differences among mitochondrial SNPs. The web site is useful for identification of mtSNPs associated with age-related conditions, such as longevity, Parkinson's disease, and Alzheimer's disease; as well as those associated with energy metabolism, such as obesity, thinness, and type-2 diabetes, or with atherosclerosis. The main part of this mtSNP database includes the entire mitochondrial genome sequences of individuals belonging to seven different groups (with 96 individuals in each group): namely, centenarians, patients with Parkinson's disease, patients with Alzheimer's disease, young obese males, young non-obese males, and type-2 diabetes patients with or without severe vascular involvement. The purpose of this mtSNP website is to provide various means to evaluate the functional differences between individuals carrying different sets of mtSNPs. For this purpose, we have taken five different approaches:

- Molecular simulation of nsSNP (nonsynonymous single nucleotide polymorphism)–associated structural alterations in cytochrome c oxidase (complex IV) and ubiquinol-cytochrome c oxidoreductase (cytochrome bc_1 complex or complex III).
- Visualization of nsSNP effect with respect to the folding patterns in the mitochondrial inner membrane of mitochondrially encoded subunits of the respiratory chain enzyme complexes (complexes I–V).
- Visualization of SNP effect on the secondary structures of two ribosomal RNAs (12S and 16S rRNA) and 22 transfer RNA molecules.
- Comparison of amino acid replacements caused by nsSNPs with mammalian mitochondrial protein sequences for evaluation of functional effects.
- Dynamic systems for searching for individuals carrying specific SNPs and for comparing frequencies of SNPs among disease and control groups.

TABLE 1. Distribution of polymorphisms in the mitochondrial genome

Region	mtSNP	bp	Mutational Rate
Non-coding regions	264	1,228	21.5%
Ribosomal RNA genes	90	2,513	3.6%
Transfer RNA genes	70	1,512	4.6%
Protein-coding genes	870	11,367	7.7%
Total	1,294	16,620^a	7.8%

^{a}Total number includes overlapping regions of 51 bp.

Overview of mtSNPs Registered in the Database

Sequence analysis of entire mitochondrial genomes from 576 individuals belonging to six different groups except Alzheimer's disease group (opened to the public in January 2004) revealed a total of 1294 mtSNPs (TABLE 1, http://www.giib.or.jp/mtsnp/search_mtSNP_stat_e.html). The numbers of insertions, deletions, and substitutions in the hypervariable regions are excluded from TABLE 1, and the sequences in the regions with length polymorphisms are listed in other tables (http://www.giib.or.jp/mtsnp/search_LengthPolymorphisms_e.html). The frequency of 1,294 mtSNPs in 16,620 bp (including the overlapping regions of 51 bp) is 1 mtSNP per every 13 bp of mitochondrial genome. Because the frequency of SNPs in the nuclear genome is estimated to be 1 SNP per every 1,000 bp, the frequency of mtSNPs is at least 70 times higher than that of nuclear SNPs. Among the mitochondrial genome regions, the mtSNP frequency was highest for the non-coding regions, equivalent to a mutational rate of 21.5%. In the protein-coding genes, a total of 870 mtSNPs were detected, indicating a mutational rate of 7.7%. The mutational rates for the ribosomal RNA genes and transfer RNA genes were 3.6% and 4.6%, respectively, being lower than those for the protein-coding genes.

Difference in Frequencies of mtSAP among the Protein-Coding Genes

TABLE 2 shows the distribution of mtSNPs among the protein-coding genes. A total of 870 mtSNPs were detected in the 13 protein-coding genes including 268 mt-SAPs (mitochondrial single amino acid polymorphisms) and 602 synonymous mtSNPs. The percentage of nonsynonymous substitutions (mtSAP/mtSNP) was 30.8%. Ratios of mtSAP/mtSNP for ND3 (18.2%), CO1 (18.4%), ND4L (20.0%), and ND4 (21.9%) were lower than the average, whereas those for ND6 (34.8%), Cytb (40.1%), and ATP6 (54.1%) were higher than the average.

Prediction of Effect of mtSAP on the Three-Dimensional Structure

We have detected a total of 268 amino acid replacements (mtSAPs) in the mitochondrial genome. Because the majority of mtSAPs are detected either in several of the 96 individuals in a given group or in only 1 among 576 subjects, it is difficult to

TABLE 2. Distribution of mtSNP among protein-coding genes

Gene	bp	mtSAP	(%)	Syn	(%)	mtSNP	(%)	mtSAP/mtSNP
ND3	345	4	(1.16)	18	(5.22)	22	(6.38)	18.2%
CO1	1,539	18	(1.17)	80	(5.20)	98	(6.37)	18.4%
ND4L	294	2	(0.68)	8	(2.72)	10	(3.40)	20.0%
ND4	1,377	16	(1.16)	57	(4.14)	73	(5.30)	21.9%
ND5	1,809	34	(1.88)	91	(5.03)	125	(6.91)	27.2%
ATP8	204	6	(2.94)	16	(7.84)	22	(10.78)	27.3%
ND1	954	23	(2.41)	58	(6.08)	81	(8.49)	28.4%
CO2	681	16	(2.35)	40	(5.87)	56	(8.22)	28.6%
ND2	1,041	24	(2.31)	54	(5.19)	78	(7.49)	30.8%
CO3	783	19	(2.43)	42	(5.36)	61	(7.79)	31.1%
ND6	522	16	(3.07)	30	(5.75)	46	(8.81)	34.8%
Cytb	1,140	50	(4.39)	74	(6.49)	124	(10.88)	40.3%
ATP6	678	40	(5.90)	34	(5.01)	74	(10.91)	54.1%
Total	11,367	268	(2.36)	602	(5.30)	870	(7.65)	30.8%

DEFINITIONS: mtSAP, mitochondrial single amino acid replacement; Syn, synonymous mutations.

demonstrate their effects on the function of mitochondrial gene products by statistic epidemiological studies. For evaluation of the effects of mitochondrial single amino acid polymorphisms (mtSAPs) on the functioning of the mitochondrial respiratory chain, this mtSNP database provides a molecular simulation system that predicts the conformational changes caused by each mtSAP or a group of mtSAPs detected in each individual (http://www.giib.or.jp/mtsnp/search_mtSAP_3D.html). At present, the three-dimensional structures of bovine cytochrome c oxidase, and cytochrome bc_1 complex (ubiquinol-cytochrome c oxidoreductase) are known. All of the subunits, both mitochondrially encoded and nuclearly encoded, of the bovine enzymes were replaced by those of human enzymes, and then the structures of the human enzymes were calculated by using a molecular dynamic simulation system. For each mtSAP or group of mtSAPs that occur in each individual, the altered structures can be predicted by the simulation system at GiiB. These predicted structures can be sent as protein database format (PDB) files by e-mail to each client who has requested them through the "3D Structure" section in this mtSNP database.

Location of mtSAPs in the Secondary Structure of Mitochondrially Encoded Subunits

On the request of each client, locations of mtSAPs on the two-dimensional (2D) structures of subunits encoded by mtDNA are displayed on the webpage of the "2D Structure" section of this mtSNP database (http://www.giib.or.jp/mtsnp/search_mtSAP_2D_e.html). The 2D structures of NADH dehydrogenase subunits

(MTND1-6, MTND4L), subunits of cytochrome c oxidase (CO1-3), and cytochrome b protein (MTCYB) were based on the prediction of their hydrophobic transmembrane domains by the SOSUI system (Classification and Secondary Structure Prediction of Membrane Proteins by Mitaku Group, Department of Biotechnology, Tokyo University of Agriculture and Technology; sosui@proteome.bio.tuat.ac.jp; http://sosui.proteome.bio.tuat.ac.jp/sosuiframe0.html). Coordinates of the amino acid residues were predicted by the SOSUI system. For ND1-6 and ND4L, as well as for subunits of ATP synthase (MTATP6 and MTATP8), the three-dimensional (3D) structures of which are unknown, the amino (N) terminus of each subunit is located in the upper left corner. For the MTCO1, MTCO2, and MTCO3 subunits and the cytochrome b protein, the 3D structures of which are known for bovine enzymes, the orientation and folding of each subunit has been adjusted according to the crystal structure. The upper part corresponds to the intermembrane space side; and the lower part, to the matrix side.

Location of mtSNPs in the Secondary Structure of tRNA or rRNA

In the database web site, the location of nucleotide replacements can be displayed (http://www.giib.or.jp/mtsnp/search_RNA_2D_e.html). The structure of transfer RNA is displayed in the conventional cloverleaf structure by use of the program SstructView (http://smi-web.stanford.edu/projects/helix/sstructview/home.html). The secondary structure of the 12S ribosomal RNA molecule was obtained by use of the program SstructView according to the model of 12S rRNA (http://rrna.uia.ac.be/ssu/index.html) and that of 16S rRNA (http://rrna.uia.ac.be/lsu/index.html) predicted by Dr. Jan Wuyts and Prof. Rupert De Wachter [Department of Biomedical Sciences, Antwerp University (UIA), Belgium].

mtSAPs in Evolution

To examine whether a certain amino acid residue is conserved among species, one can compare a mtSAP with the amino acid sequences of 61 mammalian species (http://www.giib.or.jp/mtsnp/search_mtSAP_evaluation_e.html). The entire sequences of the mitochondrial genome for these 61 mammalian species are registered in an organelle genome database (http://www.ncbi.nlm.nih.gov/PMGifs/Genomes/40674.html).

GENETIC "GOLDEN MEAN" ASSOCIATED WITH LONGEVITY

In his *Doctrine of the Mean*, Confucius (551–479 BC) said: "The Superior Man actualizes the mean, the inferior man goes against it." In *Ethica Nicomachea*, Aristotle (384–322 BC) also discussed the importance of moderation. We make no claims for the superiority of the "golden mean" in politics or ethics, but we do wonder whether the golden mean is the best genetic path for longevity. Population genetics has revealed that most mitochondrial amino acid polymorphisms within a species are mildly deleterious.[4,5] We hypothesize that centenarians are free from deleterious mitochondrial variations and thereby represent the genetic golden mean for the human mitochondrial genome. To test this hypothesis, we analyzed amino acid variations in the cytochrome b molecule of 64 Japanese centenarians, 96 young obese adults, and 96 patients with Parkinson's disease by directly sequencing the entire cytochrome *b* gene in mitochondrial DNA from each subject.[6] Although the frequencies of some variations, such as N260D and G251S, differed significantly between centenarians and patients with Parkinson's disease, the most striking feature of centenarian cytochrome *b* was the rareness of amino acid variations in contrast to the variety of amino acid replacements in patients with Parkinson's disease. These results suggest that centenarians are genetically hitting the "golden mean" (less variation from the consensus cytochrome *b* sequence or less mismatch with other subunits). Our findings have at least three implications. First, the absence of certain variations in centenarians and their presence in patients with Parkinson's disease indicate that these variations are not beneficial for long-term survival but do predispose individuals to adult-onset diseases and that centenarians are genetically hitting the golden mean. Second, the multiple variations present in young adults are unlikely to greatly influence either individual survival before maturation or the transmission of the genome to subsequent generations. Third, a multiplex detection system for mildly deleterious mitochondrial variations in combination with genetic tests for longevity-associated genotypes will be useful both for predicting longevity and for assessing the risk of age-related diseases.

PHYLOGENETIC ANALYSIS OF HAPLOGROUPS IN THE JAPANESE

Mitochondrial haplogroups in Japanese people can be classified into two macrohaplogroups, N and M. The macrohaplogroup N differs from the macrohaplogroup M by 9 mtSNPs. The macrohaplogroup N includes haplogroups A, B, F, N9, U, and T, whereas the macrohaplogroup M includes haplogroups D, G, M7, M8, M9, M10, M11, and Z. For example, the haplogroup D differs from the macrohaplogroup M by 5 mtSNPs. We previously reported that the haplogroup D was frequently detected in Japanese centenarians. As haplogroup D, which is represented by Mt5178A, can be detected not only in Japanese individuals but also in the Siberian Inuit,[7] this haplogroup seems to be one of those that have adapted to a cold climate. There are large personal and ethnic differences in the degree of obesity.[8] According to the National Recommendation, the average daily energy requirement for Canadians (19–24 years old, male) is 3,000 kcal, while that for the Japanese (18–29 years old, male) is only 2,300 kcal. Caucasoids may have required thermogenesis and a large body weight to cope with cold climates,[9] whereas the Japanese may have diminished heat radiation by having short extremities in adaptation to cold climates. The differences between Caucasoids and Japanese in daily energy requirements may be attributed at least in part to functional differences in the oxidative phosphorylation systems defined by their mitochondrial haplogroups. In addition, environmental factors may be important. For example, the incidences of obesity and diabetes among Japanese-Americans living in Hawaii and in the Los Angeles area are about threefold higher than those of Japanese in Japan.[10] It can be hypothesized that the Japanese individuals with 5178C might be more susceptible to myocardial infarction than those with 5178A, possibly due to metabolic alterations associated with the recent westernization of dietary habits of Japanese. For evaluation of this hypothesis, further longitudinal large-scale molecular epidemiological studies are required.

CHARACTERISTICS AND PHENOTYPES OF MITOCHONDRIAL HAPLOGROUPS

Association with Longevity and Antiatherosclerotic Effect of Haplogroup D

Because sequence variations of mtDNA are extensive among human populations,[11] we hypothesized that rates of age-associated accumulation of mtDNA mutations may differ among various mtDNA haplotypes and that a certain haplotype of mtDNA might thus be associated with longevity. To answer this question, we determined the entire mtDNA sequences of 11 Japanese centenarians. The majority of these centenarians carried five nucleotide substitutions representing haplogroup D4 (Mt5178C→A, Mt8414→T, Mt4883C→T, Mt14668C→T, and Mt3010G→A),[1,12] which were more frequently found in Japan than in other countries,[13] in addition to nine nucleotide substitutions representing macrohaplogroup M (Mt8701A→G, Mt10398A→G, Mt9540T→C, Mt10400C→T, Mt10873T→C, Mt12705C→T, Mt14783T→C, Mt15043G→A, Mt15301G→A), which were frequently found in the general Asian population.[13,14]

Among these polymorphisms identified in Japanese centenarians, we focused on Mt5178C→A, causing a Leu→Met substitution at amino acid 237 of the ND2 gene. ND2 is one of 7 mtDNA-encoded subunits included among the approximately 41 polypeptides of respiratory complex I (NADH:ubiquinone oxidoreductase, EC 1.6.5.3).[15] Complex I accepts electrons from NADH, transfers them to ubiquinone, and uses the energy released to pump protons across the mitochondrial inner membrane. Since allelic variants of the ND2 gene have been associated with Leber's hereditary optic neuropathy,[16] the ND2 subunit is assumed to play an important role in the function of complex I.

In diabetic subjects, Matsunaga and colleagues[17] observed that the mean intima-media thickness (IMT) and plaque formation in the bilateral carotid arteries were significantly less in the Mt5178A group than in the Mt5178C group, suggesting that Mt5178A may have an anti-atherogenic effect, at least in type 2 diabetic individuals. Kokaze and colleagues[18] reported that serum concentrations of high-density lipoprotein (HDL) cholesterol were significantly higher in men carrying 5178A than in those carrying 5178C and that serum concentrations of triglycerides (TG) were significantly lower in women carrying 5178A than in those carrying 5178C, suggesting that the antiatherogenic effect of Mt5178A may be in part attributable to favorable lipid metabolism. Kokaze and colleagues[19] examined whether Mt5178 A/C polymorphism influenced the effects of habitual smoking on serum protein fraction levels in 321 healthy Japanese men. In Mt5178C genotype men, alpha-1 and alpha-2 globulin levels were higher in smokers than in nonsmokers, whereas in Mt5178A genotype men no significant difference was observed in alpha-1 or alpha-2 globulin levels between smokers and nonsmokers. These results suggest that longevity-associated Mt5178 A/C polymorphism may influence the effects of cigarette smoking on serum protein fraction levels in healthy Japanese men.

We have examined whether the Mt5178C→A (Leu237Met) polymorphism in the mitochondrial ND2 gene is associated with a low prevalence of myocardial infarction (MI) in a case control study (Takagi and colleagues, submitted for publication). Multivariate logistic regression analysis with adjustment for age, gender, body mass index, smoking status, hypertension, diabetes mellitus, hypercholesterolemia, and hyperuricemia revealed that the frequency of the Mt5178A genotype was significantly higher in controls than in subjects with MI. These results suggest that the 5178A genotype of mitochondrial ND2 gene polymorphism is protective against MI. This protective effect would explain, at least in part, its contribution to longevity.

To examine whether a certain amino acid residue is conserved among species, one can compare a mtSAP with the amino acid sequences of 61 mammalian species. The entire sequence of the mitochondrial genomes for these 61 mammalian species are registered in an organelle genome database (http://www.ncbi.nlm.nih.gov/PMGifs/Genomes/40674.html). A Met residue is found at amino acid position 237 in MTND2 of *Balaenoptera physalus* (finback whale), *Equus caballus* (horse), *Pteropus dasymallus* (Ryukyu flying fox), *Pteropus scapulatus* (little red flying fox), *Sus scrofa* (pig), and *Thryonomys swinderianus* (greater cane rat). Therefore, we tentatively nicknamed this longevity-associated mtSNP as "whale-type" mitochondria. However, it is unknown whether or not this Met^{237} residue is related to the putatively high efficiency of the oxidative phosphorylation system of fin whale, horse, or flying fox. Because haplogroup D, characterized by 5178A, is detected in the Siberian Inuit[7] and people with this haplogroup had migrated across the present Bering Strait (the land Beringia at interglacial periods) into North and South America, this haplogroup can be considered to have adapted to cold climates.

mtSNP in MTCYB Associated with Obesity

We focused on the mtSNP 15498G→A, causing a Gly251Ser change, in MTCYB, which alteration was detected in 6 of 96 patients with Parkinson's disease. We previously reported that this mtSNP was not detected in 64 centenarians, but further research revealed that this mtSNP was detected in 1 of 107 centenarians. Because Gly^{251} is highly conserved among mammalian species, this amino acid residue seems to be functionally important. The neighboring substitution 15497G→A, causing Gly251Asp, was reported to be a pathogenic mutation in an infantile patient with histiocytoid cardiomyopathy. Although the Gly251Ser replacement may be less deleterious than the Gly251Asp replacement, we hypothesized that this mtSAP is a slightly deleterious mutation and that this mtSAP predisposes individuals to age-related neurodegenerative diseases, such as Parkinson's disease and Alzheimer's disease. To examine this hypothesis, we surveyed the contribution to the physical statures of individuals with a mtSNP as a part of the NILS-LSA study. There were no associations between the 15498G→A change and the cognitive functions or the family history of Parkinson's disease. However, the population-based association study suggested that the Mt15497 polymorphism is associated with obesity-related variables and lipid metabolism (Ohkura and colleagues, submitted for publication). The 15497G→A mtSNP was detected in 3.5% ($N=60$) of all subjects (825 women and 906 men): 2.8% ($N=23$) among women and 4.1% ($N=37$) among men. After adjusting for age and smoking, we found that body weight, body mass index, waist and hip circumferences, fat mass, fat-free mass, intra-abdominal fat, and triglycerides were significantly greater in women with the A allele than in those with the G allele ($P=0.001$–0.025). For men, the waist to hip ratio was significantly greater ($P=0.032$), and waist circumference, intra-abdominal fat, and triglycerides tended to be greater ($P=0.062$–0.087) in subjects with the A allele than in those with the G allele. Given that the 15497G→A replacement is one of the representative mtSNP's for haplogroup G3 and that haplogroup G can be detected in people living in the Arctic Zone, this mtSNP can be considered to be a result of adaptation to cold climates. It is of great interest that the haplogroups G and A, which had been selected during the glacial period, are associated with obesity in westernized Japan.

Haplogroup A Associated with High Performance in Endurance Running

We also identified a mtSNP that is associated with high performance in endurance running. We sequenced the MTATP6 and MTATP8 genes of 10 elite runners, including those of a winner of the Tokyo International Women's marathon, and detected a 8794C→T transition, causing a His90Tyr replacement, in MTATP6 in 5 of the 10 runners. Mitochondrial ATP synthase (Complex V) comprises 10–16 subunits encoded by nuclear DNA and 2 subunits (ATPase 6 and ATPase 8) encoded by mtDNA. ATP synthase subunit 6 is located in the vicinity of the bundle of subunit c's that forms a proton channel. The 8993T→G transversion in MTATP6, causing a Leu153Arg replacement, is a pathogenic mutation that can be found in patients with Leigh's syndrome or NARP (neurogenic muscle atrophy associated with ataxia and pigmentary retinopathy) disease. The Leu153Arg replacement, which inhibits proton translocation through the Fo portion of ATP synthase, is located in the fourth transmembrane domain of MTATP6, whereas the His90Tyr replacement is located in the second transmembrane domain. The His^{90} residue is frequently found in mammalian species, whereas the Tyr^{90} is frequently found in the MTATP6 of birds, amphibians, and reptiles. Therefore, we tentatively named the mitochondria with haplogroup A as "bird-type" or "dove-type" mitochondria. Because haplogroup A can be detected in the Chukchi people, who live in Northeast Siberia as well as in native Americans, this haplogroup is considered to have adapted to cold climates. Further studies are necessary to elucidate the functional differences between the MTATP6 with Tyr^{90} and that with His^{90} by use of cybrids carrying the mitochondrial genome with haplogroup A or others.

Haplogroup M7b2 Detected in a Winner of Marathon Race Is Associated with Obesity

We previously determined the entire sequence of the mitochondrial genome of a winner of a male marathon race, and his mtSNPs included 4048G→A (ND1, Asp248Asn). A series of five mtSAPs were detected in 7 of 96 young obese males, whereas these mtSNPs were found in only 7 of 96 young non-obese males. These five mtSAPs represent haplogroup M7b2. Because these mtSAPs are tightly linked to one another, we cannot determine which mtSNPs are functionally important. These results suggest that individuals with haplogroup M7b2 can exhibit extremely high performance in endurance running due to the high capacity of their oxidative phosphorylation system. However, when such individuals with this haplogroup experience sedentary life, they can easily become obese because of the high efficiency of their mitochondrial energy conservation system (Fuku and colleagues, manuscript in preparation).

Haplogroup F1a Associated with Thinness Is Distributed in Southeast Asia

We also identified a group of mtSNPs associated with thinness. These mtSNPs, representing haplogroup F1a, were detected in 11 of 96 young non-obese adults, whereas these were found in only 3 of 96 young obese adults (Fuku and colleagues, manuscript in preparation). Because the frequency of haplogroup F has been reported to be high in Southeast Asia, this haplogroup may be related to adaptation to hot climates. The association of haplogroup F1a with thinness may also be relevant to mitochondrial functions adapted to hot climates.

Haplogroups B4c2 and B4c3 Predispose toward Obesity and Resistance against Obesity

When we compared the frequencies of mtSNPs between obese young males and patients with type-2 diabetes, we found a series of mtSNPs representing subhaplogroups B4c2 and B4c3, which were found in 5 of 96 young obese males but not detected in 96 patients with type-2 diabetes. The prevalence of extreme obesity (MBI>30) is much lower in Japan (2.4%) than in the United States (22.5%), whereas the prevalence of type-2 diabetes in Japan (575 per 10,000) is much higher than that in the United States (318 per 10,000). These differences suggest that the Japanese are less prone to obesity but more susceptible to type-2 diabetes than the people in the United States. For an individual to become extremely obese, insulin secretion from the pancreatic islet beta cells must be sustained at a high level for a certain period. Because haplogroup B can be detected in Polynesians in the South Pacific area, and because the prevalence of obesity is high in this area, this haplogroup may be associated with efficient mitochondrial function to maintain a high level of insulin secretion from the pancreatic islet beta cells. We can hypothesize that Japanese young males with subhaplogroups B4c2 and B4c3 are predisposed to extreme obesity. In addition, if their body weights are appropriately controlled during middle or old age, they may have a lower genetic risk for type-2 diabetes. Most Japanese with haplogroup B are protected against type-2 diabetes because of the relatively low average caloric intake in Japan.

Haplogroup M8a1 and Susceptibility to Type-2 Diabetes

We also identified a series of mtSNPs that predispose individuals toward type-2 diabetes. These mtSNPs, representing haplogroup M8a1, were detected in 5 of 96 patients with type-2 diabetes; but they were not found in 96 young obese males (Guo and colleagues, manuscript in preparation). In patients with a 3243A→G mutation in the tRNA-Leu(UUR) gene, causing diabetes and hearing loss, insulin secretion is impaired and their past histories indicate that they had never been obese. Similarly, we can hypothesize that these individuals with haplogroup M8a1 are not prone to obesity but are predisposed to type-2 diabetes. Further studies are necessary to understand the mechanisms underlying development of diabetes in patients with haplogroup M8a1.

Haplogroup D4b2 Predisposes toward Resistance against Parkinson's Disease

Among the mtSNPs detected in Japanese centenarians, we found that a series of mtSNPs representing haplogroup D4b2 were more frequently detected in centenarians than in patients with Parkinson's disease (Takeyasu and colleagues, manuscript in preparation). These longevity-associated mtSNPs seem to predispose one toward resistance against Parkinson's disease. We previously hypothesized that a haplogroup D–specific mtSAP (Leu237Met in MTND2) predisposes individuals to resistance against reactive oxygen species, because methionine can serve as a scavenger of hydroxyl radicals by forming methionine sulfoxide. In the case of subhaplogroup D4b2, because these mtSNPs cause no significant amino acid replacements, this subhaplogroup may be characterized by absence of deleterious mutations. We hypothesize that haplogroup D4b2 might be associated with a lower production of reactive oxygen species from mitochondria because of the absence of deleterious amino acid replacements in the mitochondrial oxidative phosphorylation system. Conversely, various mtSAPs detected in patients with Parkinson's disease might be regarded as slightly deleterious mutations, which would increase the vulnerability of mitochondrial proteins to reactive oxygen species. Alternatively, these putative deleterious mutations might increase the production of reactive oxygen species by the respiratory chain.

STUDIES ON mtSNPs PAVE THE WAY TO HEALTHY LONGEVITY

By investigating mtSNPs of elite marathon runners, we can elucidate the characteristics of the mitochondrial oxidative phosphorylation system having high efficiency. As a second step, further studies are needed to understand the role of these mtSNPs in the life of ordinary people who usually experience sedentary lives. Such mtSNPs identified in elite marathon runners might predispose individuals with these mtSNPs to obesity if they have sedentary life styles and do not have the habit of daily aerobic exercise.

The conventional haplogroups A, B, C, D, etc. have been deduced from the combined analyses of the D-loop sequences and the restriction fragment length polymorphism (RFLP). Because of the limited resolution of these analyses, the conventional classification of haplogroups would probably not be sufficient for molecular epidemiological studies according to the mtSNPs in the coding regions. For example, haplogroup D, characterized by 5178C→A, can be divided into two groups, namely, subhaplogroup D5 without 3010G→A and subhaplogroup D4 with 3010G→A. Subhaplogroup D4 in Japan can be further classified into subsubhaplogroups D4a, D4b2, D4c, D4e, etc.

HIGHER ORDER BRAIN FUNCTION AND MITOCHONDRIA

Recently, in the NILS-LSA study, we identified two different mtSNPs associated with high cognitive function of the brain. First, a haplogroup A–specific mtSNP in MTATP6 was shown to be associated with high cognitive function in middle-aged to old males (40–80 years old). Because this mtSNP associated with high IQ in males was originally found in a winner of the Tokyo International Women's Marathon, we interpret these findings to mean that an efficient mitochondrial oxidative phosphorylation system is essential for the high cognitive function of the brain. In addition, a subsubhaplogroup D4i-specific mtSNP in MTCYB was demonstrated to be associated with high cognitive function, also in males. Because this mtSNP was originally found in centenarians, this mtSNP seems to be associated with both longevity and high cognitive function of the brain. Brain function is dependent on both blood glucose and oxygen for oxidative phosphorylation. Consciousness is easily impaired by hypoglycemia or anoxia, and inhibition of cytochrome c oxidase by carbon monoxide results in drowsiness, coma, and death. Therefore, it is obvious that the neuronal activity is totally dependent on the mitochondrial function. Our findings that two mtSNPs are associated with high cognitive function suggest that evolution of the mitochondrial genome is involved in the development of human intelligence. The average lifespan of humans is much longer than that of the chimpanzee or gorilla. We speculate that intelligence and longevity in humans have been achieved in parallel with the evolution of the mitochondrial genome.

CONCLUSION AND PERSPECTIVES

By extensive sequence analysis of the mitochondrial genome, we have established phylogenetic relationships among the Japanese people, and we have classified mtSNPs into subhaplogroup-specific mtSNPs. Present targets of our research are the following: (1) selection of one or two mtSNPs that represent each subhaplogroup, (2) utilization of these mtSNPs for large-scale epidemiological studies, and (3) construction of diagnostic systems. For large-association studies, it is essential to select mtSNPs that have frequencies higher than 5–10%. By selection of such mtSNPs according to their frequencies, most of them are subhaplogroup-specific mtSNPs, and others represent homoplasmic mutations that occur independently in different haplogroups. In contrast, it is rather difficult to demonstrate the etiological or pathological roles of mtSNPs with low frequencies. As previously mentioned, we have developed a molecular simulation system to evaluate the structural and functional effects of these infrequent mtSNPs. In addition, we can develop a diagnostic microsequencing system for calculating the sum of slightly deleterious mutations in the mitochondrial genome of each individual.

We previously developed a diagnostic system for predicting genetic risks for coronary heart diseases on the basis of previously reported polymorphisms in the nuclear genome.[20] We are currently conducting a large-scale association study to identify genetic risk factors for cardiac infarction, brain infarction, hypertension, and type-2 diabetes in collaboration with three major hospitals in Gifu Prefecture. By integrating studies on both nuclear SNPs and mtSNPs in this collaboration, we will be able to develop more valuable and economical genetic tests that can be used for people of this prefecture and Japan as well as for people in North East Asia. Japan is confronted with various problems of an unprecedented ultra-aging society, where good health in the aged is essential for the vitality of the society in terms of both economy and welfare. Several of the mtSNPs reported here are tightly associated with longevity and others are related to common metabolic disorders, such as obesity and type-2 diabetes. We expect that longevity in good health can be achieved by most individuals if he/she chooses his/her own lifestyle according to his/her genetic characteristics defined by mtSNPs and nuclear SNPs.

REFERENCES

1. TANAKA, M. et al. 1998. Mitochondrial genotype associated with longevity. Lancet **351:** 185–186.
2. NIEMI, A.K. et al. 2003. Mitochondrial DNA polymorphisms associated with longevity in a Finnish population. Hum. Genet. **112:** 29–33.
3. CHASMAN, D. & R.M. ADAMS. 2001. Predicting the functional consequences of non-synonymous single nucleotide polymorphisms: structure-based assessment of amino acid variation. J. Mol. Biol. **307:** 683–706.
4. WEINREICH, D.M. & D.M. RAND. 2000. Contrasting patterns of nonneutral evolution in proteins encoded in nuclear and mitochondrial genomes. Genetics **156:** 385–399.
5. RAND, D.M., D.M. WEINREICH & B.O. CEZAIRLIYAN. 2000. Neutrality tests of conservative-radical amino acid changes in nuclear- and mitochondrially-encoded proteins. Gene **261:** 115–125.
6. TANAKA, M. 2002. Mitochondrial genotypes and cytochrome b variants associated with longevity or Parkinson's disease. J. Neurol. **249** (Suppl. 2): II11–18.
7. INGMAN, M. et al. 2000. Mitochondrial genome variation and the origin of modern humans. Nature **408:** 708–713.
8. YOSHIIKE, N. et al. 1998. Descriptive epidemiology of body mass index in Japanese adults in a representative sample from the National Nutrition Survey 1990–1994. Int J. Obes. Relat. Metab. Disord. **22:** 684–687.
9. FLOREZ-DUQUET, M. & R.B. M DONALD. 1998. Cold-induced thermoregulation and biological aging. Physiol. Rev. **78:** 339–358.
10. HARA, H., G. EGUSA & M. YAMAKIDO. 1996. Incidence of non-insulin-dependent diabetes mellitus and its risk factors in Japanese-Americans living in Hawaii and Los Angeles. Diabet. Med. **13:** S133–142.
11. TANAKA, M. & T. OZAWA. 1994. Strand asymmetry in human mitochondrial DNA mutations. Genomics **22:** 327–335.
12. GONG, J.-S. et al. 1998. Mitochondrial genotype frequent in centenarians predisposes resistance to adult-onset diseases. J. Clin. Biochem. Nutr. **24:** 105–111.
13. UMETSU, K. et al. 2001. Multiplex amplified product-length polymorphism analysis for rapid detection of human mitochondrial DNA variations. Electrophoresis **22:** 3533–3538.
14. TANAKA, M. et al. 2000. Mitochondrial genotype associated with longevity and its inhibitory effect on mutagenesis. Mech. Ageing Dev. **116:** 65–76.
15. WALKER, J.E., J.M. SKEHEL & S.K. BUCHANAN. 1995. Structural analysis of NADH: ubiquinone oxidoreductase from bovine heart mitochondria. Methods Enzymol. **260:** 14–34.
16. BROWN, M.D. et al. 1992. Mitochondrial DNA complex I and III mutations associated with Leber's hereditary optic neuropathy. Genetics **130:** 16373.
17. MATSUNAGA, H. et al. 2001. Antiatherogenic mitochondrial genotype in patients with type 2 diabetes. Diabetes Care **24:** 500–503.
18. KOKAZE, A. et al. 2001. Association of the mitochondrial DNA 5178 A/C polymorphism with serum lipid levels in the Japanese population. Hum. Genet. **109:** 521–525.
19. KOKAZE, A. et al. 2003. Longevity-associated mitochondrial DNA 5178 A/C polymorphism influences effects of cigarette smoking on serum protein fraction levels in Japanese men. Mech. Ageing Dev. **124:** 765–770.
20. YAMADA, Y. et al. 2002. Prediction of the risk of myocardial infarction from polymorphisms in candidate genes. N. Engl. J. Med. **347:** 1916–1923.

Mitochondrial DNA Variation in the Aboriginal Populations of the Altai-Baikal Region

Implications for the Genetic History of North Asia and America

ILIA A. ZAKHAROV,[a] MIROSLAVA V. DERENKO,[b] BORIS A. MALIARCHUK,[b] IRINA K. DAMBUEVA,[c] CHODURAA M. DORZHU,[d] AND SERGEY Y. RYCHKOV[a]

[a]*Vavilov Institute of General Genetics, Russian Academy of Sciences, Moscow, 119991 Russia*

[b]*Institute of Biological Problems of the North, Russian Academy of Sciences, Magadan, 685000 Russia*

[c]*Institute of General and Experimental Biology, Russian Academy of Sciences, Ulan-Ude, 670047 Russia*

[d]*Tuva State University, Kyzyl, 667035 Russia*

> ABSTRACT: The discovery of mtDNA types common to Asians and Amerinds (types A, B, C, and D) forced investigators to search for those nations of Asia which, though not considered the ancestors of the Amerinds, have retained a close genetic resemblance with them. We collected samples and studied the gene pools of the Turkic-speaking nations of South Siberia: Altaians, Khakassians, Shorians, Tuvinians, Todjins, Tofalars, Sojots, as well as Mongolian-speaking Buryats. The data indicate that nearly all Turkic-speaking nations of Siberia and Central Asia, as well as the Buryats, have types A, B, C, and D in their gene pool. The highest total frequency of these types is observed in the Tuvinians and Sojots. They, as well as the Buryats, also have the lowest frequency of the europeoid types. The most mixed Asian-Europeoid gene pool examined turned out to be that of the Shorians. An important finding was the presence of type X in the Altaians, which had not yet been detected in Asia. As shown by computer analysis, this DNA sequence is not a late European admixture. Rather, the Altai variant X is ancient and can be close to the ancestral form of the variants of contemporary Europeans and Amerinds. The presented results prove that of all nations in Asia, the Turkic-speaking nations living between Altai and Baikal along the Sayan mountains are genetically closest to the Amerinds.
>
> KEYWORDS: mtDNA; gene pool; haplogroups; human populations; Amerinds

---

This work supported by grants from the Russian Foundation for Basic Research (Grant Nos. 00-15-97777 and 99-06-80430).

Address for correspondence: Ilia A. Zakharov, Vavilov Institute of General Genetics, Russian Academy of Sciences, Moscow, 119991 Russia. Fax: 7-095-1328962.
zakharov@vigg.ru

INTRODUCTION

It is generally accepted that the first colonization of America came from Asia during the last glaciation, through a Bering land bridge connecting the two continents. However, the timing, place of origin, and the number of waves of migration remain controversial questions debated by geneticists, anthropologists, archaeologists, and linguists. The affinities between Asians and Native Americans (Amerinds) are well established, and are based on cultural, morphological, and genetic similarities between populations of the New World and Siberia. However, no particular population in Siberia has been proposed as direct descendant from ancient groups related to the American founder populations.

The peoples of the Altai-Baikal region are extremely variable anthropologically and are of considerable interest from this point of view. According to anthropological evidence, the modern population of this region belongs to three local races (subraces) within the Mongoloid race. The local races differ in the extent to which they manifest Mongoloid traits. The West Siberian group of the Ural race is represented by North Altaic peoples, Shorians, and several Khakassians ethnographic groups; the South Siberian race is represented by Khachin Khakassians. Tuvinians, Southern Altaians, some Khakassian groups, and Buriats are assigned to the Central Asian type of the North Asian race, which have the most complete complex of Mongoloid traits. Paleoanthropologic evidence indicates that the formation of the anthropological composition of the modern Altai-Baikal ethnic groups was largely based on ancient Mongoloid elements, although—due to an ancient Caucasoid admixture—the Mongoloid component was significantly reduced in indigenous peoples residing west of the Yenisey River.

However, the use of anthropological data for more detailed analysis aimed at investigating the origin and ethnic history of certain populations is hampered by their incompleteness, methodical incompatibility, and, in some cases, by the absence of paleoanthropological data. In contrast, modern molecular genetic methods offer the opportunity to reconstruct the process of gene pool development based on their present-day state. Mitochondrial DNA (mtDNA), which is inherited maternally and without recombination, is a very informative genetic system for describing gene pool structure and history. To date it has been established that human mitochondrial gene pools are represented by combinations of mtDNA type groups, each of which stems from a single founder. Phylogenetic analysis of mtDNA types in populations of different ethnic and racial origins made it possible to establish continental specificity for the majority of mtDNA type groups. Caucasoid gene pools mainly comprise ten mtDNA groups (HV, H, V, J, T, U, K, I, W, and X), while in Mongoloid populations ten other groups (A, B, C, D, E, F, G, Y, Z, and M*) prevail. Gene pools of Amerind populations are represented by five groups [A, B, C, D, and X (very rare)].[1,2] Thus, the data on mtDNA polymorphisms can be utilized for precise identification of female lineage, which, in turn, enables us to determine the ratio between genetic components of different origins in mixed population.

Here, we present evidence on the structure and diversity of the gene pools of several ethnic groups of the Altai-Baikal highlands based on mtDNA restriction polymorphism analysis and sequencing of the control region hypervariable segment I (HVS-I).

MATERIALS AND METHODS

Tissue samples (hair follicles and blood) were collected from 527 individuals representing eight ethnic populations of Central Asia during expeditions in 1997–2001. Judging by questionnaire data, all individuals examined were unrelated at least for three generations.

Total DNA was extracted from tissue samples using standard techniques.[3] Screening for polymorphic sites to determine the main haplogroups of mtDNA types distributed in the populations of Eurasia was conducted through the analysis of mtDNA fragments in polymerase chain reaction with the primers described previously.[7] Restriction fragments were separated by electrophoresis in 8% polyacrylamide gel. Gels were stained with ethidium bromide and DNA fragments were visualized by UV light. Polymorphism was scored by the presence or absence of restriction endonuclease recognition sites.

A total of 401 bp [from nucleotide position 16000 through up 16400 of the Cambridge reference sequence, encompassing the control region hypervariable segment I (HVS-I)] were sequenced.[8] For each mtDNA, the HVS-I was read in both directions by direct sequencing of PCR products.

Diversity of mtDNA types (h) was calculated according to Nei and Tajima[9]:

$$h = (1 - \Sigma x^2) N / (N-1),$$

where x is the population frequency of each mtDNA type and N is the sample size. Phylogenetic analysis was carried out using the PHYLIP 3.5c software package.[10] Phylogenetic trees were constructed using the NJ (neighbor-joining) method.[11] Materials and methods are described in more detail in other publications.[12–15]

RESULTS

We present data on mtDNA variation in 527 individuals from eight indigenous populations inhabiting the Altai-Baikal region of South Siberia (FIG. 1) and belonging to different linguistic groups of the Altaic linguistic family. Altaians, Khakassians, Shorians, Tuvinians, Todjins, Tofalars, and Sojots belong to Turkic-speaking group, while Buryats are Mongolic-speaking.

The prevalence of mtDNA groups in the Siberian populations examined is demonstrated in TABLE 1. As shown in the table, mitochondrial gene pools of the populations examined were characterized by different ratios between Mongoloid (M*, C, D, E, G, A, B, and F) and Caucasoid (H, I, J, K, T, U, and X) mtDNA lineages. All populations studied carried the marked Mongoloid component, whose maximum frequency was observed in Buryats and Sojots (90.2% and 88.3%, respectively). More than the half of Mongoloid mtDNA lineages in each population belongs to different haplogroups of macrohaplogroup M, and haplogroups C and D of this macrohaplogroup are the most prevalent. Maximum frequency of C (62.1%) mtDNA lineage was observed in Tofalars. Tuvinian and Todjins are also characterized by high frequencies of this haplogroup (more than 40%). mtDNA types belonging to haplogroup D were found in all populations except Tofalars. Moreover, the frequency of D mtDNA in Buryats was about 30%. mtDNA types belonging to haplogroups A, B, and F from macrohaplogroup R and distributed predominantly among Mon-

TABLE 1. Prevalence (in %) of mtDNA groups and genetic diversity of Altai-Baikal ethnic populations

mtDNA Group	Altaians (N=110)	Shorians (N=42)	Khakassians (N=54)	Tuvinians (N=90)	Todjins (N=48)	Sojots (N=34)	Buryats (N=91)	Tofalars (N=58)
C	19.1	7.1	35.2	47.8	47.9	17.6	28.6	62.1
D	15.5	9.5	9.3	17.8	4.2	50.0	33.0	0
E	0.9	0	0	2.2	0	9.0	14.3	1.7
G	0.9	0	0	4.4	18.8	0	0	0
M*	11.8	2.4	0	1.1	4.2	0	4.4	5.2
A	0	0	3.7	1.1	4.2	8.8	2.2	5.2
B	3.6	2.4	5.6	7.8	4.2	2.9	6.6	3.5
F	8.2	43.0	22.0	2.2	2.1	0	1.1	0
H	6.4	21.4	3.7	1.1	2.1	0	2.2	6.9
U	16.4	0	11.1	3.3	6.2	5.9	1.1	0
K	0	2.4	0	0	0	0	0	0
J	3.6	11.9	1.9	5.6	0	0	2.2	8.6
T	0.9	0	1.9	1.1	0	0	1.1	5.2
I	1.8	0	0	0	0	0	1.1	0
X	2.7	0	0	0	0	0	0	0
Other	8.2	0	5.6	4.4	6.2	5.9	2.2	1.7
h	0.886	0.758	0.811	0.733	0.735	0.717	0.789	0.603

NOTE: h is the diversity of mtDNA types (see text for details).

TABLE 2. The frequency of race-specific mtDNA groups in the gene pools of Altai-Baikal ethnic populations

Population	Gene pool component, frequency (%)			
	Mongoloid	"American" (A, B, C, D, in total)	Caucasoid	Unidentified
Altaians	60.0	38.2	31.8	8.2
Shorians	64.4	19.0	35.7	0
Khakassians	75.8	53.8	18.6	5.6
Tuvinians	84.4	74.5	11.1	4.4
Todjins	85.6	60.5	8.3	6.2
Sojots	88.3	79.3	5.9	5.9
Buryats	90.2	70.4	7.7	2.2
Tofalars	77.7	72.5	20.7	1.7

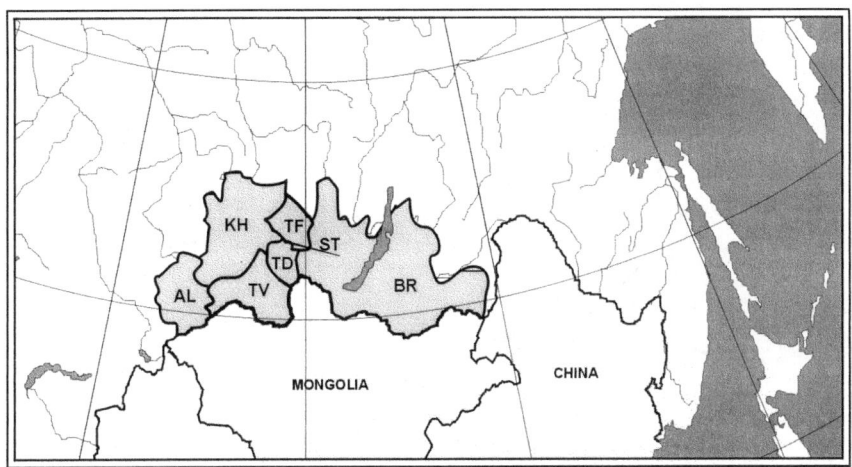

FIGURE 1. Geographic locations of populations studied. AL, Altaians; KH, Khakassians; TV, Tuvinians; TD, Todjins; TF, Tofalars; ST, Sojots; and BR, Buryats.

goloid populations were discovered in most of the populations studied, with a frequency of no more than 10% (except for the Shorians and the Khakassians, in whom group F was very frequent).

The fact that the mtDNA groups mentioned above form the basis of the Asian Mongoloid gene pools makes it possible to quantitate the proportion of the Mongoloid component in the gene pools of the Altai-Baikal populations.

As shown in TABLE 2, all populations examined were characterized by the high frequency of Mongoloid mtDNA lineages. The highest frequencies of these lineages were observed in Buryats, Sojots, and Tuvinians (90.2%, 88.3%, and 84.4% respectively), typical representatives of Central Asian Mongoloids. However, in Shorians, Altaians, and to lesser extent in Khakassians, there was a marked Caucasoid component, represented by mtDNA types of groups H, U, K, J, T, and I. Among these groups, H and J were detected in the gene pools of nearly all populations, while mtDNA types of group U were found with rather high frequency in Altaians (16.4%) and in Khakassians (11.1%). The presence of group K mtDNA lineages was specific to the gene pools of Shorians. In addition, in Khakassians, Altaians, and Buryats sporadic mtDNA types belonging to groups T and I were detected.

In general, maximum frequency of the Caucasoid component was observed in Shorians (35.7%), while maximum diversity of the lineages determining the Caucasoid component was observed in Altaians, in which mtDNA types belonged to five groups.

Types of mtDNA not belonging to any of the known mtDNA groups and defined as "other" (TABLE 1) represent a yet-unidentified component, the proportion of which in the gene pools of the ethnic groups studied is low and varies from 0–8.2% (TABLE 2).

The presence of the X mtDNA type in the gene pool of Altaians deserves special attention. Earlier, it was shown that haplogroup X was characterized by mosaic dis-

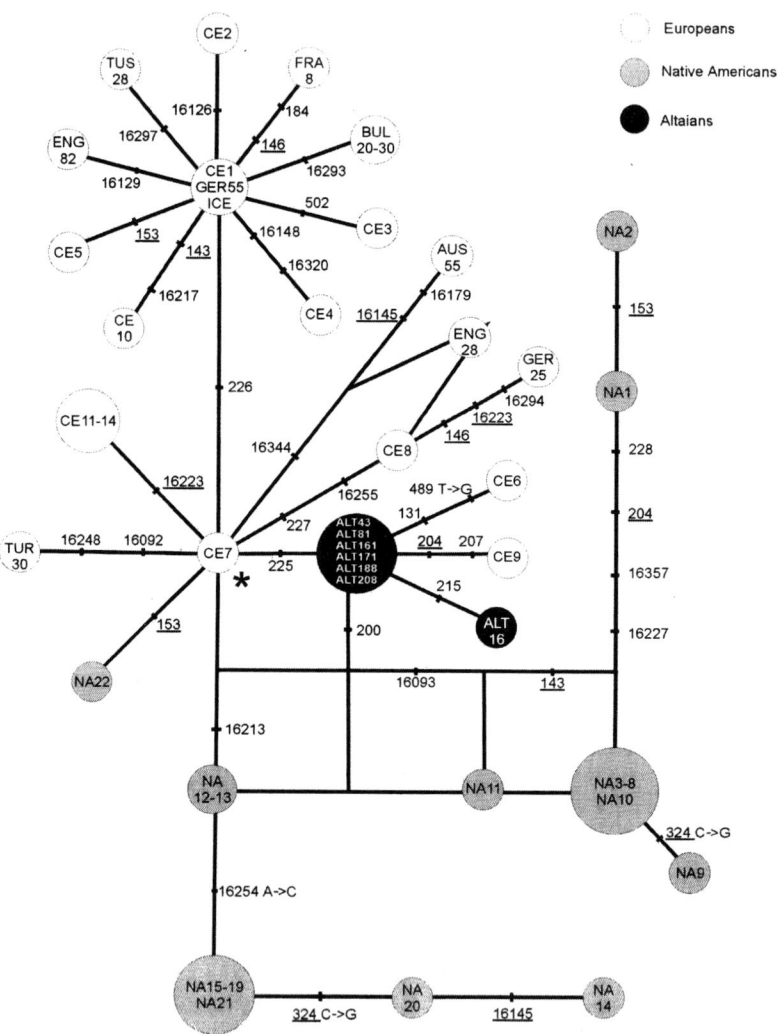

FIGURE 2. Reduced median network of haplogroup X mtDNAs as defined by HVSI and HVSII variation. Sample origin is indicated by the scheme (see key). mtDNA types are represented by *circles*, with areas proportional to number of individuals. Lines are labeled by HVSI and HVSII mutations. Nucleotide variants correspond to transitions, while transversions are further specified. The node marked with an *asterisk* matches the haplogroup X basic motif: −1715C, +14465S, 16189C, 16223T, 16278T, +16517E/16519C, 153A, 195C, and 225A.

tribution. While these mtDNAs were found in the populations of Europe, the Caucasus, and Western Asia, they were also present in the populations of North American Indians.

Comprehensive phylogenetic analysis of the X mtDNAs control region sequences has demonstrated the common ancestry of Caucasoid and American X mtDNA lineages. The time of their divergence was estimated as 12,000 to 36,000 years.[16]

Thus, according to modern ideas, haplogroup X is one of the major founding mitochondrial haplogroups (A, B, C, and D), which formed the gene pools of Native Americans. It should be noted in this respect, that the absence of haplogroup X in the populations of Eastern Asia has been a major obstacle to the reconstruction of the process of peopling of America. However, the discovery of haplogroup X lineages in the populations of Northern and Southern Altaians[17] suggests that the presence of this rare mtDNA haplogroup in South Siberia populations is evidence for the involvement of these populations in the colonization of America, since, according to the data of phylogenetic analysis, X mtDNA variants found in Altaians occupy an intermediate position between Caucasoid and North American X lineages (FIG. 2).

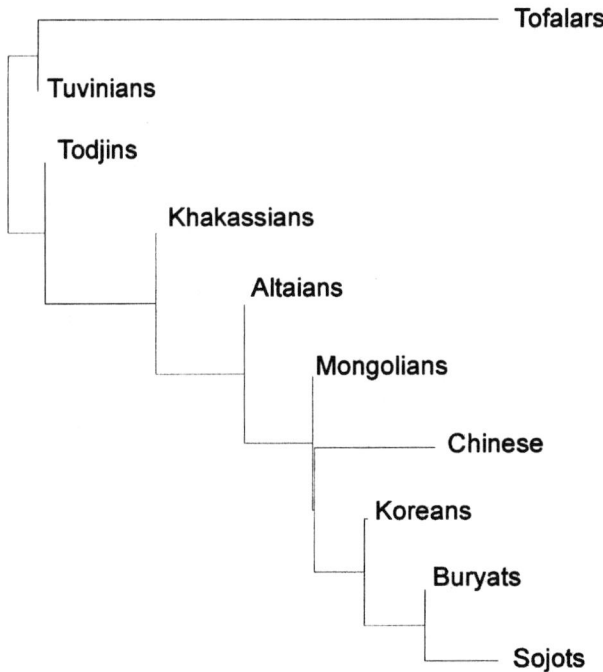

FIGURE 3. mtDNA neighbor-joining tree of several Asian populations. Tofalars, Tuvinians, Todjins, Khakassians, Altaians, Buryats and Sojots are from the present study; Mongolians are from Kolman and colleagues;[18] Koreans and Chinese are from Horai and colleagues.[19]

All populations of the Altai-Baikal region are characterized by a high level of genetic diversity, maximally in the population of Altaians (TABLE 1). Despite the fact that the gene pools of the populations examined were represented by a similar set of mtDNA type groups, the population of the Altai-Baikal region cannot be considered as genetically homogeneous as there were statistically significant differences in the frequencies of mtDNA type groups ($P<0.01$) between all population pairs compared.

The results of phylogenetic analysis of mtDNA variations in 10 populations from East and Central Asia, and from South Siberia are presented in FIGURE 3. The NJ tree shows that the populations examined differentiate into several groups.

The first group comprises two linguistically related Siberian populations (Tuvinians and Tofalars), while the last one includes Buryats and Sojots, who cluster together with Koreans. Generally, the distribution of the populations on the phylogenetic tree pointed to the presence of certain geographic trends, which probably reflected the existing ethnic relationships and the gene migrations associated with them.

DISCUSSION

For Native Americans, extensive RFLP and HVS-1 sequence analysis has unambiguously identified four major founding mtDNA haplogroups (A, B, C, and D), which, together, account for ~97% of modern Native American mtDNAs.[1] The distribution of the four founding lineage haplotypes (A, B, C, and D) in Native American populations (both contemporary and ancient) shows that all four lineages were present in the New World prior to European contact,[1,20,21] thus indicating that all Native American mtDNAs are apparently descendent from these four founding lineages. A striking example of the presence of nonhaplogroup A–D genotypes in Native Americans is haplogroup X mtDNAs, which represents a minor founding lineage restricted in distribution to northern Amerindian groups.[16]

Detection of all four haplogroups (A, B, C, and D) in an Asian population is a first criterion in the identification of a possible New World founder. The location and identification of the population that was the source for the founding lineages of the New World is a question of considerable interest. Markers defining all four founding haplogroups were found in Chinese (Han),[22] Tibetans,[5] and Mongolians.[18] It has been proposed[18] that Mongolian populations exhibit a higher percentage (48%) of New World haplogroups than any Asian population that carries all four haplogroups, and therefore appear to be the strongest candidates for New World founders. The lack of haplogroup B mtDNA in Siberian populations is considered a reason to exclude Siberians from possible New World founders.[4,18] However, our analysis has shown that haplogroup B mtDNAs are present in Buryats and Tuvinians[12,13,23] as well as in Altaians.[14,15,24] It appears that haplogroup B was part of the ancestral gene pool for South Siberian populations and that these populations played an important role in the passage of this mtDNA lineage into the New World.

TABLES 1 and 2 displays the frequencies of the four founding lineages in Siberian and Central Asian populations that carry all four haplogroups. The Buryats, Tuvinians and Sojots studied here carry all four founding haplogroups with total frequencies of 70.4%, 74.5%, and 79.3%, respectively, exceeding those reported previously for Mongolians (48%), Chinese (45%), and Tibetans (31%).[5,18,22] The candidate source population for Native American mtDNA haplotypes therefore may include the populations originating in the region to the southwest of Lake Baikal, including the Altai Mountain region, where high frequencies of A–D lineages were also observed.[14,24]

On the basis of the last accumulated data in the Laboratory of Human Genetics at the Vavilov Institute of General Genetics of the Russian Academy of Sciences (Moscow, Russia), we undertook an analysis of the geographic distribution of frequencies of HVR I motifs associated with haplogroups A, B, C, and D of mtDNA.[4,25–27] The distribution of haplogroups on the territory is quite heterogeneous (FIGS. 4–7). For example, the motif 16298C, 16223T, 16327T (or haplogroup C) is most widely spread in the north of the territory, and motif 16217C, and 16189C (haplogroup B) in the south. The map legends in turn point out the wide range of variability of fre-

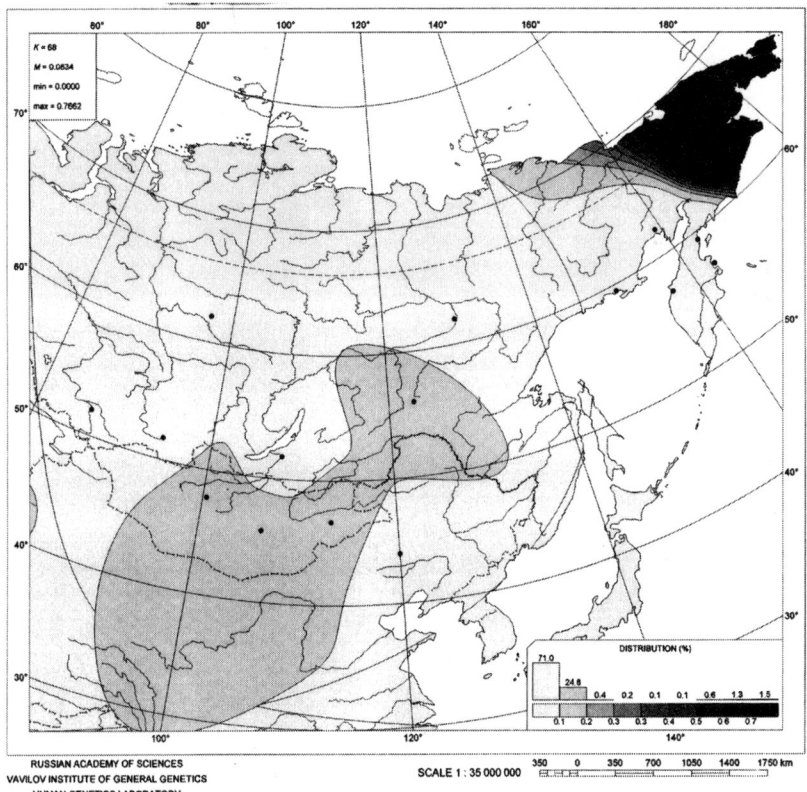

FIGURE 4. Map of geographic distribution of HVS I motif [16223 T–16290 T–16319 A] associated with mtDNA haplogroup A.

quency of mtDNA motifs (from 0 to 70–90%) in different populations. Despite the evident diversity of the genetic landscape, the frequency of considered markers shows a common latitude-oriented direction of geographic variability, although expressed in different degrees (from 30–70% of frequency correlation with geographic latitude). On the one hand, this correlates with the division of native Asian inhabitants into northern and southern Asian populations. On the other hand, it agrees with the hypothesis that the colonization of this part of the continent took place precisely in a south-north direction. Thus one ought to expect the maximum variability (i.e., the presence in the gene pool of a full set of haplogroups each having the meaningful frequency) in the south of the considered territory. To verify this proposition, we represent here the integral map, some generalized landscapes of distribution of four considered mtDNA motifs—markers of "Asian-American" haplogroups (FIG. 8). Having being created by association of interpolation maps of single signs, this map

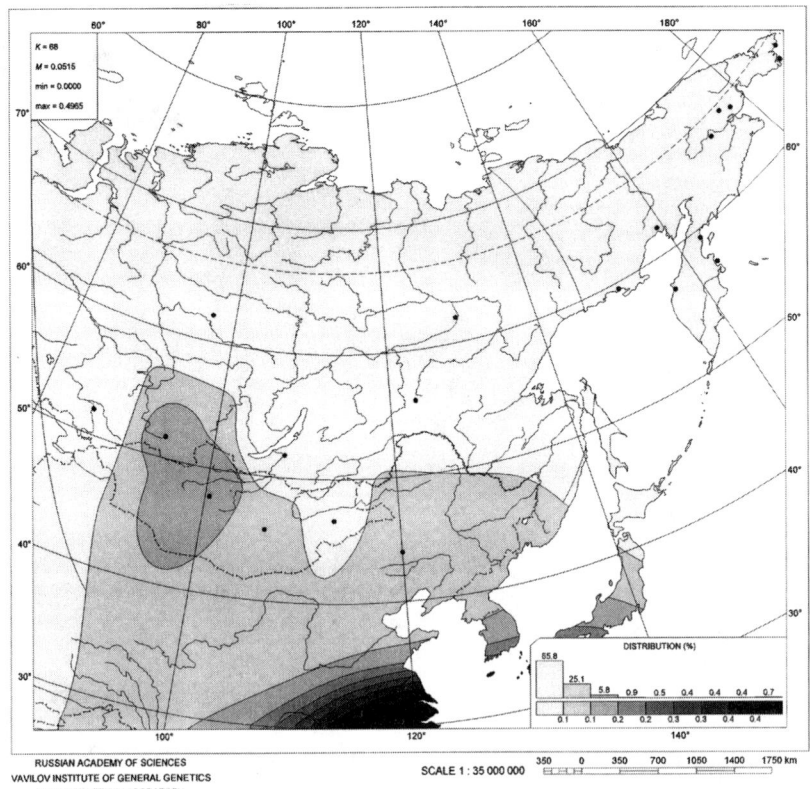

FIGURE 5. Map of geographic distribution of HVS I motif [16217C–16189C] associated with mtDNA haplogroup B.

shows those regions where all haplogroups are present with frequencies substantially different from zero. As seen on the map, these regions are Central Asia and South Siberia. In South Siberia, the Sayan Mountains are most remarkable—it is the ethnic area of the Tuvinians whose gene pool has the richest set of "Asian-American" mtDNA haplogroups. Taking into account that South Siberia as a whole and the Altai-Sayan region in particular were colonized by humans much earlier than the rest of Siberia, it is conceivable that in this particular region the mtDNA diversity represented by four haplogroups was formed and spread further, to the north of Asia and to the New World.

Moreover, it has been proposed that the source of the Y chromosome founders in the New World was a population occupying the general area including Lake Baikal, the Lena River headwaters, the Angara and Yenisey River basins, the Altai Mountain foothills, and the region south of the Sayan Mountains (including Tuva and western

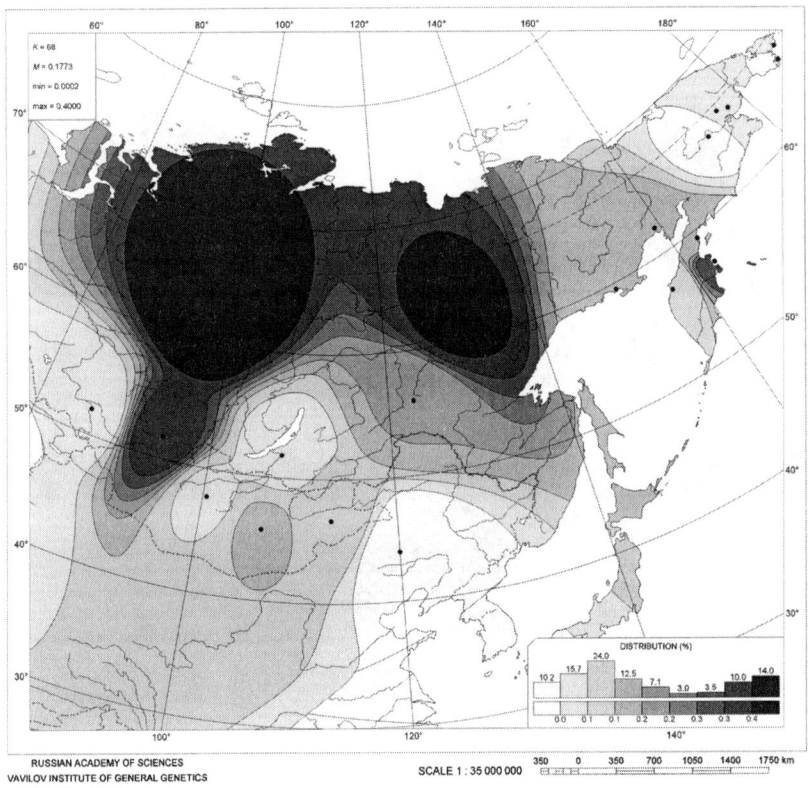

FIGURE 6. Map of geographic distribution of HVS I motif [16298C–16223T–16327T] associated with mtDNA haplogroup C.

Mongolia).[28] Thus, the results of our study demonstrate a concordance between inferences based on maternally inherited genetic systems and those based on paternally inherited genetic systems.

The earliest traces of human presence in the Altai (Denis's cave) date back to 282 ± 56 thousand years ago; the transfer from middle-paleolithic to late-paleolithic industry, i.e., the appearance of *Homo sapiens* there, dates to 50–60 to 45–35 thousand years ago. The age of the most ancient site in Sayan (Malaya Syya) is 34,500 ± 450 years.[29] Of importance is the evidence of archeologists that the Ustkarakol industrial tradition established in the Altai spread from this region to northeastern Asia as far as Beringia.[29]

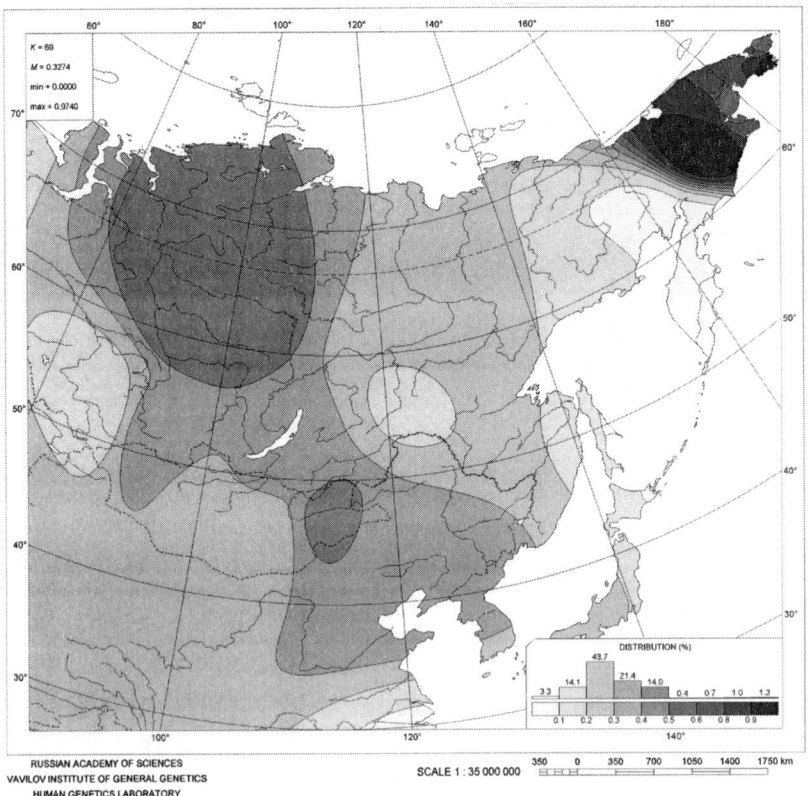

FIGURE 7. Map of geographic distribution of HVS I motif [16362C–16223T] associated with mtDNA haplogroup D.

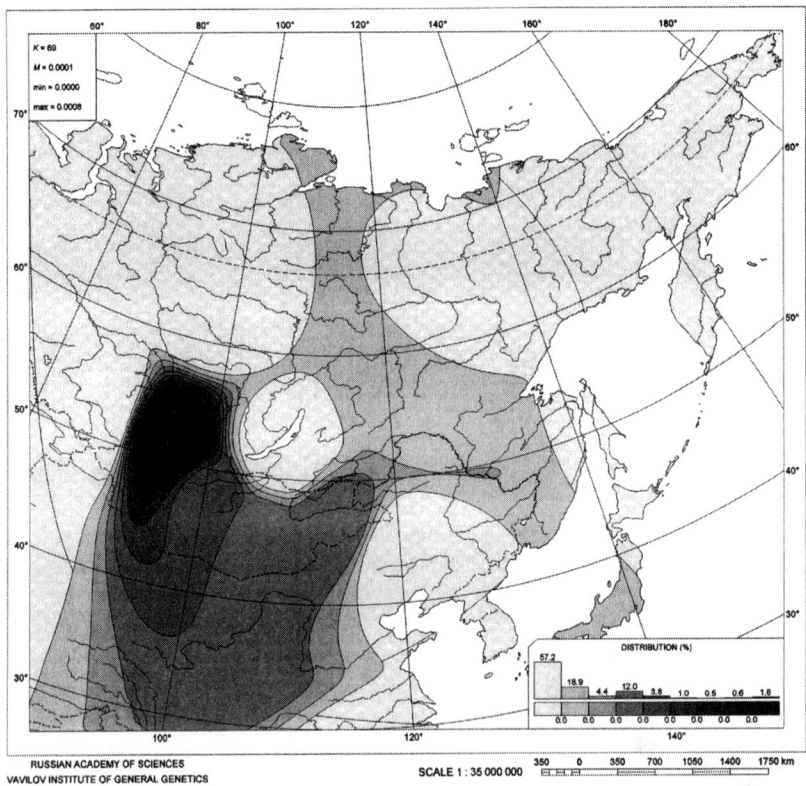

FIGURE 8. Generalized landscape of distribution of frequencies of HVS I motifs associated with haplogroups A, B, C, and D.

REFERENCES

1. WALLACE, D.C. 1995. Mitochondrial DNA variation in human evolution, degenerative disease and aging. Am. J. Hum. Genet. **57:** 201–223.
2. MACAULAY, V., M. RICHARDS, E. HICKEY, et al. 1999. The emerging tree of West Eurasian mtDNAs: A synthesis of control-region sequences and RFLPs. Am. J. Hum. Genet. **64:** 232–249.
3. WALSH, P.S., D.A. METZGER & R. HIGUCHI. 1991. Research report: Chelex 100 as a medium for simple extraction of DNA for PCR-based typing from forensic material. BioTechniques **10:** 506–513.
4. TORRONI, A., R.I. SUKERNIK, T.G. SCHURR, et al. 1993. mtDNA variation of aboriginal Siberians reveals distinct genetic affinities with Native Americans. Am. J. Hum. Genet. **53:** 591–608.
5. TORRONI, A., J.A. MILLER, L.G. MOORE, et al. 1994. Mitochondrial DNA analysis in Tibet: implications for the origin of the Tibetan population and its adaptation to high altitude. Am. J. Phys. Anthropol. **93:** 189–199.
6. TORRONI, A., K. HUOPONEN, P. FRANCALACCI, et al. 1996. Classification of European mtDNAs from an analysis of three European populations. Genetics **144:** 1835–1850.

7. FINNILA, S., I. HASSINEN, L. ALA-KOKKO, et al. 2000. Phylogenetic network of the mtDNA haplogroup U in Northern Finland based on sequence analysis of the complete coding region by conformation-sensitive gel electrophoresis. Am. J. Hum. Genet. **66:** 1017–1026.
8. SANGER, F., S. NICKLEN & A.R. COULSON. 1977. DNA sequencing with chain-terminating inhibitors. Proc. Natl. Acad. Sci. USA **74:** 5463–5467.
9. NEI, M. & F. TAJIMA. 1981. DNA polymorphism detectable by restriction endonucleases. Genetics **97:** 145–163.
10. FELSENSTEIN, J. 1989. PHYLIP: Phylogeny Inference Package (version 3.2). Cladistics **5:** 164–166.
11. SAITOU, N. & M. NEI. 1987. The neighbor-joining method: a new method for reconstructing phylogenetic trees. Mol. Biol. Evol. **4:** 406–425.
12. DERENKO, M.V., I.K. DAMBUEVA, B.A. MALIARCHUK, et al. 1999. Structure and diversity of the mitochondrial gene pool of the aboriginal population of Tuva and Buriatia from restriction polymorphism data. Genetika **35:** 1706–1712.
13. DERENKO, M.V., B.A. MALYARCHUK, I.K. DAMBUEVA, et al. 2000. Mitochondrial DNA variation in two South Siberian aboriginal populations: Implications for the genetic history of North Asia. Human Biol. **72:** 945–973.
14. DERENKO, M.V., G.A. DENISOVA, B.A. MALYARCHUK, et al. 2001. The structure of the gene pools of the ethnic populations of Altai-Sayan region based on of mitochondrial DNA polymorphism data. Genetika **37:** 1402–1410.
15. DERENKO, M.V., B.A. MALYARCHUK, G.A. DENISOVA, et al. 2002. Molecular genetic differentiation of the ethnic populations of South and East Siberia based on mitochondrial DNA polymorphism. Genetika **38:** 1409–1416.
16. BROWN, M.D., S.H. HOSSEINI, A. TORRONI, et al. 1998. mtDNA haplogroup X: an ancient link between Europe/Western Asia and North America? Am. J. Hum. Genet. **63:** 1852–1861.
17. DERENKO, M.V., T. GRZYBOWSKI, B.A. MALYARCHUK, et al. 2001. The presence of mitochondrial haplogroup X in Altaians from South Siberia. Am. J. Hum. Genet. **69:** 237–241.
18. KOLMAN, C.J., N. SAMBUUGHIN & E. BERMINGHAM. 1996. Mitochondrial DNA analysis of Mongolian populations and implications for the origin of New World founders. Genetics **142:** 1321–1334.
19. HORAI, S., K. MURAYAMA, K. HAYASAKA, et al. 1996. MtDNA polymorphism in East Asian populations, with special reference to the peopling of Japan. Am. J. Hum. Genet. **59:** 579–590.
20. LALUEZA, C., A. PEREZ-PEREZ, E. PRATS, et al. 1997. Lack of founding Amerindian mitochondrial DNA lineages in extinct Aborigines from Tierra del Fuego-Patagonia. Hum. Mol. Genet. **6:** 41–46.
21. STONE, A.C. & M. STONEKING. 1998. MtDNA analysis of a prehistoric Oncota population: implications for the peopling of the New World. Am. J. Hum. Genet. **62:** 1153–1170.
22. BALLINGER, S.W., T.G. SCHURR, A. TORRONI, et al. 1992. Southeast Asian mitochondrial DNA analysis reveals genetic continuity of ancient mongoloid migrations. Genetics **130:** 139–152.
23. DERENKO, M., B. MALYARCHUK, I. DAMBUEVA, et al. 1998. Buryat and Tuva populations from South Siberia exhibit the highest percentage of New World mtDNA haplogroups. Am. J. Hum. Genet. **63:** A211.
24. SUKERNIK, R.I., T.G. SCHURR, Y.B. STARIKOVSKAYA, et al. 1996. Mitochondrial DNA variation in native Siberians, with special reference to the evolutionary history of American Indians: studies on restriction endonuclease polymorphism. Genetika **32:** 432–439.
25. TORRONI, A., T.G. SCHURR, C.C. YANG, et al. 1992. Native American mitochondrial DNA analysis indicates that the Amerind and the Nadene populations were founded by two independent migrations. Genetics **130:** 153–162.
26. TORRONI, A., T.G. SCHURR, M.F. CABELL, et al. 1993 Asian affinities and continental radiation of the four founding Native American mtDNAs. Am. J. Hum. Genet. **53:** 563–590.

27. WARD, R.H., B.L. FRAIZER, K. DEW-JAGER, et al. 1991. Extensive mitochondrial diversity within a single Amerindian tribe. Proc. Nat. Acad. Sci. USA **88:** 8720–8724.
28. KARAFET, T.M., S.L. ZEGURA, O. POSUKH, et al. 1999. Ancestral Asian source(s) of New World Y-chromosome founder haplotypes. Am. J. Hum. Genet. **64:** 817–831.
29. DEREVIANKO, A.P. 2001. The transfer from the middle-palaeolith to the late-palaeolith in the Altai. Archeol. Ethnogr. Anthropol. Eurasia **7:** 70–103.

Mitochondrial ALDH2 Deficiency as an Oxidative Stress

SHIGEO OHTA,[a] IKUROH OHSAWA,[a] KOUZIN KAMINO,[a,c] FUJIKO ANDO,[b] AND HIROSHI SHIMOKATA[b]

[a]*Department of Biochemistry and Cell Biology, Institute of Development and Aging Sciences, Graduate School of Medicine, Nippon Medical School, Kosugi, Kawasaki, Kanagawa, 211-8533 Japan*

[b]*Department of Epidemiology, National Institute for Longevity Sciences, Obu, Aichi, 474-8522 Japan*

ABSTRACT: Mitochondrial aldehyde dehydrogenase 2 (ALDH2) plays a major role in ethanol metabolism. It is involved in acetaldehyde detoxification. A polymorphism of the ALDH2 gene is specific to North-East Asians. Sensitivity to ethanol is highly associated with this polymorphism (*ALDH2*2* allele), which is responsible for a deficiency of ALDH2 activity. We first show that this deficiency influences the risk for late-onset Alzheimer's disease (LOAD) by a case-control study in a Japanese population. In a comparison of 447 patients with sex, age, and region-matched non-demented controls, the genotype frequency for the *ALDH2*2* allele was significantly higher in the patients than in the controls ($P=0.001$). Next, we examined the combined effect of the *ALDH2*2* and the apolipoprotein E4 allele (*APOE-ε4*), which has been confirmed to be a risk factor for LOAD. The *ALDH2*2* allele more significantly affected frequency and age at onset in patients with *APOE-ε4* than in those without it. These results indicate that the ALDH2 deficiency is a risk factor for LOAD, acting synergistically with the *APOE-ε* allele. Next, to elucidate the molecular mechanism involved, we obtained *ALDH2*-deficient cell lines by introducing mouse mutant ALDH2 cDNA into PC12 cells. We speculate that ALDH2 may act to oxidize toxic aldehyde derivatives. Then, we found that the *ALDH2*-deficient transfectants were highly vulnerable to exogenous 4-hydroxy-2-nonenal, an aldehyde derivative generated from peroxidized fatty acids. In addition, the *ALDH2*-deficient transfectants were sensitive to oxidative insult induced by antimycin A, accompanied by an accumulation of proteins modified with 4-hydroxy-2-nonenal. Mitochondrial ALDH2 functions as a protector against oxidative stress.

KEYWORDS: aldehyde dehydrogenase; ethanol metabolism; Alzheimer's disease; oxidative stress; 4-hydroxy-nonenal; peroxide

---

Address for correspondence: Shigeo Ohta, Department of Biochemistry and Cell Biology, Institute of Development and Aging Sciences, Graduate School of Medicine, Nippon Medical School, Kosugi, Kawasaki, Kanagawa, 211-8533 Japan. Voice: +81-44-733-9267; fax: +81-44-733-9268.

ohta@nms.ac.jp

[c]Present address: Division of Psychiatry and Behavioral Proteomics, Department of Post-Genomics and Diseases, Osaka University Graduate School of Medicine, Suita, Osaka, 565-0871 Japan.

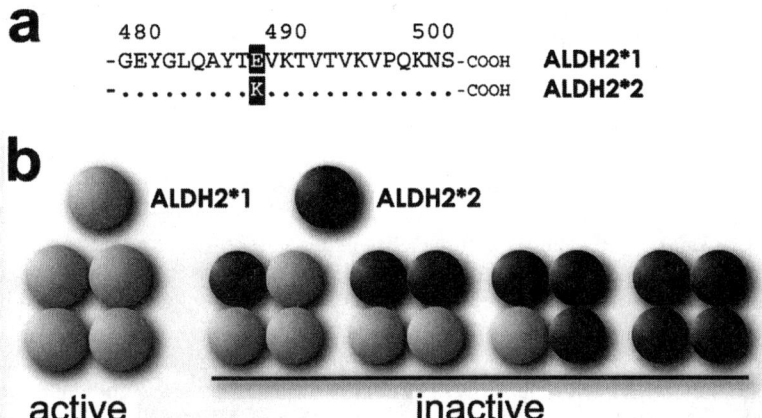

FIGURE 1. A polymorphism specific to North-East Asians in the ALDH2 gene. (a) C-terminal amino acid sequences of the active and inactive subunits termed *ALDH2*1* and *ALDH2*2*. (b) Schematic representation of a homotetrameric enzyme, ALDH2. All tetramers containing at least one *ALDH2*2* subunit are inactive.

INTRODUCTION

A Polymorphism of Aldehyde Dehydrogenase 2 Specific to North-East Asians

Mitochondrial aldehyde dehydrogenase 2 (ALDH2) is located in the matrix of mitochondria and plays a major role in metabolizing acetaldehyde produced from ethanol into acetate. A mutant allele, *ALDH2*2*, has a single point mutation (G/A) in exon 12 of the active *ALDH2* gene and is confined to North-East Asians. The mutation results in the substitution of glutamic acid 487 with lysine (E487K), acting in a dominant negative fashion (FIG. 1). Individuals with the *ALDH2*2* allele exhibit the alcohol flushing syndrome, attributable to an elevated blood acetaldehyde level. The *ALDH2*2* allele has been also reported to affect the metabolism of other aldehydes such as benzaldehyde, which is a metabolite of toluene, and chloroacetaldehyde, which is generated during the metabolism of vinyl chloride. However, the risks have been mainly associated with alcohol consumption. We directed our attention to the genetic role of ALDH deficiency to help us understand the physiological role of ALDH2.

ASSOCIATION OF ALZHEIMER'S DISEASE WITH ALDH2 DEFICIENCY

A Large-Scale Case-Control Study on Alzheimer's Disease with ALDH2 Deficiency

Late-onset Alzheimer's disease (LOAD) is a complex disease caused by multiple genetic and environmental factors upon aging. It was pointed out that alcohol intake could affect the development of LOAD, because ethanol and its metabolite, acetaldehyde, are directly neurotoxic, and patients with a history of alcohol abuse show alterations in neurotransmitting molecules in the brain, such as the muscarinic cholinergic receptor and serotonin. On the other hand, epidemiological studies have provided conflicting results, which may be explained by genetic factors that modify ethanol metabolism and potentially influence alcohol-drinking behavior.

To understand the genetic effect of *ALDH2*2*, we performed a large-scale case-control study in patients with LOAD by examining the frequency of *ALDH2*2*. Patients with LOAD and controls in three areas of Japan (447 patients and as many controls) were examined to find the effect of the *ALDH2*2* allele on the risk for LOAD.[1] The controls were strictly selected to match the patients in age, gender, and area. Since the ALDH2 deficiency appears in a dominant-negative fashion, homozygous and heterozygous carriers of the allele were combined in evaluating the risk for LOAD. The frequency of carriers with the *ALDH2*2* allele (*1/2* and *2/2*) was significantly higher in the patients than in the controls [odds ratio (O.R.)=1.6, $P=0.001$]. This trend was evident in both males (O.R.=1.9, $P=0.01$) and females (O.R.=1.4, $P=0.02$).

Synergistic Effect by APOE-ε4

To confirm the effect of *ALDH2*2* on LOAD, we examined the interaction between the *APOE-ε4* and *ALDH2*2* alleles.[1] Since *APOE-ε4* has been established as a risk of LOAD, a synergistic effect of the two genes would strongly support that the *ALDH2*2* allele is also a risk factor. Harboring of the *ALDH2*2* allele synergistical-

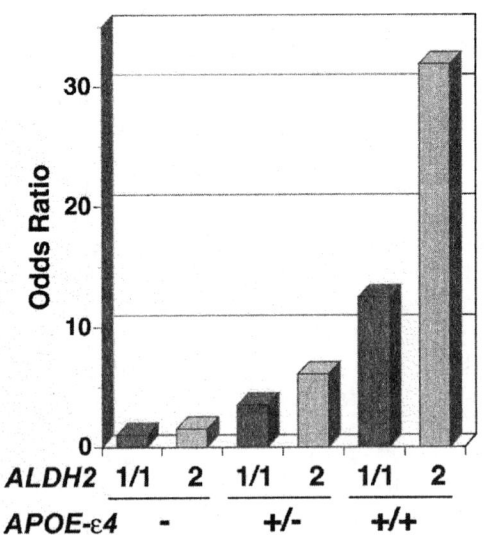

FIGURE 2. Synergistic effect on onset of LOAD by *ALDH2*2* with *APOE-ε4*. Relative risks of LOAD were estimated by the frequencies of patients ($N=447$) and controls ($N=447$) with each allele of the *APOE4* and *ALDH2* gene. ALDH2 1/1, carrier of homozygous *ALDH2*1*; *ALDH2*2*, carrier of homozygous and heterozygous *ALDH2*2*; *APOE ε4* –, no *APOE ε4*; +/–, heterozygous *APOE ε4*; and +/+, homozygous *APOE ε4*.

ly increased the odd ratios of patients with the *APOE-ε4* allele (FIG. 2), supporting that the *ALDH2*2* allele is indeed a risk factor for LOAD.

Next, the effect of these alleles on age at onset was examined. In all patients with LOAD, the difference in age at onset was independent of the *APOE-ε4* allele. In contrast, those patients with *ALDH2*2* (*1/2* or *2/2*) and homozygous for *APOE-ε4* showed a significantly earlier onset than other patients. In addition, a dosage effect of the *ALDH2*2* allele on age at onset showed a significant trend in patients homozygous for the *APOE-ε4* allele by regression analysis ($P=0.028$).

Since logistic regression analysis indicates a significant effect of the *ALDH2*2* allele ($P=0.002$), the allele is an independent risk factor for LOAD from the *APOE-ε4* allele. Therefore, we conclude that the *ALDH2*2* allele is an independent risk for LOAD and shows synergistic effects with *APOE-ε4* in affecting not only the frequency of LOAD, but also the age at onset of Alzheimer's disease.

PHENOTYPE OF INDIVIDUALS WITH THE ALDH2 DEFICIENCY

Geriatric diseases, including LOAD, are associated with many factors; genetic, life-style–related, physiological, medical, nutritional, and psychological. Thus, it is important to clarify the contributions of genetic factors and other basic background factors. In 1997, we started gene-related investigations into various geriatric diseases in the National Institute for Longevity Sciences, Longitudinal Study of Aging (NILS-LSA).[2] The subjects numbered 2,259. They were community-dwelling males and females aged 40–79 years randomly selected from the area around NILS.

We examined the association of the *ALDH2*-deficient genotype with various other factors evaluated in NILS-LSA.[3] In addition to biochemical analyses of blood and urine, renal and liver functions, serum proteins and lipids, and a complete blood count, lipid peroxide (LPO), and geriatric disease markers were also examined. Several serum proteins, lipids, and LPO levels showed differences between the non-defective (*ALDH2*1/1*) and defective (*ALDH2*1/2* and *ALDH2*2/2*) ALDH2 individuals. However, these biochemical evaluations are notoriously affected by alcohol-drinking behavior. Indeed, subjects with the *ALDH2*1/1* genotype drank alcohol more frequently than those with *ALDH2*1/2* and *2/2*. Thus, we excluded the effects of alcohol-drinking behavior from the association of the *ALDH2*-deficient genotype with the evaluation. Data were analyzed with an adjustment for alcohol consumption by the least squares method in a general linear model. We found that the concentration of LPO in females differed significantly according to *ALDH2* genotype. The concentration was higher in females carrying at least one *ALDH2*2* allele (2.922 nmol/mL) than in those carrying *ALDH2*1/1* (2.781 nmol/mL; $P=0.003$), raising the possibility that oxidative stress increases in *ALDH2*-deficient individuals.[3]

ALDH2 AS A PROTECTOR AGAINST OXIDATIVE STRESS

Model for Explaining the Role of the ALDH2 Deficiency

Oxidative stress and lipid peroxidation caused by reactive oxygen species (ROS) are reported to play an important role in the pathogenesis of neurodegenerative dis-

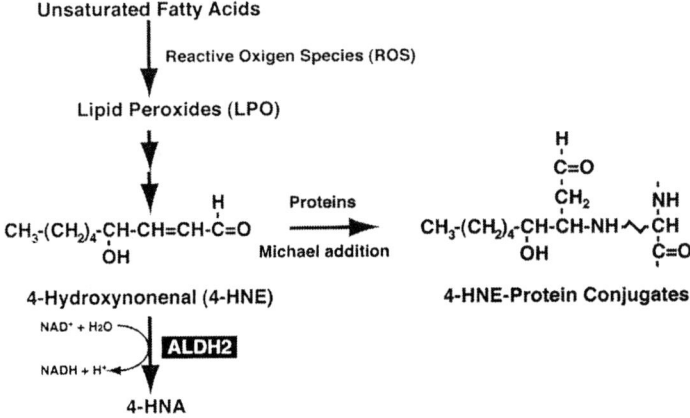

FIGURE 3. Involvement of ALDH2 in the metabolism of 4-HNE.

eases, including Alzheimer's disease. A major source of ROS is the mitochondrially derived superoxide anion radical, which induces membrane lipid peroxidation, thereby generating reactive aldehydes, including malondialdehyde (MALD) and trans-4-hydroxy-2-nonenal (4-HNE). A strong electrophile, 4-HNE, has the ability to readily adduct cellular proteins and may damage the proteins by interacting with lysine, histidine, serine, and cysteine residues.

Thus, we hypothesized that ALDH2 is involved in antioxidant defense through the oxidation of toxic aldehyde derivatives and its deficiency enhances oxidative stress (FIG. 3).

Construction of ALDH2-Deficient Cell Lines

To verify this hypothesis, we obtained ALDH2-deficient PC12 cells by transfection with a dominant-negative form of the mouse *Aldh2* gene.[4] Then, we examined the toxic effect of 4-HNE and found that exposure to 4-HNE resulted in more rapid decrease of viable cells in the ALDH2-deficient population than in control cells (FIG. 4). Exposure to 10 μM 4-HNE for about 2 h resulted in the appearance of round cells. At that time, the percentage of living ALDH2-deficient cells (K6 and K11) was 37% and 35%, whereas that of control cells (PC12, V, and E) was 99%, 85%, and 102%, respectively. Time-course study revealed that one day after exposure to 10 μM 4-HNE, the survival of ALDH2-deficient cells decreased rapidly, whereas that of control cells decreased gradually. The sensitivity of ALDH2-deficient cells to 4-HNE was dose dependent. These findings clearly show that ALDH2-deficient cells are less resistant to exogenous 4-HNE.

Effect of Generation of Superoxide on Cytotoxicity

Next, we tried to generate superoxide anion through exposure to an external insult. Partial inhibition of the mitochondrial electron transport at complex III by low concentrations of antimycin A induces the production of ROS and cell death. To in-

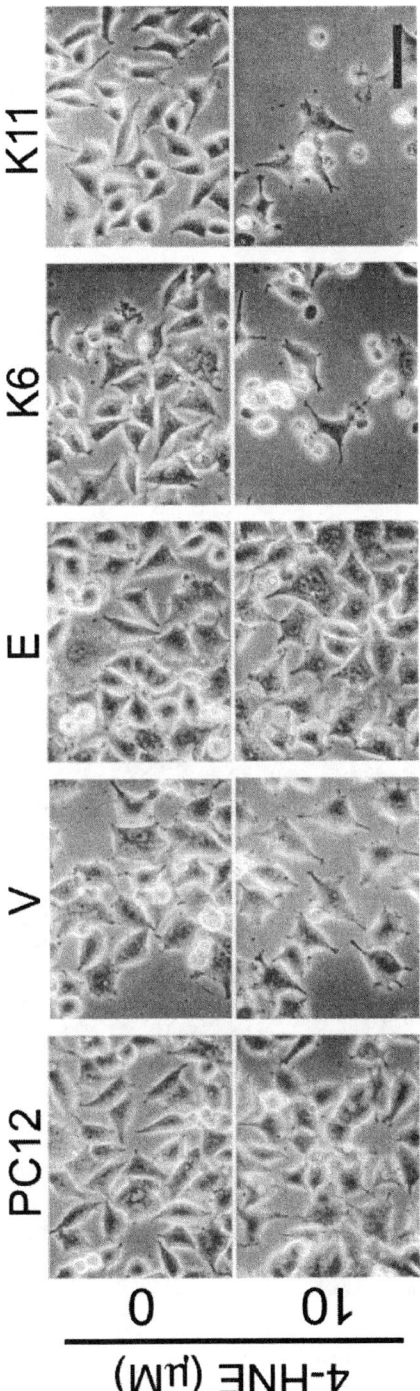

FIGURE 4. Rapid cell death of ALDH2-deficient PC12 transfectants after treatment with 4-HNE. PC12 or each transfectant (V, E, K6, or K11) was treated with 10 μM 4-HNE or ethanol (1/1,000 volume of medium) as a control (0 μM). One day after treatment, cells were observed under a phase-contrast microscope (×200). Bar=50 μm.

vestigate the effect of ALDH2 deficiency on cell vulnerability induced by oxidative stress, we examined the cellular toxicity of antimycin A in the ALDH2-deficient and parental cells of PC12.[4] In this experiment, we confirmed that the generation of ROS did not depend upon the type of transfectant. Then, we examined whether the accumulation of 4-HNE induced by the ROS differed between the ALDH2-deficient and normal cells. The accumulation after the exposure to antimycin A was measured with an anti–4-HNE antibody in immunocytochemical assays. A day after treatment with antimycin A (3 or 10 μg/mL), cellular 4-HNE immunoreactivity increased only in ALDH2-deficient cells, K6 and K11, but not in control cells (FIG. 5). These results strongly suggest that the ALDH2 deficiency caused the intracellular accumulation of 4-HNE, resulting in cell death.

ALDH2 deficiency was found to contribute to risks of diabetes,[5] cancer,[6] hypertension,[7,8] and myocardial infarction.[9] However, the risks have been mainly attributed to the association with alcohol consumption and the increase in the acetaldehyde concentration. In contrast, this study proposes that ALDH2 can contribute to the pathogenesis of various geriatric diseases by an alternative pathway, that is, the detoxification of cytotoxic products of lipid peroxidation.

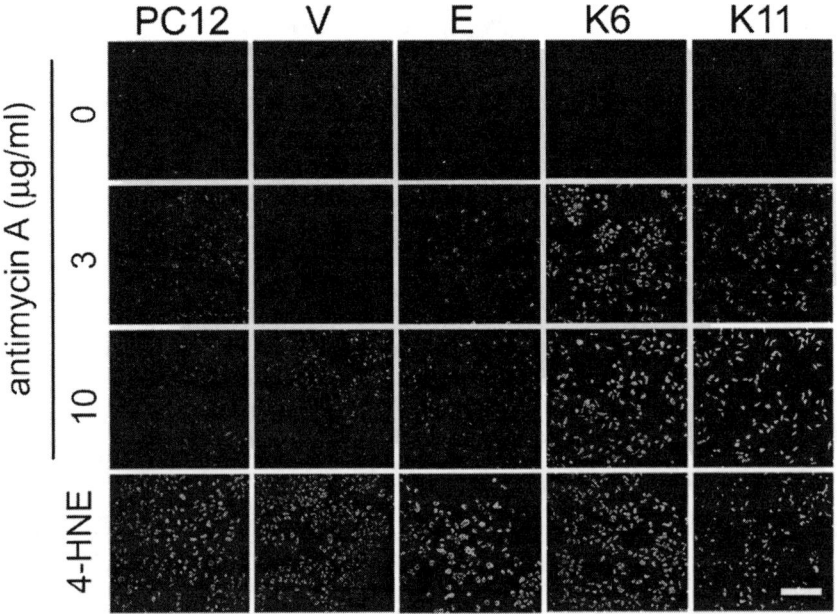

FIGURE 5. Accumulation of 4-HNE by superoxide. Cells were treated with the indicated concentration of antimycin A or 1 μM 4-HNE, and incubated for 24 h. After fixation, cells were stained with anti-4-HNE antibody. Bar=200 μm.

DISCUSSION OF THE ROLE OF ALDH2 DEFICIENCY IN OXIDATIVE STRESS

It has been shown that patients with Alzheimer's disease homozygous for *APOE-ε4* have greater 4-HNE adduct immunoreactivity associated with neurofibrillary tangles than those with other APOE genotypes. Studies of the interactions of *APOE* proteins with 4-HNE showed that the isoforms differ in the amount of 4-HNE they can bind, with the order ε2>ε3>ε4.[10] This correlated with the different abilities of APOE isoforms to protect against apoptosis induced by 4-HNE in cultured neurons. Our case-control study has revealed that ALDH2 deficiency is a risk factor for LOAD in a Japanese population, synergistically acting with *APOE-ε4*.[1] When compared with carriers of the *APOE-ε3/ε3* genotype, the risk for LOAD in Japanese subjects with the *APOE-ε4* allele is twice that in Caucasian subjects. The increased risk can partly be explained by the effect of the *ALDH2*2* allele, since this allele is very rare in non-Asian populations. Therefore, we suggest the possibility that in LOAD an enhancement of 4-HNE accumulation in Alzheimer's disease brain caused by ALDH2 deficiency may act synergistically with a weaker activity of APOE-ε4 to protect against neuronal cell death induced by 4-HNE. However, as Japanese patients with Alzheimer's disease are less numerous than Caucasian patients, other risks must overcome that posed by ALDH2 deficiency.

Taken together, our results suggest that mitochondrial ALDH2 functions to protect against oxidative stress. Thus, the metabolism of aldehyde including ALDH2 could be a preventive and therapeutic target in Alzheimer's disease and other neurodegenerative disorders.

REFERENCES

1. KAMINO, K., K. NAGASAKA, M. IMAGAWA, *et al.* 2000. Deficiency in mitochondrial aldehyde dehydrogenase increases the risk for late-onset Alzheimer's disease in the Japanese population. Biochem. Biophys. Res. Commun. **273:** 192–196.
2. SHIMOKATA, H., Y. YAMADA, M. NAKAGAWA, *et al.* 2000. Distribution of geriatric disease-related genotypes in the National Institute for Longevity Sciences, Longitudinal Study of Aging (NILS-LSA). J. Epidemiol. **10:** S46–55.
3. OHSAWA, I., K. KAMINO, K.NAGASAKA, *et al.* 2003. Genetic deficiency of a mitochondrial aldehyde dehydrogenase increases serum lipid peroxides in community-dwelling population. J. Hum. Genet. **48:** 404–409.
4. OHSAWA, I., K. NISHIMAKI, C. YASUDA, *et al.* 2003. Deficiency in a mitochondrial aldehyde dehydrogenase increases vulnerability to oxidative stress in PC12 cells. J. Neurochem. **84:** 1110–1117.
5. SUZUKI, Y., T. MURAMATSU, M. TANIYAMA, *et al.* 1996. Association of aldehyde dehydrogenase with inheritance of NIDDM. Diabetologia **39:** 1115–1118.
6. YOKOYAMA, A., T. MURAMATSU, T. OHMORI, *et al.* 1998. Alcohol-related cancers and aldehyde dehydrogenase-2 in Japanese alcoholics. Carcinogenesis **19:** 1383–1387.
7. TAKAGI, S., S. BABA, N. IWAI, *et al.* 2001. The aldehyde dehydrogenase 2 gene is a risk factor for hypertension in Japanese but does not alter the sensitivity to pressor effects of alcohol: the Suita study. Hypertens. Res. **24:** 365–370.
8. AMAMOTO, K., T. OKAMURA, S. TAMAKI, *et al.* 2002. Epidemiologic study of the association of low-Km mitochondrial acetaldehyde dehydrogenase genotypes with blood pressure level and the prevalence of hypertension in a general population. Hypertens. Res. **25:** 857–864.

9. TAKAGI, S., N. IWAI, R. YAMAUCHI, *et al.* 2002. Aldehyde dehydrogenase 2 gene is a risk factor for myocardial infarction in Japanese men. Hypertens. Res. **25:** 677–681.
10. PEDERSEN, W.A., S.L. CHAN & M.P. MATTSON. 2000. A mechanism for the neuroprotective effect of apolipoprotein E: isoform-specific modification by the lipid peroxidation product 4-hydroxynonenal. J. Neurochem. **74:** 1426–1433.

Mitochondrial Reactive Oxygen Species Generation and Calcium Increase Induced by Visible Light in Astrocytes

MEI-JIE JOU,[a] SHUO-BIN JOU,[b] MEI-JIN GUO,[a] HONG-YUEH WU,[a] AND TSUNG-I PENG[c]

[a]*Department of Physiology and Pharmacology, Chang Gung University, Tao-Yuan, Taiwan*

[b]*Department of Neurology, China Medical College Hospital, Taichung, Taiwan*

[c]*Department of Neurology, Lin-Kou Medical Center, Chang Gung Memorial Hospital, Tao-Yuan, Taiwan*

ABSTRACT: Mitochondria contain photosensitive chromophores that can be activated or inhibited by light in the visible range. Rather than utilizing light energy, however, mitochondrial electron transport oxidation-reduction reaction and energy coupling could be stimulated or damaged by visible light. Our previous work demonstrated that reactive oxygen species (ROS) were generated in cultured astrocytes after visible laser irradiation. With confocal fluorescence microscopy, we found that ROS were generated mostly from mitochondria. This mitochondrial ROS (mROS) formation plays a critical role in photoirradiation–induced phototoxicity and apoptosis. In this study, we measured changes of mitochondrial calcium level ($[Ca^{2+}]_m$) in cultured astrocytes (RBA-1 cell line) irradiated with blue light and examined the association between mROS formation and $[Ca^{2+}]_m$ level changes. Changes of intracellular ROS and $[Ca^{2+}]_m$ were visualized using fluorescent probes 2′,7′-dichlorodihydrofluorescein (DCF), and rhod-2 After exposure to visible light irradiation, RBA-1 astrocytes showed a rapid increase in ROS accumulation particularly in the mitochondrial area. Increase in $[Ca^{2+}]_m$ was also induced by photoirradiation. The levels of increase in DCF fluorescence intensity varied among different astrocytes. Some of the cells generated much higher levels of ROS than others. For those cells that had high ROS levels, mitochondrial Ca^{2+} levels were also high. In cells that had mild ROS levels, mitochondrial Ca^{2+} levels were only slightly increased. The rate of increase in DCF fluorescence seemed to be close to the rate of rhod-2 fluorescence increase. There is a positive and close correlation between mitochondrial ROS levels and mitochondrial Ca^{2+} levels in astrocytes irradiated by visible light.

KEYWORDS: mitochondria; calcium; reactive oxygen species; photoirradiation

Address for correspondence: Tsung-I Peng, M.D., Ph.D., Department of Neurology, Chang Gung Memorial Hospital, Lin-Kou Medical Center, No. 5 Fu-Shin Street, Gwei-Shan, Tao-Yuan, 333 Taiwan. Voice: 886 3 328 1200, ext. 8347; fax: 886 3 328 8849.
tipeng@adm.cgmh.org.tw

INTRODUCTION

Like the electron transport chains on the thylakoid membrane in chloroplasts, the electron transport chains on the inner membrane of mitochondria also contain photosensitive chromophores (porphyrins, cytochromes). Light in the visible range can be absorbed by these mitochondrial chromophores (porphyrin ring, flavinic or pyridinic rings). Rather than utilizing light energy, however, mitochondrial electron transport redox reaction and energy coupling can be stimulated or damaged by visible light, depending on wavelength, intensity, and duration of light exposure.

Treating mitochondria with light in 400–500 nm range, Aggarwal and colleagues[1] found the temporal sequence of changes of respiration to be stimulatory at first and inhibitory later on. Loss of respiration was principally due to inactivation of dehydrogenases. Of the components of dehydrogenase systems, flavins and quinones were most susceptible to illumination, whereas the iron-sulfur centers were remarkably resistant to being damaged. Succinate dehydrogenase was inactivated before choline and NADH dehydrogenases. Redox reactions of cytochromes and cytochrome c oxidase activity were unaffected. Similar findings were also obtained by Ramadan-Talib and colleagues[2] in their studies on the effect of visible light on respiratory activity of isolated mitochondria from *Neurospora crassa*. Photosensitive sites in the flavoprotein dehydrogenases were also discovered.

Aggarwal and colleagues[1] found that inactivation of flavin-containing dehydrogenases by photoirradiation was O_2-dependent and could be prevented by anaerobiosis or by the presence of substrates for the dehydrogenases. They suggested that activated species of oxygen could be involved in the flavin-photosensitized reaction. Hockberger and colleagues[3] found that blue light (450–490 nm) stimulated H_2O_2 production in cultured mouse, monkey, and human cells. In cells overexpressing flavin-containing oxidases, H_2O_2 production was enhanced.[3] They suggested that photoreduction of flavin, which activates flavin-containing oxidases, underlies blue light–induced production of H_2O_2. Generation of intracellular reactive oxygen species (ROS) in human fibroblasts induced by laser irradiation was also reported by Alexandratou and colleagues.[4]

In our previous work,[5] generation of ROS in cultured astrocytes after brief visible laser irradiation (argon, 488 nm line) was visualized and monitored by confocal fluorescence microscopy. We found that ROS was generated mostly from mitochondria. This mitochondrial ROS generation could be detected shortly after photoirradiation. Its level intensified progressively and remained high long after the laser irradiation. The irradiated cells eventually died from apoptosis. Antioxidants, melatonin, and vitamin E largely attenuated the laser irradiation–induced mitochondrial ROS formation in astrocytes and prevented subsequent apoptosis.

Other than ROS generation, alterations in intracellular calcium, pH, and membrane potential resulting from photoirradiation has been reported.[4–9] However, these experiments were performed by lasers with longer wavelength (helium-neon, 630 nm). The photoirradiation–induced alterations reported in these studies were rather variable and dependent on the power of laser used. The association between ROS generation and changes in intracellular calcium, pH, and membrane potential remained obscure. In this study we visualized and monitored changes of mitochondrial calcium level ($[Ca^{2+}]_m$) in cultured astrocytes (RBA-1 cell line) irradiated briefly with blue light (450–490 nm) or argon laser (488 nm). The association between mitochondrial ROS formation and $[Ca^{2+}]_m$ level changes was examined in depth.

MATERIALS AND METHODS

Cell Preparation

Normal rat brain astrocytes (RBA-1) cell line was originally established through a continuous passage of primary astrocytes in culture isolated from 3-day-old JAR-2, F51 rat brains by Dr. Teh-Cheng Jou. All cells were grown in medium consisting of Dulbecco's modified Eagle's medium (Life Technologies, Grand Island, NY, USA) supplemented with 10% (vol/vol) fetal bovine serum. All cells were plated onto No. 1 glass coverslips for fluorescent measurement (Model No. 1, VWR Scientific, San Francisco, CA).

Chemicals and Fluorescent Probes

All fluorescent probes were purchased from Molecular Probes Inc. (Eugene, OR) and chemicals were obtained from Sigma (St. Louis, MO). Intracellular ROS was detected using an intracellular ROS dye, DCF. The nonfluorescent DCF is oxidized by intracellular ROS to form the highly fluorescent DCF. Mitochondrial calcium was detected using a specific mitochondrial calcium fluorescent dye, rhod-2. Loading concentrations of fluorescent probes used were 1 mM DCF and 1 mM rhod-2. Fluorescent probes were all loaded at room temperature for 30 minutes. DCF and rhod-2 loading required 30 additional minutes of incubation to allow intracellular deacetylation of the ester forms of the dyes. After loading, cells were rinsed three times with HEPES-buffered saline (140 mM NaCl, 5 mM KCl, 1 mM $MgCl_2$, 2 mM $CaCl_2$, 10 mM glucose, 5 mM HEPES, pH 7.4).

Conventional and Confocal Imaging Microscopy

Conventional fluorescence images were obtained using a Zeiss inverted microscope (AxioVert 200M) equipped with a mercury lamp (HBO 103), a cool CCD camera (Coolsnap FX), and Zeiss objectives (Plan-NeoFluor 100X, N.A. 1.3 oil). The filter used to detect DCF was No. 10 (Exi: BP 450–490 nm; Emi: BP 515–565 nm) and that used to detect rhod-2 was No. 15 (Exi: BP 546/12 nm; Emi: LP 590 nm). Confocal fluorescence images were obtained using a Leica SP2 MP (Leica-Microsystems, Wetzlar, Germany) fiber coupling system equipped with an argon laser system. During fluorescence imaging, the illumination light was reduced to the minimal level by using a neutral density filter (3%) to prevent the photosensitizing effect from the interaction of light with fluorescent probes. All images were processed and analyzed by use of MetaMorph software (Universal Imaging Corp., West Chester, PA).

Visible Light Irradiation

For visible light-irradiation on conventional fluorescence microscope, cells were continuously exposed to visible light from the epi-illumination port of the objective of a Zeiss inverted microscope coupled with a 100 mW mercury lamp. An excitation band pass filter (450–490 nm) was used and no neutral density filter was employed. The irradiation time was 2 minutes. For visible laser irradiation, cells were continuously exposed to a 488-nm Ar laser from the epi-illumination port of the objective of a Leica microscope. The irradiation time was 1 minute. No neutral density filter was used. The size of the pinhole during laser irradiation was set to 100% open. The power density of the laser irradiation measured by a Coherent Model 210 power meter at the epi-illumination port of the objective was 1.7 mW/cm^2. Possible heat effects generated from the visible light or visible laser irradiation were avoided by perfusion of cells with fresh HEPES solution during the experiment.

RESULTS AND DISCUSSION

Blue Light–Induced Mitochondrial ROS Generation in Cultured Astrocytes

To visualize the generation of ROS induced by blue light in cultured astrocytes, we loaded the cells with dichlorodihydrofluorescein (H_2DCF, DCF in brief), a commonly used ROS-detecting fluorescent probe for living cells. Generation of ROS has been detected using DCF with conventional as well as multiphoton imaging system (see Peng and colleagues, this volume). Both techniques demonstrate a punctate increase (in contrast to the homogeneous increase due to autooxidation of DCF, also see Peng and colleagues, this volume) in DCF fluorescent intensity after the cells received blue-light irradiation. As shown in FIGURE 1, DCF fluorescence could be observed in some cells shortly after light exposure. While some cells became brightly fluorescent 1 minute after light exposure, some cells remained only faintly fluorescent. Two minutes after light exposure, all cells became brightly fluorescent. As clearly demonstrated in FIGURE 1, the bright DCF fluorescence signals induced by 2-min blue-light exposure did not distribute evenly inside the cells, but was highly localized in punctate spots. These discrete spots colocalized well with signals from Mito Red, a mitochondria-specific marker (data not shown), suggesting that blue light–induced ROS in cultured astrocytes originated largely from mitochondria. The same findings have been reported in our work published earlier.[5]

In addition to mitochondria, peroxisomes have been found by Hockberger and colleagues[3] to be another source of H_2O_2 production induced by violet-blue light in cultured mouse, monkey, and human cells. The difference in the sources of ROS production induced by blue light in different types of cells is not clear. We suspect that the amount of flavin-containing oxidases in peroxisomes may vary in different types of cells. The numbers of peroxisomes contained in astrocytes may not be high. On the other hand, it has been well established that the major site of oxidation-reduction reaction in cells is mitochondria and ROS production inside the cells mainly comes from mitochondria. Likewise, light-induced generation of ROS may arise mainly from mitochondria.

Although it has been suggested that activation of flavin-containing oxidases underlies blue light–induced production of ROS, the exact mechanisms of mitochondrial ROS generation induced by light await further exploration. It is not clear how photoreduction of flavin in mitochondria leads to generation of ROS. Whether the photo-activation process takes place merely in the electron transport chain or goes beyond the inner membrane requires further investigation. Exploring other alterations in mitochondria that are also induced by light may provide valuable information in the understanding of light-induced mitochondrial ROS generation.

Several extracellular or intracellular signaling molecules have been associated with mitochondrial ROS production. One such molecule that has drawn lots of attention is calcium. Using electron paramagnetic resonance, Dyken[10] reported that mitochondria isolated from adult rat cerebral cortex and cerebellum generate ex-

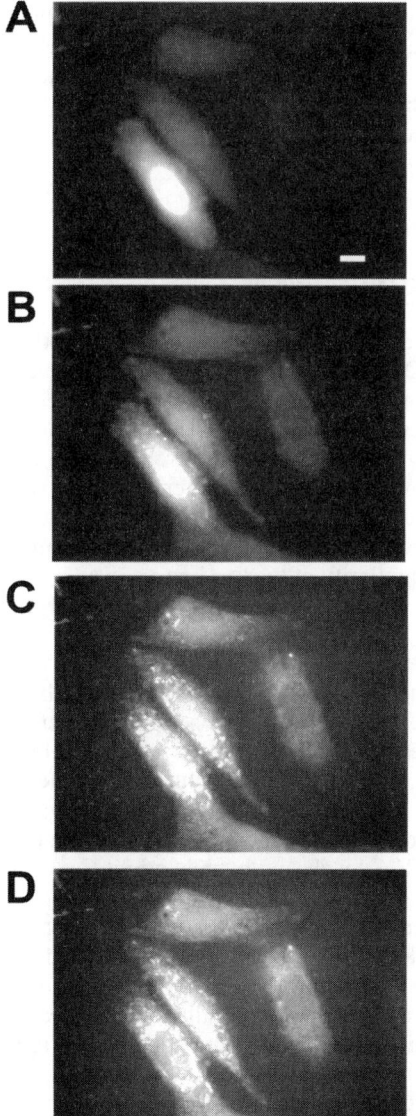

FIGURE 1. ROS generation induced by blue light irradiation in RBA-1 astrocytes. Astrocytes were irradiated with blue light (450–500 nm). DCF images of the irradiated cells were taken every 30 seconds (**A–D**). The non-fluorescent DCF became fluorescent when ROS was generated. The bright DCF signals localized in punctate spots. These discrete spots colocalized well with signals from Mito Red, a mitochondrial specific marker suggesting that blue light–induced ROS generation in cultured astrocytes originated largely from mitochondria. Bar=10 μm.

tremely reactive hydroxyl (·OH) radicals, plus ascorbyl and other carbon-centered radicals when exposed to 2.5 mM Ca^{2+}, 14 mM Na^+, plus elevated ADP under normoxic conditions. Using a variety of optical techniques, Territo and colleagues[11] reported that Ca^{2+} stimulates oxidative phosphorylation and ATP production in isolated porcine heart mitochondria by activating the Ca^{2+}-sensitive dehydrogenases and the F_0F_1-ATPase. These findings suggest that Ca^{2+} uptake into mitochondria may incite ROS production. Whether or not this is also true in blue light–induced mitochondrial ROS generation awaits exploration.

To study the association between mitochondrial Ca^{2+} change and mitochondrial ROS generation, we tried to visualize these two variants in the same cells during blue light irradiation. Astrocytes were loaded with two fluorescent probes, DCF to detect ROS and rhod-2 to detect mitochondrial Ca^{2+}. We have been successful in visualizing mitochondrial Ca^{2+} change in intact living cells by using rhod-2.[12–14] We have found rhod-2 to be sensitive, fast and rather selective in dynamically monitoring mitochondrial Ca^{2+} changes in living cells.

Mitochondrial Calcium Increases Are Highly Associated with Mitochondrial ROS Generation

DCF and rhod-2 images of astrocytes exposed to blue light are shown in FIGURE 2. ROS generation, reflected by increased DCF fluorescence (FIG. 2A) and mitochondrial Ca^{2+} changes (FIG. 2B) reflected by variations in rhod-2 fluorescence are both shown. These two images taken with different filter sets looked very similar, indicating that most of the fluorescence emitted from these two probes inside the cells arose from the same sites. However, there is one distinct difference in these two images: nucleoli could be stained by rhod-2 and not by DCF. This distinctive feature of rhod-2 has been noted in our previous reports[13] and could be used to distinguish the rhod-2 image from the DCF image. The similarity of the two images in area outside the nuclei could not be artifactual but resulted from similar localization of the fluorescence emitted by the probes.

As already demonstrated in FIGURE 1, DCF signals lit up in spots or lines that colocalized well with rhod-2 signals, indicating that ROS generation induced by blue light occurred mostly in mitochondrial areas. The levels of increase in DCF fluorescence intensity among different astrocytes were variable. Some cells had much higher levels of ROS generation than others (FIG. 2A). In those cells that had high ROS levels, mitochondrial Ca^{2+} levels were also high (FIG. 2B). In cells that had mild ROS levels, mitochondrial Ca^{2+} levels were only slightly increased. Thus, there is a positive correlation between mitochondrial ROS levels and mitochondrial Ca^{2+} levels in astrocytes irradiated by visible light.

To verify that the rhod-2 fluorescence increase observed in this study did not result from DCF fluorescence increase, we loaded astrocytes with rhod-2 alone and acquired the images before and after photoirradiation. FIGURE 3 shows the confocal images of rhod-2 loaded astrocytes before and after photoirradiation. As clearly demonstrated in this figure, rhod-2 signals lit up after photoirradiation. The bright rhod-2 fluorescent signals were also in discrete punctate pattern appearing as dot or rod areas. This is the first direct evidence ever reported of blue light-induced mitochondrial Ca^{2+} increase. An accelerated Ca^{2+} uptake by the mitochondria after low power helium-neon irradiation was suggested earlier by Breitbart and colleagues.[8] However, that study measured radiolabeled $^{45}Ca^{2+}$ uptake in digitonin-treated mammalian sperm cells.

Due to a limitation of the device used in this study, we could not acquire DCF and rhod-2 images simultaneously. The exact kinetics of DCF and rhod-2 fluorescence intensity changes was not obtained. However, from the rough temporal profiles of DCF and rhod-2 fluorescence intensity changes acquired in this study (data not shown), the rate of increase in DCF fluorescence seemed to be close to the rate of rhod-2 fluorescence increase. The time difference in the increase rates may be too small to be revealed by our imaging system. At this point, we do not know whether

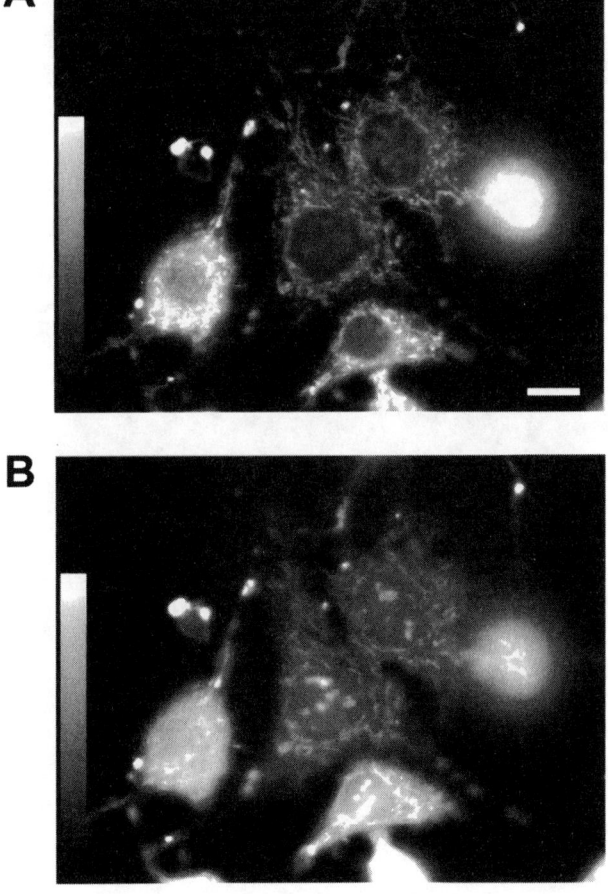

FIGURE 2. Mitochondrial calcium increases are highly associated with mitochondrial ROS generation in astrocytes irradiated with blue light. Astrocytes were loaded with two fluorescent probes, DCF and rhod-2. (**A**) DCF image of astrocytes after photoirradiation. (**B**) Rhod-2 image of the same cells after light exposure. The bright signal areas on both images localized in nearly identical areas. Bar=10 µm.

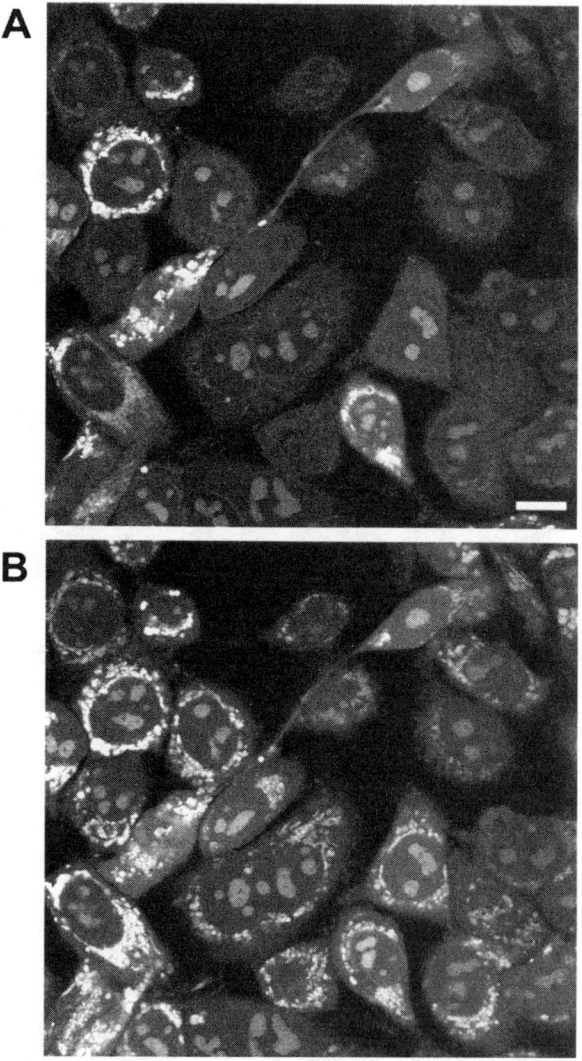

FIGURE 3. Increased mitochondrial calcium induced by light irradiation in RBA-1 astrocytes. Confocal rhod-2 images of astrocytes before (**A**) and after (**B**) laser light irradiation. Rhod-2 signals localized in discrete spots or lines, which lit up after light irradiation. Bar=10 µm.

the increase in rhod-2 fluorescence intensity preceded or followed the rise in DCF fluorescence intensity. The exact cause-effect relationship between mitochondrial ROS generation and mitochondrial Ca^{2+} increase in photoirradiated astrocytes awaits further exploration.

There are published reports of mitochondrial Ca^{2+}-elicited ROS generation. On the other hand, results from several studies have indicated that ROS could induce intracellular Ca^{2+} increase. The temporal ordering of changes in cytoplasmic ($[Ca2+]c$) and mitochondrial Ca^{2+} ($[Ca^{2+}]_m$) levels in relation to mitochondrial reactive oxygen species (ROS) accumulation was examined by Kruman and Mattson[15] in cultured neural cells exposed to either an apoptotic (staurosporine; STS) or a necrotic (aldehyde 4-hydroxynonenal; HNE) insult. STS and HNE each induced an early increase of $[Ca^{2+}]_c$, followed by a delayed increase of $[Ca^{2+}]_m$, then by mitochondrial ROS accumulation. A critical role of mitochondrial Ca^{2+} uptake for mitochondrial production of ROS was also found by Reynolds and Hastings[16] in cultured forebrain neurons treated with excitotoxic concentration of glutamate.

Alternatively, mitochondrial Ca^{2+} increase may be induced by ROS. Pei and colleagues[17] reported that Ca^{2+}-permeable channels in the plasma membrane of *Arabidopsis* guard cells are activated by hydrogen peroxide. Wang and colleagues[18] reported that H_2O_2 induced extracellular Ca^{2+} influx through sodium and calcium channels that may be directly or indirectly attributed to thiol oxidation. Although direct detection of mitochondrial Ca^{2+} changes in response to H_2O_2 was not done in these studies, increase of mitochondrial Ca^{2+} may follow cytosolic Ca^{2+} increase. A large body of evidence has shown that mitochondria function as an important Ca^{2+} buffer by taking up cytosolic Ca^{2+} through a specific transport mechanism that employs the huge electrochemical gradient across the mitochondrial membrane as a driving force.

To verify that ROS can initiate mitochondrial Ca^{2+} increase, we treated astrocytes with H_2O_2 and acquired confocal rhod-2 images to visualize mitochondrial Ca^{2+} changes. As shown in FIGURE 4, rhod-2 signals brightened up after 10 mM H_2O_2 treatment for 5 minutes. Similar to the findings in FIGURE 3, the bright rhod-2 fluorescent signals in this figure were also confined to discrete dotted areas. However, the high signal areas in FIGURE 4B were larger and more rounded than those in FIGURE 3B. In addition to taking up a large amount of Ca^{2+}, mitochondria also became swollen after H_2O_2 treatment. Opening of the mitochondrial permeability transition pore by the combined effect of Ca^{2+} and ROS may underlie the swelling of mitochondria seen here. Note that rhod-2 fluorescent signals in nucleoli also intensified after H_2O_2 treatment. In our experience with rhod-2 imaging, increased nucleolar rhod-2 signals can frequently be observed when there are increases of cytosolic Ca^{2+}. The faintly increased rhod-2 fluorescent signals in cytosol areas seen in FIGURE 4B also suggest increases of cytosolic Ca^{2+} induced by H_2O_2 treatment. In FIGURE 3B, there were very mild increases of rhod-2 fluorescent signals in nucleoli and cytosol areas, suggesting that cytosolic Ca^{2+} may increase after light exposure. Yet, it will be difficult to verify this issue by using cytosolic Ca^{2+} probes because many of these probes will be easily bleached by strong light.

Could the light-induced mitochondrial Ca^{2+} increase be a direct effect of light and independent from mitochondrial ROS generation? We address this question by exploring whether mitochondrial Ca^{2+} increase can still be induced by light in the absence of ROS production. We treated astrocytes with antioxidants before photo-

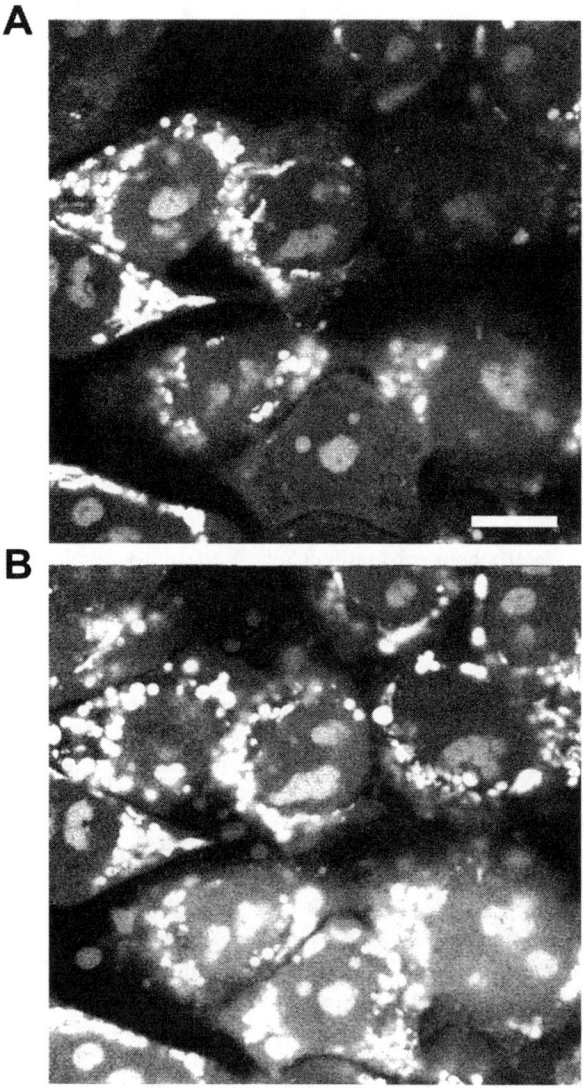

FIGURE 4. Increased mitochondrial calcium induced by H_2O_2 in RBA-1 astrocytes. Confocal rhod-2 images of astrocytes before (**A**) and after (**B**) 1 mM H_2O_2 treatment. Rhod-2 signals localized in discrete spots or lines, which lit up after H_2O_2 treatment. Bar = 10 μm.

irradiation and found that mitochondrial Ca^{2+} did not increase after light exposure in the presence of antioxidants (data not shown). This finding suggests that the light-induced mitochondrial Ca^{2+} increase resulted from the light-elicited mitochondrial ROS production.

ACKNOWLEDGMENTS

This research was supported by grants CMRP 1009 (to M-J.J.) and CMRP 930 (to T-I.P.) from the Chang Gung Medical Research Foundation, grants NSC 89-2320-B-182-089 and NSC 90-2315-B-182-006 (to M-J.J.), and NSC 91-2314-B-182A-021 (to T-I.P.) from the National Science Council, Taiwan, and grant DMR-91-009 (to S-B.J.) from China Medical College, Taiwan.

REFERENCES

1. AGGARWAL, B.B., A.T. QUINTANILHA, R. CAMMACK, et al. 1978. Damage to mitochondrial electron transport and energy coupling by visible light. Biochim. Biophys. Acta **502**: 367–382.
2. RAMADAN-TALIB, Z. & J. PREBBLE. 1978. Photosensitivity of respiration in Neurospora mitochondria. A protective role for carotenoid. Biochem. J. **176**: 767–775.
3. HOCKBERGER, P.E., T.A SKIMINA, V.E. CENTONZE, et al. 1999. Activation of flavin-containing oxidases underlies light-induced production of H_2O_2 in mammalian cells. Proc. Natl. Acad. Sci. USA **96**: 6255–6260.
4. ALEXANDRATOU, E., D. YOVA, P. HANDRIS, et al. 2002. Human fibroblast alterations induced by low power laser irradiation at the single cell level using confocal microscopy. Photochem. Photobiol. Sci. **1**: 547–552.
5. JOU, M.J., S.B. JOU, H.M. CHEN, et al. 2002. Critical role of mitochondrial reactive oxygen species formation in visible laser irradiation-induced apoptosis in rat brain astrocytes (RBA-1). J. Biomed. Sci. **9**: 507–516.
6. VACCA, R.A., L. MORO, V.A. PETRAGALLO, et al. 1997. The irradiation of hepatocytes with He-Ne laser causes an increase of cytosolic free calcium concentration and an increase of cell membrane potential, correlated with it, both increases taking place in an oscillatory manner. Biochem. Mol. Biol. Intl. **43**: 1005–1014.
7. LUBART, R., H. FRIEDMANN, M. SINYAKOV, et al. 1997. Changes in calcium transport in mammalian sperm mitochondria and plasma membranes caused by 780 nm irradiation. Lasers Surg. Med. **21**: 493–499.
8. BREITBART, H., T. LEVINSHAL, N. COHEN, et al. 1996. Changes in calcium transport in mammalian sperm mitochondria and plasma membrane irradiated at 633 nm (HeNe laser). J. Photochem. Photobiol. Biol. **34**: 117–121.
9. LUBART, R., H. FRIEDMANN, T. LEVINSHAL, et al. 1992. Effect of light on calcium transport in bull sperm cells. J. Photochem. Photobiol. Biol. **15**: 337–341.
10. DYKENS, J.A. 1994. Isolated cerebral and cerebellar mitochondria produce free radicals when exposed to elevated Ca^{2+} and Na^+: implications for neurodegeneration. J. Neurochem. **63**: 584–591.
11. TERRITO, P.R., S.A. FRENCH, M.C. DUNLEAVY, et al. 2001. Calcium activation of heart mitochondrial oxidative phosphorylation: rapid kinetics of mVO2, NADH, and light scattering. J. Biol. Chem. **276**: 2586-2599.
12. JOU, M.J., T.I. PENG & S.S. SHEU. 1996. Histamine induces oscillations of mitochondrial free Ca^{2+} concentration in single cultured rat brain astrocytes. J. Physiol. **497**: 299–308.
13. PENG, T.I., M.J. JOU, S.S. SHEU, et al. 1998. Visualization of NMDA receptor-induced mitochondrial calcium accumulation in striatal neurons. Exp. Neurol. **149**: 1–12.

14. PENG, T.I. & J.T. GREENAMYRE. 1998. Privileged access to mitochondria of calcium influx through N-methyl-D-aspartate receptors. Mol. Pharmacol. **53:** 974–980.
15. KRUMAN, I.I. & M.P. MATTSON. 1999. Pivotal role of mitochondrial calcium uptake in neural cell apoptosis and necrosis. J. Neurochem. **72:** 529–540.
16. REYNOLDS, I.J. & T.G. HASTINGS. 1995. Glutamate induces the production of reactive oxygen species in cultured forebrain neurons following NMDA receptor activation. J. Neurosci. **15:** 3318–3327.
17. PEI, Z.M., Y. MURATA, G. BENNING, *et al.* 2000. Calcium channels activated by hydrogen peroxide mediate abscisic acid signalling in guard cells. Nature **406:** 731–734.
18. WANG, H. & J.A. JOSEPH. 2000. Mechanisms of hydrogen peroxide-induced calcium dysregulation in PC12 cells. Free Rad. Biol. Med. **28:** 1222–1231.

Radical Metabolism Is Partner to Energy Metabolism in Mitochondria

JIAN-XING XU

National Laboratory of Biomacromolecule, Center for Molecular Biology, Institute of Biophysics, Chinese Academy of Sciences, Beijing 100101, China

ABSTRACT: It has been shown that cytochrome *c* plays a role in scavenging $O_2^-\cdot$ and H_2O_2 in mitochondria through two electron leak pathways of the respiratory chain. Based on the two electron leak pathways and the superoxide metabolic routes in nature, it is suggested that the concept of radical metabolism should be added as a partner to the energy metabolism of mitochondria. A total of four reactive pathways of superoxide anion have been identified, collected, and linked to electron leaks in the respiratory chain. A view is presented that envisions mitochondria working on ATP synthesis by means of transferring electrons inside the respiratory chain while at the same time working on ROS generation and elimination as electrons leak out of the chain.

KEYWORDS: mitochondria; radical metabolism; superoxide anion; hydrogen peroxide; cytochrome *c*

INTRODUCTION

Mitochondria have been the objects of more and more attention since Wang and colleagues found that cytochrome *c* is involved in cell apoptosis.[1] Basic research on mitochondria has also been promoted in recent years. Reviewing the history of mitochondria research, it becomes clear that it has been known for a long time that mitochondria not only produce ATP but also produce H_2O_2.[2] Although research on mitochondria was mostly focused on ATP synthesis during the past 50 years, mitochondrial production of H_2O_2 was not appreciated until Chance developed a method for detecting H_2O_2 generated by mitochondria.[3] Chance discovered that about 2% of the oxygen consumed in mitochondria ended up in H_2O_2 production in normal physiological conditions.[4] Ever since, increasing attention has been directed to H_2O_2 production in mitochondria. It was known that the precursor of H_2O_2 is the $O_2^-\cdot$, and that $O_2^-\cdot$ is generated through a single electron reduction of O_2 by the electrons leaked from the substrate-side of the respiratory chain.[5,6] It has been established that complexes I and III of the respiratory chain are responsible for leaking electrons and generating $O_2^-\cdot$.[7,8]

The H_2O_2 generation through electron leaks in the respiratory chain has been well studied, but the disposal of H_2O_2 in mitochondria is not well understood. This ques-

---

Address for correspondence: Jian-Xing Xu, Institute of Biophysics, Chinese Academy of Sciences, Beijing 100101, China. Voice: 86-10-64888504; fax: 86-10-64871293.
    xujx@sun5.ibp.ac.cn

tion is important because the $O_2^-\cdot$ and H_2O_2 will cause damage if they are not disposed in time. With this consideration in mind, I proposed a hypothesis in 1995 suggesting that the respiratory chain has the ability to remove H_2O_2 through a way of leaking electrons from cytochrome c to it. I pointed out that a radical metabolic path of $O_2^-\cdot \rightarrow H_2O_2 \rightarrow H_2O$ is formed as a result of electron leakage in the respiratory chain.[9,10] This hypothesis was based on the following three facts.

(1) In yeast, the enzyme cytochrome c peroxidase catalyzes the reduction of H_2O_2 by ferrocytochrome c.[11]
(2) The reduction of H_2O_2 by ferrocytochrome c is a thermodynamically favorable reaction based on redox potential values.
(3) In 1988, I showed that the exterior H_2O_2 can be reduced continuously by cytochrome c if cytochrome c is kept in its reduced state by adding trace amounts of purified succinate-cytochrome c reductase and feeding electrons by succinate.[12]

To support this hypothesis it is important to prove that the H_2O_2 generated by the electron leak of complexes I and III of the respiratory chain can be eliminated by the electron leak of cytochrome c, and to further prove that this cytochrome c–mediated electron leak pathway is operational *in vivo*.

In a recent paper,[13] we presented evidence documenting that the respiratory chain–generated $O_2^-\cdot$ and H_2O_2 can be eliminated by a cytochrome c–mediated electron leak pathway. We use the Keilin-Hartree heart muscle preparation (HMP) as the material in which observe the generation of $O_2^-\cdot$ and H_2O_2 while adding substrate to it. Two generative peaks can be observed by adding NADH as the substrate. The first peak, formed by the electron leak of complex I, is a sharp peak that appears soon after the addition of NADH. The second peak, formed by the electron leak of complex III, is a broader peak that appears later. If succinate is added as the substrate, only the second peak appears. When we prepared a cytochrome c–depleted HMP (c-dHMP), we discovered that the production of $O_2^-\cdot$ and H_2O_2 was 7–8 times higher than that in normal HMP. Reconstitution of cytochrome c to the c-dHMP causes an exponential decay of the production of $O_2^-\cdot$ and H_2O_2. The least squares analysis of the decay curves show that the relationship of $O_2^-\cdot$ or H_2O_2 generation to the amount of reconstituted cytochrome c can be described as a formula of $P(X) = C + Ae^{-ax} + Be^{-bx}$. Based on this formula, three active roles of the reconstituted cytochrome c can be deduced. First, cytochrome c facilitates the electron transfer of the respiratory chain. Second, cytochrome c scavenges the generated $O_2^-\cdot$ when it is in the oxidized state. Third, cytochrome c scavenges the generated H_2O_2 when it is in the reduced state. The biological function of the cytochrome c–mediated electron leak pathway is to keep the $O_2^-\cdot$ and H_2O_2 at lower physiological levels in mitochondria. A dramatic increase of $O_2^-\cdot$ and H_2O_2 in mitochondria will occur once the cytochrome c moves out of the respiratory chain, and this is the very early event when the cell enters programmed death. A similar behavior can also be observed in the cytochrome c–depleted mitochondria and cytochrome c–depleted submitochondrial particles (unpublished data).

The superoxide metabolic nature of the cytochrome c–mediated electron leak pathway of $O_2^-\cdot \rightarrow H_2O_2 \rightarrow H_2O$ suggests that a radical metabolism is operating in mitochondria. Four reactive pathways initiated by $O_2^-\cdot$ have been identified from published data and linked to the electron leak of respiratory chain. In addition to the pathway $O_2^-\cdot \rightarrow H_2O_2 \rightarrow H_2O$, three other pathways are (1) the $O_2^-\cdot$ reduces ferri-

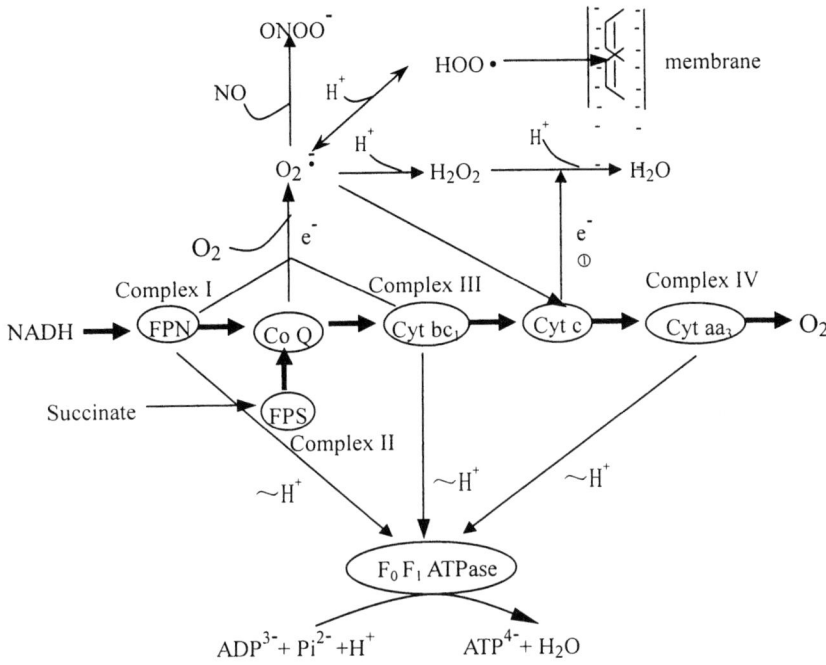

FIGURE 1. The electron leak–linked radical metabolism (*upper*) and electron transfer–coupled energy metabolism (*lower*) in mitochondria. The cycle linked by arrows (*middle*) is the respiratory chain enzymes. The upper part shows the generation of $O_2^{-}\cdot$ and its four reactive pathways. The lower part shows the coupled ATP synthesis.

cytochrome *c* directly[14]; (2) the $O_2^{-}\cdot$ protonates to HOO· and then moves into the membrane to react with unsaturated fatty acids[15]; and (3) the $O_2^{-}\cdot$ changes to $ONOO^{-}$ by combining with NO. All four superoxide reaction pathways were defined as the radical metabolism of mitochondria, as shown in the FIGURE 1.

CONCLUSION

A new view of the function of mitochondria can be obtained by assuming that radical metabolism is a partner to energy metabolism.

(1) Two pathways of oxygen consumption in mitochondria can be obtained according to FIGURE 1. One is the O_2 consumed by the electrons transferred inside the chain, which is used for ATP synthesis. The other is the O_2 consumed by the electrons leaked out of the chain, which is used in the radical metabolism. The two pathways of O_2 consumption have different KCN sensitivity. The former is KCN inhibitable and the latter is KCN insensitive. The KCN-insensitive respiration may have pathophysiological importance, but previous studies of mitochondria respiration have never distinguished this KCN-sensitive oxygen consumption from the total oxygen consumption.

(2) Radical metabolism in mitochondria has more pathophysiological effects. It is more responsible for the dysfunction of mitochondria. The potential negative effects of $O_2^-\cdot$ and H_2O_2 predicate that the mitochondria may be hurt by ROS when the electron leak of the respiratory chain increases.

(3) The four metabolic superoxide anion pathways have different biological effects. The two cytochrome c–mediated electron leak pathways of respiratory chain have more protective effects on the damages by ROS. The other two pathways may have more detrimental effects, because the reaction of HOO· with unsaturated fatty acids is a heat-release reaction[15] and NO is well known to have many physiological roles.

ACKNOWLEDGMENTS

This work is supported by the fund from National Science Foundation and the Ministry of Science and Technology of China.

REFERENCES

1. LIU, X., C.N KIM, J. YANG, *et al.* 1996. Induction of apoptotic program in cell-free extracts requirement for dATP and cytochrome c. Cell **86:** 147–157.
2. KEILIN, D. 1966. The History of Cell Respiration and Cytochrome.: 230. Cambridge University Press. Cambridge, UK.
3. BOVERIS, A. & B. CHANCE. 1973. The mitochondrial generation of hydrogen peroxide general properties and effects of hyperbaric oxygen. Biochem. J. **134:** 707–716.
4. CHANCE, B., H. SIES & A. BOVERIS. 1979. Hydroperoxide metabolism in mammalian organs. Physiol. Rev. **59:** 527–605.
5. LOSCHEN, G., A. AZZI & L. FLOHE. 1973. Mitochondrial H_2O_2 formation: relationship with energy conservation. FEBS Lett. **33:** 84–88.
6. LOSCHEN, G., A. AZZI, C. RICHTER, *et al.* 1974. Superoxide radicals as precursors of mitochondrial hydrogen peroxide. FEBS Lett. **42:** 68–72.
7. TURRENS, J.F. & A. BOVERIS. 1980. Generation of superoxide anion by the NADH dehydrogenase of bovine heart mitochondria. Biochem. J. **191:** 421–427.
8. ZHANG, L., L. YU & C.A. YU. 1998. Generation of superoxide anion by succinate-cytochrome c reductase from bovine heart mitochondria. J. Biol. Chem. **273:** 33972–33976.
9. XU, J-X. 1995. The involvement of mitochondria in the metabolism of active oxygen radicals. Prog. Biochem. Biophys. [Chinese] **22:** 179–180.
10. XU, J-X., X. LI, Y-X. ZHANG, *et al.* 1996. Mitochondrial respiratory chain: a self-defense system against oxygen toxicity. *In* Proceedings of the International Symposium on Native Antioxidants: Molecular Mechanism and Health Effects. L. Packer, M.G.Traber & W.J. Xin, Eds.: 530. AOCS Press. Champaign, IL.
11. YONETANI, T. 1967. Cytochrome c peroxidase (Bakers' yeast). Methods Enzymol. **10:** 336.
12. XU, J-X. 1988. New function of cytochromes in mitochondria. *In* Abstracts of the Second Japan-China Bilateral Symposium on Biophysics. May 16–20, 1988. Kyoto, Japan. pp. 79–80.
13. ZHAO, Y., Z.-B. WANG & J.-X. XU. 2003 Effect of cytochrome c on the generation and elimination of $O_2^-\cdot$ and H_2O_2 in mitochondria. J. Biol. Chem. **278:** 2356–2360.
14. FORMAN, H.J. & I. FRIDOVICH. 1973. Superoxide dismutase: a comparison of rat constants. Arch. Biochem. Biophys. **158:** 396-400.
15. BIELSKI, B.H., R.L. ARUDI & M.W. SUTHERLAND. 1983. A study of the reactivity of HO2/O2- with unsaturated fatty acids. J. Biol. Chem. **258:** 4759–4761.

Mitochondrial Nucleoid and Transcription Factor A

TOMOTAKE KANKI,[a] HIROSHI NAKAYAMA,[b] NARIE SASAKI,[c] KOJI TAKIO,[b] TANFIS ISTIAQ ALAM,[a] NAOTAKA HAMASAKI,[a] AND DONGCHON KANG[a]

[a]*Department of Clinical Chemistry and Laboratory Medicine, Kyushu University Graduate School of Medical Sciences, Fukuoka 812-8582, Japan*

[b]*Biomolecular Characterization Division, RIKEN (The Institute for Physical and Chemical Research), Wako 351-0198, Japan*

[c]*Department of Biology, Faculty of Science, Ochanomizu University, Tokyo 112-8610, Japan*

ABSTRACT: Nuclear DNA is tightly packed into nucleosomal structure. In contrast, human mitochondrial DNA (mtDNA) had long been believed to be rather naked because mitochondria lack histone. Mitochondrial transcription factor A (TFAM), a member of a high mobility group (HMG) protein family and a first-identified mitochondrial transcription factor, is essential for maintenance of mitochondrial DNA. Abf2, a yeast counterpart of human TFAM, is abundant enough to cover the whole region of mtDNA and to play a histone-like role in mitochondria. Human TFAM is indeed as abundant as Abf2, suggesting that TFAM also has a histone-like architectural role for maintenance of mtDNA. When human mitochondria are solubilized with non-ionic detergent Nonidet-P40 and then separated into soluble and particulate fractions, most TFAM is recovered from the particulate fraction together with mtDNA, suggesting that human mtDNA forms a nucleoid structure. TFAM is tightly associated with mtDNA as a main component of the nucleoid.

KEYWORDS: mitochondria; mitochondrial DNA; nucleoid; TFAM; transcription; HMG

INTRODUCTION

Maintenance of mitochondrial DNA (mtDNA) integrity is essential for normal function of the respiratory chain responsible for aerobic ATP production. This is clearly exemplified by many patients with encephalomyopathies caused by mtDNA mutations. mtDNA is subject to damage in part because it is under stronger oxidative stress than is nuclear DNA.[1] Naturally, mitochondria are equipped with systems for protecting their own genome.[1] Recently, the role of mitochondrial transcription factor A (TFAM) in the mtDNA-maintaining systems has gained increasing interest.

Address for correspondence: Dongchon Kang, Department of Clinical Chemistry and Laboratory Medicine, Kyushu University Graduate School of Medical Sciences, 3-1-1 Maidashi, Higashi-ku, Fukuoka 812-8582, Japan. Voice: +81-92-642-5749; fax: +81-92-642-5772.
kang@mailserver.med.kyushu-u.ac.jp

TFAM was purified and cloned as a transcription factor for mtDNA.[2,3] TFAM indeed enhances mtDNA transcription in a promoter-specific fashion in the presence of mitochondrial RNA polymerase and transcription factor B.[4] Because replication of mammalian mtDNA is to be initiated following the transcription,[5] TFAM is also thought to be crucial for replication of mtDNA. Consistent with this notion, targeted disruption of the mouse TFAM gene is an embryonic lethal mutation causing mtDNA depletion.[6] However, TFAM is virtually a DNA-binding protein with no sequence specificity, although it shows a relatively higher affinity to the light and heavy strand promoters, LSP and HSP, respectively.[2,3] TFAM is a member of the high mobility group (HMG) proteins and contains two HMG-boxes. Many HMG-family proteins bind, wrap, bend, and unwind DNA regardless of DNA sequence.[7–9] Abf2, a TFAM homologue of *Saccharomyces cerevisiae*, is also an HMG-family protein. Abf2 is abundant in yeast mitochondria: roughly one molecule to every 15 bp of mtDNA.[10] Unlike mammalian TFAM, Abf2 is not essential for either transcription initiation of mtDNA or mtDNA replication.[10] *Saccharomyces cerevisiae* devoid of Abf2 loses mtDNA when cultured in the presence of fermentable carbon sources,[11] but this mtDNA depletion is rescued by a bacterial histone-like protein HU,[12] suggesting that Abf2 maintains mtDNA as an architectural factor. Because TFAM can substitute for Abf2 as well,[3] human TFAM at least potentially retains common properties to Abf2 and HU. In this article, we show that human mtDNA is tightly associated and wrapped with TFAM, and is thus far from naked, although it is still widely believed to be naked. We propose that TFAM functions not only as a transcription factor but also as a main constitutive factor of an mtDNA/protein complex.

TFAM IN MITOCHONDRIAL NUCLEOID

mtDNA is postulated to have a nucleosome-like structure, i.e., nucleoid, in lower eukaryotes such as *Saccharomyces cerevisiae*[13] and *Physarum polycephalum*.[14] However, its structure is poorly elucidated at a molecular level. Human mtDNA was also proposed to take on a nucleoid structure, mainly based on a dotted pattern of mtDNA staining.[15,16] The human mtDNA nucleoid has not been isolated until recently[17,18] and its structure is still poorly characterized. We have found that human TFAM is two orders of magnitude more abundant than previously considered.[19] Its amount is quite comparable to that of yeast Abf2. Assuming that most TFAM molecules are bound to mtDNA, the amount of TFAM is sufficient for covering the entire length of mtDNA. The issue is whether this assumption is a fact. When human mitochondria are solubilized with a non-ionic detergent Nonidet-P40 (NP-40), most TFAM is recovered from a particulate fraction along with mtDNA,[18] indicating that few free TFAM molecules exist in mitochondria. In the NP-40–insoluble fraction (P2 fraction) of mitochondria of a human cell line of Jurkat, the TFAM/mtDNA ratio is about 900. These results suggest that about 900 molecules of TFAM are bound to one molecule of mtDNA on average. This amount of TFAM could cover the mtDNA entirely. Hence, we can no longer say that human mtDNA is naked. LSP and HSP are promoters for transcription of the light and heavy strands, respectively.[20] TFAM would bind first and leave last these two promoters, because TFAM has a higher af-

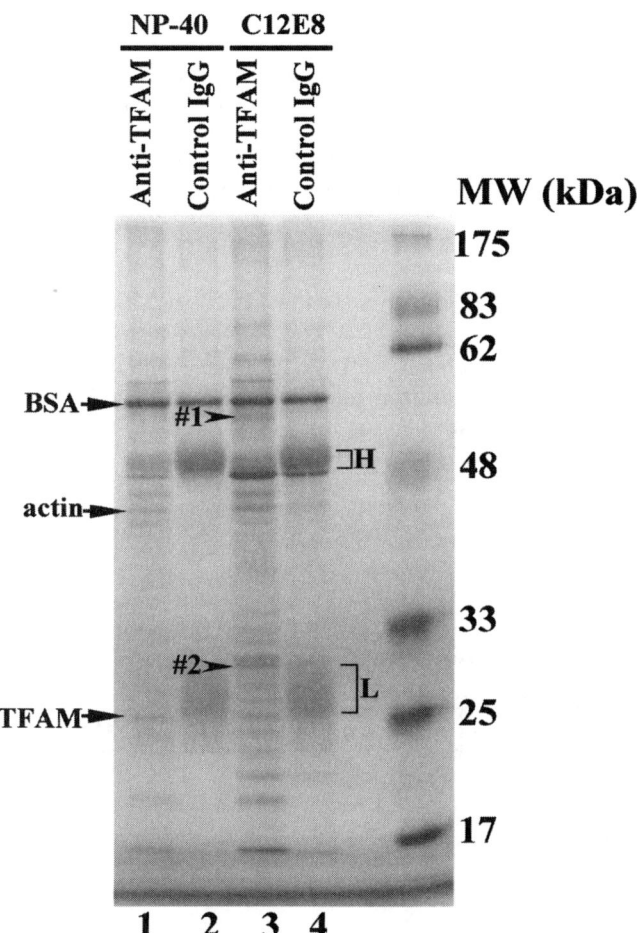

FIGURE 1. Analysis of immunoprecipitated proteins. Mitochondria were solubilized with 0.5% NP-40 or C12E8. The proteins immunoprecipitated by anti-TFAM were resolved on SDS-PAGE. Each band was excised and analyzed by mass spectrometry after lysyl endopeptidase digestion. BSA, which was used for blocking the beads, was released from the beads. #1, monoamine oxidase; #2, adenine nucleotide translocator 2; H, IgG heavy chain; L, IgG light chain.

finity for these two regions than for the rest of the mtDNA molecule.[3,21,22] Therefore, under conditions where about 900 molecules of TFAM are bound to mtDNA, LSP and HSP are likely to be persistently occupied by TFAM. Given that binding of TFAM to LSP or HSP is required to initiate transcription of mtDNA, TFAM itself would be abundant enough to constitutively and fully activate the mtDNA transcription in a normal state. Consistent with this, the gene expression of mtDNA largely depends on its copy number.[23] Thus TFAM may be essential for transcription initiation, but it seems unlikely that the transcription rate is regulated by the amount of TFAM. This idea is compatible with a recent finding that TFAM levels can be substantially reduced without significant inhibition of mtDNA transcription in insect cells.[24] In addition, TFAM alone is not able to initiate transcription from the specific promoters with purified recombinant RNA polymerase.[25,26] Recently, the human homologue of yeast mitochondrial transcription factor B has been cloned[27] and shown to be required for the promoter-specific transcription by the RNA polymerase.[17]

ASSOCIATION OF mtDNA WITH MEMBRANES

It has been proposed that human mtDNA associates with the mitochondrial inner membrane.[28] When we solubilized mitochondria with octaethylene glycol monododecyl ether (C12E8), a weaker non-ionic detergent than NP-40, more monoamine oxidase and ATP/ADP translocator, abundant outer and inner membrane proteins, respectively, were co-immunoprecipitated by anti-TFAM (FIG. 1), a result

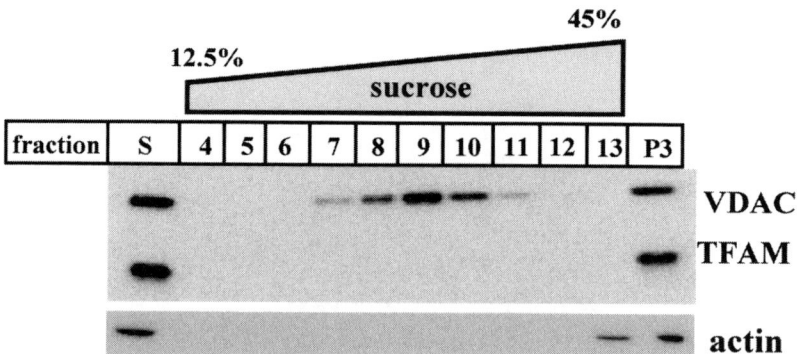

FIGURE 2. Separation of an NP-40–insoluble particulate fraction by sucrose-density gradient centrifugation. An NP-40–insoluble particulate fraction (P2) was mixed with an equal volume of buffer TEN (10 mM Tris-HCl, pH 7.0, 1 mM EDTA, and 150 mM NaCl) containing 80% sucrose (w/vol) and placed at the tube bottom. Then, stepwise sucrose gradients (5~30% in TEN buffer, 1 mL each) were layered on top. The gradient steps were serially reduced by 2.5% from the bottom to top. The tube was centrifuged at 200,000 × g for 15 h at 4°C. After centrifugation, 1-mL samples were collected from the top down. The pellet at the bottom was designated P3. Each fraction was collected from the top down and analyzed by Western blotting. S, starting sample; P3, pellet at the bottom.

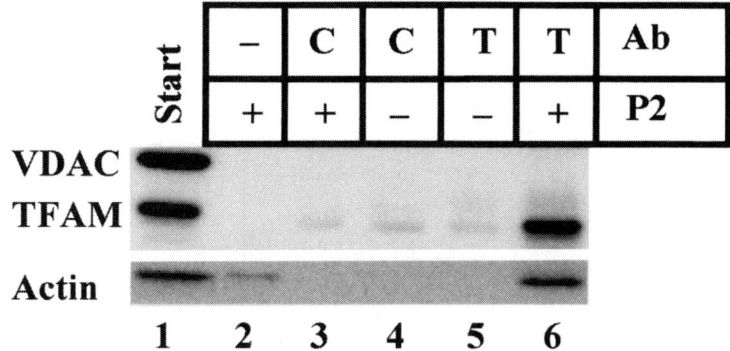

FIGURE 3. Immunoprecipitation by anti-TFAM. An NP-40–insoluble particulate fraction was incubated with rotation in the presence of antibody-immobilized magnetic beads. The beads were pelleted using a magnet. The resulting supernatants were removed. After washing three times, proteins were eluted from the beads by heating. VDAC, TFAM, and actin were detected by Western blotting. –, unconjugated magnetic beads; C, control IgG; T, anti-TFAM; VDAC, voltage-dependent anion channel.

compatible with the association of a TFAM/mtDNA complex with the membranes. It is assumed that nuclear DNA is looped into domains by attachment to some underlying skeleton (e.g., a matrix or scaffold[29]). Compared to nuclear DNA, however, little is known about how mtDNA associates with membranes. Plasma membranes harbor a special domain, a so-called raft, that is insoluble with TritonX-100. The typical rafts are recovered from around 20% sucrose fractions by sucrose-density centrifugation.[30] It is unlikely that TFAM is localized in a similar special membrane domain of mitochondria because TFAM was recovered from the bottom of the sucrose gradient (FIG. 2, fraction P3).

Our mitochondrial preparation contained actin, a major component of the cytoskeleton. The cytoskeleton is resistant to cold solubilization with non-ionic detergents, and the NP-40 insoluble fractions of the mitochondria indeed retained actin (FIG. 2, fraction S). The insoluble cytoskeleton is recovered from the bottom of the sucrose gradient,[30] and we detected actin in the bottom fraction (FIG. 2, fraction P3). Furthermore, actin was co-immunoprecipitated along with mtDNA and TFAM (FIG. 3). Actin is known to associate with mitochondria in yeast.[31] Recently it was reported that tubulin, a component of microtubules, is attached to human mitochondria.[32] It may be safe to assume that actin also associates with human mitochondrial membranes. There is a surprising report that vimentin, an intermediate filament protein, is directly cross-linked to mtDNA in mouse and human cells.[33] Intermediate filaments are interconnected to microfilaments and microtubules.[34] Taken together, it seems an attractive hypothesis that mtDNA is anchored to membranes via cytoskeletal components irrespective of whether its putative cytoskeletal association is indirect or direct. Such attachment would provide mtDNA a scaffold for transcription, replication, and repair, as in the case of nuclear DNA. Elucidation of such a relationship could also give new insights into the segregation of mtDNA.

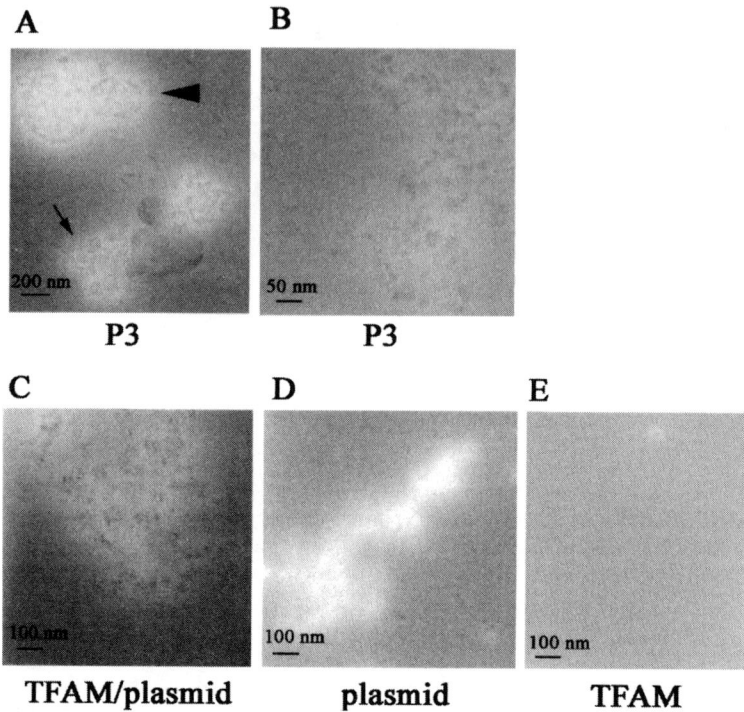

FIGURE 4. Electron microscopic structures. Four nM circular plasmids (pGL-MHC-SvpA, 9.2 kb)[35] and 2 μM recombinant human TFAM (TFAM/plasmid=500) were incubated for 10 min at 25°C in buffer containing 10 mM HEPES-NaOH, pH 7.4, 1 mM EDTA, and 150 mM NaCl. Carbon-coated grid was rendered hydrophilic by glow-discharge at low pressure in air. Samples were adsorbed to the carbon-coated grid for 1 min, and immediately stained with 1% uranyl acetate. The samples were observed in a JEOL JEM 2000 EX electron microscope operated 100 kV acceleration voltage and 60,000× magnification. (**A**) The P3 fraction was observed by electron microscopy. A typical vesicle is indicated by an arrowhead. An arrow indicates a string-like structure. (**B**) A more magnified image of a string-like structure in another area. TFAM/plasmid complexes (**C**), plasmid alone (**D**), and TFAM alone (**E**) were similarly observed by electron microscopy.

DNA WRAPPING WITH TFAM

We examined the P3 fraction by electron microscopy. Vesicle-like bodies were mainly observed in the P3 fraction (FIG. 4A, arrowhead). We also observed some bodies in which a string-like structure could be noticed (FIG. 4A, arrow, and 4B). Some mtDNA nucleoids might be exposed by removal of membrane lipids in the P3. These observations suggest that mtDNA is precipitated along with mitochondrial membranes which were not solubilized with NP-40. When we incubated circular plasmids with TFAM, we found a rosary-like structure (FIG. 4C). In the case of plasmids alone, we observed fine threads only (FIG. 4D). TFAM alone showed few structural elements (FIG. 4E). Hence, the rosary-like structure may reflect a TFAM-plasmid complex, suggesting that TFAM in fact has the ability to cover and wrap the entire region of closed circular DNA.

CONCLUDING REMARKS

Human mtDNA has been postulated to take on a nucleoid structure, mostly on the basis of morphological observations that mtDNA stains in a punctate pattern in cells.[15,16] Our results have provided the first biochemically substantial bases for this contention. We propose that naked mtDNA is unstable and that formation of the mtDNA nucleoid structure is one of the major functions of TFAM for maintaining mtDNA, based on the following evidence provided by several groups including ours: (1) TFAM indeed can wrap circular plasmids entirely (FIG. 4); (2) TFAM is able to substitute for Abf2, a TFAM homologue of yeast, which is supposed to mainly play a histone-like role for maintaining mtDNA;[3] (3) TFAM is abundant enough to cover mtDNA;[19] (4) most mtDNA molecules are associated with TFAM;[18] (5) most TFAM is associated with mtDNA;[18] and (6) the 50% reduction in TFAM in $TFAM^{+/-}$ mice decreases mtDNA by roughly 50%.[6]

ACKNOWLEDGMENTS

This work was supported in part by Uehara Memorial Foundation and Grants-in-Aid for Scientific Research from the Ministry of Education, Science, Technology, Sports, and Culture of Japan.

REFERENCES

1. KANG, D. & N. HAMASAKI. 2002. Maintenance of mitochondrial DNA integrity: repair and degradation. Curr. Genet. **41:** 311–322.
2. FISHER, R.P. & D.A. CLAYTON. 1988. Purification and characterization of human mitochondrial transcription factor 1. Mol. Cell. Biol. **8:** 3496–3509.
3. PARISI, M.A. *et al.* 1993. A human mitochondrial transcription activator can functionally replace a yeast mitochondrial HMG-box protein both *in vivo* and *in vitro*. Mol. Cell. Biol. **13:** 1951–1961.
4. FALKENBERG, M. *et al.* 2002. Mitochondrial transcription factors B1 and B2 activate transcription of human mtDNA. Nat. Genet. **31:** 289–294.
5. SHADEL, G.S. & D.A. CLAYTON. 1997. Mitochondrial DNA maintenance in vertebrates. Annu. Rev. Biochem. **66:** 409–435.
6. LARSSON, N.G. *et al.* 1998. Mitochondrial transcription factor A is necessary for mtDNA maintenance and embryogenesis in mice. Nat. Genet. **18:** 231–236.
7. BUSTIN, M. 1999. Regulation of DNA-dependent activities by the functional motifs of the high-mobility-group chromosomal proteins. Mol. Cell. Biol. **19:** 5237–5246.
8. WOLFFE, A.P. 1994. Architectural transcription factors. Science **264:** 1100–1103.
9. WOLFFE, A.P. 1999. Architectural regulations and HMG1. Nat. Genet. **22:** 215–217.
10. DIFFLEY, J.F.X. & B. STILLMAN. 1992. DNA binding properties of an HMG1-related protein from yeast mitochondria. J. Biol. Chem. **267:** 3368–3374.
11. DIFFLEY, J.F.X. & B. STILLMAN. 1991. A close relative of the nuclear, chromosomal high-mobility group protein HMG1 in yeast mitochondria. Proc. Natl. Acad. Sci. USA **88:** 7864–7868.

12. MEGRAW, T.L. & C.B. CHAE. 1993. Functional complementarity between the HMG-like yeast mitochondrial histone HM and bacterial histone-like protein HU. J. Biol. Chem. **268:** 12758–12763.
13. MIYAKAWA, I. *et al.* 1987. Isolation of morphologically intact mitochondrial nucleoids from the yeast, *Saccharomyces cerevisiae*. J. Cell Sci. **88:** 431–439.
14. SASAKI, N. *et al.* 1998. DNA synthesis in isolated mitochondrial nucleoids from plasmodia of *Physarum polycephalum*. Protoplasma **203:** 221–231.
15. SATOH, M. & T. KUROIWA. 1991. Organization of multiple nucleoids and DNA molecules in mitochondria of a human cell. Exp. Cell Res. **196:** 137–140.
16. SPELBRINK, J.N. *et al.* 2001 Human mitochondrial DNA deletions associated with mutations in the gene encoding Twinkle, a phage T7 gene 4-like protein localized in mitochondria. Nat. Genet. **28:** 223–231.
17. GARRIDO, N. *et al.* 2003. Composition and dynamics of human mitochondrial nucleoids. Mol. Biol. Cell. **14:** 1583–1596.
18. ALAM, T.I. *et al.* 2003. Human mitochondrial DNA is packaged with TFAM. Nucleic Acids Res. **31:** 1640–1645.
19. TAKAMATSU, C. *et al.* 2002. Regulation of mitochondrial D-loops by transcription factor A and single-stranded DNA-binding protein. EMBO Rep. **3:** 451–456.
20. CLAYTON, D.A. 1991. Replication and transcription of vertebrate mitochondrial DNA. Annu. Rev. Cell Biol. **7:** 453–478.
21. FISHER, R.P. *et al.* 1991. A rapid, efficient method for purifying DNA-binding proteins. J. Biol. Chem. **266:** 9153–9160.
22. OHNO, T. *et al.* 2000. Binding of human mitochondrial transcription factor A, an HMG box protein, to a four-way DNA junction. Biochem. Biophys. Res. Commun. **271:** 492–498.
23. WILLIAMS, R.S. 1986. Mitochondrial gene expression in mammalian striated muscle. J. Biol. Chem. **261:** 12390–12394.
24. GOTO, A. *et al.* 2001. *Drosophila* mitochondrial transcription factor A (d-TFAM) is dispensable for the transcription of mitochondrial DNA in Kc167 cells. Biochem. J. **354:** 243–248.
25. NAM, S.C. & C. KANG. 2001. Expression of cloned cDNA for the human mitochondrial RNA polymerase in *Escherichia coli* and purification. Protein Expr. Purif. **21:** 485–491.
26. PRIETO-MARTIN, A. *et al.* 2001 A study on the human mitochondrial RNA polymerase activity points to existence of a transcription factor B-like protein. FEBS Lett. **503:** 51–55.
27. MCCULLOCH, V. *et al.* 2002. A human mitochondrial transcription factor is related to RNA adenine methyltransferases and binds S-adenosylmethionine. Mol. Cell. Biol. **22:** 1116–1125.
28. JACKSON, D.A. *et al.* 1996. Sequences attaching loops of nuclear and mitochondrial DNA to underlying structures in human cells: the role of transcription units. Nucleic Acids Res. **24:** 1212–1219.
29. GETZENBERG, R.H. *et al.* 1991. Nuclear structure and the three-dimensional organization of DNA. J. Cell. Biochem. **47:** 289–299.
30. BROWN, D.A. & J.K. ROSE. 1992 Sorting of GPI-anchored proteins to glycolipid-enriched membrane subdomains during transport to the apical cell surface. Cell **68:** 533–544.
31. YAFFE, M.P. 1999. The machinery of mitochondrial inheritance and behavior. Science **283:** 1493–1497.
32. CARRE, M. *et al.* 2002. Tubulin is an inherent component of mitochondrial membranes that interacts with the voltage-dependent anion channel. J. Biol. Chem. **277:** 33664–33669.
33. TOLSTONOG, G.V. *et al.* 2001. Isolation of SDS-stable complexes of the intermediate filament protein vimentin with repetitive, mobile, nuclear matrix attachment region, and mitochondrial DNA sequence elements from cultured mouse and human fibroblasts. DNA Cell Biol. **20:** 531–554.
34. HERRMANN, H. & U. AEBI. 2000. Intermediate filaments and their associates: multi-talented structural elements specifying cytoarchitecture and cytodynamics. Curr. Opin. Cell. Biol. **12:** 79–90.
35. KUBOTA, T. *et al.* 1997. Cardiac-specific overexpression of tumor necrosis factor-alpha causes lethal myocarditis in transgenic mice. J. Card. Fail. **3:** 117–124.

Regulation of Mitochondrial Transcription Factor A Expression by High Glucose

YON SIK CHOI,[a] KI-UP LEE,[b] AND YOUNGMI KIM PAK[a]

[a]*Asan Institute for Life Sciences, University of Ulsan, Seoul, 138-736, Korea*
[b]*Division of Endocrinology, Department of Internal Medicine, University of Ulsan, Seoul, 138-736, Korea*

> ABSTRACT: Mitochondrial transcription factor A (Tfam, previously mtTFA) is a key regulator of mitochondrial DNA (mtDNA) transcription and replication. We have reported that overexpression of nuclear respiratory factor-1 (NRF-1) and high concentration (50 mM) of glucose increased the promoter activity of the rat Tfam in L6 rat skeletal muscle cells. In this study, we investigated the mechanism of high glucose–induced Tfam transactivation. The addition of 50 mM glucose for 24 h increased Tfam promoter activity up to twofold. The glucose-induced Tfam expression was dose-dependent and cell-type specific. Glucose increased the Tfam promoter-driven transactivity in L6 (skeletal muscle), HIT (pancreatic beta-cell), and CHO (ovary) cells, but not in HepG2 (hepatoma), HeLa, and CV1 (kidney) cells. Among various monosaccharides, only glucose and fructose increased the Tfam promoter activity. Oxidative stress might not be involved in glucose-induced Tfam expression since treatment with antioxidants such as vitamin C, vitamin E, probucol, or α-lipoic acid did not suppress the induction. None of the inhibitors of protein kinase C, MAP kinase, and PI3 kinase altered the glucose-induced Tfam promoter activity, suggesting that general phosphorylation is involved in its signaling. However, a dominant negative mutant of NRF-1, in which 200 amino acids of C-terminus were truncated, completely suppressed the glucose-induced Tfam induction. It was concluded that high glucose–induced Tfam transcription in L6 cells might be mediated by NRF-1.
>
> KEYWORDS: mitochondria; Tfam; diabetes; NRF-1

INTRODUCTION

Mitochondrial transcription factor A (Tfam; previously mtTFA) is a nuclear-encoded protein of 246 amino acids (25 kDa) with a mitochondrial targeting presequence of 42 amino acids.[1] Tfam was originally identified as a transcriptional activator of mitochondrial DNA (mtDNA) but later it was found to be essential for mtDNA replication, since fragments of RNA transcripts from mtDNA were required for priming mtDNA replication.[2] Therefore, Tfam is considered a key regulator of

---

Address for correspondence: Youngmi Kim Pak, Asan Institute for Life Sciences, University of Ulsan, Songpa-Ku Pungnap-Dong 388-1, Seoul, 138-736, Korea. Voice: +82-2-3010-4191; fax: +82-2-3010-4182.
 ymkimpak@amc.seoul.kr

mtDNA transcription and replication. There are also several lines of evidence suggesting that Tfam regulates the transcription and replication of mtDNA *in vivo*. First, disruption of the Tfam gene in mice causes major cellular dysfunction, embryonic lethality, and mitochondrial diabetes resulting from mtDNA depletion and loss of oxidative phosphorylation capacity.[3,4] Second, Tfam levels are responsive to the amounts of mtDNA in the cell, since it is present in low amount in rho zero cells lacking mtDNA.[5–7] Third, differences in mitochondrial transcriptional activity and mtDNA synthesis correlate with the relative amounts of Tfam.[8,9]

The expression of Tfam has been coordinated by a limited set of transcription factors. Nuclear respiratory factors (NRF-1 and 2) are two major trans-acting factors that play a key role in the transcription of the human Tfam gene.[10,11] Peroxisome proliferator–activated receptor-γ coactivator-1 (PGC-1) stimulates the expression of NRF-1 and coactivates the transcription function of NRF-1 on the promoter of Tfam.[12] Furthermore, genomic footprinting studies in tumor cells showed that high level of Sp1 binding to the promoter upregulated Tfam.[13] Although some important factors involved in Tfam expression have been identified, the upstream signaling pathway of these factors and information on other relating factors in various cellular conditions remain largely undetermined.

Mitochondrial DNA (mtDNA)[14] and mitochondrial dysfunction[15] might have important roles in developing insulin resistance and type 2 diabetes. Lee and colleagues reported that a decrease of mtDNA copy number preceded type 2 diabetic development.[16] We reported that depletion of mtDNA impaired glucose metabolism including glucose uptake and hexokinase activity in hepatoma cells.[17] These data suggest that the decrease of mtDNA content may be one of the pathogenic causes of type 2 diabetes. Conversely, we also observed that the amount of mtDNA in leukocytes from diabetic patients increased gradually as diabetes worsened (unpublished observation). This observation leads us to postulate that glucose may regulate mtDNA replication by modulating Tfam transcriptional activity.

In the present study, we tested if glucose regulated Tfam promoter activity in L6 cells using the rat Tfam promoter ligated to the luciferase reporter vector,[18] and investigated putative signaling factors regulating Tfam expression.

MATERIALS AND METHODS

Cell Culture

L6 (skeletal muscle myoblasts, rat), CV-1 (kidney, African green monkey), and CHO (ovary, Chinese hamster) cell lines were grown in Dulbecco's modified Eagle's medium, 10% fetal bovine serum. HepG2 (hepatocellular carcinoma, human) cell line was grown in Eagle's minimal essential medium, 10% fetal bovine serum. HeLa (epitheloid carcinoma, cervix, human) and HIT-T15 (β cell, Syrian hamster) cell lines were grown in RPMI 1640, 10% fetal bovine serum.

Plasmid Constructs

The rat Tfam-luciferase reporter vectors (pTfam-luc) containing −461 to +50 rat promoter region (Genbank accession number, AF264733) were utilized for Tfam promoter activity analysis as described previously.[18]

To construct the NRF-1 dominant negative expression vector, the sequence of human NRF-1 cDNA (Genbank accession number, L22454) from 1 to 304 amino acids lacking the transactivation domain, was amplified by PCR from human NRF-1 expression vector using 5'-CGC GGT ACC ATG GAG GAA CAC GGA GTG ACC-3' as a forward primer and 5'-CGC GGT ACC AGT CTG TGA TGG TAC AAG ATG-3' as a reverse primer, which harbors *Kpn*I sites, respectively. The 924 kb PCR fragments were cloned into a TA-type vector and digested by *Kpn*I. The final fragments were ligated with *Kpn*I restricted pcDNA3.1, a mammalian expression vector. The insert orientation in vector was determined by digestion using appropriate restriction enzymes and sequencing.

Transfection

The plasmid constructs were transfected into the cells by calcium phosphate precipitation method with a β-galactosidase expression vector, BG2-β-gal (Invitrogen, Co., San Diego, CA). After transfection, the cells were grown in culture medium supplemented with 10% fetal bovine serum for 24 h, treated with the various compounds and then collected by scraping. Luciferase and β-galactosidase activities were assayed as described.[19]

Luciferase Assay

The transfected cells were harvested and the cell extracts were assayed for luciferase activity using the luciferase assay kit (Promega, Madison, WI, USA) and a luminometer (Berthold, Badwildbad, Germany). The measured luciferase activity was normalized to β-galactosidase activity.

RESULTS

Induction of Tfam Expression by High Glucose in Various Cells

Since liver, skeletal muscle, and pancreatic islets are glucose-responding tissues, we first examined whether the exposure of various concentrations of glucose affected the Tfam promoter activity in L6 rat skeletal muscle cells. Exposure of glucose increased the Tfam-driven transactivation in a dose-dependent manner after transient transfection of Tfam reporter vectors (pTfam-luc). (FIG. 1A). In order to examine whether the glucose-induced transactivation is tissue specific or not, the Tfam promoter activity was monitored using various cells, such as HepG2 (hepatocellular carcinoma, human), HIT-T15 (pancreatic β cell, Syrian hamster), HeLa (epitheloid carcinoma, cervix, human), CV-1 (kidney, African green monkey), and CHO (ovary, Chinese hamster) cells. As shown FIGURE 1, 24-h incubation with glucose stimulated Tfam transactivation in skeletal muscle cells (L6), ovary cells (CHO), or pancreatic β cells (HIT-T15) in a concentration-dependent manner, but not in HepG2, HeLa, or CV-1 cells, indicating that glucose responsiveness of the Tfam promoter was cell- or tissue-specific.

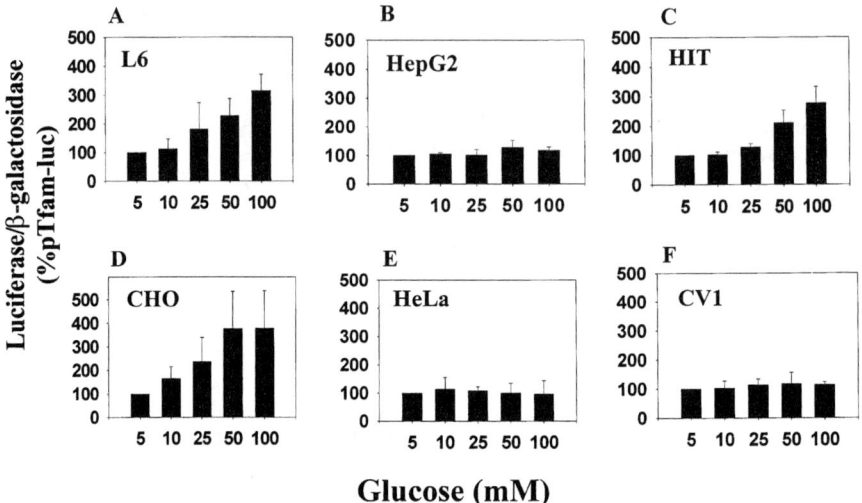

FIGURE 1. Cell-specific induction of Tfam promoter activity by various concentrations of glucose. The cells (**A**, L6; **B**, HepG2; **C**, HIT; **D**, CHO; **E**, HeLa; **F**, CV1) were transfected with pTfam-luc (2 μg) along with the β-galactosidase expression vector, BG2-β-gal, and then treated with various concentrations of glucose for 24 hours. The luciferase activities of the cell lysates were determined and normalized by β-gal activity. Values are mean ± SD of three independent duplicate assays.

Induction of Tfam Promoter by Various Monosaccharides in L6 Skeletal Muscle Cells

To exclude the possibility that the stimulatory effect of high glucose resulted from osmotic shock and to compare the effect of other monosaccharides, we tested several monosaccharides for 24 h in pTfam-luc-transfected L6 cells. Treatment with 50 mM glucose or fructose increased the activity of Tfam promoter 2.5-fold, whereas treatment with other monosaccharides, such as galactose, mannose or mannitol, did not increase the activity (FIG. 2).

Effect of Antioxidants and Other Protein Kinase Inhibitors on High Glucose–Induced Tfam Promoter

It is known that glucose can generate reactive oxygen species (ROS) as byproducts of oxidative phosphorylation or other oxidative reactions. Whether generation of ROS might be involved in high glucose-induced stimulation was tested using various antioxidants. Interestingly, strong antioxidants, such as ascorbic acid, α-tocopherol, probucol, and α-lipoic acid, did not reduce the stimulatory effects of high glucose on Tfam promoter in L6 cells (FIG. 3), implying that ROS do not play a role in the effect of glucose on Tfam transcription. Similarly, none of the inhibitors of protein kinase C (staurosporine, GF 109203X), MAP kinase (PD98059, SB202190), and PI3 kinase (wortmannin) altered the glucose-induced Tfam promot-

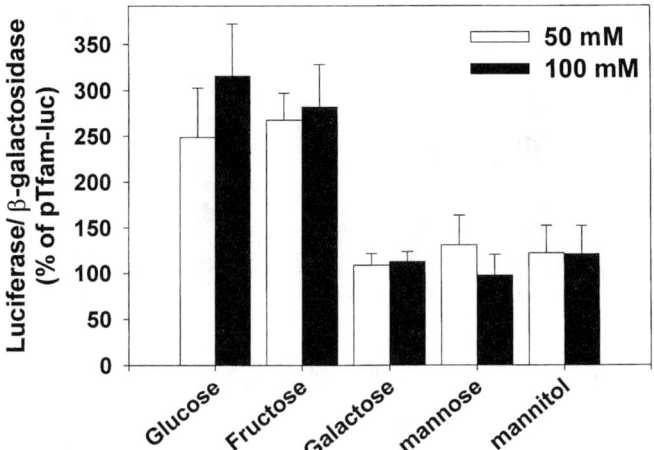

FIGURE 2. Effects of different monosaccharides on the Tfam promoter activity in L6 skeletal muscle cells. The pTfam-luc–transfected cells were incubated in the presence of 50 mM (*open bar*) or 100 mM (*closed bar*) of glucose, fructose, galactose, mannose, or mannitol for 24 h. The luciferase activities of the cell lysates were determined and normalized by β-gal activity. Values are mean ± SD of three independent duplicate assays.

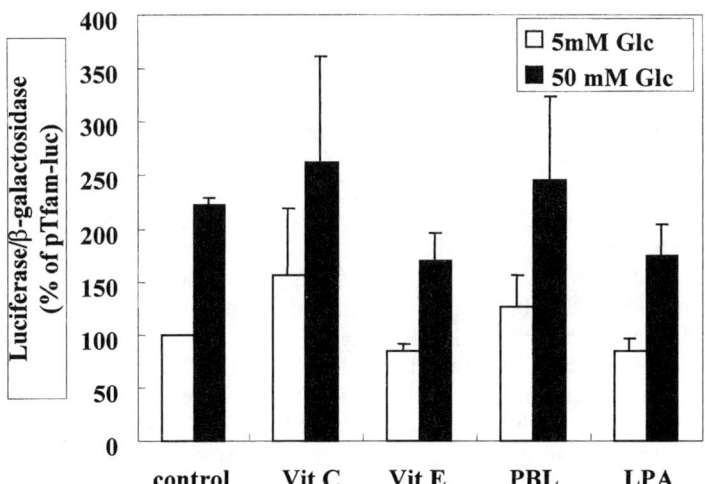

FIGURE 3. Antioxidants did not block the high glucose–induction of Tfam promoter activity in L6 skeletal muscle cells. The pTfam-luc–transfected cells were incubated with vehicle (control), vitamin C (Vit C, 100 μM), vitamin E (Vit E, 1 μM), probucol (PBL, 1 μM), or α–lipoic acid (LPA, 1 μM) for 24 h in the presence of 5 mM (*open bar*) or 50 mM glucose (*closed bar*). The luciferase activities of the cell lysates were determined and normalized by β-gal activity. Values are mean±SD of three independent duplicate assays.

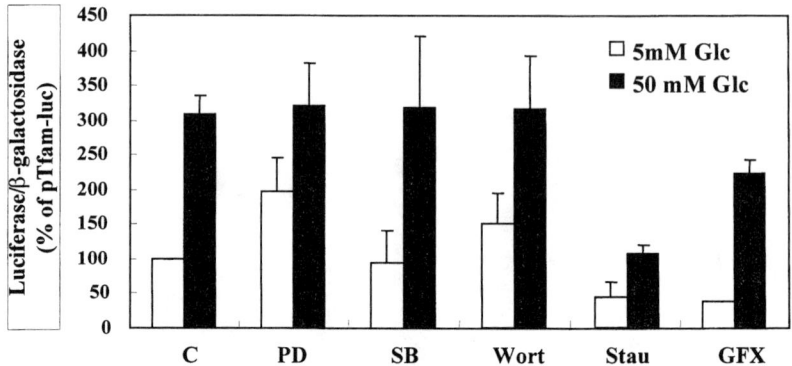

FIGURE 4. No effect of various kinase inhibitors on the high glucose–induction of Tfam promoter activity of in L6 skeletal muscle cells. The pTfam-luc–transfected cells were incubated with the indicated inhibitors of protein kinase C (Stp, 0.1 μM staurosporine; GFX, 6 μM GF109203X), MAP kinase (PD, 2 μM PD98059; SB, 0.3 μM SB202190), and PI3 kinase (Wort, 0.2 μM wortmannin) for 24 h in the presence of 5 mM (*open bar*) or 50 mM glucose (*closed bar*). The luciferase activities of the cell lysates were determined and normalized by β-gal activity. Values are mean ± SD of three independent duplicate assays.

er activity (FIG. 4). These results suggest that PKC, MAPK, and PI3K are not involved in the signaling pathway to increase Tfam transcription by glucose.

Induction of Tfam Promoter by Glucose Is NRF-1 Dependent in L6 Skeletal Muscle Cells

We had shown that overexpression of NRF-1 increased the Tfam promoter activity.[14] Although rat Tfam promoter does not contain an obvious consensus sequence for NRF-1 binding, the −112 to +49 region of the promoter competed with the DNA binding to NRF-1 oligonucleotides by EMSA, demonstrating that the proximal region of the Tfam promoter could bind to NRF-1. To investigate if the glucose-induced Tfam transactivation is mediated through NRF-1, we constructed a dominant negative mutant form of NRF-1 (NRF-1/DN) in which the transactivation domain was deleted but the DNA binding/dimerization domain was maintained. NRF-1/DN could inactivate/suppress all endogenous NRF-1, possibly by forming heterodimer with endogenous NRF-1 or by binding to the NRF-1 binding site on the Tfam promoter. The overexpression of NRF-1 did not further increase the glucose-induced Tfam promoter activity, while overexpression of NRF-1/DN suppressed it completely (FIG. 5). The result strongly suggests that high glucose may induce Tfam promoter expression through NRF-1 transactivation.

DISCUSSION

In the present study, we demonstrated that *in vitro* exposure to high glucose (over 25 mM) in various cells induced Tfam promoter expression. Glucose-induced Tfam

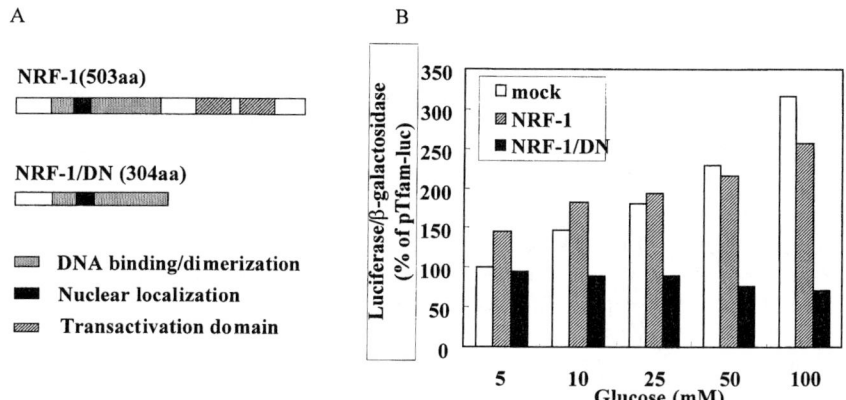

FIGURE 5. High glucose–induced Tfam promoter activity was NRF-1-dependent in L6 skeletal muscle cells. (**A**) A schematic representation of the domains of NRF-1 and NRF-1 dominant-negative protein (NRF-1/DN). There is no transactivation domain in NRF-1/DN, which suppresses wild-type NRF-1 activity by heterodimerization with NRF-1 or other coregulators. (**B**) The cells were co-transfected with Tfam-luc and either of mock (*open bar*), NRF-1 (*hatched bar*), or NRF-1/DN (*closed bar*) expression vectors. The transfected cells were treated with indicated concentrations of glucose for 24 h and harvested. The luciferase activities of the cell lysates were determined and normalized by β-gal activity. Values are mean ± SD of three independent duplicate assays.

expression was mediated by NRF-1, but not by PKC, MAPK, PI3K, or ROS. Since Tfam plays a pivotal role in mitochondrial DNA replication and transcription, the effect of glucose may explain the correlation between diabetes and dysfunctions of mitochondrial biogenesis and/or oxidative phosphorylation.

High glucose conditions, routinely observed in diabetes, have been shown to cause many changes in cellular and molecular aspects. These changes include the induction of intercellular adhesion molecule-1 expression,[20] inhibition of glucose-6-phosphate dehydrogenase in endothelial cells,[21] and apoptosis in endothelial,[22] neuron,[23] or mesangial[24] cells. Hyperglycemia is known to induce the generation of ROS, and sequential activation of c-Jun NH(2)-terminal kinase and caspase-3. Furthermore high concentration of glucose is a potent stimulator for the activation of PKC isozymes and nuclear factor-kappa B (NF-κB). Although the signaling pathways from glucose to these molecules are sometimes different among cell systems, most of them are involved in glucose-induced oxidative stress, glucotoxicity, and eventual diabetic complications. We first hypothesized that high glucose might generate ROS, such as hydrogen peroxide (H_2O_2), and decrease mitochondrial biogenesis. Surprisingly, we observed that high glucose increased, but H_2O_2 decreased Tfam transactivation in L6 skeletal muscle cells,[18] which strongly suggested that H_2O_2 did not mediate the glucose effect in L6 cells. To support this, antioxidants or inhibitors against signaling molecules involved in glucotoxicities did not alter the glucose-induced Tfam promoter activity.

Skeletal muscle has the remarkable capacity to consume glucose as energy source and to vary mitochondrial content and function.[25] Hyperglycemia activates glucose transport in skeletal muscle cell, which is distinct from the insulin-signaling pathway.[26] Thus, the induction of Tfam expression by high glucose may provide some clue to the change of mitochondria metabolism of skeletal muscle in the presence of high glucose. The concentration of glucose in the present study was relatively high compared to previous studies. This might explain the difference in mitochondrial metabolism of physiological *in vivo* conditions, such as hyperglycemia or diabetes.

To explain how Tfam promoter activity is induced by high glucose, we tested many possible inhibitors or known transcription regulators. None of them influenced the glucose effect in L6 cells except NRF-1. It is known that 25 mM glucose activates hexosamine pathway, increases O-glycosylation of Sp1, and consequently induces Sp1-dependent expression, including TGF β1 and PAI-1.[27] Because the Tfam promoter contains three putative Sp1 binding sites, glucose activation of hexosamine pathway might increase Sp1 activity. Analysis of cDNA microarray after glucose treatment of primary cultured human skeletal muscle cells demonstrated that cAMP phosphodiesterase was suppressed to 40% of control by exposure to 50 mM glucose, whereas glutathione peroxidase was induced twofold (unpublished data). These results suggest that cAMP-dependent signaling might be one of the induction mechanisms, but this must be elucidated by further study.

ACKNOWLEDGMENTS

This study was supported by a grant from the Good Health 21 Program, Ministry of Health and Welfare Grant 02-PJ1-PG1-CH04-0001 and an intramural grant from Asan Institute for Life Sciences, University of Ulsan.

REFERENCES

1. FISHER, R.P. & D.A. CLAYTON. 1985. A transcription factor required for promoter recognition by human mitochondrial RNA polymerase. J. Biol. Chem. **260:** 11330–11338.
2. CLAYTON, D.A. 2000. Transcription and replication of mitochondrial DNA. Hum. Reprod. **15**(Suppl. 2)**:** 11–17.
3. LARSSON, N.G., J. WANG, H. WILHELMSSON, *et al.* 1998. Mitochondrial transcription factor A is necessary for mtDNA maintenance and embryogenesis in mice. Nat. Genet. **18:** 231–236.
4. SILVA, J.P., M. KOHLER, C. GRAFF, *et al.* 2000. Impaired insulin secretion and beta-cell loss in tissue-specific knock out mice with mitochondrial diabetes. Nat. Genet. **26:** 336–340.
5. LARSSON, N.G., A. OLDFORS, E. HOLME, *et al.* 1994. Low levels of mitochondrial transcription factor A in mitochondrial DNA depletion. Biochem. Biophys. Res. Commun. **200:** 1374–1381.
6. DAVIS, A.F., P.A. ROPP, D.A. CLAYTON, *et al.* 1996. Mitochondrial DNA polymerase gamma is expressed and translated in the absence of mitochondrial DNA maintenance and replication. Nucleic Acid Res. **24:** 2753–2759.
7. POULTON, J., K. MORTEN, C. FREEMAN-EMMERSON, *et al.* 1994. Deficiency of the human mitochondrial transcription factor h-mtTFA in infantile mitochondrial myopathy is associated with mtDNA depletion. Hum. Mol. Genet. **3:** 1763–1769.
8. MONTOYA, J., A. PEREZ-MARTOS, H.L. GARSTKA, *et al.* 1997. Regulation of mitochondrial transcription by mitochondrial transcription factor A. Mol. Cell. Biochem. **174:** 227–230.

9. GENSLER, S., K. WEBER, W.E. SCHMITT, et al. 2001. Mechanism of mammalian mitochondrial DNA replication. Nucleic Acids Res. **29:** 3657–3663.
10. SCARPULLA, R.C. 1997. Nuclear control of respiratory chain expression in mammalian cells. J. Bioenerg. Biomembr. **29:** 109–119.
11. VIRBASIUS, J.V. & R.C. SCARPULLA. 1994. Activation of the human mitochondrial factor a gene by nuclear respiratory factors. Proc. Natl. Acad. Sci. USA **91:** 1309–1313.
12. WU, Z., P. PUIGSERVER, U. ANDERSON, et al. 1999. Mechanisms controlling mitochondrial biogenesis and respiration through the thermogenic coactivator PGC-1. Cell **98:** 115–124.
13. DONG, X., K. GHOSHAL, S. MAJUMDER, et al. 2002. Mitochondrial transcription factor A and its downstream targets are upregulated in a rat hepatoma. J. Biol. Chem. **277:** 43309–43318.
14. ANTONETTI, D.A., C. REYNET & C.R. KAHN. 1995. Increased expression of mitochondrial-encoded genes in skeletal muscle of humans with diabetes mellitus. J. Clin. Invest. **95:** 1383–1388.
15. PETERSEN, K.F., D. BEFROY, S. DUFOUR, et al. 2003. Mitochondrial dysfunction in the elderly: possible role in insulin resistance. Science **300:** 1140–1142.
16. LEE, H.K., J.H. SONG, C.S. SHIN, et al. 1998. Decreased mitochondrial DNA content in peripheral blood precedes the development of non-insulin-dependent diabetes mellitus. Diabetes Res. Clin. Pract. **42:** 161–167.
17. PARK, K.S., K.J. NAM, J.W. KIM, et al. 2001. Depletion of mitochondrial DNA alters glucose metabolism in SK-Hep1 cells. Am. J. Physiol. Endocrinol. Metab. **280:** E1007–1014.
18. CHOI, Y.S., H.K. LEE & Y.K. PAK. 2002. Characterization of the 5′-flanking region of the rat gene for mitochondrial transcription factor A (Tfam). Biochim. Biophys. Acta **1574:** 200–204.
19. ALAM, J. & J.L. COOK. 1990. Reporter genes: application to the study of mammalian gene transcription. Anal. Biochem. **188:** 245–254.
20. KADO, S., T. WAKATSUKI, M. YAMAMOTO, et al. 2001. Expression of intercellular adhesion molecule-1 induced by high glucose concentrations in human aortic endothelial cell. Life Sci. **68:** 727–737.
21. ZHANG, Z., K. APSE, J. PANG, et al. 2000. High glucose inhibits glucose-6-phosphate dehydrogenase via camp in aortic endothelial cells. J. Biol. Chem. **275:** 40042–40047.
22. HO, F.M., S.H. LIU, C.S. LIAU, et al. 2000. High glucose-induced apoptosis in human endothelial cells is mediated by sequential activations of c-Jun NH(2)-terminal kinase and caspase-3. Circulation **101:** 2618–2624
23. RUSSEL, J.W., D. GOLOVOY, A.M. VINCENT, et al. 2002. High glucose-induced oxidative stress and mitochondrial dysfunction in neurons. FASEB J. **16**(13)**:** 1738–1748.
24. KANG, B.P., S. FRENCHER, V. REDDY, et al. 2003. High glucose promotes mesangial cell apoptosis by oxidant-dependent mechanism. Am. J. Physiol. Renal Physiol. **284:** F455–466.
25. DUGUEZ, S., L. FÉASSON, C. DENIS, et al. 2002. Mitochondrial biogenesis during skeletal muscle regeneration. Am. J. Physiol. Endocrinol. Metab. **282:** E802–E809.
26. KAWANO, Y., J. RINCON, A. SOLER, et al. 1999. Changes in glucose transport and protein kinase Cbeta(2) in rat skeletal muscle induced by hyperglycemia. Diabetologia **42:** 1071–1079.
27. DU, X.L., D. EDELSTEIN, L. ROSSETTI, et al. 2000. Hyperglycemia-induced mitochondrial superoxide overproduction activates the hexosamine pathway and induces plasminogen activator inhibitor-1 expression by increasing Sp1 glycosylation. Proc. Natl. Acad. Sci. USA **97:** 12222–12226.

Regulation and Role of the Mitochondrial Transcription Factor in the Diabetic Rat Heart

YOSHIHIKO NISHIO,[a] AKIO KANAZAWA,[a] YOSHIO NAGAI,[a] HIDETOSHI INAGAKI,[b] AND ATSUNORI KASHIWAGI[a]

[a]*Division of Endocrinology and Metabolism, Department of Medicine, Shiga University of Medical Science, Otsu, Shiga, Japan*

[b]*National Institute of advanced Industrial Science and Technology, Tsukuba, Ibaragi, Japan*

ABSTRACT: To clarify the mechanism of abnormalities in mitochondrial expression and function in diabetic rat heart, we have studied the transcriptional activities of mitochondrial DNA using isolated intact mitochondria from the heart of either diabetic or control rats. The transcriptional activity of cardiac mitochondria isolated from diabetic rats decreased to 40% of the control level ($P<0.01$). Consistently, in the heart of diabetic rats, the content of cytochrome *b* mRNA encoded by mitochondrial DNA was reduced to 50% of control ($P<0.01$). This abnormal transcriptional activity of mitochondrial DNA could not be explained by mRNA or protein contents of mitochondrial transcription factor (mtTFA), but mtTFA binding to the promoter sequence of mitochondrial DNA, assessed by gel-shift assay, was attenuated in diabetic rats. In contrast, the mRNA expression of nuclear-encoded mitochondrial genes, such as ATP synthase-β, was not affected by diabetes. Although O_2 consumption of the mitochondria from diabetic rats was decreased, H_2O_2 production in these rats was increased compared with the control. Insulin treatment reversed all the abnormalities found in diabetic rats. These results clearly indicate that an impairment of binding activity of mtTFA to the promoter sequence has a key role in the abnormal mitochondrial gene expression, which might explain the mitochondrial dysfunction found in diabetic heart.

KEYWORDS: mtTFA; PGC-1; transcription; mitochondrial genome; oxidative stress

INTRODUCTION

Abnormalities in mitochondrial function have been found in diabetics and thought to be a possible cause of oxidative stress in cardiovascular tissues. To clarify the mechanism of abnormalities in mitochondrial expression and function in the heart of diabetic rat, we have studied the transcriptional activities of mitochondrial DNA using isolated intact mitochondria from the heart of either diabetic or control

Address for correspondence: Yoshihiko Nishio, Division of Endocrinology and Metabolism, Department of Medicine, Shiga University of Medical Science, Tsukinowa-cho, Seta, Otsu, Shiga 520-2192, Japan. Voice: +81-77-548-2223; fax:+81-77-543-3858.
nishio@belle.shiga-med.ac.jp

Ann. N.Y. Acad. Sci. 1011: 78–85 (2004). © 2004 New York Academy of Sciences.
doi: 10.1196/annals.1293.009

rats. The transcriptional activity of cardiac mitochondria isolated from diabetic rats decreased to 40% of the control level ($P<0.01$). Consistently, in the heart of diabetic rats, the content of cytochrome b mRNA encoded by mitochondrial DNA was reduced to 50% of control ($P <0.01$). This abnormal transcriptional activity of mitochondrial DNA could not be explained by mRNA or protein contents of mitochondrial transcription factor (mtTFA), but mtTFA binding to the promoter sequence of mitochondrial DNA, assessed by gel-shift assay, was attenuated in diabetic rats. In contrast, the mRNA expression of nuclear-encoded mitochondrial genes, such as ATP synthase-β, was not affected by diabetes. Although O_2 consumption of the mitochondria from diabetic rats was decreased, H_2O_2 production in these rats was increased compared with the control. Insulin treatment reversed all the abnormalities found in diabetic rats. These results clearly indicate that an impairment of binding activity of mtTFA to the promoter sequence has a key role in the abnormal mitochondrial gene expression, which might explain the mitochondrial dysfunction found in diabetic heart.

Transcription of mitochondrial DNA is initiated at two different promoters, heavy- and light-strand promoters, located within the D-loop region, and requires mitochondrial transcription factor A (mtTFA)[1] and RNA polymerase.[2] The mtTFA is a nuclear-encoded DNA binding protein containing two high-mobility group domains. Previous reports[3] showed that the gene expression of mtTFA is regulated by ligand-activated nuclear hormone receptors, such as peroxisome proliferator–activated receptors (PPARs) and their coactivator PGC-1 (FIG. 1). Recently, Wang and colleagues[4] generated heart specific mtTFA knockout mice. In this animal model, the level of mitochondrial transcription was markedly decreased, and the level of mitochondrial-encoded respiratory chain subunits were also decreased, suggesting

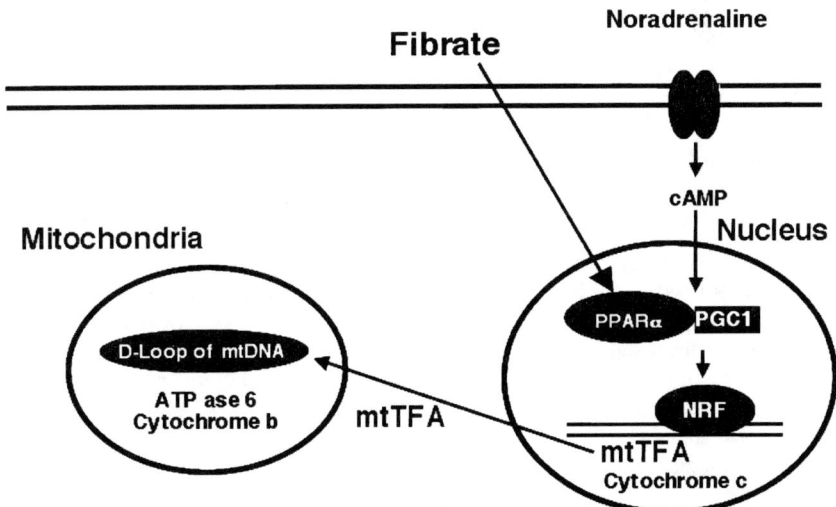

FIGURE 1. Regulation of mitochondrial genes. PPAR, peroxisome proliferator–activated receptor; PGC-1, peroxisome proliferator–activated receptor gamma coactivator 1; and NRF, nuclear respiration factor.

that mtTFA plays a major role in mitochondrial transcription and its functions. Furthermore, it was reported that the mRNA contents of mitochondrial-encoded genes were decreased in the pancreas of diabetic rats.[5] However, the mechanisms underlying the decreased mitochondrial gene expression in diabetes are not clear, and little is known about mitochondrial gene regulation in the diabetic heart.

An interaction between mitochondrial dysfunction and oxidative stress has been reported.[6] We[7] previously reported elevated oxidative stress in the hearts of diabetic rats, and Nishikawa and colleagues[8] reported that high glucose levels increased the production of superoxide anions in mitochondria of cultured bovine aortic endothelial cells. Thus, the pathological roles of mitochondria as a source of reactive oxygen species have been very well established. Conversely, oxidative stress itself causes mitochondrial disorders such as functional impairment in respiration, accumulation of mutated mitochondrial DNA,[9] and decreased mitochondrial gene expression.[10] In turn, reduced mitochondrial function causes oxidative stress in the cells.

On the basis of these observations, we suggest that mtTFA may be a clue to elucidating the mechanisms for mitochondrial gene regulation and reduced mitochondrial function in diabetes mellitus and that diabetes-induced oxidative stress may affect mitochondrial transcriptional activity in the heart. In the present study, we investigated the expression and activity of mtTFA and mitochondrial gene regulation in the hearts of diabetic rats.

MATERIALS AND METHODS

Animals

Male Sprague-Dawley rats (Japan SLC, Shizuoka, Japan) weighing 180–200 g were randomly separated into control and experimental groups (diabetic and insulin-treated diabetic rats). The experimental animals were given an intravenous injection of streptozotocin (50 mg/kg) in 0.05 mM citrate buffer. These animals were maintained on a laboratory diet and water ad libitum for 4 weeks after the streptozotocin injection. Insulin pellets (Linshin, Scarborough, Canada) were implanted subcutaneously 3 days after the streptozotocin injection to normalize blood glucose levels.

Northern Blot Analysis

Northern blot analysis was performed as described previously.[11] In brief, Total RNA was extracted from the cardiac ventricle by the acid guanidinium thiocyanate-phenol-chloroform method and fractionated by denaturing 1% formaldehyde agarose gel electrophoresis. Hybridization and washing of the membrane were carried out with α-^{32}P-labeled complementary DNA (cDNA) probes for rat ATP synthase subunit 6, cytochrome *b*, cytochrome *c*, PGC-1, and mtTFA, as reported previously. To normalize the loading differences, the same membrane was rehybridized with 36B4 (human acidic ribosomal phosphoprotein) cDNA.[12]

In Vitro *Transcription Assay*

Mitochondria were isolated from rat hearts by the protease method,[13] and the integrity of isolated mitochondria was confirmed by evaluating respiratory control ratio (>4.5) and ADP/O ratio (1.8), which were measured using an oxygen electrode with 10 mM succinate as the substrate. The activity of mitochondrial transcription was measured as reported previously.[14]

Electrophoretic Gel Shift Assay

A radioactive probe containing the nucleotide sequence of the heavy-strand promoter was prepared by annealing paired oligonucleotides with the sequences 5'-TTTCCTCCTAACTAAACCCTCTTTAC-3' and 5'-GTAGGCAAGTAAAGAGGGTTTAGTTA-3' and was labeled using [α-^{32}P]dATP (New England Nuclear) and DNA polymerase (Takara, Shiga, Japan). The protein-DNA binding protein reaction was performed as described previously.[15]

Western Blot Analysis of mtTFA

Anti-rat mtTFA serum was prepared as described previously.[16] For Western blotting, total heart homogenate and isolated mitochondria were suspended in ice-cold lysis buffer containing 20 mM Tris/HCl (pH 7.5), 50 mM sodium pyrophosphate, 50 mM sodium fluoride, 1 mM EDTA, 140 mM NaCl, 1% Nonidet P-40, 1 mM sodium orthovanadate, 1 mM phenylmethylsulfonyl fluoride, 50 μM aprotinin, 5 μg/mL leupeptin, and 2 mM benzamidine. After centrifugation at 17,000g at 4°C for 20 min, the supernatant (30 μg of protein) was resolved on 12% SDS-polyacrylamide gel, electrotransferred to an Immobilon P membrane (Millipore, Bedford, MA), and blotted with anti-rat mtTFA serum. Bound antibodies were detected with horseradish peroxidase–conjugated anti-IgG and visualized with an enhanced chemiluminescence detection system (ECL, Amersham Pharmacia Biotech, Piscataway, NJ).

RESULTS

Animal Characteristic

Diabetic rats showed significantly higher plasma glucose levels (28.7±1.2 vs. 9.1±0.3 mM, $P<0.01$). The levels of plasma glucose were not changed between control and insulin-treated diabetic rats (7.6±1.2 mM).

mRNA Contents of ATP Synthase Subunit 6 and Cytochrome b in the Hearts of Diabetic Rats

The mRNA contents of ATP synthase subunit 6 and cytochrome b were significantly decreased by 40% ($P<0.05$) in the hearts of diabetic rats. Insulin treatment prevented the diabetes-induced decrease in these mRNA levels in the hearts.

mRNA Contents of PGC-1, mtTFA, and Cytochrome c in the Hearts of Diabetic Rats

FIGURE 2 presents the results of Northern blot analysis of the PGC-1 in the heart of diabetic rats. In contrast to the results of gene expression of mitochondrial genome, the mRNA contents of the PGC-1, mtTFA, and cytochrome c in diabetic rats were not different from those of control rats.

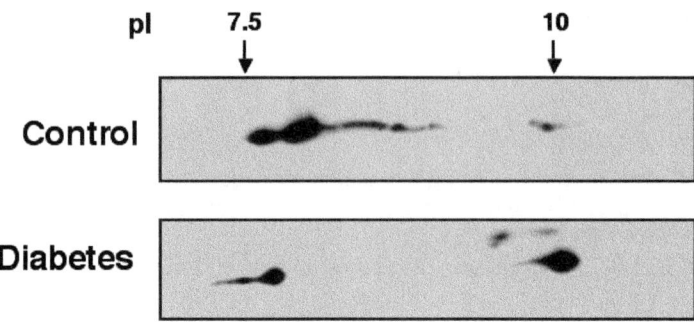

FIGURE 2. (**A**) Northern blot analysis of PGC-1 by use of 20 µg of total RNA isolated from the hearts of control and diabetic rats. The same blots were reprobed with 36B4 cDNA as a loading control. (**B**) Quantification of Northern blot analysis for PGC-1. Values are means ± SEM. $N=6$.

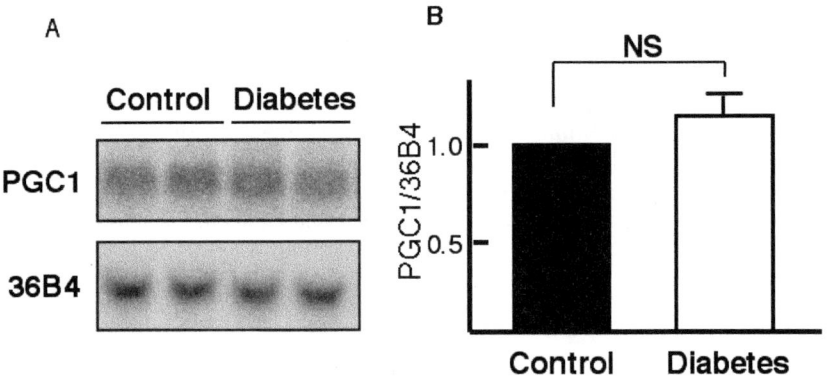

FIGURE 3. Western blot analysis of mtTFA with two-dimensional gel electrophoresis. A sample of 100 µg mitochondrial protein was isolated from the hearts of control or diabetic rat and analyzed.

Protein Contents of mtTFA in the Hearts of Diabetic Rats

Western blot analysis of the mtTFA demonstrated that there was no significant difference in the mtTFA protein contents in the total heart homogenate and isolated mitochondria between control and diabetic rats. FIGURE 3 shows Western blot analysis of mtTFA from either control or diabetic rats' hearts with two-dimensional gel electrophoresis, showing the different pattern of isoelectric focusing of mtTFA between control and diabetic rats.

In Vitro *Transcription Assay in the Hearts of Diabetic Rats*

The incorporation of ^{32}P-labeled UTP into RNA in mitochondria isolated from the heart of diabetic rats was significantly decreased during the 30-min assay period as compared with that of control (3,900±820 vs. 9,900±1,400 cpm/mg protein,

$P<0.01$) and insulin treatment completely prevented this diabetes-induced decrease in transcriptional activity ($10,800\pm1,200$ cpm/mg protein). However, the UTP transport rate into mitochondria was not different between control and diabetic rats.

Binding Activity of mtTFA to the D-Loop Region in the Hearts of Diabetic Rats

We investigated the binding activity of the ^{32}P-labeled oligonucleotide containing the heavy-strand promoter sequence within the D-loop to mitochondrial extracted protein from the hearts of control and diabetic rats using an electrophoretic gel mobility shift assay. The specificity of the shifted band was confirmed by the finding that the band disappeared in the presence of a 50-fold molar excess of unlabeled oligonucleotide. The binding activity of mtTFA to oligonucleotide containing the heavy-strand promoter sequence was decreased in the hearts of diabetic rats, and insulin treatment prevented this decrease.

DISCUSSION

The present study demonstrated that the mRNAs of the mitochondrial-encoded ATP synthase subunit 6 and cytochrome *b* were decreased in the hearts of streptozotocin-induced diabetic rats. In agreement with this finding, the transcriptional activity in mitochondria isolated from hearts of diabetic rats was also reduced compared to control. However, the mRNA expression of PGC-1, which regulates nuclear-encoded mitochondrial genes, such as cytochrome *c* and mtTFA, through the activity of nuclear respiration factors (FIG. 1), was unchanged in the hearts of diabetic rats. Although the mRNA expression of mtTFA was not altered, we found that both the binding activity of mtTFA and the transcriptional activity of mitochondria were reduced in the hearts of diabetic rats. Insulin treatment completely normalized these abnormalities, indicating that these changes were the result of the diabetic state and not a toxic effect of streptozotocin itself. Recently, some studies have reported that mtTFA plays a critical role in the regulation of mitochondrial gene expression.[17] mtTFA is imported into the mitochondrial matrix and activates mitochondrial transcription by binding to the promoter region within the D-loop of mitochondrial DNA. Therefore, the decreased transcriptional activity in mitochondria in diabetes may be explained by abnormality of mtTFA function, such as reduced mtTFA transport into mitochondria, modification of mtTFA protein, or mutations in the binding site of mitochondrial DNA. Western blot analysis of mtTFA in this study showed that the content of mtTFA in mitochondria isolated from the hearts of diabetic rats was not significantly different from that in control hearts, indicating that mtTFA transport into mitochondria was not impaired in diabetic rats. However, the binding activity of mtTFA to the promoter region of mitochondrial DNA was significantly decreased in diabetic rats. Furthermore, the results of two-dimensional gel electrophoresis of mtTFA suggested that some modification of mtTFA may have caused reduced binding activity. This modification of mtTFA and the decrease in its binding to the D-loop of mitochondrial DNA may explain the reduced transcriptional activity in mitochondria found in the heart of diabetic rats. We found an increased basal production of hydrogen peroxide and an increased lipid peroxide content in the mitochondria isolated from diabetic rat hearts.[18] Furthermore, hydrogen peroxide

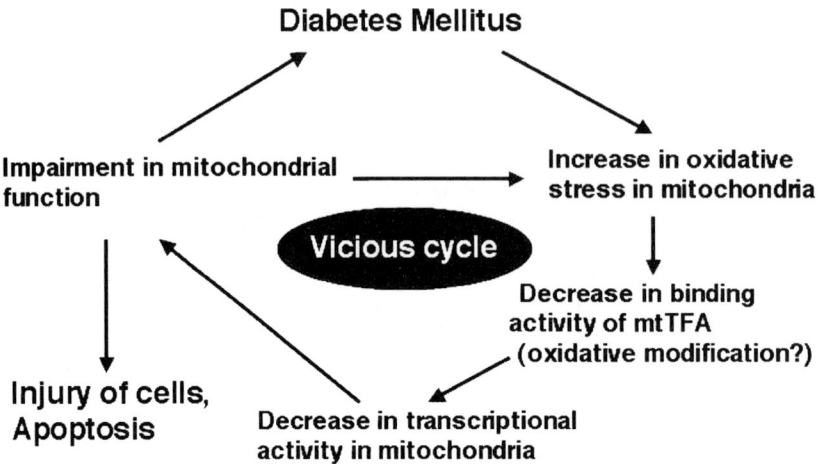

FIGURE 4. Interaction between mitochondrial dysfunction and oxidative stress in mitochondria.

treatment of mitochondria isolated from rat hearts decreased their transcriptional activity.[18] These findings may suggest that mtTFA exposed to the elevated oxidative stress induced by diabetes may undergo additional modifications, such as protein oxidation. The finding of our previous study and those of the present study suggest that the reduced binding activity of mtTFA might contribute to mitochondrial dysfunction in the heart of diabetic rats. In turn, excessive generation of reactive oxygen species by mitochondrial dysfunction might further decrease the binding activity of mtTFA. As shown in FIGURE 4, this idea implies that the reduced binding activity of mtTFA in diabetes might be part of a vicious cycle. The mtTFA may play a key role connecting oxidative stress and mitochondrial dysfunction in this cycle.

REFERENCES

1. PARISI, M.A. & D.A. CLAYTON. 1991. Similarity of human mitochondrial transcription factor 1 to high mobility group proteins. Science **252:** 965–969.
2. FISHER, R.P. & D.A. CLAYTON. 1985. A transcription factor required for promoter recognition by human mitochondrial RNA polymerase. Accurate initiation at the heavy- and light-strand promoters dissected and reconstituted in vitro. J. Biol. Chem. **260:** 11330–11338.
3. WU, Z. et al. 1999. Mechanisms controlling mitochondrial biogenesis and respiration through the thermogenic coactivator PGC-1. Cell **98:** 115–124.
4. WANG, J.H. et al. 1999. Dilated cardiomyopathy and atrioventricular conduction blocks induced by heart-specific inactivation of mitochondrial DNA gene expression. Nat. Genet. **21:** 133–137.
5. SERRADAS, P. et al. 1995. Mitochondrial deoxyribonucleic acid content is specifically decreased in adult, but not fetal, pancreatic islets of the Goto-Kakizaki rat, a genetic model of non-insulin-dependent diabetes. Endocrinology **136:** 5623–5631.
6. MELOV, S. et al. 1999. Mitochondrial disease in superoxide dismutase 2 mutant mice. Proc. Natl. Acad. Sci. USA **96:** 846–851.

7. NISHIO, Y. *et al.* 1988. Altered activities of transcription factors and their related gene expression in cardiac tissues of diabetic rats. Diabetes **37:** 1181–1187.
8. NISHIKAWA, T. *et al.* 2000. Normalizing mitochondrial superoxide production blocks three pathways of hyperglycaemic damage. Nature **404:** 787–790.
9. WALLACE, D.C. 1999. Mitochondrial diseases in man and mouse. Science **283:** 1482–1488.
10. BALLINGER, S.W. *et al.* 2000. Hydrogen peroxide- and peroxynitrite-induced mitochondrial DNA damage and dysfunction in vascular endothelial and smooth muscle cells. Circ. Res. **86:** 960–966.
11. NISHIO, Y. *et al.* 1995. Identification and characterization of a gene regulating enzymatic glycosylation which is induced by diabetes and hyperglycemia specifically in rat cardiac tissue. J. Clin. Invest. **96:** 1759–1767.
12. NISHIO, Y. *et al.* 1994. Glucose induced genes in bovine aortic smooth muscle cells identified by mRNA differential display. FASEB J. **8:** 103–106.
13. MELA, L. & S. SEITZ. 1979. Isolation of mitochondria with emphasis on heart mitochondria from small amounts of tissue. Methods Enzymol. **55:** 39–46.
14. ENRIQUEZ, J.A. *et al.* 1996. An organellar RNA synthesis system from mammalian liver and brain. Methods Enzymol. **264:** 50–57.
15. NAGAI, Y. *et al.* 2002. Amelioration of high fructose-induced metabolic derangements by activation of PPARa. Am. J. Physiol. Endocrinol. Metab. **282:** E1180–1190.
16. INAGAKI, H. *et al.* 2000. Isolation of rat mitochondrial transcription factor A (r-Tfam) cDNA. DNA Seq. **11:** 131–135.
17. INAGAKI, H. *et al.* 1998. Inhibition of mitochondrial gene expression by antisense RNA of mitochondria transcription factor A (mtTFA). Biochem. Mol. Biol. Int. **45:** 567–573.
18. KANAZAWA, A. *et al.* 2002. Reduced activity of mtTFA decreases the transcription in mitochondria isolated from diabetic rat heart. Am. J. Physiol. Endocrinol. Metab. **282:** E778–785.

The Mitochondrial Production of Reactive Oxygen Species in Relation to Aging and Pathology

MARIA LUISA GENOVA, MILENA MERLO PICH, ANDREA BERNACCHIA, CRISTINA BIANCHI, ANNALISA BIONDI, CARLA BOVINA, ANNA IDA FALASCA,[a] GABRIELLA FORMIGGINI, GIOVANNA PARENTI CASTELLI, AND GIORGIO LENAZ

Dipartimento di Biochimica "G. Moruzzi," University of Bologna, 40126 Bologna, Italy

[a]*Dipartimento di Scienze Farmacologiche, Biologiche e Chimiche Applicate, Università di Parma, 43100 Parma, Italy*

ABSTRACT: Mitochondria are known to be strong producers of reactive oxygen species (ROS) and, at the same time, particularly susceptible to the oxidative damage produced by their action on lipids, proteins, and DNA. In particular, damage to mtDNA induces alterations to the polypeptides encoded by mtDNA in the respiratory complexes, with consequent decrease of electron transfer, leading to further production of ROS and thus establishing a vicious circle of oxidative stress and energetic decline. This deficiency in mitochondrial energetic capacity is considered the cause of aging and age-related degenerative diseases. Complex I would be the enzyme most affected by ROS, since it contains seven of the 13 subunits encoded by mtDNA. Accordingly, we found that complex I activity is significantly affected by aging in rat brain and liver mitochondria as well as in human platelets. Moreover, due to its rate control over aerobic respiration, such alterations are reflected on the entire oxidative phosphorylation system. We also investigated the role of mitochondrial complex I in superoxide production and found that the one-electron donor to oxygen is most probably the Fe-S cluster N2. Short chain coenzyme Q (CoQ) analogues enhance ROS formation, presumably by mediating electron transfer from N2 to oxygen, both in bovine heart SMP and in cultured HL60 cells. Nevertheless, we have accumulated much evidence of the antioxidant role of reduced CoQ_{10} in several cellular systems and demonstrated the importance of DT-diaphorase and other internal cellular reductases to reduce exogenous CoQ_{10} after incorporation.

KEYWORDS: mitochondria; oxidative stress; bioenergetic defects; aging; complex I; superoxide; coenzyme Q

Address for correspondence: Prof. Giorgio Lenaz, Dipartimento di Biochimica "G. Moruzzi," Via Irnerio 48, 40126 Bologna, Italy. Voice: +39 051 2091229; fax: +39 051 2091217.
lenaz@biocfarm.unibo.it

INTRODUCTION

Mitochondria are known to be involved in the etiology and pathogenesis of a variety of diseases and in aging.[1–4] This is a consequence of some peculiar aspects of the function of these cellular organelles. The main function of mitochondria is oxidative phosphorylation, by means of which the energy released by electron transfer in the respiratory chain is conserved in the form of ATP.

The mitochondrial respiratory chain is composed of a series of enzyme complexes embedded in the lipid bilayer of the inner mitochondrial membrane. Electrons deriving from oxidation of metabolic substrates in the mitochondrial matrix flow through a series of redox carriers from NADH or other reducing substrates to molecular oxygen. The most accepted view of the chain consists of a series of enzymes independently dissolved in the membrane and connected through mobile components by means of random diffusion of small connecting molecules, coenzyme Q and cytochrome c, in the lipid bilayer.[5,6] The main complexes (I, III, IV) exploit the free energy of oxidation to translocate protons from the matrix to the intermembrane space: the electrochemical potential deriving from the proton movement is used to drive ATP synthesis from ADP and Pi by the ATP synthase (F_0F_1) complex.[7]

Most of the mitochondrial proteins are synthesized in the cytoplasm on information by nuclear genes and then transferred to the mitochondria.[8] However, mitochondrial DNA (mtDNA) encodes for 13 hydrophobic subunits belonging to the four complexes involved in proton translocation, in particular seven subunits for complex I, 1 for complex III, 3 for complex IV, and 2 for ATP synthase. In addition, mtDNA encodes for the tRNA and rRNA involved in mitochondrial protein synthesis.[9]

The special features of mtDNA have been the reason for the postulation of the mitochondrial theory of aging.[10,11] Mitochondria are known to be at once strong producers of reactive oxygen species (ROS) and particularly susceptible to damage by their action on lipids, proteins, and DNA.[4,12,13] In particular, damage to mtDNA would induce damage to the polypeptides encoded by mtDNA in the respiratory complexes, with consequent decrease of electron transfer, leading to further production of ROS, and thus establishing a vicious circle of oxidative stress and energetic decline.[14,15] This decline of mitochondrial energetic capacity is considered the cause of aging and age-related degenerative diseases.[2,4] This vicious circle might be broken by agents capable of preventing the chain reaction of ROS formation and damage.

MITOCHONDRIAL COMPLEX I AND THE ORGANIZATION OF THE RESPIRATORY CHAIN

According to the mitochondrial theory of aging,[10,11] complex I would be the enzyme most affected by ROS, since it contains seven out of the 13 subunits encoded by mtDNA. Complex I (NADH coenzyme Q oxidoreductase)[16] is the most intricate of mitochondrial respiratory complexes, and its molecular organization has not been solved yet.[17,18] Complex I catalyzes the oxidation of matrix NADH by coenzyme Q (CoQ) dissolved in the lipid bilayer.[19] It contains 43 protein subunits and a large number of prosthetic groups having redox activity: FMN, seven iron-sulfur clusters, and some molecules of bound CoQ. The mechanism of electron transfer is not known with certainty; it may, however, involve the consecutive reduction of a bound quinone and of a further quinone molecule from the pool by a FeS center, followed by dismutation of the formed semiquinones and release of ubiquinol to the pool in the lipid bilayer.[20] Several inhibitors are known that interrupt electron transfer at different levels in the complex and can be used to functionally dissect the electron pathway within the enzyme.[20]

The seven subunits encoded by mtDNA are located in the hydrophobic core of the complex in correspondence of the lipid bilayer of the membrane and are probably involved in CoQ binding and in proton translocation. They are denoted as ND1, ND2, ND3, ND4, ND4L, ND5, ND6, and are homologous to the main subunits of bacterial complex I.[21] They are also the site for the binding of several inhibitors considered as CoQ antagonists.[20]

An eventual change in complex I activity is significant only if the enzyme controls the entire oxidative phosphorylation system. The extent of control of an enzyme on a pathway is studied by flux control analysis.[22] The flux control coefficient represents the fractional change of total activity of a pathway induced by a fractional change of the individual enzyme in the pathway. Such a change is usually obtained by stepwise addition of a specific inhibitor. A flux control coefficient approaching one means that the change of the individual step induces the same change on the whole pathway, so the enzyme is totally rate limiting. A flux control coefficient approaching zero means that the enzyme has no control on the whole pathway and its activity is in large excess.

A more evident way of showing flux control is by means of threshold plots,[23] where the residual activity of the total pathway is plotted as a function of the extent of inhibition of the individual step.

We have applied flux control analysis on each segment of the respiratory chain using specific inhibitors of each complex. In open bovine heart submitochondrial particles (SMP),[24,25] where the system is simplified by the absence of membrane potential and ATP synthesis and by the lack of requirement of the carrier systems for substrates, we surprisingly found that both complex I and complex III are almost completely controlling aerobic NADH oxidation[26] (FIG. 1). The most obvious explanation is that complex I and complex III behave as a single enzyme,[27] forming a supercomplex with metabolic channeling of the connecting intermediate, CoQ. This is contrary to the common view of electron transfer;[6] however, it is in line with recent structural findings by Schägger and Pfeiffer;[28] applying digitonin solubilization and blue-native polyacrylamide gel electrophoresis they found evidence for association of complexes I, III, and IV in fixed stoichiometric relations. At difference with them, however, in our study only complexes I and III appear to form a supercomplex, while complex IV seems to behave independently and not to be rate limiting. On the other hand, complex II is rate limiting over succinate oxidation, but there appears to be no channeling between complexes II and III (not shown). The arrangement of the respiratory chain in the inner membrane according to our results is shown in FIGURE 2.

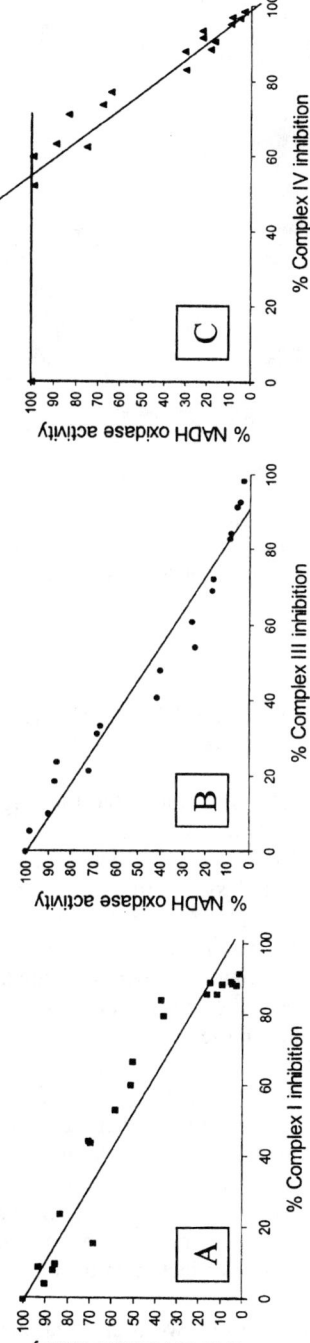

FIGURE 1. Threshold plots of NADH oxidase activity in bovine heart SMP after stepwise inhibition with rotenone, for complex I (**A**); mucidin, complex III (**B**); and KCN, complex IV (**C**). Each point represents the percent rate of NADH oxidase activity as a function of the percent inhibition of the indicated respiratory complexes for the same inhibitor concentration.

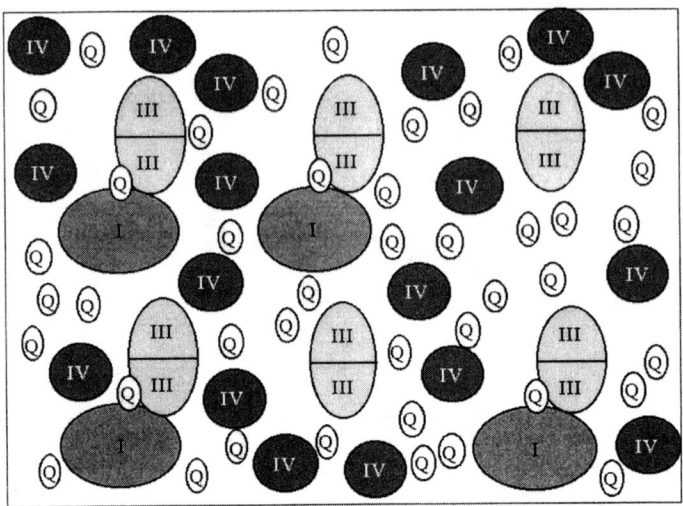

FIGURE 2. Schematic drawing of the arrangement of mitochondrial complexes in the lipid bilayer as viewed from the membrane surface. The presence of specific aggregates is depicted according to our results. For simplicity, only complexes I, III, and IV and CoQ are indicated.

EXPERIMENTAL SUPPORT TO THE MITOCHONDRIAL THEORY OF AGING

The mitochondrial theory of aging states that the original damage to mtDNA is induced by the continuous generation of ROS and other toxic species; thus it is not necessary that an increase of ROS generation occurs in aging, since it is the damage that would accumulate. Nevertheless, an increase of oxidative stress was indeed found in hepatocytes from aged rats.[29,30] By flow cytometric analysis of peroxide production detected by dichlorofluorescein diacetate (DCFDA) labeling, hepatocytes from old rats were found to have a higher peroxide level than hepatocytes from young animals; in addition, the peroxide production after an oxidative stress induced by adriamycin was much higher in the old animals.

We have applied flux control analysis to complex I in aerobic respiration in coupled liver mitochondria from young and aged rats of 32 months.[31] In this system, in comparison with open SMP as cited above, there are other steps that partly control total respiration, such as the substrate carriers and the ATP synthesis, so that the control coefficients for complex I are lower than in SMP. The results show that complex I has little control in young rats but very high control in the old animals (see also Figure 3 provided earlier in this series[32]). This means that aging induces a profound alteration of complex I that is reflected on the entire oxidative phosphorylation system.

In agreement with this observation, we found that aerobic state 3 respiration is decreased in aged rats.[31,32] The alteration of complex I is also documented by the small, albeit significant, decrease of rotenone sensitivity of the enzyme and of whole respiration, documented by the increase of I_{50}, the inhibitor concentration inducing half inhibition of the activities. A similar increase in I_{50} for rotenone was also found in the brain cortex of aged rats.[33]

We have also investigated complex I activity in human platelets from young and aged individuals. In this case the most striking result was the decrease of rotenone sensitivity;[34] the class distribution of rotenone titers for half-inhibition was within a narrow range in the young, but was highly scattered in the old, with many individuals having high rotenone titers for half-inhibition.

Contrary to many tissues both in rats and in humans, platelets do not exhibit the so-called common deletion of 5 kb in mtDNA[35] that is considered as a marker of aging.[36] The deletion is not apparent either in young or in old individuals, as shown by PCR analysis.

The lack of the mtDNA common deletion is not incompatible with the mitochondrial theory of aging. In fact point mutations could accumulate in mtDNA. In particular, Attardi has recently[37] found that the D-loop region of mtDNA, involved in control of duplication and transcription, is heavily mutated only in old individuals.

Since the D-loop controls transcription, we are investigating the steady-state level of mitochondrial RNA transcripts of subunits ND1 and ND5 of complex I using Real Time RT PCR. Preliminary data show that the levels of complex I transcripts are increased in aging (FIG. 3). It must be noted that in rat tissues, mtDNA transcription

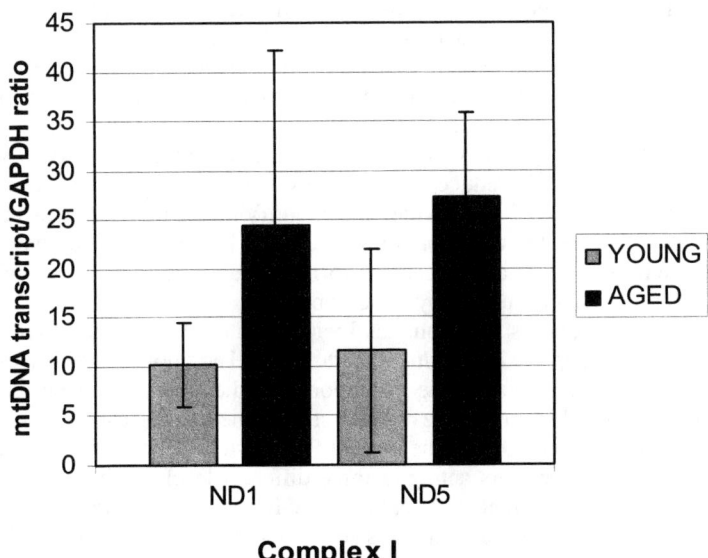

FIGURE 3. Level of the mtDNA transcripts for ND1 and ND5 subunits of complex I in platelets from young and old individuals. RNA was isolated from platelets. Drastic DNase treatment was performed to have pure RNA, then cDNAs for mtDNA-encoded complex I subunits were generated by reverse transcription-PCR amplification. Quantification analysis of templates was performed through Real Time Taqman PCR.

was unchanged or decreased in aging, however complex I transcripts were not investigated.[38]

The unexpected increase of complex I transcripts in aging may be considered as a compensatory mechanism to cope with the functional alteration of the enzyme, due to its rate control over the oxidative phosphorylation system. It is not known whether the protein expression of these genes is also increased.

SUPEROXIDE RADICAL PRODUCTION BY COMPLEX I

It is calculated that up to 4–5% of oxygen in mitochondria is not converted to water by complex IV, but is reduced by a single electron transfer to superoxide radical by the complexes located upstream in the chain.[39] In particular, the main sites of ROS production by the respiratory chain are considered to be located within complex I and complex III.[40] We have studied the role of complex I in superoxide production in bovine heart submitochondrial particles (SMP), where the matrix antioxidant enzymes (Mn-superoxide dismutase and glutathione peroxidase), that could interfere with the results, have been removed by the sonication procedure.

In order to functionally isolate complex I from the rest of the chain we need to inhibit complex III. Since antimycin induces by itself superoxide production by complex III,[41] we have used mucidin, which we have shown does not induce ROS production at the complex III site.[42] This way, the only superoxide detected would derive from complex I. Use of the different inhibitors, as previously indicated, allows the functional dissection of complex I and the detection of the redox group in the complex responsible for superoxide production. The production of superoxide has been assayed by the oxidation of epinephrine to adrenochrome by the superoxide radical.[43] The reaction is fully sensitive to superoxide dismutase, showing that the assay only detects the superoxide radical.

The epinephrine assay is however a potential source of artifacts: the oxidation product, adrenochrome, is reduced by complex I to a semiquinone intermediate, which is a further source of superoxide, thus amplifying the reaction in an autocatalytic fashion.[44] In fact, the reaction is not linear but becomes progressively faster. However, the initial rate of adrenochrome formation appears to be a real indication of the true superoxide production by the complex.

FIGURE 4 (upper panel) shows complex I with the sites of action of different inhibitors. Their effect on superoxide production after NADH addition is shown in the lower panel. Mucidin enhances superoxide formation over the control, as expected by the block of the chain resulting in a more reduced state of the electron carriers; p-hydroxymercuribenzoate, which inhibits at the level of FeS clusters, decreases superoxide production, whereas all inhibitors acting at three different levels at the CoQ reduction sites, either singly or in combination, enhance the formation of the radical.[42]

TABLE 1 shows that different quinones added in presence of mucidin elicit a large increase of superoxide production. The short chain quinones are used as acceptors in place of the physiological CoQ_{10},[19] which is too hydrophobic to be used in the assay system. They are either lower homologs or analogues of the native quinone. Our results mean that under normal conditions of assay of complex I, there is production of superoxide radical. The effect is further potentiated by rotenone and other complex I inhibitors. Idebenone is shown to be the most powerful enhancer of super-

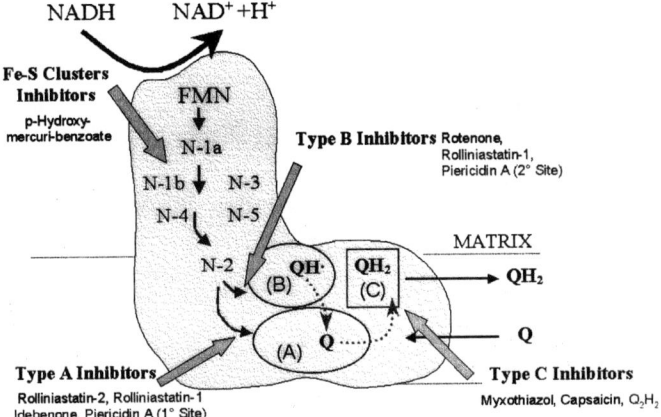

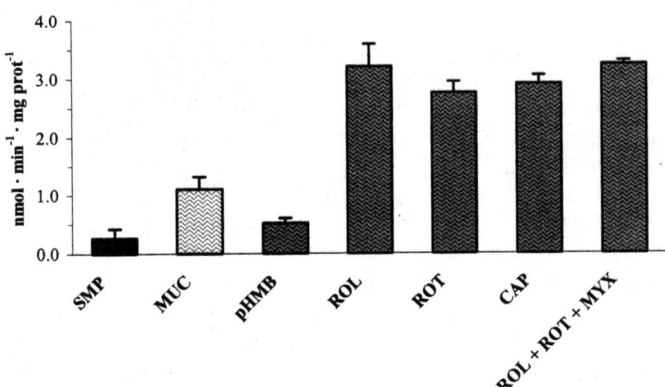

FIGURE 4. Superoxide production by complex I in bovine heart SMP. (*Top*) Scheme of complex I with the redox centers, the path of electrons, and the inhibitors used in this study. (*Bottom*) Superoxide production after addition of 125 µM NADH and different inhibitors, in absence of quinone acceptors. Mucidin (MUC, 1.8 µM) was added to inhibit complex III in all samples except in the control (SMP). Inhibitor concentrations: *p*-hydroxymercuribenzoate (pHMB), 59 µM; rolliniastatin-2 (ROL), 0.2 nmol/mg protein; rotenone (ROT), 0.2 nmol/mg protein; capsaicin (CAP), 4 µmol/mg protein; myxothiazol (MYX), 230 nmol/mg protein. Inhibitor classes are according to the nomenclature of Degli Esposti.[20]

TABLE 1. Effect of different quinones and enzyme inhibitors on superoxide production by complex I in bovine heart SMP

Quinone Acceptor	Inhibitor	Superoxide Production (nmol min^{-1} mg prot^{-1})
–	–	1.10 ± 0.21
–	rotenone	2.76 ± 0.19
DB	–	2.36 ± 1.71
DB	rotenone	5.71 ± 1.46
DB	rotenone + rolliniastatin-2 + myxothiazol	8.56 ± 0.27
CoQ$_1$	–	7.56 ± 1.52
CoQ$_1$	rotenone	12.47 ± 2.76
CoQ$_2$	–	4.23 ± 1.80
CoQ$_2$	rotenone	6.39 ± 0.64
Idebenone	–	17.55 ± 1.21
Idebenone	rotenone	19.10 ± 1.32

The activity was assayed in mucidin-inhibited SMP by monitoring the oxidation of epinephrine to adrenochrome by the superoxide radical produced after addition of 125 μM NADH, 60 μM acceptor (either CoQ lower homologues or decylubiquinone, DB), and inhibitors as indicated. Inhibitor concentrations were the same as in the legend of FIGURE 4. Idebenone was 2 μmol/mg protein.

oxide production, reaching levels 20-fold or more greater than the basal ones; the effect of idebenone is of special interest, since it has a clinical use in some mitochondrial pathologies.[45] Although reduced idebenone is a good electron donor to complex III, it was shown to inhibit complex I and to induce ROS production.[46] This finding suggests caution and further studies *in vivo* before considering the use of this compound in clinics.

The results overall suggest that the site of superoxide production is located between the *p*-hydroxy-mercuribenzoate and the rotenone inhibition sites,[42] thus excluding bound CoQ$_{10}$ as the source of the radical. In fact, CoQ is bound at the site of rotenone and other similar inhibitors.

We have produced further evidence by studying superoxide production by CoQ-extracted and reconstituted particles.[42,47] In all conditions studied, either in presence or absence of inhibitors and short chain quinones, the two types of particles behave similarly with a conspicuous superoxide production, suggesting that the endogenous CoQ$_{10}$ has no part in the univalent reduction of oxygen.

Due to the side reactions of the epinephrine-adrenochrome assay system, we have decided to test our results with an alternative method. Acetylated cytochrome *c*[48] is reduced in mucidin- and cyanide-inhibited SMP in the presence of NADH. The reaction is largely sensitive to SOD, showing that the reducing agent is superoxide. The superoxide production detected by this assay is qualitatively similar to that detected by the epinephrine system. FIGURE 5, showing the flux of electrons through the redox groups in complex I, illustrates our interpretation of the results.[40,42] The main electron donor to oxygen is FeS cluster N2, in view of its proximity to the water phase and its role of reducing bound CoQ in the complex.[49] In addition to direct re-

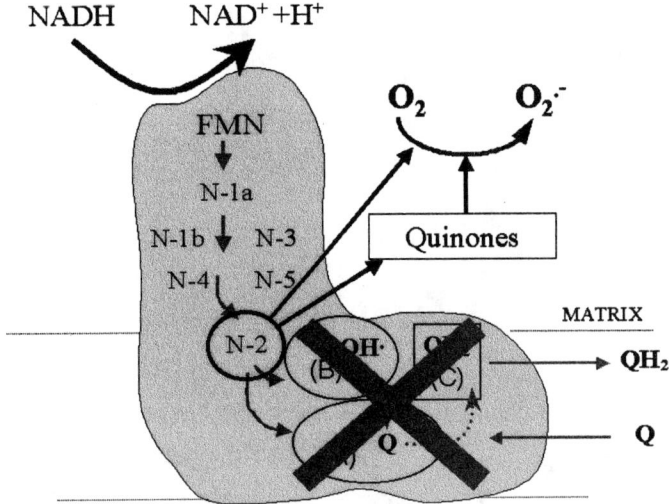

FIGURE 5. Schematic drawing of the postulated site of superoxide formation in complex I according to the results of this study. The effect of different inhibitors and acceptors (see text) suggests Fe-S cluster N2 as the source of one electron to oxygen, directly, or to exogenous quinones that, in turn, can reduce oxygen monoelectronically.

action with oxygen, N2 may give one electron to a water-soluble quinone, and the semiquinone thus formed then reduces molecular oxygen. Bound CoQ_{10} does not appear to have any role in ROS production in complex I.

We have also started studying ROS production in cultured HL60 cells. In the cell, the mitochondrial Mn-SOD and the cytosolic Cu,Zn-SOD convert superoxide to hydrogen peroxide,[40] which can be detected by the probe dichlorofluorescein diacetate (DCFDA) by an increase of its fluorescence upon oxidation. Our preliminary data show that quinones incubated with these cells enhance ROS formation, presumably acting at the mitochondrial level. Nevertheless, CoQ_{10} decreases ROS production (FIG. 6).

We have also shown that short chain quinones are toxic for platelets; in fact they appear to damage the oxidative phosphorylation system, as shown by the enhanced production of lactate after incubation in presence of glucose.[50]

The production of ROS by complex I and other mitochondrial enzymes may be related to the pathogenesis of several diseases.[1] A first example may be represented by some hereditary disorders due to point mutations in mtDNA. Leber's hereditary optic neuropathy is due to three main types of mutations affecting ND subunits of complex I.[51] A puzzling observation is that the most severe form of the disease, due to a mutation at nucleotide 11778 leading to an Arg-to-His change in subunit ND4, is not accompanied by any change of electron transfer rate in the complex.[52] Similarly to what happens in other mutations in nuclear DNA and affecting complex I,[53] or in experimental complex I alterations,[54] we are testing the hypothesis that the al-

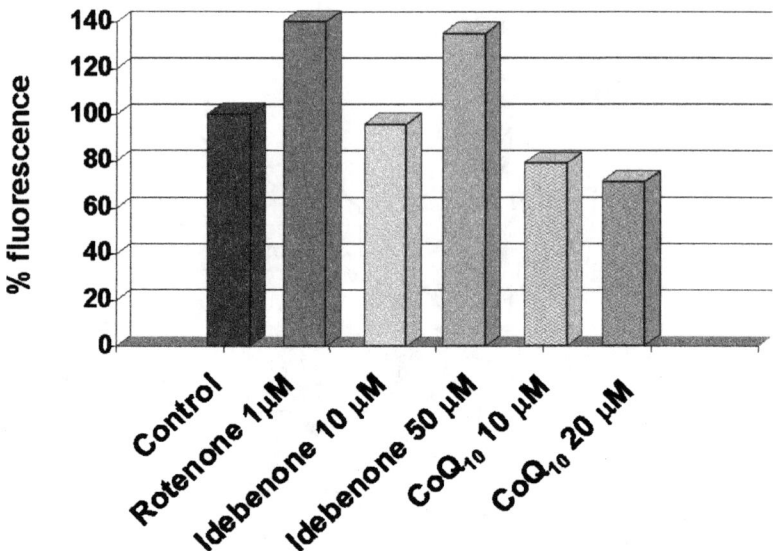

FIGURE 6. Peroxide production by HL60 cells under different conditions. Cells (150 × 10^3) were incubated in a final volume of 0.15 mL (in 96-well plates) for 1 h in presence of 5 μM DCFDA and in presence of rotenone or idebenone or CoQ$_{10}$ as indicated in the picture. Fluorescence was measured at 37°C with a Wallace Victor plate reader (excitation at 485 nm and emission at 520 nm).

teration in the enzyme may be a cause of enhanced ROS production, so that the pathological changes leading to optic nerve degeneration are the result of a secondary oxidative stress.[55]

Similarly, neuropathy, ataxia with retinitis pigmentosa (NARP), which is due to a mutation at nucleotide 8993 leading to a Leu-to-Arg change in subunit ATPase-6 of the membrane sector of ATP synthase,[56] besides a primary defect in ATP synthesis,[57] may be accompanied by secondary alterations due to ROS production, as a consequence of diminished electron flow caused by respiratory control for the lack of ATP synthesis.

The number of pathologies where free radicals have been considered to be heavily involved is overwhelming: we mention here neurodegenerative diseases,[2,58] diabetes,[59] atherosclerosis,[60] and ischemic diseases,[61] failure of organ transplantation,[62] cancer,[63] sepsis,[64] etc.

In addition, the toxic effects of several drugs and pollutants may also be ascribed to ROS: we are conducting a study on the mitochondrial toxicity of asbestos fibers, where the toxic affect of asbestos appears to be correlated with the iron content, as a source of free radicals by the Fenton reaction.[65]

ANTIOXIDANT FUNCTION OF COENZYME Q AND ROLE OF COENZYME Q REDUCING SYSTEMS

It is well known that CoQ_{10} in the reduced form is a powerful antioxidant.[13,66] We have accumulated a lot of evidence of its antioxidant role in cellular systems. For example, in rat hepatocytes, adriamycin induces an oxidative stress, as shown by increase of peroxide levels and decrease of mitochondrial membrane potential.[67] At the same time, CoQ becomes more oxidized with respect to the control. Incubation with CoQ_{10} protects cells from the increase of peroxides and maintains the membrane potential. Both CoQ_9 (the major quinone in the rat) and exogenously added CoQ_{10} are largely reduced by internal cellular systems. There are several cellular systems that can be used to reduce CoQ: besides cytosolic DT diaphorase,[67] a NAD(P)H quinone reductase[68] and plasma membrane NADH oxidase (PMOR)[69] may fulfill this function. In the rat hepatocytes we demonstrated the importance of DT-diaphorase to reduce exogenous CoQ_{10} after incorporation.[67]

We have demonstrated that CoQ_{10} behaves as an antioxidant in human platelets too.[50] Reduced CoQ_{10} protects platelets from an oxidative stress induced by an azido–free radical initiator, shown by using a test based on the Pasteur effect.[70] The rationale of this test is that the enhancement of production of lactate by platelets after antimycin A inhibition of electron flow is proportional to the mitochondrial energetic capacity that has been lost. In that case we had to add reduced CoQ_{10} because, contrary to hepatocytes, platelets appear to have insufficient ability to reduce the quinone after incorporation. In fact, oxidized CoQ_{10} can protect platelets from the oxidative stress, provided they are incubated with purified DT-diaphorase.[50]

The plasma membrane oxidoreductase (PMOR) is a trans-membrane system that can reduce several exogenous compounds including the ascorbate free radical to regenerate the reduced antioxidant form of ascorbate.[69] The system uses CoQ and may also reduce exogenous quinones.

It has been proposed that PMOR may be overexpressed in conditions of oxidative stress, together with plasma membrane-bound DT-diaphorase.[71] We have performed a study in patients with insulin-dependent diabetes[72] and found in lymphocytes a large increase of both the PMOR and DT-diaphorase, this latter shown by the extent of dicoumarol-sensitive activity. The effect is probably related to the oxidative stress that occurs in diabetic patients. In particular DT-diaphorase is almost absent in the plasma membrane of lymphocytes from normal individuals, but is largely increased in most diabetics. We believe the assay can be exploited as a sensitive biomarker of oxidative stress.

ACKOWLEDGMENTS

The studies reported here were supported by grants from the Ministero dell'Istruzione dell'Università e della Ricerca (Rome), from the International CoQ_{10} Association, and from the University of Bologna. The lactate assay is part of U.S. Patent No. US 6,261,796 B1 (July 17, 2001).

REFERENCES

1. BEAL, M.F. 1992. Does impairment of energy metabolism result in excitotoxic neuronal death in neurodegenerative illnesses? Ann. Neurol. **31:** 119–130.
2. DIPLOCK, A.T. 1994. Antioxidants and disease prevention. Mol. Aspects Med. **15:** 293–376.
3. GÖTZ, M.E. *et al.* 1994. Oxidative stress: free radical production in neural degeneration. Pharmacol. Ther. **63:** 37–122.
4. LENAZ, G. 1998. Role of mitochondria in oxidative stress and ageing. Biochim. Biophys. Acta **1366:** 53–67.
5. GREEN, D.E. 1966. The mitochondrial electron transfer system. *In* Comprehensive Biochemistry. M. Florkin & E.H. Stotz, Eds.: 309–327. Elsevier. Amsterdam.
6. HACKENBROCK, C.R., B. CHAZOTTE & S.S. GUPTE. 1986. The random collision model and a critical assessment of diffusion and collision in mitochondrial electron transport. J. Bioenerg. Biomembr. **18:** 331–368.
7. MITCHELL, P. 1968. Chemiosmotic Coupling and Energy Transduction. Glynn Research Inc. Bodmin, UK.
8. PASCHEN, S.A. & W. NEUPERT. 2001. Protein import into mitochondria. IUBMB Life **52:** 101–112.
9. DI MAURO, S. & D.C. WALLACE, EDS. 1993. Mitochondrial DNA in Human Pathology. Raven Press. New York.
10. MIQUEL, J. *et al.* 1980. Mitochondrial role in cell aging. Exp. Gerontol. **15:** 575–591.
11. LINNANE, A.W. *et al.* 1989. Mitochondrial DNA mutations as an important contributor to ageing and degenerative diseases. Lancet **i:** 642–645.
12. ERNSTER, L. 1993. Lipid peroxidation in biological membranes: mechanisms and implications. *In* Active Oxygen, Lipid Peroxides and Antioxidants. K. Yagi, Ed.: 1–38. CRC Press. Boca Raton, FL.
13. ERNSTER, L. & G. DALLNER. 1995. Biochemical, physiological and medical aspects of ubiquinone function. Biochim. Biophys. Acta **1271:** 195–204.
14. LENAZ, G. *et al.* 1999. Mitochondria, oxidative stress, and antioxidant defences. Acta Biochim. Polon. **41:** 1–21.
15. OZAWA, T. 1997. Genetic and functional changes in mitochondria associated with aging. Physiol. Rev. **77:** 425–464.
16. BRANDT, U. 1997. Proton-translocation by membrane-bound NADH:ubiquinone-oxidoreductase (complex I) through redox-gated ligand conduction. Biochim. Biophys. Acta **1318:** 79–91.
17. BRANDT, U. (Ed.). 1998. Structure and function of complex I. Biochim. Biophys. Acta **1364:** 285–296.
18. VINOGRADOV, A.D. & V.G. GRIVENNIKOVA. 2001. The mitochondrial complex I: progress in understanding of catalytic properties. IUBMB Life **52:** 129–134.
19. LENAZ, G. 1998. Quinone specificity of complex I. Biochim. Biophys. Acta **1364:** 207–221.
20. DEGLI ESPOSTI, M. 1998. Inhibitors of NADH-ubiquinone reductase: an overview. Biochim. Biophys. Acta **1364:** 222–235.
21. YAGI, T. *et al.* 1998. Procaryotic complex I (NDH-1), an overview. Biochim. Biophys. Acta **1364:** 125–133.
22. KACSER, A. & J.A. BURNS. 1979. Molecular democracy: who shares the controls? Biochem. Soc. Trans. **7:** 1149-1160.
23. ROSSIGNOL, R. *et al.* 1999. Threshold effect and tissue specificity. Implication for mitochondrial cytopathies. J. Biol. Chem. **274:** 33426–33432.
24. HANSEN, M. & A.L. SMITH. 1964. Biochim. Biophys. Acta **81:** 214–222.
25. FATO, R. *et al.* 1993. Steady-state kinetics of ubiquinol-cytochrome c reductase in bovine heart submitochondrial particles: diffusional effects. Biochem. J. **290:** 225–236.
26. GENOVA, M.L., C. BIANCHI & G. LENAZ. 2003. Structural organization of the mitochondrial respiratory chain. Ital. J. Biochem. **52:** 58–61.
27. LENAZ, G. 2001. A critical appraisal of the mitochondrial coenzyme Q pool. FEBS Lett. **509:** 151–155.

28. SCHÄGGER, H. & K. PFEIFFER. 2001. The ratio of oxidative phosphorylation complexes I-V in bovine heart mitochondria and the composition of respiratory chain supercomplexes. J. Biol. Chem. **276:** 37861–37867.
29. CAVAZZONI, M. *et al.* 1999. The effect of aging and an oxidative stress on peroxide levels and the mitochondrial membrane potential in isolated rat hepatocytes. FEBS Lett. **449:** 53-56.
30. BAROGI, S. *et al.* 2000. Effect of the oxidative stress induced by adriamycin on rat hepatocyte bioenergetics during ageing. Mech. Ageing Dev. **113:** 1–21.
31. VENTURA, B. *et al.* 2002. Control of oxidative phosphorylation by Complex I in rat liver mitochondria: implications for aging. Biochim. Biophys. Acta **1553:** 249–260.
32. LENAZ, G. *et al.* 2002. Role of mitochondria in oxidative stress and aging. Ann. N.Y. Acad. Sci. **959:** 199–213.
33. GENOVA, M.L. *et al.* 1997. Decrease of rotenone inhibition is a sensitive parameter of complex I damage in brain non-synaptic mitochondria of aged rats. FEBS Lett. **410:** 467–469.
34. MERLO PICH, M. *et al.* 1996. Inhibitor sensitivity of respiratory complex I in human platelets: a possible biomarker of ageing. FEBS Lett. **380:** 176–178.
35. BIAGINI, G. *et al.* 1998. Mitochondrial DNA in platelets from aged subjects. Mech. Ageing Dev. **101:** 269–275.
36. SCHON, E.A. *et al.* 1996. Mitochondrial DNA mutations and aging. *In* Cellular Aging and Cell Death. N.J. Holbrook, G.R. Martin & R.A. Lockshin, Eds.: 19–34. J. Wiley & Sons Inc. New York.
37. MICHIKAWA, Y. *et al.* 1999. Aging-dependent large accumulation of point mutations in the human mtDNA control region for replication. Science **286:** 774–779.
38. BARAZZONI, R., K.R. SHORT & K.S. NAIR. 2000. Effects of aging on mitochondrial DNA copy number and cytochrome c oxidase gene expression in rat skeletal muscle, liver, and heart. J. Biol. Chem. **275:** 3343–3347.
39. RICHTER, C. 1992. Reactive oxygen and DNA damage in mitochondria. Mutat. Res. **275:** 249–255.
40. LENAZ, G. 2001. The mitochondrial production of reactive oxygen species: mechanisms and implications in human pathology. IUBMB Life **52:** 159–164.
41. STANIEK, K. & H. NOHL. 2000. Are mitochondria a permanent source of reactive oxygen species? Biochim. Biophys. Acta **1460:** 268–275.
42. GENOVA, M.L. *et al.* 2001. The site of production of superoxide radical in mitochondrial Complex I is not a bound ubisemiquinone but presumably iron-sulfur cluster N2. FEBS Lett. **505:** 364–368.
43. BOVERIS, A. 1984. Determination of the production of superoxide radicals and hydrogen peroxide in mitochondria. Methods Enzymol. **105:** 429–435.
44. BINDOLI, A. *et al.* 1990. Direct and respiratory chain-mediated redox cycling of adrenochrome. Biochim. Biophys. Acta **1016:** 349–356.
45. GILLIS, J.C., P. BENFIELD. & D. MCTAVISH. 1994. Idebenone. A review of its pharmacodynamic and pharmacokinetic properties, and therapeutic use in age-related cognitive disorders. Drugs Aging **5:** 133–152.
46. DEGLI ESPOSTI, M. *et al.* 1996. The interaction of Q analogs, particularly hydroxydecyl benzoquinone (idebenone), with the respiratory complexes of heart mitochondria. Arch. Biochem. Biophys. **330:** 395–400.
47. GENOVA, M.L. *et al.* 2003. Mitochondrial production of oxygen radical species and the role of coenzyme Q as an antioxidant. Exp. Biol. Med. **228:** 506–513.
48. AZZI, A., C. MONTECUCCO & C. RICHTER. 1975. The use of acetylated ferricytochrome c for the detection of superoxide radicals produced in biological membranes. Biochem. Biophys. Res. Commun. **65:** 597–603.
49. YANO, T. & T. OHNISHI. 2001. The origin of cluster N2 of the energy-transducing NADH-quinone oxidoreductase: comparisons of phylogenetically related enzymes. J. Bioenerg. Biomembr. **33:** 213–221.
50. MERLO PICH, M. *et al.* 2002. Ubiquinol and a coenzyme Q reducing system protect platelet mitochondrial function of transfusional buffy coats from oxidative stress. Free Radical Res. **36:** 429–436.

51. BROWN, M.D. *et al.* 1992. Mitochondrial DNA complex I and III mutations associated with Leber's hereditary optic neuropathy. Genetics **130:** 163–173.
52. DEGLI ESPOSTI, M. *et al.* 1994. Functional alterations of the mitochondrially encoded ND4 subunit associated with Leber's hereditary optic neuropathy. FEBS Lett. **352:** 375–379.
53. ROBINSON, B.H. 1998. Human complex I deficiency: clinical spectrum and involvement of oxygen free radicals in the pathogenicity of the defect. Biochim. Biophys. Acta **1364:** 271–286.
54. BARRIENTOS, A. & C.T. MORAES. 1999. Titrating the effects of mitochondrial complex I impairment in the cell physiology. J. Biol. Chem. **274:** 16188–16197.
55. CARELLI, V., F.N. ROSS-CISNEROS & A.A. SADUN. 2002. Optic nerve degeneration and mitochondrial dysfunction: genetic and acquired optic neuropathies. Neurochem. Int. **40:** 573–584.
56. HOLT, I.J. *et al.* 1990. A new mitochondrial disease associated with mitochondrial DNA heteroplasmy. Am. J. Hum. Genet. **46:** 428–433.
57. BARACCA, A. *et al.* 2000. Catalytic activities of mitochondrial ATP synthase in patients with mitochondrial DNA T8993G mutation in the ATPase 6 gene encoding subunit a. J. Biol. Chem. **275:** 4177–4182.
58. BEAL, M.F. 1998. Mitochondrial dysfunction in neurodegenerative diseases. Biochim. Biophys. Acta **1366:** 211–223.
59. LOW, P.A., K.K. NICKANDER & H.J. TRITSCHLER. 1997. The roles of oxidative stress and antioxidant treatment in experimental diabetic neuropathy. Diabetes **46:** 838–842.
60. MASHIMA, R., P.K. WITTING & R. STOCKER. 2001. Oxidants and antioxidants in atherosclerosis. Curr. Opinion Lipidol. **12:** 411–418.
61. RICE-EVANS, C.A. & A.T. DIPLOCK. 1993. Current status of antioxidant therapy. Free Radical Biol. Med. **15:** 77–96.
62. JAESCHKE, H. 1991. Reactive oxygen and ischemia/reperfusion injury of the liver. Chem. Biol. Interact. **79:** 115–136.
63. BIANCHI, N.O., M.S. BIANCHI & S.M. RICHARD. 2001. Mitochondrial genome instability in human cancers. Mutation Res. **488:** 9–23.
64. TAYLOR, D.C., A. GHIO & C.A. PIANTADOSI. 1995. Reactive oxygen species produced by liver mitochondria of rats in sepsis. Arch. Biochem. Biophys. **316:** 70–76.
65. KAMP, D.W. *et al.* 2002. Asbestos-induced alveolar epithelial cell apoptosis: role of mitochondrial dysfunction caused by iron-derived free radicals. Mol. Cell. Biochem. **234/235:** 153–160.
66. BEYER, R.E. & L. ERNSTER. 1990. The antioxidant role of Coenzyme Q. *In* Highlights of Ubiquinone Research. G. Lenaz, O. Barnabei, A. Rabbi & M. Battino, Eds.: 191–213. Taylor & Francis. London.
67. BEYER, R.E. *et al.* 1996. The role of DT-diaphorase in the maintenance of the reduced antioxidant form of coenzyme Q in membrane systems. Proc. Natl. Acad. Sci. USA **93:** 2528–2532.
68. TAKAHASHI, T. *et al.* 1995. Reduction of ubiquinone in membrane lipids by rat liver cytosol and its involvement in the cellular defence system against lipid peroxidation. Biochem. J. **309:** 883–890.
69. VILLALBA, J.M. *et al.* 1995. Coenzyme Q reductase from liver plasma membrane: purification and role in trans-plasma-membrane electron transport. Proc. Natl. Acad. Sci. USA **92:** 4887–4891.
70. D'AURELIO, M. *et al.* 2001. Decreased Pasteur effect in platelets of aged individuals. Mech. Ageing Dev. **122:** 823–833.
71. NAVARRO, F. *et al.* 1998. Vitamin E and selenium deficiency induces expression of the ubiquinone-dependent antioxidant system at the plasma membrane. FASEB J. **12:** 1665–1673.
72. LENAZ, G. *et al.* 2002. Enhanced activity of the plasma membrane oxidoreductase in circulating lymphocytes from insulin-dependent diabetes mellitus patients. Biochem. Biophys. Res. Commun. **290:** 1589–1592.

Biological Significance of the Defense Mechanisms against Oxidative Damage in Nucleic Acids Caused by Reactive Oxygen Species

From Mitochondria to Nuclei

YUSAKU NAKABEPPU, DAISUKE TSUCHIMOTO, AKIMASA ICHINOE, MIZUKI OHNO, YASUHITO IDE, SEIKI HIRANO, DAISUKE YOSHIMURA, YOHEI TOMINAGA, MASATO FURUICHI, AND KUNIHIKO SAKUMI

Division of Neurofunctional Genomics, Medical Institute of Bioregulation, Kyushu University, and Core Research for Evolutional Science and Technology (CREST), Japan Science and Technology Agency (JST), Fukuoka, 812-8582, Japan

ABSTRACT: In mammalian cells, more than one genome in a single cell has to be maintained throughout the entire life of the cell, namely, one in the nucleus and the other in the mitochondria. The genomes and their precursor nucleotides are highly exposed to reactive oxygen species, which are inevitably generated as a result of the respiratory function in mitochondria. To counteract such oxidative damage in nucleic acids, cells are equipped with several defense mechanisms. Modified nucleotides in the nucleotide pools are hydrolyzed, thus avoiding their incorporation into DNA or RNA. Damaged bases in DNA with relatively small chemical alterations are mainly repaired by the base excision repair (BER) system, which is initiated by the excision of damaged bases by specific DNA glycosylases. MTH1 protein hydrolyzes oxidized purine nucleoside triphosphates, such as 8-oxo-dGTP, 8-oxo-dATP, and 2-hydroxy (OH)-dATP to the monophosphates, and MTH1 are located in the cytoplasm, mitochondria, and nucleus. We observed an increased susceptibility to spontaneous carcinogenesis in *Mth1*-deficient mice and an alteration of MTH1 expression along with the accumulation of 8-oxo-dG in patients with various neurodegenerative diseases. Enzymes for the BER pathway, namely, 8-oxoG DNA glycosylase (OGG1), 2-OH-A/adenine DNA glycosylase (MUTYH), and AP endonuclease (APEX2) are also located both in the mitochondria and in the nuclei, and the expression of mitochondrial OGG1 is altered in patients with various neurodegenerative diseases. We also observed increased susceptibilities to spontaneous carcinogenesis in OGG1 and MUTYH-deficient mice. The increased occurrence of lung tumor in OGG1-deficient mice was completely abolished by the concomitant disruption of the *Mth1* gene.

KEYWORDS: oxidative damage; DNA repair; carcinogenesis; neurodegeneration; nucleotide pool; mitochondria

Address for correspondence: Yusaku Nakabeppu, D.Sc., Division of Neurofunctional Genomics, Medical Institute of Bioregulation, Kyushu University, 3-1-1 Maidashi, Higashi-Ku, Fukuoka, 812-8582, Japan. Voice: 81-92-642-6800; fax: 81-92-642-6791.
yusaku@bioreg.kyushu-u.ac.jp

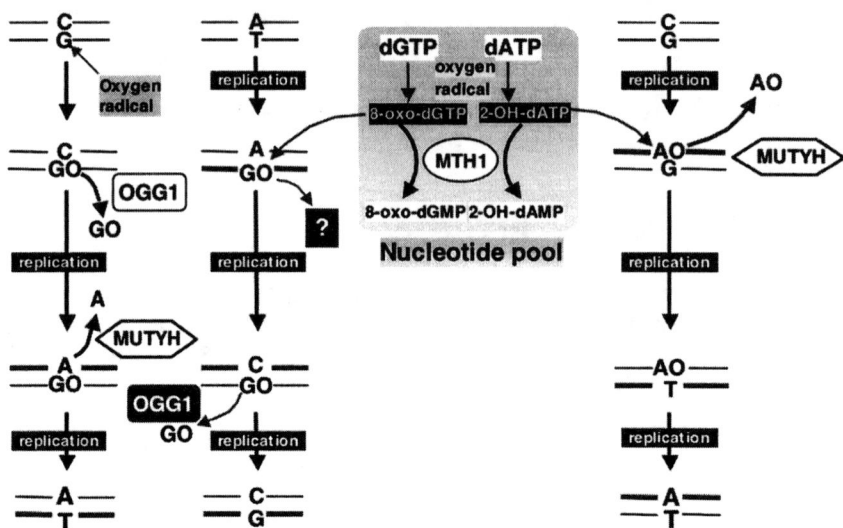

FIGURE 1. Mutagenesis caused by the oxidation of nucleic acids and error-avoiding mechanisms in mammals. 8-OxoG accumulates in DNA, as a result of the incorporation of 8-oxo-dGTP from nucleotide pools or because of direct oxidation of DNA, increases the occurrence of A:T to C:G or G:C to T:A transversion mutation, whereas 2-OH-A is mainly derived from of the incorporation of 2-OH-dATP from nucleotide pools. The accumulation of 8-oxoG or 2-OH-A in DNA is minimized by coordinated actions of MTH1, OGG1, and MUTYH. See text for details. GO, 8-oxoguanine (8-oxoG); AO, 2-hydroxyadenine (2-OH-A). Grayish thick lines represent newly synthesized strands of DNA.

INTRODUCTION

Among many classes of DNA damage caused by reactive oxygen species (ROS), an oxidized form of guanine base, 8-oxoguanine (8-oxoG), is considered to play a role in mutagenesis and carcinogenesis Because 8-oxoG can pair with adenine as well as cytosine with almost equal efficiency during DNA replication, it thus has the potential to cause a high frequency of mutations.[1] The accumulation of 8-oxoG in DNA, as a result of the incorporation of 8-oxo-dGTP from nucleotide pools or because of direct oxidation of DNA, increases the occurrence of A:T to C:G or G:C to T:A transversion mutation, respectively (FIG. 1).[2]

Studies on mutator mutants have shown that *Escherichia coli* has several error-avoiding mechanisms, which minimize the deleterious effects of 8-oxoG, and in which MutT, MutM(FPG), and MutY proteins play important roles.[3,4] MutT protein hydrolyses 8-oxo-dGTP to 8-oxo-dGMP and pyrophosphate, thus avoiding the spontaneous occurrence of A:T to C:G transversion mutation during DNA synthesis, the rate of which in a *mutT*-deficient strain increases 100-fold to 1,000-fold compared with the wild type. MutM protein, originally identified as formamidopyrimidine (Fapy) DNA glycosylase, removes the 8-oxoG paired with cytosine. MutY protein with its DNA glycosylase activity excises adenine paired with guanine or 8-oxoG.

The resulting base-loss sites, called apurinic or apyrimidinic (AP) sites, are further processed through the base excision repair (BER) pathway, in which AP endonuleases such as XTH and NFO play critical roles in the initiation of repair replication. The rate of spontaneous occurrence of G:C to T:A transversion mutation in double mutants of *mutM* and *mutY* is equivalent to that of the *mutT* mutant. In contrast, double mutants of *xth* and *nfo* exhibit an increased sensitivity toward various damaging agents such as hydrogen peroxide.[5]

It has been established recently that mammalian cells are also equipped with error-avoiding mechanisms similar to those found in prokaryotes.[2,6] We have identified and extensively characterized four enzymes, MTH1 as a MutT homologue, OGG1 as a functional homologue for MutM, MUTYH, or MYH as MutY homologue, and APEX2 as the second member of XTH homologue in mammals. In mammalian cells, more than one genome in a single cell has to be maintained throughout the entire life of the cell, one in the nucleus and the other in mitochondria. Genomes in mitochondria are likely to be more susceptible to ROS-induced oxidative damage as oxygen metabolism is high.[7] We previously reported that all four enzymes, namely, MTH1, MUTYH, OGG1, and APEX2, which are considered to minimize oxidative DNA damage in mammalian cells, are located both in nuclei and in mitochondria of mammalian cells.[8-10] Here, we describe the biochemical and biological functions of these enzymes in mammalian cells.

MAMMALIAN ENZYMES INVOLVED IN ERROR-AVOIDING MECHANISMS AGAINST OXIDATIVE DAMAGE IN NUCLEIC ACIDS

The occurrence of 8-oxoG in DNA is derived through two pathways, incorporation of an oxidized precursor, 8-oxo-dGTP, into DNA during DNA synthesis, and direct oxidation of a guanine base in DNA.[2,3,11] To prevent the use of the former pathway, organisms possess the oxidized purine nucleoside triphosphatase encoded by the *MTH1* gene, that degrades 8-oxo-dGTP into 8-oxo-dGMP and pyrophosphate.[12,13] MTH1 efficiently hydrolyzes two forms of oxidized dATP, 2-hydroxy (OH)-dATP and 8-oxo-dATP, as well as 8-oxo-dGTP, and MTH1 also hydrolyzes oxidized ribonucleotides, 2-OH-ATP, 8-oxoATP, and 8-oxo-GTP. The MTH1 protein therefore is designated as an oxidized purine nucleoside triphosphatase (TABLE 1). The substrate specificity of MTH1 for oxidized purine nucleoside triphosphates was investigated by mutation analyses based on sequence comparison with the *Escherichia coli* homologue, MutT, which hydrolyzes only 8-oxo-dGTP and 8-oxoGTP but not oxidized forms of dATP or ATP. Neither a replacement of the phosphohydrolase module of MTH1 with that of MutT nor deletions of the C-terminal region of MTH1, which is unique for MTH1, altered the substrate specificity of MTH1. In contrast, the substitution of residues at position W117 and D119 of MTH1, which showed apparent chemical shift perturbations with 8-oxo-dGDP in NMR analyses but are not conserved in MutT, affected the substrate specificity. W117 is essential for MTH1 to recognize both 8-oxo-dGTP and 2-OH-dATP, whereas D119 is only essential for recognizing 2-OH-dATP, thus suggesting that the origins of the substrate-binding pockets for MTH1 and MutT are different.[14]

Once 8-oxoG comes into existence in DNA, 8-oxoG DNA glycosylase, encoded by *OGG1* gene, removes the oxidized base from the DNA.[2,15,16] The DNA glycosylase activity of OGG1 preferentially excises 8-oxoG or Fapy opposite cytosine and to a lesser extent, thymine, but not guanine and adenine, and OGG1 also possesses AP lyase activity. OGG1 contains a functional motif that is common to members of a superfamily of DNA glycosylases, such as *E. coli* endonuclease III and AlkA. Members of this superfamily are found in organisms ranging from prokaryotes to mammals, and they repair a wide variety of base lesions resulting from, for example, DNA oxidation, alkylation, hydration, and deamination. The hallmark of these proteins is a helix-hairpin-helix structural element followed by a Gly/Pro-rich loop and a conserved aspartic acid (HhH-GPD motif).[17]

DNA polymerases may insert adenine opposite 8-oxoG in template strand during DNA replication, and excision of 8-oxoG from the A: 8-oxoG pair may result in G:C to T:A transversion mutation. Thus, a DNA glycosylase encoded by *MUTYH* gene but not OGG1 excises adenine.[18] MUTYH protein has an ability to excise adenine opposite 8-oxoG and 2-OH-A incorporated opposite any normal base in template, and it also lacks any AP lyase activity. MUTYH has a PCNA binding motif, and the MUTYH repair activity for adenine misincorporated opposite 8-oxoG in plasmid DNA transfected into cultured cells is likely to be dependent on PCNA.[19,20] MUTYH thus plays an important role to avoid mutagenesis caused by 8-oxoG and 2-OH-A during postreplicative BER in association with PCNA.

After a base excision by DNA glycosylase, the incision of DNA 5' to the AP sites is the essential step to generate accessible 3'-OH termini before repair synthesis by DNA polymerase, and the incision is catalyzed by class II AP endonuclease. In mammalian cells, APEX1 (APE1/HAP1/REF-1) is a major class II AP endonuclease in the nucleus. Recently, we and others reported a second AP endonuclease, APEX2 (APE2) in human cells,[9,10,21] and we further demonstrated that the human APEX2 binds to PCNA in the nucleus through its functional PCNA binding motif. APEX1 does not have a PCNA binding motif; as a result, APEX2 is most likely responsible for the postreplicative BER which is PCNA dependent.

MITOCHONDRIAL LOCALIZATION OF MTH1, OGG1, MUTYH, AND APEX2

The human *MTH1* gene located on chromosome 7p22 consists of five major exons. There are two alternative exon 1 sequences, namely, exon 1a and 1b, and three contiguous segments (exon 2a, 2b, and 2c) in exon 2, which are alternatively spliced.[22,23] The *MTH1* gene thus produces seven types (type 1, 2A, 2B, 3A, 3B, 4A, and 4B) of mRNAs.[24] The B-type mRNAs with exon 2b-2c segments direct synthesis of three forms of MTH1 polypeptides, MTH1b (p22), MTH1c (p21), and MTH1d (p18) by an alternative initiation of translation, whereas the others encode only MTH1d. In human cells, MTH1d, the major form is mostly localized in the cytoplasm with approximately 5% in the mitochondrial matrix and lesser extent in nuclei. A single cell of Jurkat and HeLa lines contains approximately 2×10^5 or 4×10^5 molecules, respectively, of the hMTH1d; thus, 1 to 2×10^4 molecules of hMTH1d are present in mitochondria of each cell. The other two forms are mostly

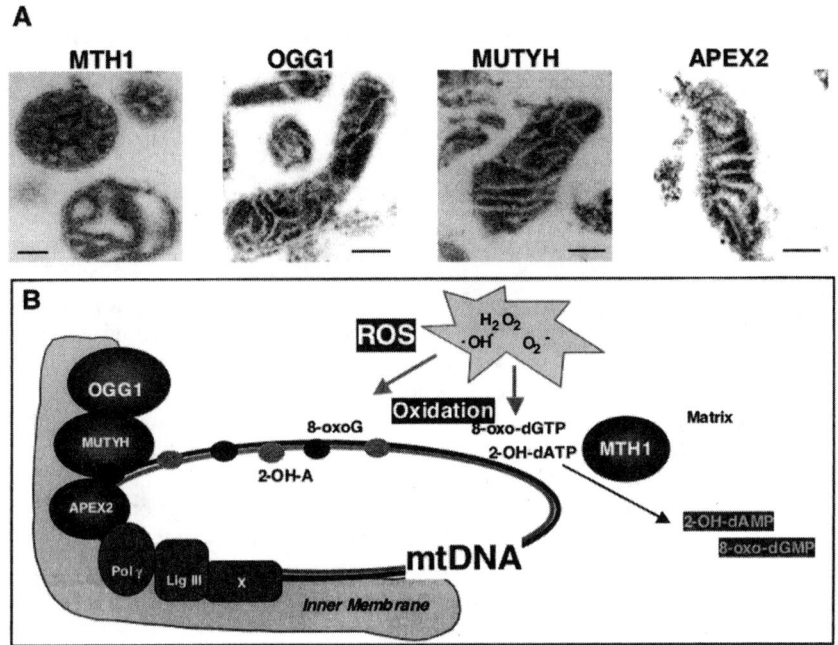

FIGURE 2. Error-avoiding mechanisms against oxidative damage of nucleic acids in mitochondria. (**A**) Electron microscopic immunocytochemistry for MTH1, MTH1, OGG1, MUTYH, and APEX2 protein in HeLa cells. (**B**) Hypothetical representation of MTH1, OGG1, MUTYH, and APEX2 proteins located in mitochondria. See text for details. 8-oxoG, 8-oxoguanine; 2-OH-A, 2-hydroxyadenine.

localized in the cytoplasm. In mitochondria, MTH1d is located in the matrix, as a soluble component (FIG. 2).[15,25]

The human *OGG1* gene located on chromosome 3p25 has eight major exons. More than seven alternatively spliced forms of *OGG1* mRNAs are produced, which are classified into two types based on their last exons (type 1 with exon 7: 1a and 1b; type 2 with exon 8: 2a to 2e).[15,26] Types 1a and 2a mRNAs are major *OGG1* transcripts in various human tissues. All forms of OGG1 polypeptides carry a relatively poor mitochondrial targeting sequences (MTSs), consisting of residues 9 to 26 at the common N-terminal region and the MTS is processed at residue 23 (W) after being translocated into the mitochondria. Among all the known OGG1 polypeptides, only OGG1-1a has a nuclear localization signal (NLS) at the C-terminal end.

A 40-kDa polypeptide corresponding to a processed form of hOGG1-2a was detected in the mitochondria of Jurkat cells. Electron microscopic immunocytochemistry and subfractionation of the mitochondria showed OGG1-2a to be located on the innermembrane of the mitochondria, in contrast with the MTH1d present in the mitochondrial matrix. Deletion mutant analyses showed that the unique C terminus of hOGG1-2a and its MTS are essential for mitochondrial localization, whereas the nuclear localization of hOGG1-1a depends on NLS at its C terminus. Because the unique C-terminal region of hOGG1-2a consists of two distinct regions—namely, the N-terminal sided acidic region (amino acids from Ile^{345} to Asp^{381}) and the C-terminal sided hydrophobic region (the last 20 residues)—we speculate that the latter mediates the hydrophobic interaction between OGG1-2a and innermembrane of the mitochondria (FIG. 2).[15]

The human *MUTYH* gene is located on the short arm of chromosome 1 between p32.1 and p34.3 and consists of 16 exons.[27] There are three alternative exons 1 for *MUTYH* gene, and three major *MUTYH* transcripts, namely, *MUTYH*α,β,γ, are produced and each transcript is alternatively spliced, thus forming more than 10 mature transcripts.[18,28] In human cells, Western blots showed authentic MUTYH proteins in the nuclei and in mitochondria and that the molecular masses differed, p52/53 and p57, respectively. A major transcript, *MUTYH*α3 encodes two polypeptides, p54 and p53, the former translated from the first AUG and the latter from second AUG. The amino-terminal sequence of the MUTYH translated from the first AUG in the *MUTYH*α3 mRNA functions as a mitochondrial targeting signal. The *MUTYH*α1 transcript has a 33-nucleotide insertion into *MUTYH*α3, thus encoding a polypeptide with an expected molecular mass of 60,031; this may be the p57 detected in the mitochondria. Moreover, the nuclear form of MUTYH, p52 partially purified from Jurkat cells, most likely corresponds to the p53 translated from the second AUG of the *MUTYH*α3 transcript, and *MUTYH*β1, β3, and γ2 transcripts which are missing the first AUG and may encode the nuclear form of p52 MUTYH. Electron microscopic immunocytochemistry showed that MUTYH in the mitochondria is associated with the innermembrane structure, as is hOGG1-2a (FIG. 2).[8]

The human *APEX2* gene is located on chromosome X (Xp11.21) and consists of six exons and encodes only one transcription/translation product.[9,21] The APEX2 protein is mostly localized in the nucleus of human cells and also to some extent in the mitochondria.[9] Its N-terminal 15–amino acid residues are functional as MTS. There is a repair complex containing the entire machinery essential for BER in the mitochondria—namely, DNA glycosylases (OGG1-2a, MUTYH), AP endonuclease (APEX2), DNA polymerase (Polγ), and DNA ligase [Lig III])—and it is likely that they are associated with the innermembrane and/or mitochondrial DNAs (FIG. 2).

CANCER SUSCEPTIBILITY INCREASED IN *MTH1*, *OGG1*, AND *MUTYH*-DEFICIENT MICE

In *Mth1*-deficient mice, the incidence of spontaneous carcinogenesis in the liver, and to a lesser extent in the lung and stomach, increased to severalfold of that observed in wild-type mice (TABLE 1). Furthermore, the increased accumulation of 8-oxo-dG in human cancerous tissues including brain tumors generally is coincidental with the increased expression of MTH1 protein (TABLE 1). G:C to T:A transversion mutations that can be caused by 8-oxo-dGTP and 2-OH-dATP frequently is observed in p53 gene in human cancer, especially in the lung and liver, and these findings agree with our observations in *Mth1*-deficient mice.[13,29]

We found that lung adenoma/carcinoma spontaneously developed in *Ogg1* knockout mice at approximately 1.5 years after birth, in which 8-oxoG was found to accumulate in their genomes.[30] The mean number of tumors per mouse was 0.71 for the *Ogg1* knockout mice, which was five times higher than that observed in wild-type mice (0.14). Although the accumulation of 8-oxoG also was confirmed in the

TABLE 1. Mechanisms protecting the genomic integrity in mammalian cells from oxidation of nucleic acids by reactive oxygen species and their implications for carcinogenesis and neurodegeneration

Enzyme	8-oxoG DNA Glycosylase	2-OH-A/Adenine DNA Glycosylase	AP endonuclease	Oxidized Purine Nucleoside Triphosphatase
Function	C:GO ↓ C:■ + GO	AO:G ↓ AO + ■:G A:GO ↓ A + ■:GO	C:■ ↓ C + ■	2-OH-dATP / 2-OH-ATP 8-oxo-dATP (8-oxo-ATP) 8-oxo-dGTP (8-oxo-GTP) 8-Cl-dGTP ↓ dNMP/NMP + PPi
Gene	*OGG1*	*MUTYH(MYH)*	*APEX2*	*MTH1*
Localization	Nucleus Mitochondria	Nucleus Mitochondria	Nucleus Mitochondria	(Nucleus) Cytoplasm Mitochondria
Expression	Brain > Thymus,Testis, Kidney, Spleen, Ovary	Thymus > Brain, Testis, Kidney, Spleen, Ovary	Thymus, Kidney > Liver, Lung, Brain	Thymus,Testis, Kidney, Spleen, Ovary > Brain
Human Disease	PD, AD, ALS, SAH	PD, Hypoxia	ND	PD, AD, ALS, BT
KO Mice	Lung cancer	Cancers	ND	Liver cancer

GO, 8-oxoguanine; AO, 2-hydroxyadenine; PD, Parkinson's disease; AD, Alzheimer's disease; ALS, amyotrophic lateral sclerosis; SAH, subarachnoid hemorrhage; BT, brain tumor; ND, not done.

Ogg1, *Mth1* double knockout mice, we found no tumor in the lungs of these mice. This observation suggests that *Mth1* gene disruption resulted in a suppression of the tumorigenesis caused by an *Ogg1* deficiency.

In various human cancers, a high level of MTH1 expression is commonly observed along with an increased accumulation of 8-oxoG in cellular DNA,[31–33] thus suggesting that MTH1 may protect cancer cells from oxidative damage. We recently have obtained evidence that MTH1 efficiently protects cells from oxidative damage, and we speculate that the simultaneous loss of OGG1 and MTH1 may enhance cell death when a large amount of 8-oxo-dGTP in a nucleotide pool or 8-oxoG in cellular genome is accumulated. In such circumstances, damaged cells may not survive to produce progenitors with mutations in protooncogenes or tumor suppressor genes, thus suppressing carcinogenesis in the *Ogg1*, *Mth1* double knockout mice.

MUTYH and APEX2-deficient mice are now available in our laboratory, and, at least in MUTYH-deficient mice, we found increased occurrence of cancers in various tissues approximately 1.5 years after birth.

ALTERERD EXPRESSION OF MTH1 AND OGG1 IN VARIOUS NEURODEGENERATIVE DISEASES

We recently found the regional accumulation of 8-oxoG and an altered expression of MTH1 in patients with various neurodegenerative diseases (TABLE 1). The most typical cases are patients with Parkinson's disease, in which we found a significant increase of 8-oxoG accumulated in either the cytoplasm or mitochondria with a coincidentally elevated expression of MTH1 in the substantia nigral neurons.[34] In postmortem tissue specimens from patients with Alzheimer's disease, the expression levels of MTH1 in the entorhinal cortex also were elevated, whereas the levels of MTH1 apparently decrease in the stratum lucidum at CA3 corresponding to mossy fiber synapses, where hMTH1 is highly expressed in control subjects.[35]

We examined brain tissue specimens from autopsy cases with Alzheimer's disease and controls using antibodies for OGG1-2a, the mitochondrial form of OGG1. OGG1-2a is mainly expressed in the neuronal cytoplasm in both Alzheimer's disease and control cases in regionally different manners (TABLE 1).[36] The expression of OGG1-2a decreases in orbitofrontal gyrus and the entorhinal cortex in Alzheimer's disease in comparison with that in the control cases. Immunoreactivity for OGG1-2a occasionally is associated with neurofibrillary tangles, dystrophic neurites, and reactive astrocytes in Alzheimer's disease. Our results indicate that the repair enzyme for the oxidative damage in mitochondrial DNA may not appropriately exert its function in Alzheimer's disease, and thus oxidative DNA damage in the mitochondria may be involved in the first step of the pathomechanism of Alzheimer's disease.

Furthermore, we also investigated the expression of OGG1-2a by comparing the 8-oxoG accumulation observed in the large motor neurons of the lumbar spinal cord in seven cases of adult-onset sporadic amyotrophic lateral sclerosis (ALS), four cases of subarachnoid hemorrhage (SAH), and in four control cases (TABLE 1).[37] The 8-oxoG immunoreactivity increased in many large motor neurons in both the ALS and SAH cases. However, the mitochondrial OGG1-2a immunoreactivity varied as follows: homogenous cytoplasmic staining was noted in the control cases, a fine granular pattern was seen in the SAH cases, and either no staining or only weak staining was seen in the ALS cases, respectively. Our results thus indicate that oxidative damage accumulates in the mitochondria of motor neurons in ALS and that the repair function of OGG1 is not sufficient to efficiently overcome any damage, which therefore may lead to a loss of motor neurons in ALS.

In addition to the altered expression of OGG1 in various degenerative disorders such as neurodegenerative diseases and acute renal failure after ischemia-reperfusion,[38] we obtained evidence of another mechanism which causes an alteration of the OGG1 function, namely, a chemical modification of the OGG1 protein itself by NO or its metabolites.[39] The formation of S-nitrosothiol adducts and loss/ejection of zinc ions inhibits the OGG1 function in cells exposed to NO, thus suggesting that the inactivation of OGG1 by NO is another pathological pathway involved in neurodegeneration.

FUTURE PERSPECTIVE

Among the many oxidized bases, 8-oxoG has been intensively studied, and it also has been implicated in various diseases, such as cancer, neurodegeneration, and teratogenecity. 8-oxoG has a mutagenic or cytotoxic potential and various enzymes specifically act on the 8-oxoG lesions (TABLE 1).

The accumulation of 8-oxoG in DNA is minimized by the coordinated actions of MTH1 and OGG1, MUTYH, and APEX2. When guanine in a template strand of DNA is oxidized and adenine then is inserted opposite 8-oxoG during the first round of DNA replication, MUTYH excises adenine paired with 8-oxoG in association with PCNA at least in part, thus initiating postreplicative BER. APEX2, an AP endonuclease which binds PCNA, may be involved in the next step of BER by incising the AP site introduced by MUTYH. Cytosine then is inserted opposite 8-oxoG by DNA polymerase β or PCNA-dependent DNA polymerase δ and the ends are joined by DNA ligase. Next, OGG1 excises 8-oxoG opposite cytosine by its glycosylase activity, to initiate the final round of BER, thus resulting in a cytosine:guanine pair. Because no enzyme has been identified to excise 8-oxoG misinserted opposite adenine (FIG. 1), MTH1 may play a major role by hydrolyzing 8-oxo-dGTP or other oxidized nucleotides to prevent their misincorporation into newly synthesized strands.

MTH1, OGG1, MUTYH, and APEX2 are found in mitochondria as well as in nuclei, thus indicating that the mitochondrial genomes are efficiently protected from an attack of ROS. It is very likely that the accumulation of oxidative DNA damage either in the nuclear genome or mitochondrial genome may result in carcinogenesis or degenerative disorders; however, it remains unclear as to where such causative damage actually occurs, namely, in the nuclear genome or in the mitochondrial genome. To reduce and suppress the progression of cancer and degenerative disease during aging, it is essential to precisely evaluate the deleterious effects of oxidative DNA damage in each genome. Furthermore, we still do not know whether oxidized free nucleotides themselves, rather than those incorporated into DNA or RNA, are cytotoxic or not. We are currently trying to solve these novel issues, using both artificially manipulated proteins and cells or mice.

ACKNOWLEDGMENTS

We extend our special thanks to all other members in our laboratory and to Drs. M. Takahashi, T. Iwaki, M. Shirakawa, M. Mishima H. Kamiya, H. Kasai, H. Maki, Y. Nakatsu, and T. Tsuzuki for helpful discussions, and to Dr. B. Quinn for useful comments on the manuscript.

REFERENCES

1. SHIBUTANI, S., M. TAKESHITA & A.P. GROLLMAN. 1991. Insertion of specific bases during DNA synthesis past the oxidation-damaged base 8-oxodG. Nature **349:** 431–434.
2. NAKABEPPU, Y., Y. TOMINAGA, D. TSUCHIMOTO, et al. 2001. Mechanisms protecting genomic integrity from damage caused by reactive oxygen species: implications for carcinogenesis and neurodegeneration. Environ. Mutagen. Res. **23:** 197–209.
3. SEKIGUCHI, M. 1996. MutT-related error avoidance mechanism for DNA synthesis. Genes Cells **1:** 139–145.
4. FOWLER, R.G., S.J. WHITE, C. KOYAMA, et al. 2003. Interactions among the *Escherichia coli mutT, mutM,* and *mutY* damage prevention pathways. DNA Repair **2:** 159–173.
5. CUNNINGHAM, R.P., S.M. SAPORITO, S.G. SPITZER, et al. 1986. Endonuclease IV (*nfo*) mutant of *Escherichia coli.* J. Bacteriol. **168:** 1120–1127.
6. LU, A.L., X. LI, Y. GU, et al. 2001. Repair of oxidative DNA damage: mechanisms and functions. Cell. Biochem. Biophys. **35:** 141–170.
7. SHIGENAGA, M.K., T.M. HAGEN & B.N. AMES. 1994. Oxidative damage and mitochondrial decay in aging. Proc. Natl. Acad. Sci. USA **91:** 10771–10778.

8. NAKABEPPU, Y. 2001. Regulation of intracellular localization of human MTH1, OGG1, and MYH proteins for repair of oxidative DNA damage. Prog. Nucleic Acid Res. Mol. Biol. **68:** 75–94.
9. TSUCHIMOTO, D., Y. SAKAI, K. SAKUMI, et al. 2001. Human APE2 protein is mostly localized in the nuclei and to some extent in the mitochondria, while nuclear APE2 is partly associated with proliferating cell nuclear antigen. Nucleic Acids Res. **29:** 2349–2360.
10. IDE, Y., D. TSUCHIMOTO, Y. TOMINAGA, et al. 2003. Characterization of the genomic structure and expression of the mouse Apex2 gene. Genomics **81:** 47–57.
11. MAKI, H. & M. SEKIGUCHI. 1992. MutT protein specifically hydrolyses a potent mutagenic substrate for DNA synthesis. Nature **355:** 273–275.
12. SAKUMI, K., M. FURUICHI, T. TSUZUKI, et al. 1993. Cloning and expression of cDNA for a human enzyme that hydrolyzes 8-oxo-dGTP, a mutagenic substrate for DNA synthesis. J. Biol. Chem. **268:** 23524–23530.
13. NAKABEPPU, Y. 2001. Molecular genetics and structural biology of human MutT homolog, MTH1. Mutat. Res. **477:** 59–70.
14. SAKAI, Y., M. FURUICHI, M. TAKAHASHI, et al. 2002. A molecular basis for the selective recognition of 2-hydroxy-dATP and 8-Oxo-dGTP by human MTH1. J. Biol. Chem. **277:** 8579–8587.
15. NISHIOKA, K., T. OHTSUBO, H. ODA, et al. 1999. Expression and differential intracellular localization of two major forms of human 8-oxoguanine DNA glycosylase encoded by alternatively spliced OGG1 mRNAs. Mol. Biol. Cell **10:** 1637–1652.
16. BOITEUX, S. & J.P. RADICELLA. 1999. Base excision repair of 8-hydroxyguanine protects DNA from endogenous oxidative stress. Biochimie **81:** 59–67.
17. BRUNER, S.D., H.M. NASH, W.S. LANE, et al. 1998. Repair of oxidatively damaged guanine in *Saccharomyces cerevisiae* by an alternative pathway. Curr. Biol. **8:** 393–403.
18. OHTSUBO, T., K. NISHIOKA, Y. IMAISO, et al. 2000. Identification of human MutY homolog (hMYH) as a repair enzyme for 2-hydroxyadenine in DNA and detection of multiple forms of hMYH located in nuclei and mitochondria. Nucleic Acids Res. **28:** 1355–1364.
19. MATSUMOTO, Y. 2001. Molecular mechanism of PCNA-dependent base excision repair. Prog. Nucleic Acid Res. Mol. Biol. **68:** 129–138.
20. HAYASHI, H., Y. TOMINAGA, S. HIRANO, et al. 2002. Replication-associated repair of adenine:8-oxoguanine mispairs by MYH. Curr. Biol. **12:** 335–339.
21. HADI, M.Z. & D.M. WILSON III. 2000. Second human protein with homology to the *Escherichia coli* abasic endonuclease exonuclease III. Environ. Mol. Mutagen **36:** 312–324.
22. FURUICHI, M., M.C. YOSHIDA, H. ODA, et al. 1994. Genomic structure and chromosome location of the human mutT homologue gene MTH1 encoding 8-oxo-dGTPase for prevention of A:T to C:G transversion. Genomics **24:** 485–490.
23. ODA, H., Y. NAKABEPPU, M. FURUICHI, et al. 1997. Regulation of expression of the human MTH1 gene encoding 8-oxo-dGTPase. Alternative splicing of transcription products. J. Biol. Chem. **272:** 17843–17850.
24. ODA, H., A. TAKETOMI, R. MARUYAMA, et al. 1999. Multi-forms of human MTH1 polypeptides produced by alternative translation initiation and single nucleotide polymorphism. Nucleic Acids Res. **27:** 4335–4343.
25. KANG, D., J. NISHIDA, A. IYAMA, et al. 1995. Intracellular localization of 8-oxo-dGTPase in human cells, with special reference to the role of the enzyme in mitochondria. J. Biol. Chem. **270:** 14659–14665.
26. BOITEUX, S., C. DHERIN, F. REILLE, et al. 1998. Excision repair of 8-hydroxyguanine in mammalian cells: the mouse Ogg1 protein as a model. Free Radic. Res. **29:** 487–497.
27. SLUPSKA, M.M., C. BAIKALOV, W.M. LUTHER, et al. 1996. Cloning and sequencing a human homolog (hMYH) of the *Escherichia coli* mutY gene whose function is required for the repair of oxidative DNA damage. J. Bacteriol. **178:** 3885–3892.
28. YAMAGUCHI, S., K. SHINMURA, T. SAITOH, et al. 2002. A single nucleotide polymorphism at the splice donor site of the human MYH base excision repair genes results in reduced translation efficiency of its transcripts. Genes Cells **7:** 461–474.
29. TSUZUKI, T., A. EGASHIRA, H. IGARASHI, et al. 2001. Spontaneous tumorigenesis in mice defective in the MTH1 gene encoding 8-oxo-dGTPase. Proc. Natl. Acad. Sci. USA **98:** 11456–11461.

30. SAKUMI, K., Y. TOMINAGA, M. FURUICHI, et al. 2003. Ogg1 knockout-associated lung tumorigenesis and its suppression by Mth1 gene disruption. Cancer Res. **63:** 902–905.
31. OKAMOTO, K., S. TOYOKUNI, W.J. KIM, et al. 1996. Overexpression of human mutT homologue gene messenger RNA in renal-cell carcinoma: evidence of persistent oxidative stress in cancer. Int. J. Cancer **65:** 437–441.
32. KENNEDY, C.H., R. CUETO, S.A. BELINSKY, et al. 1998. Overexpression of hMTH1 mRNA: a molecular marker of oxidative stress in lung cancer cells. FEBS Lett. **429:** 17–20.
33. IIDA, T., A. FURUTA, M. KAWASHIMA, et al. 2001. Accumulation of 8-oxo-2′-deoxyguanosine and increased expression of hMTH1 protein in brain tumors. Neurooncology **3:** 73–81.
34. SHIMURA-MIURA, H., N. HATTORI, D. KANG, et al. 1999. Increased 8-oxo-dGTPase in the mitochondria of substantia nigral neurons in Parkinson's disease. Ann. Neurol. **46:** 920–924.
35. FURUTA, A., T. IIDA, Y. NAKABEPPU, et al. 2001. Expression of hMTH1 in the hippocampi of control and Alzheimer's disease. Neuroreport **12:** 2895–2899.
36. IIDA, T., A. FURUTA, K. NISHIOKA, et al. 2002. Expression of 8-oxoguanine DNA glycosylase is reduced and associated with neurofibrillary tangles in Alzheimer's disease brain. Acta Neuropathol. **103:** 20–25.
37. KIKUCHI, H., A. FURUTA, K. NISHIOKA, et al. 2002. Impairment of mitochondrial DNA repair enzymes against accumulation of 8-oxo-guanine in the spinal motor neurons of amyotrophic lateral sclerosis. Acta Neuropathol. **103:** 408–414.
38. TSURUYA, K., M. FURUICHI, Y. TOMINAGA, et al. 2003. Accumulation of 8-oxoguanine in the cellular DNA and the alteration of the OGG1 expression during ischemia-reperfusion injury in the rat kidney. DNA Repair **2:** 211–229.
39. JAISWAL, M., N.F. LARUSSO, K. NISHIOKA, et al. 2001. Human Ogg1, a protein involved in the repair of 8-oxoguanine, is inhibited by nitric oxide. Cancer Res. **61:** 6388–6393.

Mitochondrial Swelling and Generation of Reactive Oxygen Species Induced by Photoirradiation Are Heterogeneously Distributed

TSUNG-I PENG[a] AND MEI-JIE JOU[b]

[a]*Department of Neurology, Lin-Kou Medical Center, Chang Gung Memorial Hospital, Tao-Yuan, Taiwan*

[b]*Department of Physiology and Pharmacology, Chang Gung University, Tao-Yuan, Taiwan*

ABSTRACT: Abundant evidence has been gathered to show that overproduction of reactive oxygen species (ROS) can lead to the opening of the mitochondrial permeability transition pore (MPTP) and result in apoptosis in mammalian cells. The information regarding spatial and temporal regulation of intracellular ROS formation related to the MPTP opening, however, is relatively limited. In this study, we used a fluorescent probe, dihydro-2′,7′-dichloroforescin (DCF), to detect intracellular ROS levels in different compartments of the cell in a time-resolved manner. The roles of mitochondrial ROS (mROS) in the MPTP opening and mitochondrial membrane potential drop were investigated by using H_2DCFDA coloaded with a mitochondrial marker dye MitoTracker Red, and by a mitochondrial membrane potential dye tetramethyl rhodamine ethyl ester. We applied multiphoton laser scanning microscopy to avoid autooxidation and bleaching of DCF so that long-term visualization of intracellular ROS formation could be performed. Moreover, we noted that the resting mROS levels of different mitochondria were not homogeneous. After cells had been exposed to photoirradiation, the intracellular ROS gradually increased but the heterogeneity of mROS was maintained. Later, swelling was observed in mitochondria that contained higher levels of ROS, indicating the opening of the MPTP. In cells in which all the mitochondria swelled, they were translocated to the perinuclear area, which became the site of ROS production. At this stage, mROS reached the highest level concomitantly with a complete loss of mitochondrial membrane potential, indicating full opening of the MPTP. At the end, photoirradiation resulted in apoptotic cell death. In summary, we demonstrated by multiphoton laser scanning microscopy that photoirradiation induces heterogeneous intracellular ROS formation and mitochondrial permeability transition pore opening in single intact cells. These observations imply the existence of a microdomain in the regulation of mROS formation and subsequent opening of the MPTP.

Address for correspondence: Mei-Jie Jou, Ph.D., Department of Physiology and Pharmacology, Chang Gung University, 259 Wen-Hwa 1st Road, Yao-Yuan, Taiwan. Voice: 886-3-3283016 ext. 5251 or 5974; fax: 886-3-3283016 ext. 5251.

mjjou@mail.cgu.edu.tw

KEYWORDS: mitochondria; reactive oxygen species; mitochondrial permeability transition pore; photoirradiation; multiphoton imaging

INTRODUCTION

Abundant evidence has been gathered to suggest that oxidative insults, resulting from overproduction of reactive oxygen species (ROS), are associated with cellular signal transduction systems as well as the pathogenesis of various diseases including age-related neurodegeneration.[1–3] Mitochondria particularly play a pivotal role in ROS-induced cell injury, because they themselves are the major intracellular ROS production sites[4–6] as well the most susceptible targets of ROS.[7–11] ROS-induced oxidative damage to mitochondrial lipids or proteins can severely depolarize mitochondrial membrane potential, deplete ATP, and damage mitochondrial DNA.[12–14] Moreover, ROS can directly trigger the opening of the mitochondrial permeability transition pore (MPTP), a crucial mechanism involved in releasing apoptotic lethal proteins including cytochrome c,[15] apoptosis-inducing factor,[16] Smac/DIABLO,[17,18] and endonuclease G[19] from the intermembrane space of mitochondria and lead the cell to irreversible apoptotic death.[20–22] Information regarding spatial and temporal regulation of intracellular ROS formation related to the MPTP opening, however, is relatively limited. In the past, ROS formation has been successfully detected using fluorescent probes such as dihydrorhodamine123 and dihydro-2′,7′-dichloroforescin (H_2DCFDA).[23–25] Fluorescence increases because of the oxidation of these probes by ROS, and this can be readily detected using a flow cytometer. However, information regarding the spatial and temporal distribution of ROS in the cell is lacking. Mitochondria in single intact cells are reported to be heterogeneous for morphology, bioenergetic function, calcium regulation, and membrane potential.[26–30] In this study, we used a dihydro-2′,7′-dichloroforescin (DHDCF), to detect intracellular ROS production in different subcellular compartments including the cytosol, the mitochondria, and the nucleus.

The relationship between mitochondrially generated reactive oxygen species (mROS) and the opening of the MPTP was investigated by coloading cells with a mitochondrial marker dye, MitoTracker Red (Mito-R), or with a dye that indicates mitochondrial membrane potential, tetramethyl rhodamine ethyl ester (TMRM) concurrently with DCF. To reduce photoxidation and photobleaching of DCF caused by conventional single photon illumination during the recording, we applied Ti-Sa laser fiber coupled multiphoton laser scanning microscopy (MLSM) to provide illumination with low phototoxicity.

On the other hand, we assessed the spatial and temporal profiles of photoirradiation-induced mROS formation and their relationship with the opening of the MPTP in intact single rat brain astrocyte (RBA-1).

MATERIALS AND METHODS

Cell Preparation

A cell line of normal rat brain astrocytes (RBA-1) was established originally by repeated passage of primary astrocytes isolated from the brain of a 3-day-old JAR-2, F51 rat by Dr. Teh-Cheng Jou.[31] All cells were grown in a medium consisting of

Dulbecco's modified Eagle's medium (Life Technologies, Grand Island, NY) supplemented with 10% (v/v) fetal bovine serum. The cells were plated onto a no. 1 glass coverslip coated with poly-L-lysine for fluorescence measurement (Flow Cytometer Model No. 1; VWR Scientific, San Francisco, CA).

Chemicals and Fluorescent Dyes

All fluorescent dyes were purchased from Molecular Probes Inc. (Eugene, ORA) and chemicals were obtained from Sigma Chemical Co. (St. Louis, MO). Loading concentrations of fluorescent probes were as follows: mitochondrial fluorescent dyes: MitoTracker Red (Mito R), 100 nM; TMRM, 1 μM; carboxy-2′,7′-dichlorodihydrofluorescein diacetate (H_2DCFDA), 1 μM. Fluorescent probes were all loaded at room temperature for 30 min. After loading of the dyes, cells were rinsed three times with HEPES-buffered saline (140 mM NaCl, 5 mM KCl, 1 mM $MgCl_2$, 2 mM $CaCl_2$, 10 mM glucose, 5 mM HEPES, pH 7.4). Cells loaded with H_2DCFDA required additional 30 min of incubation after dye loading to allow intracellular deacetylation of the ester form of the dyes.

Conventional and Multiphoton Imaging Microscopy

All phase-contrast images and conventional fluorescence images were obtained using a Zeiss (Thornwood, NY) inverted microscope (AxioVert 200M) equipped with a mercury lamp (HBO 103) and a cool CCD camera (coolsnap fx) with Zeiss objectives (Plan-NeoFluar 100x, N.A. 1.3 oil). Filters used for detecting DCF was Zeiss No. 10 (exitation bandpass 450–490 nm; emission: bandpass 515–565 nm) and Mito-R and TMRM was Zeiss No. 15 (excitation: BP 546/12 nm; emission: long pass 590 nm). Multiphoton fluorescence images were obtained using a Leica SP2 MP (Leica-Microsystems, Wetzlar, Germany) fiber coupling system equipped with a Ti:Sa-Laser system (model Millenia/Tsunami; Spectra-Physics), providing pulse repetition rate at 82 MHz, laser pulse width of 1.2 ps, spectral bandwidth of 1 nm, and object pulse width of 1.3 ps). Wavelength at 800 nm with average laser power of 600 mW was selected for illumination. During fluorescence imaging, the illumination light was reduced to the minimal level by using a neutral density filter (3%) to minimize the photosensitizing effect from the interaction of light with fluorescent probes. All images were processed and analyzed using MetaMorph software (Universal Imaging Corp., West Chester, PA). Intensity levels were analyzed from the original images and graphed using Microsoft EXCEL software and Adobe Photoshop.

Visible Laser Irradiation

For visible laser irradiation, cells were continuously exposed to a 488-nm argon laser from the epi-illumination port of a Leica microscope. The irradiation time was 1 minute. No neutral density filter was used. The size of the pinhole during laser irradiation was set to 100% open. The power density of the laser irradiation measured by a Coherent Model 210 power meter at the epi-illumination port of the objective was 1.7 mW/cm^2. For preliminary observation of the effect of light-induced phototoxicity on cell viability and nuclear condensation, photoirradiation also was performed by continuously exposing cells to visible light from the epi-illumination port of the objective of a Zeiss conventional microscope coupled with a mercury lamp. An excitation bandpass filter (450–490 nm) was used and no neutral density filter was used. The irradiation time was 3 min. Possible heat effects generated from the visible laser and mercury lamp irradiation were minimized by perfusion of cells with fresh HEPES-buffered saline during the experiment.

Measurement of Intracellular ROS with Multiphoton Imaging Microscopy

Intracellular ROS was detected by monitoring DCF, the fluorescent product of intracellular H_2DCF oxidation.[23,32,33] Multiphoton DCF images were obtained and displayed with a pseudocolor scale from 0 to 255 units. In DCF images, mROS was calculated from areas based on a Mito R distribution from the same cell. Intensity values were analyzed per pixel from the cytosolic, the mitochondrial, and the nuclear area of the cells. Differences were evaluated using Student's t test. Fluorescence intensity changes in the experiments with DCF were normalized with the control.

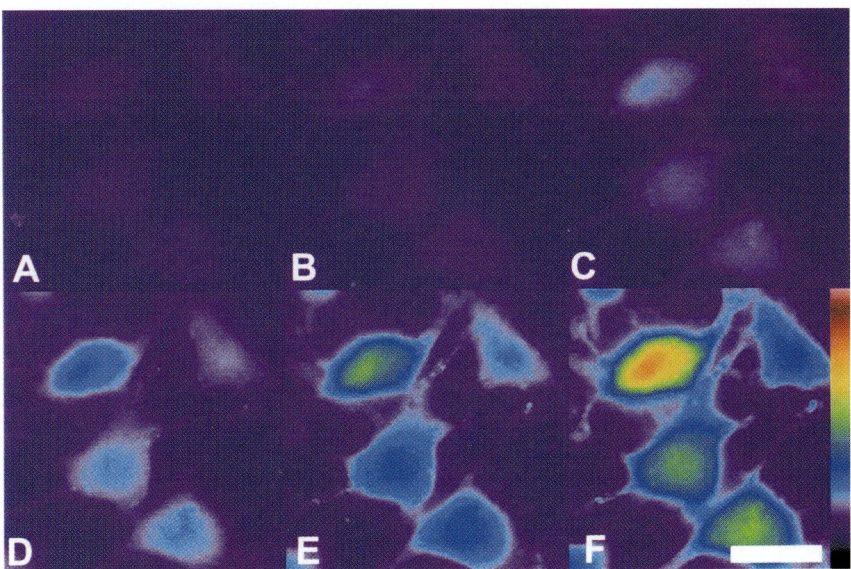

FIGURE 1. Photoirradiation-induced autooxidation of DCF. Continuous time-lapse DCF images recorded from a conventional fluorescence imaging system demonstrated a rapid autooxidation of DCF after cells had exposed to visible light. Note that the increase in DCF fluorescence intensity reached a saturated level in 2 min, and their distribution throughout the cell was relatively homogeneous as compared with the photoirradiation-induced ROS formation (see below). Images were taken at the rate of a frame per 20 s. Fluorescence intensity of DCF was displayed with a pseudocolor scale ranged between 0 and 255 units. Bar = 10 μm.

RESULTS AND DISCUSSION

Heterogeneity of Intracellular ROS Level in Single Intact Cells at the Resting Stage

To visualize whether or not there is a heterogeneous intracellular distribution of ROS in individual RBA-1 cells, we used DCF. Using DCF to detect intracellular ROS formation, we experienced severe photoinduced autooxidation and photobleaching of DCF particularly when a high concentration of DCF (>10 µM) or strong excitation light was used (e.g., without neutral density filters). As demonstrated in FIGURE 1, the DCF fluorescent signal increased swiftly and reached a saturated level within 2 min of exposure of the cells to conventional (single photon based) excitation light. The autooxidation-induced increase in DCF fluorescent intensity was relatively homogeneous among the cells as compared with the ROS generated by mitochondria induced by photoirradiation (see below). Although the increase in DCF signal caused by conventional excitation light was more than five times as compared with the control, cells experiencing autooxidation of DCF did not display mitochondrial dysfunction and survived (data not shown). We concluded that the increase of the fluorescence intensity seen during illumination of DCF was a result of light-induced oxidation of the dye and was not the result of intracellular ROS formation. Measurement of intracellular ROS formation using H_2DCF-DA was clearly impossible with the conventional single-photon technique, using short-wavelength excitation light.

In addition, light-induced photobleaching of DCF is another commonly seen undesired photodamage effect which gradually diminishes fluorescence intensity of DCF to an undetectable level after multiple light exposures (results not shown). To avoid the above artifact induced by single photon-based illumination, we applied a Ti:Sa laser fiber coupled MLSM to provide multiphoton illumination. By using two or more long wavelength photons that contain less energy and therefore less toxic to interact cooperatively to achieve excitation, we found that the MLSM drastically reduced autooxidation and bleaching of DCF during long-term recording. As shown in FIGURE 2, this DCF-loaded cell had received multiple exposure of multiphoton-based illumination light (five frames per second for 10 min), and the fluorescence of DCF was maintained. It revealed high intracellular heterogeneity of ROS distribution in the control RBA-1 cells. Three major intracellular ROS compartments corresponding to the cytoplasm, the nucleus, and the mitochondria were identified. The intracellular thread-like area of DCF-loaded cells used to define as mROS was confirmed by the colocalization of DCF distribution with a mitochondria-specific fluorescent marker, Mito R. Among these three compartments, the ROS content in the mitochondria (mROS) is the highest as displayed by the blue-to-red pseudocolor. The gradient of fluorescent intensities calculated from the three compartments is the mitochondria: the nucleus: the cytosol = 1.00: 0.29: 0.27 ($n = 5$), which is similar to our previous observation.[33] Intriguingly, the mROS level for individual mitochondrion also was found to be heterogeneous as indicated by the variation of pseudocolor among individual mitochondria. The difference between the highest mROS (red color) and lowest mROS (blue color) was approximately 2.3-fold ($n = 10$). The heterogeneity of mROS distribution in single cells seen in this study was not caused by mitochondria located on different focus planes because our three-dimensional recon-

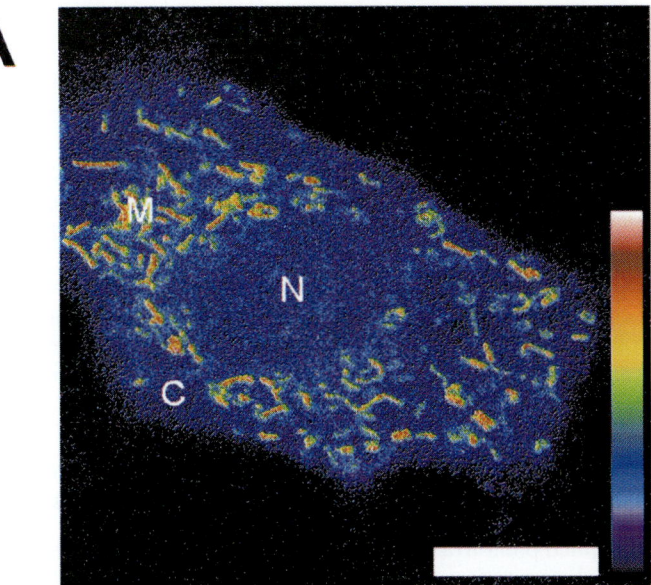

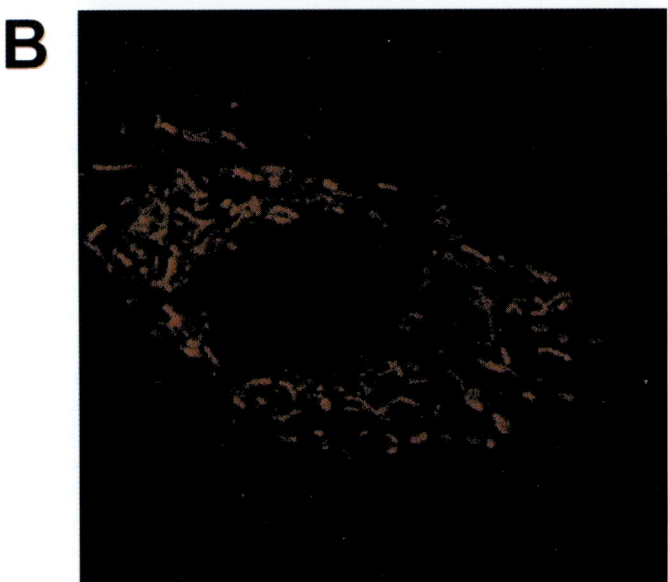

FIGURE 2. Heterogeneous distribution of intracellular ROS in a RBA-1 cell at the resting stage. (**A**) The heterogeneous distribution of intracellular ROS detected by DCF in the cytosol (**C**), the nucleus (**N**) and the mitochondria (**M**) are displayed with pseudocolor scale ranged between 0 and 255 units. Note that the mitochondrial DCF signal was much higher than that in the other two compartments indicating mROS was higher than the ROS in the other two compartments. (**B**) The area of DCF identified as mROS was confirmed by using a mitochondria-specific fluorescent marker, Mito-R. Note the identical subcellular localization of Mito-R and high DCF fluorescence area. Bar = 10 μm.

struction of DCF images showed similar results. The heterogeneous distribution of mROS among different individual mitochondria implies the existence of microdomains in regulating mROS formation. For instance, the activities and distribution of the major mROS source such as mitochondrial complexes I and III, as well as the mitochondrial antioxidant enzymes such as Mn-superoxide dismutase, may be different among various individual mitochondria. In addition, heterogeneous distribution of mitochondrial ions including calcium and proton have been reported.[28,29] It is possible that local concentration of these ions also may be involved in the regulation of mROS formation and mitochondrial function. Therefore, further investigation on the correlation of mROS formation in relation to the concentration of calcium ion and proton may provide deeper insight into how these parameters interact with one another to regulate these heterogeneous mitochondrial activities (see Jou and colleagues, this volume).

Photoirradiation Enhanced the Heterogeneity of mROS and Induced a Heterogeneous Opening of MPTP in Single Intact RBA-1 Cells

We then applied photoirradiation to specifically induce an increase of mROS and examined how heterogeneous distribution of mROS was affected by photoirradiation. As shown in FIGURE 3A and B, 5 min after photoirradiation mROS level was increased approximately 1.3-fold as compared with that of control cells ($n = 5$). However, mitochondrial morphology during induction of mROS by photoirradiation maintained thread-like, and no mitochondrial permeability transition pore (MPTP) opening was observed. The heterogeneity among individual mitochondria was apparently maintained. To explore whether elevated mROS induced by photoirradiation can trigger the opening of the MPTP, we applied stronger irradiation by doubling the irradiation time to 2 min. FIGURE 3C demonstrates that 5 min after the cells had received stronger strength of photoirradiation, mROS increased significantly in all mitochondria. Intriguingly, we observed that mitochondria containing high levels of ROS induced the opening of the MPTP, which appeared to be heterogeneous within single cells. The mROS levels of the swollen mitochondria were approximately three times higher than those of the thread-like mitochondria in the control cells. The difference in mROS levels between rod-shaped mitochondria and swollen mitochondria, however, decreased from 2.5- to 1.8-fold ($n = 5$). Ten minutes after photoirradiation, all the mitochondria swelled and mROS reached a relative homogeneous level of approximately 3.0- to 3.5-fold of that of the control cells (FIG. 3D). We often observed that the swollen mitochondria had translocated to the perinuclear area before the cells could be stained with a nuclear DNA-specific fluorescent dye, propidium iodide (FIG. 3E). Because ROS are membrane permeable, whether the translocation of mitochondria containing high levels of ROS can deliver ROS signals to the nucleus and regulate nuclear functions and subsequent fragmentation of nuclear DNA requires further investigation.

Mitochondrial swelling has been considered as an indication of the opening of the MPTP, which results in depolarization of mitochondrial membrane potential. We then recorded mitochondrial membrane potential using TMRM together with DCF. We observed that swollen mitochondria containing the highest level of ROS in concomitance with the complete loss of their membrane potential as revealed by the absence of the fluorescent signal of TMRM (FIG. 4). Complete loss of mitochondrial

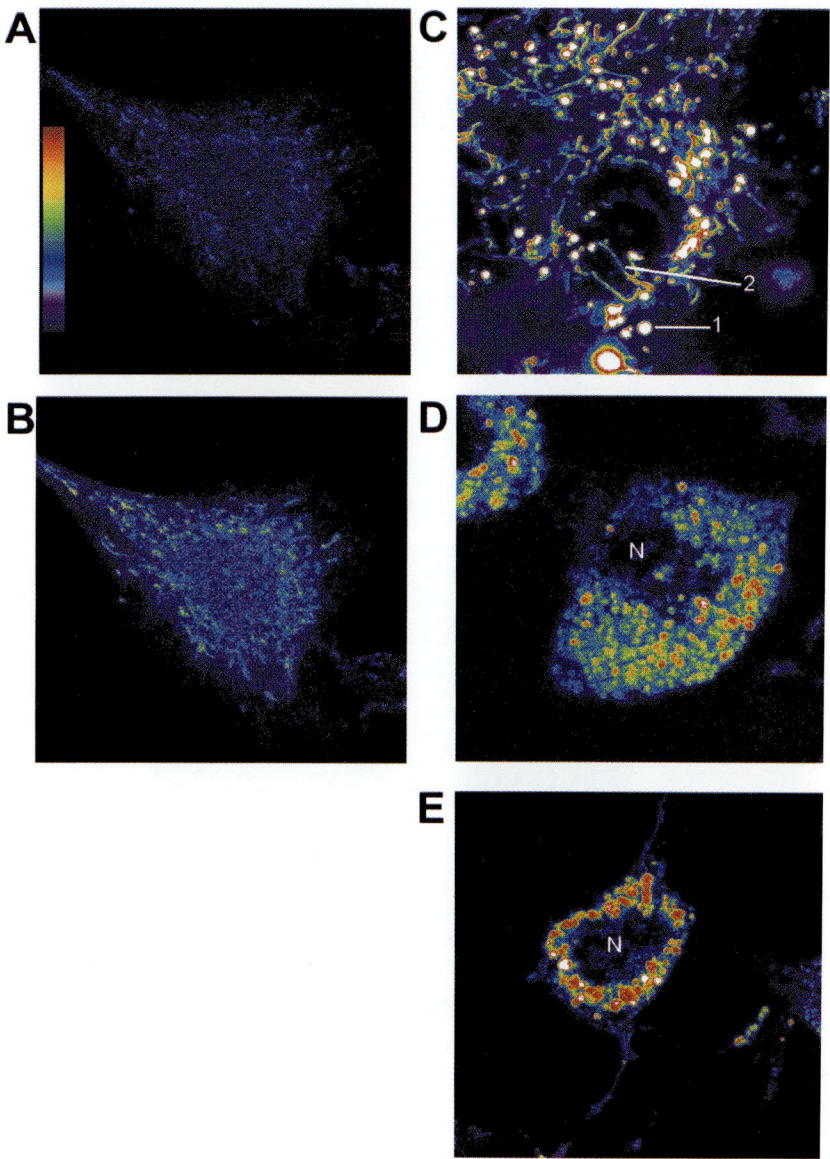

FIGURE 3. Photoirradiation enhanced the heterogeneity in mROS formation and the opening of MPTP. (**A**) mROS of a control cell. (**B**) The levels of mROS increased significantly 5 min after the cell received 1 min of 488-nm laser irradiation. The difference in the DCF fluorescence intensity analyzed from different populations of mitochondria in a photoirradiated cell was approximately twofold. (**C**) Five minutes after cells had received 2 min of 488-nm laser irradiation, two distinct mitochondrial populations and mROS formation appeared. Swollen mitochondria contained relatively high mROS and thread-like mitochondria contained relatively low level of mROS. (**D**) Ten minutes after laser irradiation, all the mitochondria swelled and process saturate levels of ROS. (**E**) Fifteen minutes after the photoirradiation, mitochondria were translocated to the perinuclear area.

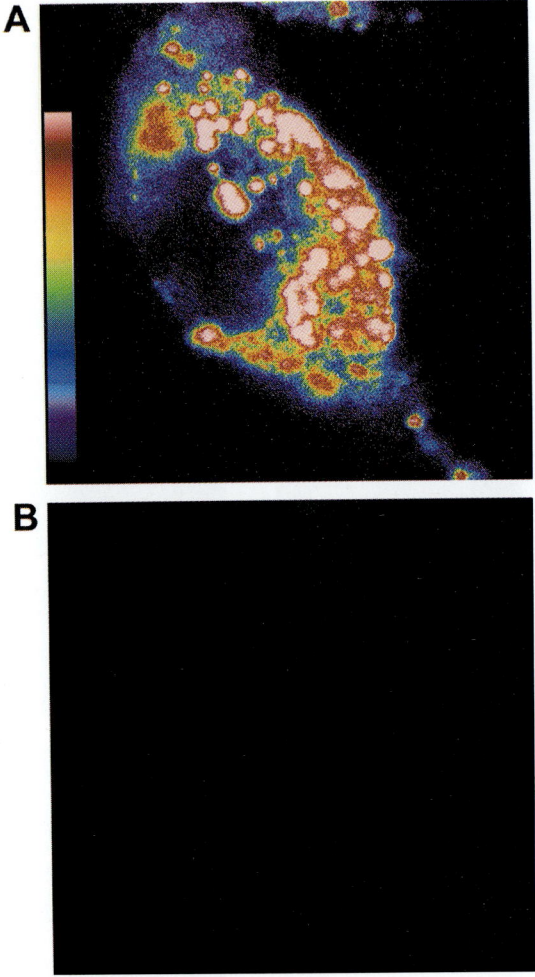

FIGURE 4. Mitochondrial membrane potential depolarization during the increase of mROS. (**A**) mROS formation reached a saturated level 20 min after photoirradiation by a laser light. (**B**) At this stage, the complete depolarization of mitochondrial membrane potential was detected by the loss of the TMRM signal.

membrane potential also was observed in swollen mitochondria of cells, which contained heterogeneous population of mitochondria (data not shown). Approximately 2–3 h after photoirradiation, DNA fragmentation of the cells was observed, indicating that photoirradiation induced apoptotic cell death as previously reported.[33–35]

Taken together, our findings of this study suggest that heterogeneous mROS formation elicited by photoirradiation can induce nonhomogeneous opening of the MPTP and decline of mitochondrial membrane potential within single cells. ROS-induced opening of the MPTP has been shown to play a crucial role in triggering the release of several proapoptotic proteins from mitochondria. Our observation of the spatial and temporal heterogeneity of mROS formation and the subsequent opening of the MPTP, therefore, implies a focal distribution of apoptotic activities within single cells. In addition, mitochondrial regulatory microdomains among individual mitochondrion as well as focal interaction of mitochondria with other intracellular compartments such as endoplasmic reticulum (or sarcoplasmic reticulum in heart muscle) via calcium ions may play a role in the regulation of heterogeneous mROS formation and subsequent opening of the MPTP.[26–30]

ACKNOWLEDGMENTS

We gratefully acknowledge Chen-Yi Hsu and Hong-Yueh Wu for assistance in the procurement and analysis of conventional fluorescent imaging data. We gratefully thank Dr. Yau-Huei Wei for critical reading of the manuscript and valuable suggestions. This research work was supported by grants CMRP 1009 (to M-J.J.) and CMRP 930 (to T-I.P.) from the Chang Gung Research Foundation, and grants NSC 90-2315-B-182-056 (to M-J.J.) and 91-2314-B-182A-021 (to T-I.P.) from the National Science Council.

REFERENCES

1. GHOSH, S.S., R.H. SWERDLOW, S.W. MILLER, *et al.* 1999. Use of cytoplasmic hybrid cell lines for elucidating the role of mitochondrial dysfunction in Alzheimer's disease and Parkinson's disease. Ann. N.Y. Acad. Sci. **893:** 176–191.
2. HARMAN, D. 1992. Role of free radicals in aging and disease. Ann. N.Y. Acad. Sci. **673:** 126–141.
3. SCHAPIRA, A.H., M. GU, J.W. TAANMAN, *et al.* 1998. Mitochondria in the etiology and pathogenesis of Parkinson's disease. Ann. Neurol. **44:** S89–S98.
4. BOVERIS, A. & B. CHANCE. 1973. The mitochondrial generation of hydrogen peroxide. General properties and effect of hyperbaric oxygen. Biochem. J. **134:** 707–716.
5. CAI, J. & D.P. JONES. 1998. Superoxide in apoptosis. Mitochondrial generation triggered by cytochrome *c* loss. J. Biol. Chem. **273:** 11401–11404.
6. TURRENS, J.F. 1997. Superoxide production by the mitochondrial respiratory chain. Biosci. Rep. **17:** 3–8.
7. ATLANTE, A., P. CALISSANO, A. BOBBA, *et al.* 2000. Cytochrome *c* is released from mitochondria in a reactive oxygen species (ROS)-dependent fashion and can operate as a ROS scavenger and as a respiratory substrate in cerebellar neurons undergoing excitotoxic death. J. Biol. Chem. **275:** 37159–37166.
8. BACKWAY, K.L., E.A. MCCULLOCH, S. CHOW & D.W. HEDLEY. 1997. Relationships between the mitochondrial permeability transition and oxidative stress during ara-C toxicity. Cancer Res. **57:** 2446–2451.
9. KOWALTOWSKI, A.J. & A.E. VERCESI. 1999. Mitochondrial damage induced by conditions of oxidative stress. Free Radical Biol. Med. **26:** 463–471.
10. KOWALTOWSKI, A.J., R.F. CASTILHO & A.E. VERCESI. 2001. Mitochondrial permeability transition and oxidative stress. FEBS Lett. **495:** 12–15.
11. LIU, S.S. 1997. Generating, partitioning, targeting and functioning of superoxide in mitochondria. Biosci. Rep. **17:** 259–272.
12. VLADIMIROV, Y.A., V.I. OLENEV, T.B. SUSLOVA & Z.P. CHEREMISINA. 1980. Lipid peroxidation in mitochondrial membrane. Adv. Lipid Res. **17:** 173–249.
13. WEI, Y.H. 1998. Oxidative stress and mitochondrial DNA mutations in human aging. Proc. Soc. Exp. Biol. Med. **217:** 53–63.

14. YAKES, F.M. & B. VAN HOUTEN. 1997. Mitochondrial DNA damage is more extensive and persists longer than nuclear DNA damage in human cells following oxidative stress. Proc. Natl. Acad. Sci. USA **94:** 514–519.
15. LIU, X., C.N. KIM, J. YANG, et al. 1996. Induction of apoptotic program in cell-free extracts: requirement for dATP and cytochrome c. Cell **86:** 147–157.
16. SUSIN, S.A., H.K. LORENZO, N. ZAMZAMI, et al. 1999. Molecular characterization of mitochondrial apoptosis-inducing factor. Nature **397:** 441–446.
17. DU, C., M. FANG, Y.LI, et al. 2000. Smac, a mitochondrial protein that promotes cytochrome c-dependent caspase activation by eliminating IAP inhibition. Cell **102:** 33–42.
18. VERHAGEN, A.M., P.G. EKERT, M. PAKUSCH, et al. 2000. Identification of DIABLO, a mammalian protein that promotes apoptosis by binding to and antagonizing IAP proteins. Cell **102:** 43–53.
19. LI, L.Y., X. LUO & X. WANG. 2001. Endonuclease G is an apoptotic DNase when released from mitochondria. Nature **412:** 95–99.
20. BERNARDI, P., V. PETRONILLI, F. DI LISA & M. FORTE. 2001. A mitochondrial perspective on cell death. Trends Biochem. Sci. **26:** 112–117.
21. LEMASTERS, J.J., A.L. NIEMINEN, T. QIAN, et al. 1998. The mitochondrial permeability transition in cell death: a common mechanism in necrosis, apoptosis and autophagy. Biochim. Biophys. Acta **1366:** 177–196.
22. ZORATTI, M. & I. SZABO. 1995. The mitochondrial permeability transition. Biochim. Biophys. Acta **1241:** 139–176.
23. LEBEL, C.P., H. ISCHIROPOULOS & S.C. BONDY. 1992. Evaluation of the probe 2′,7′-dichlorofluorescin as an indicator of reactive oxygen species formation and oxidative stress. Chem. Res. Toxicol. **5:** 227–231.
24. ROYALL, J.A. & H. ISCHIROPOULOS. 1993. Evaluation of 2′,7′-dichlorofluorescin and dihydrorhodamine 123 as fluorescent probes for intracellular H2O2 in cultured endothelial cells. Arch. Biochem. Biophys. **302:** 348–355.
25. SUREDA, F.X., C. GABRIEL, J. COMAS, et al. 1999. Evaluation of free radical production, mitochondrial membrane potential and cytoplasmic calcium in mammalian neurons by flow cytometry. Brain Res. Protoc. **4:** 280–287.
26. COLLINS, T.J., M.J. BERRIDGE, P. LIPP & M.D. BOOTMAN. 2002. Mitochondria are morphologically and functionally heterogeneous within cells. EMBO J. **21:** 1616–1627.
27. ICHAS, F., L.S. JOUAVILLE & J.P. MAZAT. 1997. Mitochondria are excitable organelles capable of generating and conveying electrical and calcium signals. Cell **89:** 1145–1153.
28. 2IZZUTO, R., M. BRINI, M. MURGIA & T. POZZAN. 1993. Microdomains with high Ca2+ close to IP3-sensitive channels that are sensed by neighboring mitochondria. Science **262:** 744–747.
29. SMILEY, S.T., M. REERS, C. MOTTOLA-HARTSHORN, et al. 1991. Intracellular heterogeneity in mitochondrial membrane potentials revealed by a J-aggregate-forming lipophilic cation JC-1. Proc. Natl. Acad. Sci. USA **88:** 3671–3675.
30. SONNEWALD, U., L. HERTZ & A. SCHOUSBOE. 1998. Mitochondrial heterogeneity in the brain at the cellular level. J. Cereb. Blood Flow Metab. **18:** 231–237.
31. JOU, T.C., M.J. JOU, J.Y. CHEN & S.Y. LEE. 1985. Properties of rat brain astrocytes in long-term culture. J. Formos. Med. Assoc. **84:** 865–881.
32. CARTER, W.O., P.K. NARAYANAN & J.P. ROBINSON. 1994. Intracellular hydrogen peroxide and superoxide anion detection in endothelial cells. J. Leukoc. Biol. **55:** 253–258.
33. JOU, M.J., S.B. JOU, H.M. CHEN, et al. 2002. Critical role of mitochondrial reactive oxygen species formation in visible laser irradiation-induced apoptosis in rat brain astrocytes (RBA-1). J. Biomed. Sci. **9:** 507–516.
34. CHURCHILL, M.E., J.G. PEAK & M.J. PEAK. 1991. Repair of near-visible- and blue-light-induced DNA single-strand breaks by the CHO cell lines AA8 and EM9. Photochem. Photobiol. **54:** 639–644.
35. TYRRELL, R.M., P. WERFELLI & E.C. MORAES. 1984. Lethal action of ultraviolet and visible (blue-violet) radiations at defined wavelengths on human lymphoblastoid cells: action spectra and interaction sites. Photochem. Photobiol. **39:** 183–189.

Mitochondrial Dysfunction via Disruption of Complex II Activity during Iron Chelation–Induced Senescence-like Growth Arrest of Chang Cells

YOUNG-SIL YOON, HYESEONG CHO, JAE-HO LEE, AND GYESOON YOON

Department of Biochemistry and Molecular Biology, Ajou University School of Medicine, Suwon 442-721, South Korea

ABSTRACT: When cells are deprived of iron, their growth is invariably inhibited. However, the mechanism involved remains largely unclear. Recently, we have reported that subcytotoxic concentration of deferoxamine mesylate (DFO), an iron chelator, specifically inhibited transition of Chang cell, a normal hepatocyte cell line, from G1 to S phase, which was accompanied by irreversible appearance of senescent biomarkers. To investigate factors responsible for the irreversible arrest, we examined mitochondrial activities because they require several irons for their proper structure and function. After exposure to 1 M DFO, total cellular ATP level was irreversibly decreased with concurrent disruption of mitochondrial membrane potential ($\Delta\Psi m$), implying that it might be one of the crucial factors involved in the arrest. DFO did not directly inhibit the mitochondrial respiratory activities *in vitro*. Among the respiratory activities, complex II activity was specifically inhibited through a down-regulation of the expression of its iron-sulfur subunit. We also observed that mitochondrial morphology was drastically changed to highly elongated form. Our results suggest that mitochondrial function is sensitive to cellular iron level and iron deprivation might be involved in inducing the senescent arrest. In addition, complex II, which is a part of both oxidative phosphorylation and the Krebs cycle, could be one of the critical factors that regulate mitochondrial function by responding to iron levels.

KEYWORDS: mitochondrial dysfunction; disruption of complex II; deferoxamine mesylate-induced senescent arrest

INTRODUCTION

Iron is essential for the maintenance of metabolic status and the growth of cells in all organisms. It participates not only as a cofactor in numerous critical biochemical activities including ATP synthesis in mitochondria, ribonucleotide reductase activity for DNA synthesis, O_2 transfer by hemoglobin, and many other

Address for correspondence: Gyesoon Yoon, Department of Biochemistry and Molecular Biology, Ajou University School of Medicine, Suwon 442-721, Korea. Voice: 82-31-219-5054; fax: 82-31-219-5059.
　ypeace@ajou.ac.kr

metalloenzymes, but also as a modulator of gene expression via a translational regulation by iron regulatory proteins.[1,2] Despite its essential role, iron overload has been implicated in several serious liver diseases such as cirrhosis and hepatocellular carcinoma, implying the importance of cellular iron homeostasis.[3–5] It also has been reported that cells deprived of iron do not proceed from the G1 to the S phase of the cell cycle.[6–8] Iron chelation therefore often has been used to inhibit the growth of cancer cells. Deferoxamine mesylate (DFO) is extremely effective at chelating intracellular iron and decreasing cell proliferation because it possesses a high binding affinity and selectivity for Fe^{3+} as a water-soluble iron chelator.[2] DFO therefore has long been investigated for its potential use in the treatment of iron overload disease such as thalassemia, and it has been used recently as a potential antineoplastic agent in the treatment of neuroblastoma and hepatocellular carcinoma.[9–13] Despite various studies on the growth inhibitory effect of DFO, the molecular mechanism remains poorly understood. Recently, we attempted to elucidate the mechanism of DFO-induced growth arrest using an immortalized human hepatocyte cell line, Chang cell. In a previous report, we described that cells deprived of iron by DFO do not proceed from G1 to S phase of the cell cycle.[14] During the arrest, we observed induction of TGF-β1 and p27^{Kip1}, which may be involved in the senescent arrest. However, blocking TGF-β1/p27^{Kip1} signaling with neutralizing antibody against TGF-β1 was not sufficient to reverse the G1 arrest, implying that other mechanisms must be involved in the irreversible arrest.

In aerobically growing cells, mitochondria play a pivotal role in generating most of the cellular energy through oxidative phosphorylation (OXPHOS). OXPHOS is conducted by the four respiratory chain complexes (I–IV) and F_1F_0-ATP synthase (complex V). The respiratory chain comprises NADH-ubiquinone reductase (NQR or complex I), succinate-ubiquinone reductase (SQR, complex II), ubiquinol-cytochrome c reductase (QCR, complex III), and cytochrome c oxidase (complex IV). The energy, released from electron transfer reaction through the complexes, is stored in the form of proton gradient across the inner membrane, the mitochondrial transmembrane potential (ΔΨm). Finally, ATP synthesis from adenosine diphosphate (ADP) and inorganic phosphate is accomplished through complex V, dissipating the ΔΨm energy.[15] These mitochondrial respiratory complexes require many iron atoms for their proper conformation and activities because irons within the complexes act as direct electron acceptors and donors in the form of hemes and iron-sulfur clusters. Therefore, the activities of mitochondrial respiratory complexes required for maintenance of ΔΨm and cellular energy level must be sensitive to cellular iron levels. Therefore, mitochondrial activity was targeted as a prime factor involved in iron chelation–induced senescent arrest.

INDUCTION OF SENESCENCE-LIKE G1 ARREST BY IRON CHELATION IN CHANG CELLS

Liver is a prime target organ for iron toxicity, because one of its functions is to take up and store excess iron in the body. Heavy iron overload has been implicated in several liver diseases, such as cirrhosis and hepatocellular carcinoma. Iron chelators often are used to eliminate the toxic effect of excess iron. Recently, we tried to explore the mechanism by which DFO induced growth arrest using an immortalized

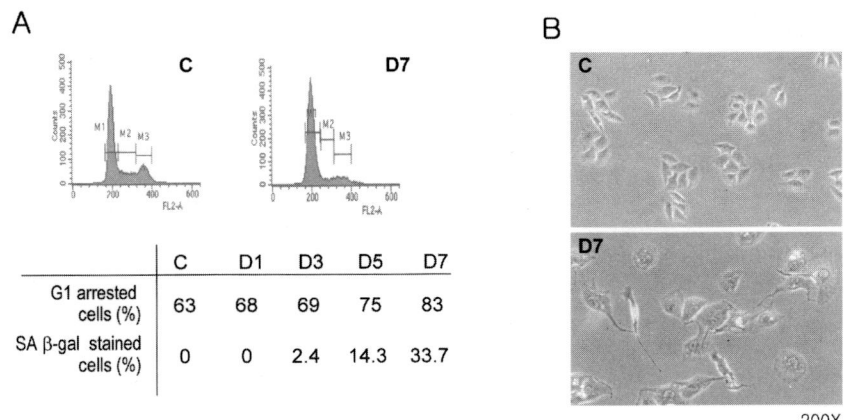

FIGURE 1. Senescent characteristics of Chang cells acquired under treatment of DFO. Two hundred thousand logarithmically growing Chang cells were seeded in six-well plates and cultured in DMEM/F12 media containing 10% FBS for 12 h and treated with or without various concentrations of DFO for the indicated periods. Cells were refreshed with the media with or without DFO every 3 days and, if necessary, passaged to keep subconfluent. (**A**) Cell cycle pattern and SA β-gal activity. Top panel is a representative pattern of cell cycle induced by DFO. Cell cycle was analyzed by flow cytometric analysis after staining cells with propidium iodide. Percentage of cells arrested at G1 was quantitated by flow cytometry. SA β-gal activity was assayed at pH 6.0 according to the method of Dimri *et al.* with a slight modification, and the number of SA β-gal–stained cells were counted.[19] Numbers in Dn indicate days after DFO treatment. (**B**) Cellular morphological changes were observed.

human normal hepatocyte cell line, Chang cell. In contrast with the low DFO concentrations (1–10 μM) required for effective growth arrest and apoptotic cell death in neuroblastoma,[13] relatively higher concentrations (>2 mM) were needed to induce apoptotic death in Chang cells.[14] Although it is not clear whether the liver has a still unknown protective mechanism against iron depletion, its function as iron storage might provide higher tolerance toward iron depletion than other organs have.

After exposure to 1 mM DFO, a subcytotoxic concentration, the cells were arrested at G1 phase in a time-dependent manner, with 83% at G1 phase after 7 days of exposure to DFO (FIG. 1A). During the growth arrest, the cells progressively displayed enlarged morphology mimicking replicative senescent cells (FIG. 1B). To confirm that DFO induces cellular senescence in Chang cells, we examined the acquisition of senescence-associated β-galactosidase activity (SA β-gal activity), a widely accepted senescence marker.[18,19] As shown in FIGURE 1, the proportion of SA β-gal–stained cells increased as the time of exposure increased. Furthermore, the arrest was proved to be irreversible.[14] Thus, DFO-induced G1 arrest could be defined as senescence-like G1 arrest. This was the first model of irreversible growth arrest representing replicative senescence induced by iron chelation. In a previous report, we demonstrated that the senescent arrest of Chang cells induced by DFO was associated with p27^{Kip1} induction through TGF-β1 expression, independent of p53. However, the TGF-β1/p27^{Kip1} signaling probably is not sufficient to arrest the cells irreversibly because clear induction was observed only after 3 days of exposure to DFO, and blocking TGF-β1/p27[Kip1] signaling with neutralizing antibody against TGF-β1 could not reverse the arrest. These results suggested that other mechanisms are involved in the irreversible arrest and prompted us to monitor mitochondrial activities as a prime target of iron chelation.

MITOCHONDRIAL ACTIVITY AS A PRIMARY TARGET FOR IRON DEPRIVATION

Most of the cellular energy is produced by mitochondrial OXPHOS. Mitochondrial ATP production depends on the mitochondrial membrane potential that is primarily generated and maintained by proton pumping through respiratory chain complexes. Each complex contains several irons that act as direct electron donor/acceptor. The irons exist in coordinated forms, such as heme or iron-sulfur cluster. This functional and structural involvement of irons in the respiratory chain complexes implies that the respiratory chain depends on cellular iron level. TABLE 1 shows various mitochondrial activities of Chang cells after exposure to DFO. In contrast with the late association of TGF-β1/p27[Kip1] signaling, the total cellular ATP level decreased to 8.7 pmole/μg lysate (53.8% of control) within 1 day after exposure to DFO, but did not decrease further, suggesting that the early ATP loss could provide an important clue to find the target of iron chelation.

To investigate how cellular ATP decreased, we first examined $\Delta\Psi_m$ by staining cells with JC-1 fluorescence dye and monitoring green fluorescence by flow cytom-

TABLE 1. Changes of mitochondrial activities of Chang cells after exposure to DFO

Time	ATP (pmol/μg lysate)	LDH (% of control)	$\Delta\Psi_m$ disrupted cells (% of total)	NCR (cyt. c μmol/ min/mg lysate)	SCR (cyt. c μmol/ min/mg lysate)	SQR initial slope
0	15.4 ± 3.0	100	0.6	10.2 ± 0.9	8.1 ± 0.8	0.22 ± 0.02
1 h	16.0 ± 4.6	101.4 ± 2.4	0.7	10.3 ± 1.1	9.4 ± 0.5	n.d.
3 h	15.9 ± 4.9	101.1 ± 4.1	0.5	9.0 ± 2.1	7.5 ± 1.9	0.24 ± 0.01
6 h	15.8 ± 3.2	110.1 ± 2.1*	0.5	11.6 ± 0.4*	4.8 ± 2.2*	0.18 ± 0.01*
12 h	14.4 ± 3.6	n.d.	1.4	10.8 ± 0.4	3.6 ± 0.2**	0.09 ± 0.01**
1 day	9.7 ± 2.9*	129.7 ± 13.9	10.1	10.7 ± 0.8	3.4 ± 0.6**	0.10 ± 0.01*
2 days	9.1 ± 1.4**	128.2 ± 4.9*	30.1	14.3 ± 1.4*	1.5 ± 0.1**	0.07 ± 0.02**
3 days	6.0 ± 0.4**	126.2 ± 2.9*	n.d.	11.3 ± 0.9	n.d.	n.d.
4 days	6.5 ± 0.3**	135.4 ± 20.3*	n.d.	n.d.	n.d.	n.d.
5 days	7.7 ± 1.0**	149.2 ± 4.9**	n.d.	10.1 ± 0.6	n.d.	n.d.
7 days	7.4 ± 0.6**	n.d.	n.d.	9.5 ± 0.7	n.d.	n.d.

NOTE: $\Delta\Psi_m$ was quantitated by monitoring green fluorescence using flow cytometry after staining cells with JC-1 fluorescence dye. LDH activity of control is 246.6 U/min/μg lysate. LDH assay was performed according to an earlier report.[20] One unit corresponds to the conversion of 1.0 μmole of pyruvate to lactate. CR, SCR, and SQR assays were performed according to an earlier report.[21] *$P < 0.05$ by Student's t test; **$P < 0.01$ by Student's t test; n.d., not defined.

etry. Green fluorescence reveals for inactive mitochondria with weak ΔΨm. The number of cells with inactive mitochondria increased after 12 h of exposure to DFO. We next monitored the respiratory chain activity to explore the cause of ΔΨm disruption. Except for a slight increase at 48 h, there were no changes, in NCR activities, but SCR activity decreased to 48% of control at 6 h and to 22% and 18% at 24 and 48 h, respectively. NCR and SCR activities reflect the activities of complex I and II, respectively, because both activities require complex III as a common factor. To ensure whether the decrease of SCR activity was caused by that of complex II, we studied the activity of succinate-ubiquinone reductase (SQR, complex II only). As expected, we confirmed the significant decrease in SQR within 12 h. From these results, we could summarize that a significant inhibition of complex II activity started 6 h after exposure to DFO and was followed by ΔΨm disruption and loss of cellular ATP level, implying that the disruption of complex II activity was the primary cause of cellular ATP loss. Moreover, LDH activity increased steadily under the situation, suggesting that the shortage of cellular energy might be partially compensated by activating glycolytic ATP generation. Taken together, these results imply that complex II might be the initial target for iron chelation by DFO and that the resulting mitochondrial dysfunction might be one important primary factor in the cellular progression to the irreversible senescence-like growth arrest of hepatocytes.

DOWN-REGULATED EXPRESSION OF IRON-SULFUR SUBUNIT RESULTED IN THE DECREASE OF COMPLEX II ACTIVITY AND WAS ASSOCIATED WITH THE IRREVERSIBLE ARREST

Next, we asked by which mechanism complex II activity was inhibited in the DFO-treated cells. No direct inhibition of the NCR and SCR activities by DFO was observed when their activities were measured at 4°C (data not shown). Therefore, we investigated gene expression of complex II by Western blot analysis. Complex II is composed of four subunits: flavoprotein (Fp), iron-sulfur protein (Ip), and two membrane-anchoring ubiquinone-binding proteins (QPs1 and QPs2), which are all encoded by nuclear DNA (FIG. 2A). Fp and Ip comprise succinate dehydrogenase (SDH), with its active site for oxidation of succinate to fumarate. The two QP subunits convert SDH to SQR by anchoring the soluble SDH fraction to the mitochondrial innermembrane. As shown in FIGURE 2B, the level of expression of flavoprotein did not show any change up to 2 days of exposure, whereas that of iron-sulfur protein started to decrease after 6 h, which agrees with the decrease of complex II activity. This result implies that the expression level of iron-sulfur subunit might be the initial limiting factor for the activity of complex II under the treatment of DFO.

To determine whether the down-regulation of complex II-Ip is associated to the irreversible arrest, we examined its expression status after treatment of Chang cells with 1 mM DFO for 3 days before replacing the culture medium with medium lacking DFO. As shown in FIGURE 2C, the down-regulation of complex II-Ip by iron chelation was not reversed even 3 days after discontinuing chelation. These data correlate well with the irreversible cellular ATP loss and with the irreversible acquisition of SA β-gal activity (FIG. 2D), suggesting that the down-regulation of complex II-Ip protein by DFO might be one of the critical factors involved in the progress to the irreversible senescence arrest.

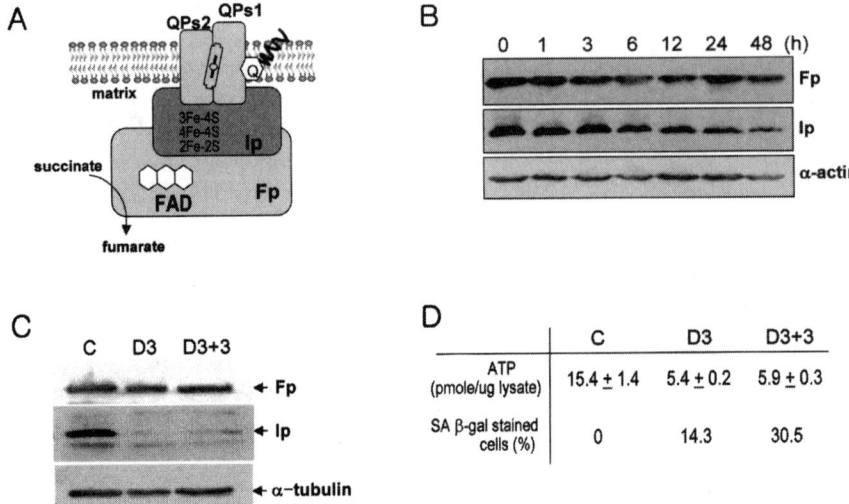

FIGURE 2. Specific down-regulation of Ip protein of complex II is associated with the irreversible arrest. (**A**) Schematic model of complex II. (**B**) Western blot analysis of flavoprotein subunit and iron-sulfur subunit of complex II. (**C**) Irreversible down-regulation of iron-sulfur subunit of complex II is confirmed. Chang cells were treated with 1 mM DFO for 3 days before being replaced with a medium lacking DFO. The cell lysates were applied to Western blot analysis. (**D**) Irreversible ATP loss and senescent arrest. Under the same condition as in **B**, the cells were applied to ATP assay or SA β-gal activity. Cellular ATP levels were measured by the bioluminescence assay according to the protocol provided with an ATP Determination Kit (Molecular Probes) following the manufacturer's instructions. SA β-gal activity was assayed according to the same procedure as that of FIGURE 1B.

ABERRANT MORPHOLOGY OF MITOCHONDRIA

Mitochondria are known to be dynamic organelle in which morphology and distribution changes readily in respose to functional changes and environmental stimuli. To examine whether the mitochondrial dysfunction in our system is accompanied by morphological changes, we investigated mitochondrial morphology by electron microscopy. As shown in FIGURE 3A, Chang cells contain mitochondria that are homogeneous in size and shape. When the cells are exposed to DFO for 2 days, most mitochondria assume highly elongated shapes that have rarely been reported (FIG. 3A). How do mitochondria become elongated? Mitochondria are dynamic organelle which maintain a balance between fusion and fission (division) in many cell types. Through these processes, there is continuous exchange of genetic content between mitochondria. Furthermore, the high capacity of mitochondria to fuse has been implicated as a defense of highly oxidative organelles against mitochondrial dysfunction caused by the accumulation of mitochondrial DNA (mtDNA) lesions with age.[16] At the present stage, it is not clear in our system whether the elongated mitochondria were formed by enhanced mitochondrial fusion or by a block of the fission process. Therefore, we first examined ROS production and mtDNA integrity. We

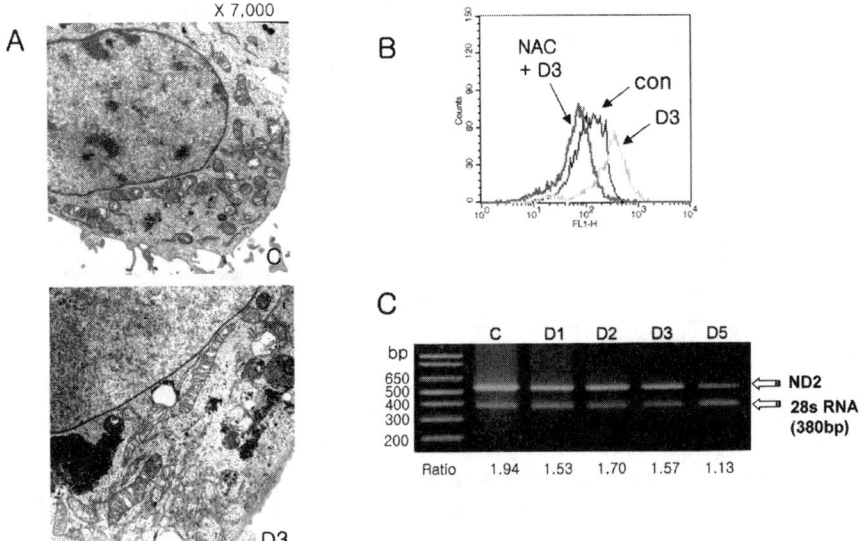

FIGURE 3. Mitochondrial morphological changes. (**A**) Electron microscopy of Chang cells with or without treatment of DFO. (**B**) ROS generation was monitored by flow cytometry after staining cells with DCF fluorescence dye. (**C**) Mitochondrial DNA quantity was examined by genomic PCR using total genomic DNA as template with primer set for mitochondrial ND2 gene and primer set for nuclear 28S RNA gene.

found that cellular ROS was continuously generated and the amount of mtDNA decreased gradually, implying that the damage in mtDNA integrity might induce the morphological changes. However, the detailed mechanism involved in this process has yet to be elucidated.

INVOLVEMENT OF MITOCHONDRIAL DYSFUNCTION IN INDUCING REPLICATIVE SENESCENCE OF HEPATOCYTES BY CELLULAR IRON DEPRIVATION

It is generally accepted that replicative senescent cells display flat cell morphology, acquire several senescent marker such as SA β-gal activity, and are irreversibly arrested during the G1 phase of cell cycle.[17–19] When we first observed replicative senescence-like growth arrest in Chang cells by depriving them of iron through exposure to DFO, we expected that cell cycle regulators would be primarily targeted to arrest the cells at G1 phase. However, the arrest of cells at G1 appeared to be a slow process, which could be observed only 3 days after exposure to DFO. Moreover, induction of TGF β1/p27^{Kip1} signaling, the only cell cycle regulator known to be associated with cellular arrest in our system, also progressed slowly. These data prompted us to screen for other primary factors that might be involved in the irreversible effect of DFO.

Metal ions are involved in many cellular processes, particularly in maintaining protein structure and in forming catalytic groups in enzymes. Iron is crucially important for oxido-reduction processes, because iron atoms can easily gain or lose electrons. The activities of the biological molecules (oxido-reductases) possessing iron in their active sites might be sensitive to cellular iron level. The oxidative respiration of mitochondria is a major site of iron requirement because all four respiratory complexes possess several irons as intracomplex electron carriers. Therefore, mitochondrial respiration was targeted in our studies as a key regulator responsive to iron deprivation. Expectedly, in DFO-treated Chang cells mitochondrial ATP production was disrupted relatively early and irreversibly. The decrease of complex II activity preceded the decrease in ATP production, implying that it was probably a primary target. The decrease of complex II activity after DFO exposure was not caused by direct inhibition, but by down-regulation of the iron-sulfur subunit (Ip), which, in turn, may be caused by translational regulation via the interaction of the iron-regulating protein and the iron-responsive element located in 5′ untranslated region of the Ip gene. Our results suggest that complex II activity is highly sensitive to cellular iron depletion and that its disruption might cause irreversible mitochondrial dysfunction, which is involved in inducing replicative senescence-like arrest. Fur-

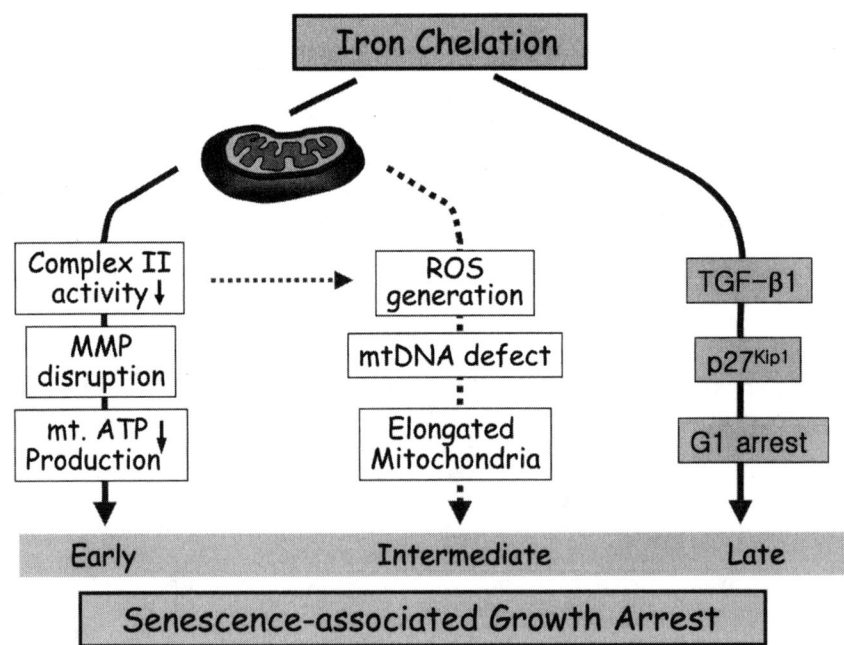

FIGURE 4. Proposed mechanism for the involvement of mitochondrial dysfunction during iron chelation–induced senescence-associated growth arrest of Chang cells. Mitochondrial dysfunction was involved before the action of TGF β1/p27^{Kip1} signaling. MMP, mitochondrial membrane potential.

thermore, there were dynamic changes of mitochondrial morphology (elongation), which might represent a defense against mitochondrial oxidative damage through interorganellar complementation. In summary, we propose a new mechanism underlying iron chelation–induced senescence-associated growth arrest: mitochondrial dysfunction via complex II disruption occurs early, followed by mitochondrial defense through elongation, and finally by involvement of cell cycle regulators (FIG. 4). Therefore, we emphasize the importance of maintenance of mitochondrial function and its integrity in cellular proliferation and the importance of mitochondrial complex II activity, which is disrupted by iron chelation and induces replicative senescent arrest. Finally, we propose our DFO-induced senescence-associated system as a good model, in which mitochondrial dysfunction and morphological changes are closely related to the replicative senescence-associated growth arrest.

ACKNOWLEDGMENT

This work was supported by Grant No. R05-2001-000-00497-0 from the Basic Research Program of the Korea Science & Engineering Foundation.

REFERENCES

1. ROBBINS, E. & T. PEDERSON. 1970. Iron: intracellular localization and possible role in cell division. Proc. Natl. Acad. Sci. USA **66:** 1244–1251.
2. BOLDT, D.H. 1999. New perspectives on iron: an introduction. Am. J. Med. Sci. **318:** 207–212.
3. DEUGNIER, Y. & B. TURLIN. 2001. Iron and hepatocellular carcinoma. J. Gastroenterol. Hepatol. **16:** 491–494.
4. ADAMS, P.C. 1998. Iron overload in viral and alcoholic liver disease. J. Hepatol. **28:** 19–20.
5. ITO, K., D. MITCHELL, T. GABATA, et al. 1999. Hepatocellular carcinoma association with increased iron deposition in the cirrhotic liver at MR imaging. Radiology **212:** 235–240.
6. NYHOLM, S., G.J. MANN, AG. JOHANSSON, et al. 1993. Role of RR in inhibition of mammalian cell growth by potent iron chelators. J. Biol. Chem. **268:** 26200–26205.
7. KICIC, A., A.C.G. CHUA & E. BAKER. 2001. Effect of iron chelators on proliferation and iron uptake in hepatoma cells. Cancer **92:** 3093–3110.
8. CHENOUFI, N., G. BAFFET, B. DRENOU, et al. 1998. Deferoxamine arrests in vitro the proliferation of porcine hepatocyte in G1 phase of the cell cycle. Liver **18:** 60–66.
9. LUCAS, J.J., A. SZEPESI, J. DOMENICO, et al. 1995. Effects of iron-depletion on cell cycle progression in normal human T lymphocytes: selective inhibition of the appearance of the cyclin A-associated component of the p33cdk2 kinase. Blood **86:** 2268–2280.
10. RENTON, F.J. & T.M. JEITNER. 1996. Cell cycle-dependent inhibition of the proliferation of human neural tumor cell lines by iron chelators. Biochem. Pharmacol. **51:** 1553–1561.
11. HOYES, K.P., R.C. HIDER & J.B. PORTER. 1992. Cell cycle synchronization and growth inhibition by 3-hydroxypyridin-4-one iron chelators in leukemia cell lines. Cancer Res. **52:** 4591–4599.
12. FAN, L., J. IYER, S. ZHU, et al. 2001. Inhibition of N-myc expression and induction of apoptosis by iron chelation in human neuroblastoma cells. Cancer Res. **61:** 1073–1079.
13. BRODIE, C., G. SIRIWARDANA, J. LUCAS, et al. 1993. Neuroblastoma sensitivity to growth inhibition by deferrioxamine: evidence for a block in G1 phase of the cell cycle. Cancer Res. **53:** 3968–3975.

14. YOON, G., H.-J. KIM, Y.S. YOON, et al. 2002. Iron chelation-induced senescence-like growth arrest in hepatocyte cell lines: association of transforming growth factor β1 (TGF β 1)-mediated p27^{Kip1} expression. Biochem. J. **366:** 613–621.
15. SARASTE, M. 1999. Oxidative phosphorylation at the fin de siecle. Science **283:** 1488–1493.
16. ONO, T., K. ISOBE, K. NAKADA & J.I. HAYASHI. 2001. Human cells are protected from mitochondrial dysfunction by complementation of DNA products in fused mitochondria. Nat. Genet. **28:** 272–275.
17. FINKEL, T. & N. HOLBROOK. 2000. Oxidants, oxidative stress and the biology of ageing. Nature **408:** 239–247.
18. COATES, P.J. 2002. Markers of senescence. J. Pathol. **196:** 371–373.
19. DIMRI, G.P., X. LEE, G. BASILE, et al. 1995. A biomarker that identifies senescent human cells in culture and in aging skin in vivo. Proc. Natl. Acad. Sci. USA **92:** 9363–9367.
20. ANDERSON, G.R. & B.K. FARKAS. 1988. The major anoxic stress response protein p34 is a distinct lactate dehydrogenase. Biochemistry **27:** 2187–2193.
21. ESPOSTI, M.D. 2001. Assessing functional integrity of mitochondria in vitro and in vivo. Methods Cell Biol. **65:** 75–96.

Mitochondrial DNA Mutation and Depletion Increase the Susceptibility of Human Cells to Apoptosis

CHUN-YI LIU,[a] CHENG-FENG LEE,[a] CHIUNG-HUI HONG, AND YAU-HUEI WEI

Department of Biochemistry, National Yang-Ming University, Taipei, Taiwan 112

ABSTRACT: Mitochondrial diseases, such as MELAS, MERRF, and CPEO syndromes, are associated with specific point mutations or large-scale deletions of mitochondrial DNA (mtDNA), which impair mitochondrial respiratory functions and result in decreased production of ATP in affected tissues. Recently, mitochondria have been recognized to act as key players in the regulation of cell death. To investigate whether a pathogenic mutation of mtDNA exerts any effect on the process of apoptosis of human cells, we constructed a series of cybrid human cells harboring different proportions of mtDNA with the A3243G or the A8344G transition, or with the 4,977-bp deletion, by cytoplasmic fusion of patients' skin fibroblasts with mtDNA-depleted ρ^0 cells of an immortal human osteosarcoma cell line (143B). We observed that the decrease in cell viability upon staurosporine treatment or exposure to ultraviolet (UV) irradiation was more pronounced in the cybrids harboring high levels of mutated mtDNA compared with the control cybrids. Using DNA fragmentation analysis, we found that the cell death induced by treatment with 100 nM staurosporine or by exposure to UV irradiation at 20 J/m^2 was caused by apoptosis, not necrosis. Moreover, we demonstrated activation of caspase 3 by Western blot and enhanced release of cytochrome *c* after 100 nM staurosporine treatment or 20 J/m^2 UV irradiation of the cybrids harboring high levels of the three mtDNA mutations. Furthermore, as compared with parental osteosarcoma 143B cells, the ρ^0 cells were found to be more susceptible to apoptosis, which was accompanied by caspase 3 activation and cytochrome *c* release. This indicates that mtDNA plays an important role in the regulation of apoptosis in human cells. Taken together, these findings suggest that mutation and depletion of mtDNA increase the susceptibility of human cells to apoptosis triggered by exogenous stimuli such as UV irradiation or staurosporine.

KEYWORDS: apoptosis; mitochondrial diseases; mitochondrial DNA mutation; osteosarcoma; UV irradiation; staurosporine

---

[a]Authors C-Y. Liu and C-F. Lee contributed equally to this work.

Address for correspondence: Professor Yau-Huei Wei, Department of Biochemistry, National Yang-Ming University, Taipei, Taiwan 112. Voice: 886-2-2826-7118; fax: 886-2-2826-4843.
joeman@ym.edu.tw

Ann. N.Y. Acad. Sci. 1011: 133–145 (2004). © 2004 New York Academy of Sciences.
doi: 10.1196/annals.1293.014

INTRODUCTION

Apoptosis is an important biological process involved in embryonic development and in several physiological and pathological events in humans and animals. A similar, or identical, process seems to be involved when cells, and in particular their DNA, are damaged beyond a certain degree by viral or oxidative attack. In the past few years, mitochondria have been established to play a critical role in the early events leading to apoptotic cell death.[1] Recent studies have clearly demonstrated that mitochondria act as a major switch and arbitrator for the initiation of apoptosis in mammalian cells. The switch is believed to involve the release of proteins shown to regulate apoptosis via interaction with the mitochondrial permeability transition pore (mtPTP). The mtPTP spans the inner and outer membranes and is composed of a set of membrane proteins including the voltage-dependent anion channel, adenine nucleotide translocase (ADP/ATP carrier), the benzodiazepine receptor, and cyclophilin D.[2] Factors influencing the opening or closing of the mtPTP may be either proapoptotic or antiapoptotic. Furthermore, the apoptotic pathway is amplified by the release of apoptogenic proteins from the mitochondrial intermembrane space, including cytochrome c, apoptosis-inducing factor (AIF, a flavoprotein), and latent forms of specialized proteases called procaspases.[3] Although cytochrome c activates the caspases to degrade cytosolic proteins, leading to the destruction of cytoplasm and nucleus, apoptosis-inducing factor translocates to the nucleus and participates in the destruction of chromatin.[4] This process can be triggered by excessive mitochondrial uptake of calcium ion (Ca^{2+}), exposure to high levels of reactive oxygen species (ROS), or decline in the cellular ATP level. Thus, marked reduction in mitochondrial energy production and chronic increase in intracellular oxidative stress, conditions that are likely to occur in mitochondria carrying specific mtDNA mutations, may activate the opening of mtPTP and initiate cell apoptosis.[5]

The mtPTP is highly sensitive to many pathological factors affecting mitochondria such as oxidative stress, membrane potential depolarization, and Ca^{2+} overload.[6] Opening of the mtPTP will lead to collapse of the proton gradient across the mitochondrial inner membrane and release of apoptogenic proteins including cytochrome c. Inhibition of mtPTP by pharmacological agents (e.g., cyclosporin A) prevents cell death by changing the expression and subcellular localization of the antiapoptotic protein Bcl-2, suggesting that mtPTP opening is one of the key events leading to cell apoptosis.[7,8]

Respiratory chain deficiency and apoptosis are associated with neurodegenerative diseases, heart failure, diabetes mellitus, and aging.[9] Several mtDNA mutations have been demonstrated to cause defects of respiratory function and to be associated with mitochondrial diseases. To investigate the relationship between mtDNA mutations and apoptosis, we have established a set of cybrid cell lines harboring different proportions of mutated mtDNA from skin fibroblasts of patients with clinically proven mitochondrial diseases.[10] We have hypothesized that the mutated mtDNA will influence the apoptotic behavior of human cells.

In this study, we tested this hypothesis by examining the molecular and cellular responses to apoptosis of cybrids harboring different proportions of mtDNA with the A3243G or the A8344G point mutation, or the common 4,977-bp deletion under condition of ultraviolet (UV) irradiation and staurosporine treatment. We examined apoptosis by measuring cell viability, DNA fragmentation, cytochrome c release from mitochondria, and caspase 3 activation. Our findings have furthered our understanding of the correlation between mtDNA mutation, mitochondrial dysfunction, and apoptosis, which is pivotal to the ultimate elucidation of the of pathogenic mechanism of mitochondrial diseases.

MATERIALS AND METHODS

Cell Culture

We used primary cultures of skin fibroblasts established from three patients with clinically proven MELAS, MERRF, and CPEO syndromes, respectively. A series of cytoplasmic hybrids (cybrids) harboring different proportions of mtDNA with either the A3243G or the A8344G point mutation, or with the 4,977-bp deletion, were made by fusing enucleated skin fibroblasts with mtDNA-less human osteosarcoma (ρ^0) cells in the presence of PEG-1500 according to the method developed by King and Attardi.[11] In brief, the enucleated skin fibroblasts of the patients with mitochondrial disease (MELAS, MERRF, or CPEO syndrome) were incubated with the ρ^0 osteosarcoma cells in medium containing PEG-1500. The resulted cybrids were grown in a selection medium of DMEM containing 5-bromo-2'-deoxyuridine (but without uridine and pyruvate). After approximately 2 weeks of subculture in the selection medium, only the successfully fused cybrids were able to grow.[12] The cybrids were selected in a medium containing 100 µg/mL of 5-bromo-2'-deoxyuridine and were grown in DMEM supplemented with 5% FBS, 100 µg/mL pyruvate and 50 µg/mL uridine and further incubated at 37°C in humidified 5% CO_2/95% air. The proportions of mtDNA with A3243G or A8344G point mutation or 4,977-bp deletion were quantified in the cybrids, by PCR-RFLP or by Southern hybridization, as established in our laboratory.[10,13]

Treatments for Induction of Apoptosis

To induce human cells to undergo apoptosis, we seeded 1×10^6 cells in each of the 60-mm dishes and incubated for 24 h, and staurosporine to the medium was added at a final concentration of 100 nM. After 6 h, the cells were harvested for the following assays. For assessment of UV irradiation-induced apoptosis, 7.5×10^5 cells were seeded in 100-mm dishes for 24 h and then subjected to short wavelength ultraviolet irradiation at 20 J/m^2. After exposure to UV irradiation, the cells were allowed to keep growing in a CO_2 incubator for 25 h before harvesting.

Cytosol Extracts for Western Blot Analysis of Cytochrome c

Cells were washed once with phosphate-buffered saline (Sigma, St. Louis, MO) and then resuspended in lysis buffer A (70 mM Tris-HCl and 250 mM sucrose, pH 7.0). Digitonin was added to provide a condition that resulted in 95% cells being stained with 0.2% trypan blue. The cells were centrifuged immediately at 600g for 10 min at 4°C, and the supernatant was collected as the cytosol fraction. The pellets were incubated at 4°C for 15 min in lysis buffer B (50 mM HEPES, 4 mM EDTA, 2 mM EGTA, and 1% Triton X-100). The samples were centrifuged at 1,700g for 15 min at 4°C, and the supernatant thus obtained was referred to as the mitochondrial fraction. The amount of protein in the cytosol and in mitochondrial fraction was measured by the Bradford method, and 40 µg of cytosol proteins and 25 µg of mitochondrial proteins were loaded into each lane of a 10% SDS-polyacrylamide gel. The separated proteins were blotted to a piece of Hybond-P$^+$ membrane (Amersham Biosciences, Buckinghamshire, UK). Nonspecific binding was blocked by incubation of the membrane in the TBS buffer containing 3% nonfat milk and 0.1% Tween 20 for 1 h at room temperature. The membrane was subjected to Western blotting using anti–cytochrome c (diluted 1:300, IMG-101; Imgenex, San Diego, CA) as first antibody for 1 h at room temperature.[14] After washing three times with TBS containing 0.1% Tween 20, the membrane was incubated with the secondary antibody, HRP-conjugated rabbit anti–mouse IgG (Amersham Biosciences, diluted 1:3,000), for 1 h at room temperature, and the signal was detected using the ECL reagent (Amersham Biosciences).

DNA Fragmentation Assay

The cells were pelleted and resuspended in 100 µL of the lysis buffer (1% NP-40, 20 mM EDTA, 50 mM Tris-HCl, pH 7.5). After incubation at 4°C for 20 s, cells were centrifuged at 900g for 5 min at 4°C. The supernatant was added with RNase A (final concentration of 5 µg/µL) and SDS (final concentration of 1%), and the mixture was incubated for 2 h at 56°C. Proteinase K was added to the supernatant (final concentration of 2.5 µg/µL) and incubated for 4 h at 37°C. After adding 2.5 volumes of ethanol and 1/10 volume of 3 M sodium acetate, the solution was stored at –20°C for more than 8 h to precipitate DNA. DNA was collected by centrifugation at 9,000g for 30 min at 4°C, and washed by 70% cool ethanol and air-dried.[15] The DNA pellet then was dissolved in distilled H$_2$O and subjected to electrophoresis on a 2% agarose gel.

Measurement of Caspase 3 Activity

Cytosol extracts were prepared by repeated cycles of freezing and thawing of the cybrids in 100 µL of extraction buffer (12.5 mM Tris-HCl, pH 7.0, 1 mM dithiothreitol, 0.125 mM EDTA, 5% glycerol, 1 mM phenylmethyl sulfonyl fluoride, 1 µg/mL leupeptin, and 1 µg/mL aprotinin). After centrifugation of the cell lysate at 9,000g for 10 min at 4°C, 50 µg of the supernatant was diluted with 500 µL of assay buffer (50 mM Tris-HCl, 1 mM EDTA, and 10 mM EGTA, pH 7.0), and incubated at 37°C for 30 min in the dark with 20 µM of a fluorescent substrate, Ac-DEVD-AFC.[16] The fluorescence of the cleaved substrate generated by caspase 3 was determined using a spectrofluorometer (Hitachi F-3000, Tokyo, Japan) set at an excitation wavelength of 380 nm and an emission wavelength of 508 nm.

Western Blot Analysis of Caspase 3 Activation

Cells were pelleted and incubated at 4°C for 15 min in the lysis buffer (50 mM HEPES, pH 7.8, 4 mM EDTA, 2 mM EGTA, 1% Triton X-100). The samples were centrifuged at 900g for 15 min at 4°C, and the supernatant was referred to as cell lysate. The amount of protein in the cell lysate was measured by the Bradford method, and 75 µg of protein was loaded into each lane of a 10% SDS-polyacrylamide gel. The separated proteins were blotted onto a Hybond-P$^+$ membrane (Amersham Biosciences). Nonspecific binding was blocked by incubation in TBS containing 3% nonfat milk and 0.1% Tween 20 for 1 h at room temperature. The membrane was subjected to Western blot using anti–caspase 3 (sc-7272, diluted 1:300; Santa Cruz Biotechnology, Santa Cruz, CA) and anti–D4-GDI (IMG-160, diluted 1:250; Imgenex), for 1 h at room temperature. After being washed three times with TBS containing 0.1% Tween 20, the membrane was incubated with the secondary antibody, HRP-conjugated rabbit anti–mouse IgG (diluted 1:3,000; Amersham Biosciences), for 1 h at room temperature, and the signal was detected using the ECL reagent (Amersham Biosciences).

Cell Viability Assay

Cell viability was measured by MTT assay. Approximately 5×10^3 cells were seeded in a 96-well plate and allowed to grow for 24 h. After treatment with 100 nM staurosporine for 6 h, cells were incubated with 3-(4,5-dimethylthiazol-2-yl)-2,5-diphenyl-2H-tetrazolium bromide (MTT; Sigma) at a final concentration of 2.0 μg/mL during the last 2 h of the incubation period. The cells were lysed with DMSO and the absorbance at 570 and 690 nm were measured concurrently.[17]

RESULTS

Genotyping of the Cybrids

After construction and subcloning of the cybrids, we determined the relative proportion of mtDNA with A3243G or A8344G point mutation by PCR-RFLP, and the proportion of mtDNA with the 4,977-bp deletion by Southern hybridization using a probe comprising the D-loop region of human mtDNA. There were two clones for each type of the cybrids fused from each of the patients with each mitochondrial disease: one was the control cybrid harboring only wild-type mtDNA, the other was the mutant cybrid harboring a high proportion of mutated mtDNA. TABLE 1 shows the proportions of mtDNA with point mutations or 4,977-bp deletions in the six cybrid clones used in this study and derived from skin fibroblasts of three patients with MELAS, MERRF, and CPEO syndromes. After subcloning, there were no obvious variations in the proportions of mutated mtDNA in these six cybrid clones.

TABLE 1. Genotyping of the cybrid clones used in this study

	A3243G point mutation		A8344G point mutation		4977-bp deletion	
Cybrid clone	Lu 02	Lu 04	D5	B2	1-3-16	51-10
Genotype	Wild type	Mutant	Wild type	Mutant	Wild type	Mutant
Proportion of mutant mtDNA (%)	Undetectable	80 ± 2.1	Undetectable	94.3 ± 1.7	Undetectable	79.8 ± 3.6

TABLE 2. Apoptotic features in the cybrids harboring 4977-bp deleted mtDNA after treatment with staurosporine or UV irradiation

		Cybrid clone	
		1-3-16	51-10
Viability (%)	STS[a]	73.5 ± 8.1	45.3 ± 8.0
	UV[b]	35.4 ± 3.4	14.8 ± 2.4
Caspase 3 activity	STS	107.7 ± 8.2	154.7 ± 27.2
	UV	78.7 ± 1.3	204.2 ± 3.9
D4-GDI cleavage	STS	+[c]	++
	UV	++	+++
DNA ladder	STS	+	++
	UV	++	+++

NOTE: In this set of experiments, each measurement was performed for all cybrid clones after treatment with 100 nM staurosporine for 6 or 25 h after exposure to UV irradiation at 20 J/m^2. [a]STS was determined by the MTT assay method. [b]UV was determined by the trypan blue exclusive method. [c]The number of "+" symbols indicates the extent of increase in that assay.

TABLE 3. Comparison of caspase 3 activities in the cybrid clones harboring mutated mtDNA after treatment with staurosporine or UV irradiation

	A3243G point mutation		A8344G point mutation		4977-bp deletion	
Cybrid clone	Lu 02	Lu 04	D5	B2	1-3-16	51-10
Staurosporine	16.2 ± 1.1	36.1 ± 3.3	59.3 ± 6.2	136.6 ± 12.2	107.7 ± 8.2	154.7 ± 27.2
UV irradiation	21.2 ± 5.7	50.7 ± 4.5	88.5 ± 1.6	146.9 ± 3.1	78.71 ± 1.3	204.2 ± 3.9

NOTE: In this set of experiments, the activities of caspase 3 were assayed for all the cybrid clones after treatment with 100 nM staurosporine for 6 or 25 h after exposure to UV irradiation at 20 J/m^2.

The Cybrids Harboring Mutated mtDNA Are More Susceptible to Apoptosis

Both staurosporine and UV irradiation caused cell death as shown by the cell viability assay, but cybrids harboring a high content of 4,977-bp deleted mtDNA (51-10) were more susceptible than the control cybrid (1-3-16) (TABLE 2). We then examined the activation of caspase 3, which is one of the key features of the cells undergoing apoptosis. After treating the cybrids with 100 nM staurosporine or exposure to UV irradiation at 20 J/m^2, we observed that the activity of caspase 3 was significantly increased in all cybrids harboring mutated mtDNA (Lu 04, B2, and 51-10) compared with the corresponding control cybrids (Lu 02, D5, and 1-3-16), respectively (TABLE 3).

DNA fragmentation is another key indicator of apoptosis, particularly fragmentation in a quantized manner into a ladder-like appearance of different sized fragments. After treating the cybrids with 100 nM staurosporine or UV irradiation at 20 J/m^2, we observed more obvious DNA laddering in the cybrids harboring higher levels of 4,977-bp deleted mtDNA (FIG. 1). To understand the upstream changes of caspase 3 in the UV-induced apoptotic processes, we monitored the release of cytochrome c from mitochondria by immunoblotting with anti–cytochrome c antibody. The results showed that cytochrome c release was induced by UV irradiation and that cytochrome c release from mitochondria occurred much earlier in the cybrids harboring high levels (80%) of mtDNA with 4,977-bp deletion compared with cybrids containing only wild-type mtDNA (FIG. 2).

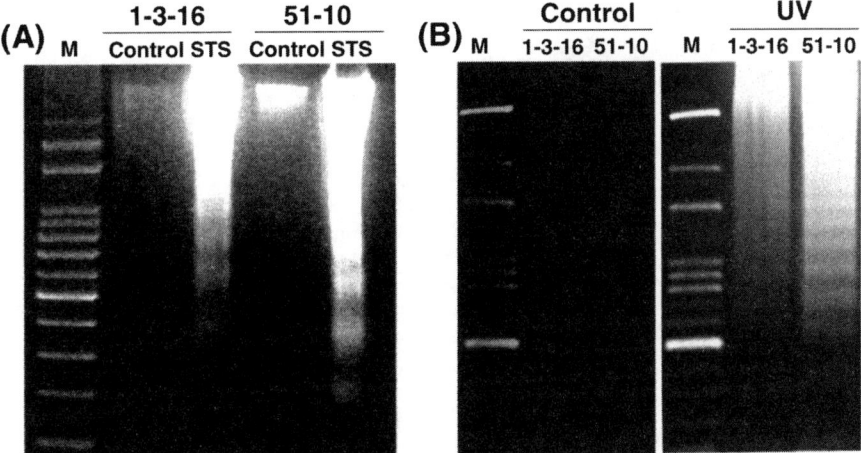

FIGURE 1. DNA fragmentation of the cybrids harboring 4,977-bp deleted mtDNA after treatment with 100 nM staurosporine for 6 or 25 h after exposure to 20 J/m^2 UV irradiation. Total cellular DNA was extracted from each of the cybrids and then subjected to electrophoresis on a 2% agarose gel. DNA fragmentation was clearly seen in cybrid clone 51-10, which contained approximately 80% of 4,977-bp deleted mtDNA, after treatment with 100 nM staurosporine for 6 (**A**) or 25 (**B**) h after exposure to 20 J/m^2 UV irradiation compared with control cybrid clone 1-3-16 harboring only wild-type mtDNA.

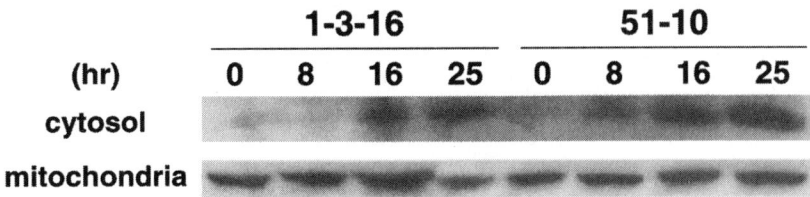

FIGURE 2. Induction of cytochrome c release from mitochondria of the cybrids by 20 J/m^2 UV irradiation. An aliquot of 40 μg of cytosolic proteins and 25 μg of mitochondrial proteins from the mutant and control cybrids were subjected to electrophoresis and Western blotting using anti–cytochrome c antibody. The results showed that cytochrome c release from mitochondria occurred much earlier in the cybrids harboring high levels of 4,977-bp deleted mtDNA than those containing only wild-type mtDNA.

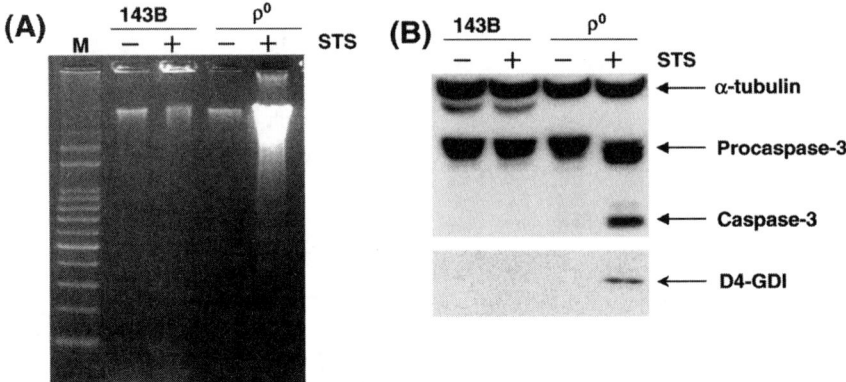

FIGURE 3. Mitochondrial DNA-less ρ^0 cells are more susceptible to staurosporine-induced apoptosis. (**A**) Total DNA was extracted from ρ^0 cells and parental 143B cells and were subjected to electrophoresis on a 2% agarose gel. The results showed that DNA fragmentation was more obvious in ρ^0 cells than in 143B cells after treatment with 100 nM staurosporine for 6 h. (**B**) Caspase 3 activation was more pronounced in ρ^0 cells than in 143B cells after staurosporine treatment. After treatment of the 143B cells and ρ^0 cells with 100 nM staurosporine for 6 h, total cellular proteins were extracted and subjected to SDS-PAGE and Western blotting. The results showed that caspase 3 activation and the D4-GDI fragment generated by proteolytic cleavage of caspase 3 was more pronounced in ρ^0 cells.

Cells Lacking Mitochondrial DNA(ρ^0 cells) Are More Susceptible to Staurosporine-Induced Apoptosis

We observed that caspase 3 activity was increased and cell viability was decreased in ρ^0 cells compared with 143B parental cells after treatment with 100 nM staurosporine for 6 h. Here, we showed more obvious DNA laddering in ρ^0 cells that had been treated with 100 nM staurosporine (FIG. 3A). We also detected enhanced proteolytic cleavage of pro-caspase 3 and appearance of the product of one of the caspase 3 downstream substrates (D4-GDI) by Western blot analysis in ρ^0 cells after staurosporine treatment (FIG. 3B). On the other hand, we examined whether the upstream changes of caspase 3 in ρ^0 cells induced by staurosporine treatment are also effected through the cytochrome c–mediated pathway. As shown in FIGURE 4, the immunoblot demonstrated that cytochrome c was released from mitochondria to cytoplasm in ρ^0 cells after treatment with 100 nM staurosporine for 6 h (FIG. 4).

DISCUSSION

Mitochondrial encephalomyopathies usually are accompanied by neuronal cell death and muscle weakness. More than 100 different point mutations and large-scale deletions of mtDNA have been associated with human diseases.[18] Mitochondrial diseases caused by mtDNA mutations are genetically and phenotypically heterogeneous. The proportion and distribution of mutated mtDNA in affected tissues cannot fully explain this clinical heterogeneity. Apoptotic features have been reported in af-

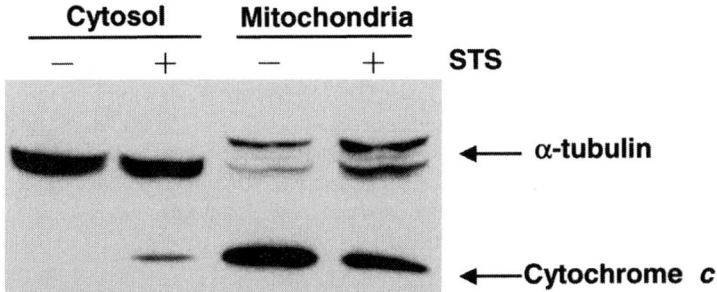

FIGURE 4. Induction of cytochrome c release from mitochondria of ρ^0 cells in apoptosis induced by staurosporine treatment. After staurosporine treatment of ρ^0 cells for 6 h, 40 μg of cytosolic proteins and 25 μg of mitochondrial proteins were subjected to electrophoresis on 10% SDS-PAGE and Western blotting using anti–cytochrome c antibody. The results showed that cytochrome c was detected in cytosolic fraction after staurosporine treatment, whereas the intensity of the immunoblotted cytochrome c band was decreased in the mitochondrial fraction.

fected cells in muscle fibers of patients carrying high proportions of single mtDNA deletions (>40%) or point mutations in mitochondrial tRNA genes (>70%).[19] It is thought that apoptosis not only removes the cells with dysfunctional mitochondria but also plays an important role in the pathogenesis of mitochondrial encephalomyopathies, because of mtDNA mutations affecting mitochondrial protein synthesis. The clinical pictures of mitochondrial diseases may be associated with or caused by deficit in energy metabolism as well as enhanced cell apoptosis of affected tissues. In this study, we used cybrid clones constructed by fusion of mtDNA-less human osteosarcoma-derived ρ^0 cells with skin fibroblasts harboring A3243G or A8344G point mutation or 4,977-bp deletion of mtDNA to investigate the effect of mtDNA mutation on apoptotic behaviors of human cells. Apoptosis of the cybrids was easily induced by 100 nM staurosporine or UV irradiation at 20 J/m^2 as shown by caspase 3 activation, cytochrome c release from mitochondria, and DNA fragmentation.

It has been widely reported that staurosporine and UV irradiation induce apoptosis in a broad range of cell lines. Several mechanisms have been proposed to explain how staurosporine induces cell death. These include inhibition of protein kinase C,[20] caspase activation,[21] and alteration of intracellular calcium ion concentration ($[Ca^{2+}]_i$),[22] but which one of these is the major pathway still remains to be investigated. On the other hand, UV irradiation may cause damage to DNA molecules, then trigger downstream signaling pathways and force the cells to enter into cell death process.[23,24] Moreover, intracellular ROS increase in cells after exposure to UV, and accumulation of ROS may further turn on stress-related genes that lead the cells to apoptosis.[25,26]

We demonstrated that cybrids harboring mutated mtDNA exhibited much higher amplitude of caspase 3 activation upon staurosporine treatment or UV irradiation (TABLE 3). Danielson et al.[27] also reported that osteosarcoma-derived cybrids carrying the most common and severe LHON-associated mtDNA mutations (11778 and 3460) were prone to undergo apoptosis. This is consistent with our previous findings that mtDNA mutations affect apoptotic behavior in the human cell.[17] Moreover, we found that caspase 3 activity was higher in the cybrids harboring 4,977-bp deleted mtDNA than in cybrids carrying mtDNA A3243G or A8344G point mutations. Interestingly, cybrids with 4,977-bp deleted mtDNA were more sensitive to staurosporine or UV irradiation than cybrids harboring A3243G or A8344G point mutation of mtDNA (TABLE 3). This finding also supports our clinical observation that a lower proportion of mtDNA with large-scale deletion was sufficient to cause mitochondrial diseases, whereas much higher levels of mtDNA with a point mutations are required for mitochondrial diseases to manifest.[28]

The results of this study suggest that apoptosis of mutant cybrids triggered by staurosporine treatment or UV irradiation was mediated by mtPTP opening, cytochrome c release from mitochondria, and activation of downstream caspases (FIGS. 2 and 4). In a separate experiment, we observed that caspase 9, but not caspase 8, was significantly activated by UV irradiation of cybrids harboring high proportions of mtDNA with 4,977-bp deletion. Moreover, we found that the percentage of cells with low mitochondrial membrane potential, as shown by the lack of JC-1 dye aggregation, was increased in cybrids harboring mutated mtDNA. These observations indicate that the alteration of mitochondrial structure and function in mutant cybrids may be responsible for the enhanced susceptibility to UV or staurosporine-induced apoptosis.

Note that the apoptotic features in cybrids harboring a high level of 4,977-bp deleted mtDNA were much more pronounced than those observed in control cybrids containing only wild-type mtDNA (TABLE 2). In another study, we observed the same trend in other series of cybrids harboring 4,366-bp deleted mtDNA, which were constructed from skin fibroblasts of another patient with CPEO syndrome. These data demonstrate that large-scale deletions of mtDNA certainly render human cells more sensitive to proapoptotic stimuli. We also observed that the extent of cytochrome c release and activation of caspase 3 in cybrids are positively correlated with the proportion of mtDNA harboring the 4,977-bp deletion (data not shown). This suggests that the mtDNA genotype indeed plays an important role in the apoptotic process. We confirmed this view by investigating apoptotic behaviors of mtDNA-less ρ^0 cells after staurosporine treatment or UV irradiation. It is controversial whether human cells lacking mtDNA are more susceptible to cell death. Wang et al.[5] and Cai et al.[29] demonstrated that ρ^0 cells were prone to undergo apoptosis induced by different stimuli in vitro, but other investigators suggested that ρ^0 cells were more resistant to cell death.[30,31] In this study, we found that DNA laddering was more obvious and caspase 3 activation was more pronounced in ρ^0 cells than in parental 143B osteosarcoma cells (FIG. 3). These results lend support to the notion that mtDNA mutation and depletion influence the life and death of human cells and suggest that increased apoptosis is an important pathogenic event in mitochondrial diseases caused by mtDNA mutations.

Cybrids are an excellent tool to investigate the alterations in mitochondrial function and phenotype of cells caused by quantitative or qualitative alteration of mtDNA, because there is no interference of the nuclear background. It has been demonstrated that the bioenergetic function is impaired in affected tissues of patients with mitochondrial encephalomyopathies as a result of defects in the structure and function of respiratory enzymes.[32] Although these biochemical defects can explain the pathology of muscle weakness and exercise intolerance of the patients, the muscle wasting, premature aging, and neurodegenerative features commonly seen in such patients have remained unclear. The results obtained in this study lead us to propose a general scheme that mutation and depletion of mtDNA indeed render human cells more vulnerable to apoptosis triggered by staurosporine treatment or UV irradiation. Although how exactly cells carrying mutated mtDNA were more susceptible to cell death remains largely unknown, we believe that apoptosis may play an important role in the manifestation of pathologies such as muscle wasting and neurological disorders in mitochondrial encephalomyopathies. In a recent study, we observed that

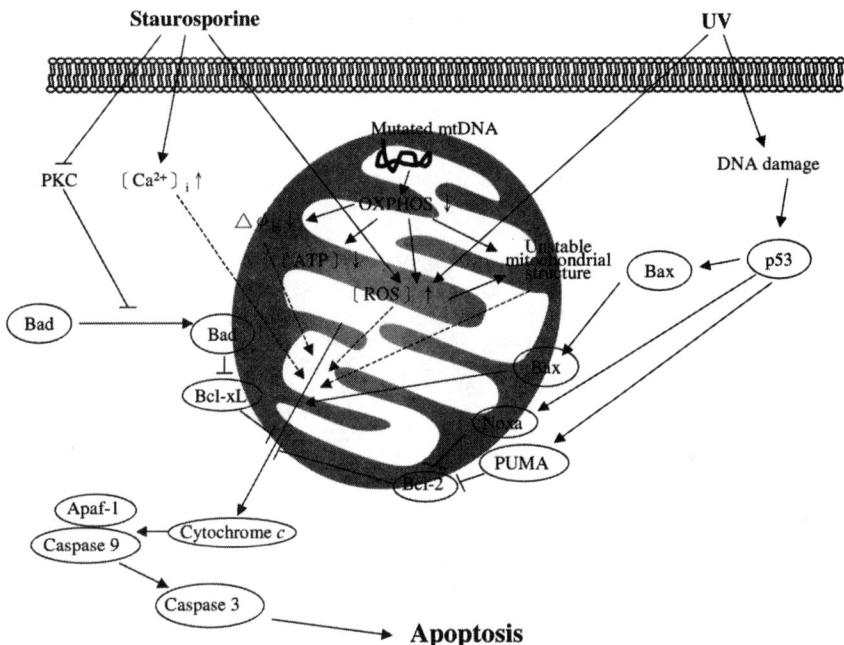

FIGURE 5. An integrated scheme illustrating the events leading to apoptosis in the cybrids harboring mutated mtDNA induced by staurosporine treatment or exposure to UV irradiation. Essentially, the mutant cybrids are more susceptible to apoptosis triggered by UV irradiation due to lower ATP level, lower membrane potential ($\Delta\psi_m$), higher oxidative damage, easier cytochrome c release, higher caspase 3 activation, and downstream apoptotic cascade. Staurosporine may induce apoptosis of the mutant cybrids through a different pathway, which involves Ca^{2+} overload to mitochondria, opening of mitochondrial permeability transition pores and cytochrome c release, and activation of the effector caspases. The roles of p53 and several Bcl-2 family proteins in the apoptotic process elicited by staurosporine treatment or exposure to UV irradiation are also depicted on the basis of findings reported recently and our own unpublished data. One of the major viewpoints of this scheme is that apoptogenic proteins (e.g., AIF and cytochrome c) are more easily released from mitochondria harboring mutant mtDNA, which are caused by unstable or poorly assembled mitochondrial structure of the mutant cybrids containing a high proportion of mutant mtDNA. The cell death pathway initiated or activated by staurosporine and that triggered by UV irradiation converge at the mitochondria and involve activation of the same downstream caspases cascade.

p53 gene expression was significantly increased in cybrids harboring high proportions (>50%) of mtDNA with the 4,977-bp deletion (data not shown). We propose on the basis of these recent findings and pertinent information gathered from previous studies an integrated scheme to explain the observations that human cells harboring mutated mtDNA are more susceptible to apoptosis (FIG. 5). The possible roles of up-regulation of p53, translocation of Bcl-2 family proteins, and dynamic change of intracellular Ca^{2+} ions in apoptosis elicited by mtDNA mutation warrant further investigation.

ACKNOWLEDGMENTS

This work was supported by a grant (NSC91-2320-B010-068) from the National Science Council and a grant (NHRI-EX92-9120BN) from the National Health Research Institutes. The authors are grateful to Dr. Chin-Chang Huang, Dr. Yuh-Jyh Jong, and Dr. Edward K. Wang for assistance in the procurement of the skin biopsies from the patients used in this study.

REFERENCES

1. LEONARD, J.V. & A.V.H. SCHAPIRA. 2000. Mitochondrial respiratory chain disorders II: neurodegenerative disorders and nuclear gene defects. Lancet **355**: 389–394.
2. TATTON, W.G. & C.W. OLANOW. 1999. Apoptosis in neurodegenerative diseases: the role of mitochondria. Biochim. Biophys. Acta **1410**: 195–213.
3. HU, Y., M.A. BENEDICT, L. DING & G. NUÑEZ. 1999. Role of cytochrome c and dATP/ATP in Apaf-1-mediated caspase-9 activation and apoptosis. EMBO J. **18**: 3586–3595.
4. GREEN, D.R. & J.C. REED. 1998. Mitochondria and apoptosis. Science **281**: 1309–1312.
5. WANG, J., J.P. SILVA, C.M. GUSTAFSSON, et al. 2001. Increased in vivo apoptosis in cells lacking mitochondrial DNA gene expression. Proc. Natl. Acad. Sci. USA **98**: 4038–4043.
6. BERNARDI, P., L. SCORRANO, R. COLONNA, et al. 1999. Mitochondria and cell death: mechanistic aspects and methodological issues. Eur. J. Biochem. **264**: 687–701.
7. LOEFFLER, M. & G. KROEMER. 2000. The mitochondrion in cell death control: certainties and incognita. Exp. Cell Res. **256**: 19–26.
8. VOEHRINGER, D.W., D.L. HIRSCHBERG, J. XIAO, et al. 2000. Gene microarray identification of redox and mitochondrial elements that control resistance or sensitivity to apoptosis. Proc. Natl. Acad. Sci. USA **97**: 2680–2685.
9. DIMAURO, S., E. BONILLA, M. DAVIDSON, et al. 1998. Mitochondria in neuromuscular disorders. Biochim. Biophys. Acta **1366**: 199–210.
10. WEI, Y.H., C.F. LEE, H.C. LEE, et al. 2001. Increase of mitochondrial mass and mitochondrial genome in association with enhanced oxidative stress in human cells harboring 4,977 bp-deleted mitochondrial DNA. Ann. N.Y. Acad. Sci. **928**: 97–112.
11. KING, M.P. & G. ATTARDI. 1989. Human cells lacking mtDNA repopulation with exogenous mitochondria by complementation. Science **246**: 500–503.
12. WEI, Y.H. 1998. Mitochondrial DNA mutations and oxidative damage in aging and diseases: an emerging paradigm of gerontology and medicine. Proc. Natl. Sci. Counc. ROC, Part B: Life Sci. **22**: 55–67.
13. LU, C.Y., D.J. TSO, T. YANG, et al. 2002. Detection of DNA mutations associated with mitochondrial diseases by Agilent 2100 bioanalyzer. Clin. Chim. Acta **318**: 97–105.
14. JIANG, S., J. CAI, D.C. WALLACE, et al. 1999. Cytochrome c-mediated apoptosis in cells lacking mitochondrial DNA. Signaling pathway involving release and caspase 3 activation is conserved. J. Biol. Chem. **274**: 29905–29911.

15. HERRMANN, M., H.M. LORENZ, R. VOLL, et al. 1994. A rapid and simple method for the isolation of apoptotic DNA fragments. Nucleic Acids Res. **22:** 5506–5507
16. SHIAH, S.G., S.E. CHUANG, Y.P. CHAU, et al. 1999. Activation of c-Jun NH_2-terminal kinase and subsequent CPP32/Yama during topoisomerase inhibitor beta-lapachone-induced apoptosis through an oxidation-dependent pathway. Cancer Res. **59:** 391–398.
17. LEE, H.C., P.H. YIN, C.Y. LU, et al. 2000. Increase of mitochondria and mitochondrial DNA in response to oxidative stress in human cells. Biochem. J. **348:** 425–432.
18. SILVA, J.P. & N.G. LARSSON. 2002. Manipulation of mitochondrial DNA gene expression in the mouse. Biochim. Biophys. Acta **1555:** 106–110.
19. MIRABELLA, M., S.D. GIOVANNI, G. SILVESTRI, et al. 2000. Apoptosis in mitochondrial encephalomyopathies with mitochondrial DNA mutations: a potential pathogenic mechanism. Brain **123:** 93–104.
20. JARVIS, W.D., A.J. TURNER, L.F. POVIRK, et al. 1994. Induction of apoptotic DNA fragmentation and cell death in HL-60 human promyelocytic leukemia cells by pharmacological inhibitors of protein kinase C. Cancer Res. **54:** 1707–1714.
21. RABKIN, S.W. 2001. Prevention of staurosporine-induced cell death in embryonic chick cardiomyocyte is more dependent on caspase-2 than caspase-3 inhibition and is independent of sphingomyelinase activation and ceramide generation. Arch. Biochem. Biophys. **390:** 119–127.
22. KRUMAN, I., Q. GUO & M.P. MATTSON. 1998. Calcium and reactive oxygen species mediate staurosporine-induced mitochondrial dysfunction and apoptosis in PC12 cells. J. Neurosci. Res. **51:** 293–308.
23. CAMPBELL, C., A.G. QUINN, B. ANGUS, et al. 1993. Wavelength specific patterns of p53 induction in human skin following exposure to UV radiation. Cancer Res. **53:** 2697–2699.
24. BRASH, D.E., A. ZIEGLER, A.S. JONASON, et al. 1996. Sunlight and sunburn in human skin cancer: p53, apoptosis, and tumor promotion. J. Invest. Dermatol. Symp. Proc. **1:** 136–142.
25. ZHUANG, L., B. WANG & D.N. SAUDER. 2000. Molecular mechanism of ultraviolet-induced keratinocyte apoptosis. J. Interferon Cytokine Res. **20:** 445–454.
26. MATSUMURA, Y. & H.N. ANANTHASWAMY. 2002. Molecular mechanisms of photocarcinogenesis. Front. Biosci. **7:** d765–d783.
27. DANIELSON, S.R., A. WONG, V. CARELLI, et al. 2002. Cells bearing mutations causing Leber's hereditary optic neuropathy are sensitized to Fas-induced apoptosis. J. Biol. Chem. **277:** 5810–5815.
28. WEI, Y.H., C.Y. LU, H.C. LEE, et al. 1998. Oxidative damage and mutation to mitochondrial DNA and age-dependent decline of mitochondrial respiratory function. Ann. N.Y. Acad. Sci. **854:** 155–170.
29. CAI, J., D.C. WALLACE, B. ZHIVOTOVSKY, et al. 2000. Separation of cytochrome c-dependent caspase activation from thiol-disulfide redox change in cells lacking mitochondrial DNA. Free Radic. Biol. Med. **29:** 334–342.
30. DEY, R. & C.T. MORAES. 2000. Lack of oxidative phosphorylation and low mitochondrial membrane potential decrease susceptibility to apoptosis and do not modulate the protective effect of Bcl-x(L) in osteosarcoma cells. J. Biol. Chem. **275:** 7087–7094.
31. SCIACCO, M., G. FAGIOLARI, C. LAMPERTI, et al. 2001. Lack of apoptosis in mitochondrial encephalomyopathies. Neurology **56:** 1070–1074.
32. CORTOPASSI, G. & E. WANG. 1995. Modelling the effects of age-related mtDNA mutation accumulation; complex I deficiency, superoxide and cell death. Biochim. Biophys. Acta **1271:** 171–176.

Resistance of ρ^0 Cells against Apoptosis

MYUNG-SHIK LEE, JA-YOUNG KIM, AND SUN YOUNG PARK

Department of Medicine, Samsung Medical Center, Sungkyunkwan University School of Medicine, 50 Irwon-dong Kangnam-ku, Seoul 135-710, Korea

ABSTRACT: Mitochondrion is one of the master players in both apoptosis and necrosis. However, most previous articles report that mitochondrial DNA-depleted cells without oxidative phosphorylation underwent apoptosis by several apoptotic effectors as efficiently as their parental cells, suggesting that intact mitochondrial function is dispensable for the progression of apoptosis. We studied the role of mitochondrial function in several apoptosis models. TRAIL, a recently identified member of the TNF family with cytotoxicity on a wide variety of transformed cells, killed SK-Hep1 cells with characteristic features of apoptosis such as DNA fragmentation, sub-G1 ploidy peak, and cytochrome c translocation. In contrast with parental cells, mitochondrial DNA-deficient SK-Hep1 ρ^0 cells were resistant to TRAIL-induced apoptosis. Dissipation of mitochondrial potential or cytochrome c translocation did not occur in ρ^0 cells after TRAIL treatment. Bax translocation also was absent in ρ^0 cells, accounting for the failure of cytochrome c release in ρ^0 cells. SK-Hep1 ρ^0 cells were resistant to other death effectors such as staurosporine. Our results indicate that apoptosis of SK-Hep1 hepatoma cells is dependent on intact mitochondrial function. Because aged cells or tumor cells have frequent mutations or deletions of mitochondrial DNA, they might acquire the ability to evade apoptosis or tumor surveillance imposed by TRAIL or other death effectors *in vivo*, accounting for the selection advantage of cancer cells and frequent development of cancer in aged individuals.

KEYWORDS: mitochondria; apoptosis; cell death; TRAIL; ROS; aging; cancer

Mitochondrion is the organelle where most cellular ATP is generated through oxidative phosphorylation. Mitochondrion is also one of the master players in both apoptosis and necrosis, two classic modes of cell death. Although morphological changes of mitochondria such as swelling and appearance of flocculent densities have been regarded as the hallmarks of necrosis,[1] mitochondria also play an important role in apoptosis by releasing cytochrome c and thereby activating caspase-9.[2,3] Mitochondrial permeability transition (PT) is observed in many apoptotic events and has been described as a point of no return of death programming.[4] However, results from most recent articles using mitochondrial DNA (mtDNA)–depleted cells (ρ^0 cells)[5] did not apparently support the indispensability of intact mitochondrial function in apoptosis.

Address for correspondence: M.-S. Lee, Department of Medicine, Samsung Medical Center, 50 Irwon-dong Kangnam-ku, Seoul 135-710, Korea. Voice: 82-2-3410-3436; fax: 82-2-3410-0388.

mslee@smc.samsung.co.kr

Ann. N.Y. Acad. Sci. 1011: 146–153 (2004). © 2004 New York Academy of Sciences.
doi: 10.1196/annals.1293.015

Most ρ^0 cells underwent apoptosis by several death signals as efficiently as their parental cells.[6–9] In contrast, some ρ^0 cells were more resistant to TNF-induced apoptosis than their parental cells, suggesting the necessity of oxidative phosphorylation in certain types of apoptosis.[10]

In an attempt to study the susceptibility of ρ^0 cells to cell death and to elucidate the role of intact mitochondrial function in various modes of cell death, we treated SK-Hep1 ρ^0 cells developed by Y.K. Pak (author in this volume)[11] with TRAIL and other apoptosis-inducing agents.

RESISTANCE OF MITOCHONDRIAL DNA-DEPLETED CELLS AGAINST CELL DEATH

We used, colorimetric assays, as a measure of death of mtDNA-depleted cells, using crystal violet or trypan blue exclusion assay because MTT assay assessing mitochondrial succinate dehydrogenase activity might be affected by the absence of mtDNA. Our experiments showed that TRAIL killed parental (ρ^+) human hepatoma cells (SK-Hep1) after incubation for 24 h (FIG. 1A). Hoechst 33342 staining of nuclei showed markedly increased nuclear condensation and fragmentation. Apoptosis by TRAIL was inhibited by a pan-caspase inhibitor (z-VAD-fmk), a caspase 3 inhibitor (Ac-DEVD-fmk), or a caspase 8 inhibitor (Ac-IETD-fmk), suggesting that TRAIL-induced ρ^+ cell death was a classic caspase-dependent apoptosis (FIG. 1B). DNA ploidy assays also showed that the death of SK-Hep1 cell by TRAIL was classic apoptosis accompanied by sub-G1 DNA peak (FIG. 1C, D). In contrast, mtDNA-deficient ρ^0 SK-Hep1 cells were resistant to TRAIL-induced apoptosis[12] (FIG. 1A). After TRAIL treatment, there was virtually no increase in the number of apoptotic cells with nuclear condensation/fragmentation identified by Hoechst 33342 staining. The difference in their susceptibility to TRAIL was not caused by a difference in the expression of TRAIL receptors, because parental SK-Hep1 cells and ρ^0 cells expressed various TRAIL receptors at similar levels. The resistance of SK-Hep1 ρ^0 cells against TRAIL was overcome by actinomycin D (FIG. 1A).

TRAIL treatment of ρ^+ cells triggered dissipation of mitochondrial potential associated by flow cytometric measurement with the fluorescence after loading of mitochondrial membrane-permeable cationic dye $DiOC_6$. However, dissipation of mitochondrial potential was not observed in TRAIL-treated ρ^0 cells. We also studied possible translocation of cytochrome c after TRAIL treatment of ρ^+ cells. Fluorescent staining using nuclear chromatin-binding Hoechst 33342 and FITC-visualized anti–cytochrome c antibody showed that cytochrome c was dispersed diffusely in the whole cytoplasm of apoptotic cells undergoing nuclear condensation/fragmentation, indicating that cytochrome c was translocated from mitochondria to cytosol in ρ^+ cells. Such cells constituted about half of the total ρ^+ cells after TRAIL treatment, whereas the number of such cells was very low before the TRAIL treatment. However, the percentage of cells showing cytochrome c translocation did not notably increase after TRAIL treatment of ρ^0 cells, indicating that TRAIL treatment did not induce cytochrome c translocation in ρ^0 cells.[12] To elucidate the mechanism of the absence of cytochrome c translocation in ρ^0 cells, we studied the translocation of Bax. Before TRAIL treatment, Bax was diffusely present in the cells. However, a large proportion of ρ^+ cells displayed a punctuate mitochondrial pattern of Bax im-

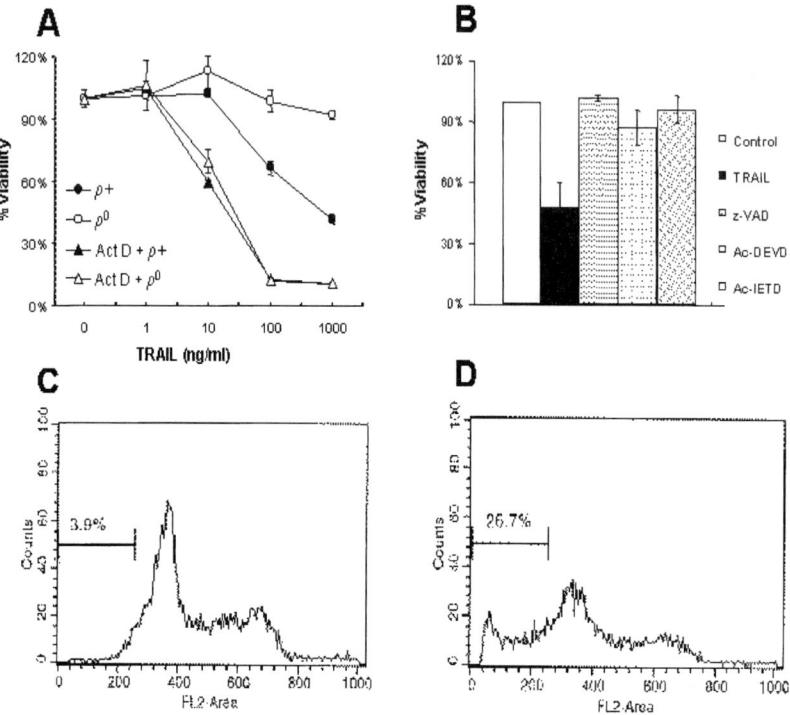

FIGURE 1. TRAIL-induced apoptosis of parental ρ^+ cells and resistance of mitochondrial DNA-deficient ρ^0 cells against TRAIL. (**A**) TRAIL treatment for 24 h induced death of SK-Hep1 hepatoma cells in a dose-dependent manner; however, mitochondrial DNA-deficient ρ^0 cells were resistant to the cell killing effect of TRAIL. The resistance of ρ^0 cells against TRAIL was overcome with actinomycin D. (**B**) Apoptosis of parental cells by TRAIL was blocked by caspase inhibitors. (**C, D**) DNA ploidy analysis. Sub-G1 peak was significantly increased after TRAIL treatment of parental SK-Hep1 cells (**D**) compared with the pretreatment level (**C**), which was not observed after TRAIL treatment of ρ^0 cells (data not shown). From Kim et al.[12] Reprinted with permission from *Oncogene*.

munostaining after TRAIL treatment, which suggests the translocation of Bax facilitating recognition by anti–Bax antibody. Triple-colored staining with a mitochondria-selective dye MitoTracker Red, Hoechst 33342, and anti–Bax antibody confirmed that the punctate perinuclear Bax was indeed localized to mitochondria of ρ^+ cells undergoing nuclear condensation/fragmentation. Such cells showing Bax translocation constituted more than half of the total ρ^+ cells after TRAIL treatment, whereas the number of such cells was negligible before TRAIL treatment. However, the percentage of such cells did not increase after TRAIL treatment of ρ^0 cells, suggesting that the absence of Bax translocation is responsible for the absence of cytochrome *c* release and also the lack of TRAIL sensitivity in ρ^0 cells. The absence of Bax translocation after TRAIL treatment was confirmed by Western blot analysis after cell fractionation. Bax expression in the cytosolic fraction was decreased, whereas it was increased in the heavy membrane fraction by TRAIL treat-

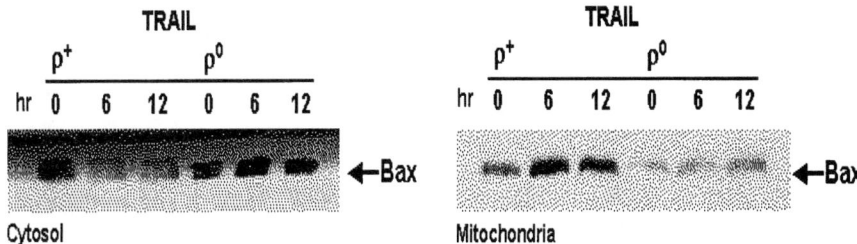

FIGURE 2. Bax translocation after TRAIL treatment assessed by cell fractionation. Bax expression in the cytosolic fraction was decreased, whereas it was increased in the heavy membrane fraction after TRAIL treatment of ρ^+ cells. However, changes in the Bax expression were not observed in the heavy membrane and cytosolic fractions after TRAIL treatment of ρ^0 cells. From Kim et al.[12] Reprinted with permission from *Oncogene*.

ment of SK-Hep1 parental cells, indicating that TRAIL induces Bax translocation from cytosolic fraction to heavy membrane fraction. However, Bax expression remained stationary in both heavy membrane and cytosolic fractions before and after TRAIL treatment of SK-Hep1 ρ^0 cells, strongly indicating that Bax translocation did not occur in ρ^0 cells[12] (FIG. 2). The mechanism of the absence of Bax translocation in SK-Hep1 ρ^0 cells is not clear. Deficient ATP production in ρ^0 cells might contribute to the failure of Bax translocation in response to apoptotic stimuli because incubation of SK-Hep1 ρ^0 cells with ADP + Pi that could increase mitochondrial ATP contents restored TRAIL susceptibility in SK-Hep1 ρ^0 cells. However, a direct effect of Pi on mitochondrial potential or on permeability transition cannot be ruled out.

APOPTOSIS OF ρ^0 CELLS BY OTHER DEATH SIGNALS

Our preliminary data suggest that SK-Hep1 ρ^0 cells are also resistant to various ROS donors such as staurosporine or hydrogen peroxide (FIG. 3). Although we have reported that SK-Hep1 ρ^0 cells and their parental cell are not different in their sensitivity to staurospoine or etoposide,[12] careful experiments using various doses of death effectors showed that they differed in their susceptibility to death effectors other than TRAIL. The resistance of SK-Hep1 ρ^0 cells to various modes of cell death might be related to the induction of antioxidant enzymes. Because superoxide radicals are generated during mitochondrial electron transfer, the interruption of mitochondrial electron transfer in ρ^0 cells might lead to the leakage of a massive amount of superoxide anion between mitochondrial respiratory enzyme complexes. If so, cells might adapt to such environmental changes by inducing antioxidant enzymes. Such enzymes might play a role in their resistance against various agents increasing external ROS as well as in the survival of ρ^0 cells despite intracellular ROS overproduction. Such stimuli entail not only direct ROS donors but also TNF family members that use ROS as a tool for intracellular signal transduction.

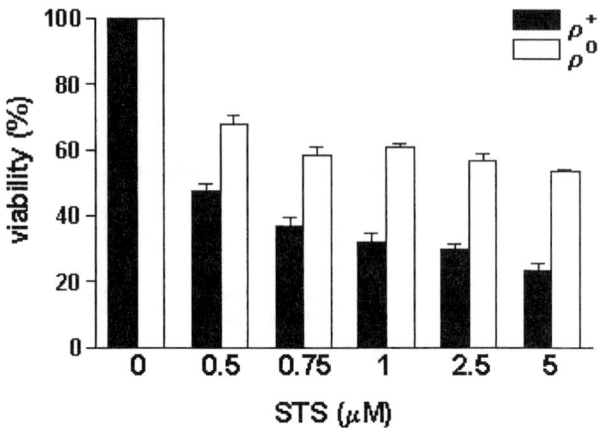

FIGURE 3. Relative resistance of SK-Hep1 ρ^0 cells against staurospoine. Careful titering of the doses of death effectors showed that SK-Hep1 ρ^0 cells were also resistant to death effectors other than TRAIL.

SIGNIFICANCE OF ρ^0 CELL RESISTANCE AGAINST CELL DEATH

The clinical or biological significance of ρ^0 cell resistance to various modes of cell death is not clear. Several studies reported prevalent mitochondrial mutations in a variety of tumors.[13,14] Then, tumor cells with mutations in mitochondrial genes may have diminished mitochondrial function and consequently decreased susceptibility to TRAIL, which might serve as a selection advantage over normal cells with intact mitochondrial function. Because TRAIL seems to be expressed on natural killer (NK) cells or monocytes upon appropriate stimuli[15,16] and TRAIL is active on the majority of tumor cells, TRAIL might be involved in tumor surveillance *in vivo*. A direct involvement of TRAIL on NK cells in the surveillance of tumor metastasis has been shown in recent *in vivo* experiments.[17,18] In addition to cancer cells, mitochondrial mutations or deletions are frequently found in aged cells.[19,20] MtDNA is more susceptible to ROS damage than genomic DNA for several reasons. First, mitochondria are the site of electron transfer, one of the most abundant source of oxygen radical in cells. Second, mtDNA is not protected by the nucleosomes such as nuclear DNA , and it exists partly in a single stranded conformation because of its mode replication, rendering mtDNA vulnerable to attack by free radicals. Third, DNA repair system for mtDNA is not as efficient as that for genomic DNA, and mtDNA replication does not stop for repair like G1 arrest. Finally, the presence of mtDNA damages could lead to altered function of the electron transfer and ROS leakage, leading to vicious cycles of ROS overproduction.[21] Because of such vulnerability of mtDNA to ROS and inefficient DNA damage repair, damage to mtDNA could accumulate over a long period of time in aged cells or individuals. The percentage of individual mutation or deletion may be low in aged cells, but the cumulative frequency of various kinds of mutations or deletions may be high. Several specific mutations have been reported to occur up to 50% in aged fibroblasts.[22] In fact, mitochondrial damage is one of the two main theories regarding the molecular mechanism of aging in addition to telomere shortening. One mechanism of the ROS overproduction in aged cells could involve accumulated mtDNA mutations. The resistance of SK-Hep1 ρ^0 cells against cell death might be related to the decreased apoptotic response of aged cells to genotoxic stress[23] and the relatively frequent development of cancer in aged individuals.

In accordance with our hypothesis that ρ^0 cell resistance against cell death may represent resistance of aged cell against cell death, ρ^0 cells showed some phenotypes resembling senescent cells such as flattened morphology and delayed growth.

FUTURE PERSPECTIVE

We showed that mtDNA-depleted ρ^0 cells are resistant to TRAIL and other death effectors, which is in contrast with other articles reporting no difference in the susceptibility of several ρ^0 cells to cell death compared with parental cells. However, our results are from only one kind of ρ^0 cells. Further studies will be necessary to see if our results from SK-Hep1 ρ^0 cells can be extended to other types of ρ^0 cells from different sources. It will be imperative to carefully reassess the sensitivity of various ρ^0 cells to death effectors over a wide range of concentrations because the different susceptibility to death effectors might manifest only at certain range of concentrations. Molecular mechanism of the resistance of ρ^0 cells against various death effectors and the detailed mechanism of the survival of ρ^0 cells in the absence of oxidative phosphorylation should be unraveled.

We suggested that our results might be related to the resistance of aged cells or cancer cells with mitochondrial mutations to apoptosis and their avoidance of host surveillance. However, our results are from an artificial system utilizing ρ^0 cells. Our hypothesis should be confirmed in a natural condition using "real" aged cells and primary cancer cells. Study of the true incidence of mtDNA mutations or deletions in natural aged cells or primary cancer cells will be very important to test our hypothesis. So far, only a few studies have been conducted regarding the incidence of mtDNA mutations or deletions particularly in cancer cells, and most studies covered only a portion of mtDNA, partly because of technical difficulties in screening the whole mtDNA. However, the recent introduction of two-dimensional gel scanning[24] as shown by Y. Suh or denaturing high-performance liquid chromatography[25] will facilitate searching of the entire mitochondrial genome. Both techniques are sensitive enough to detect minor heteroplasmy. With such technical advances and dedication of many researchers, the biological significance of the resistance of mtDNA-depleted cells to apoptosis in relation to aging and the development of cancer will be clarified.

ACKNOWLEDGMENTS

This work was supported by Health Planning Technology and Evaluation Board Grants (02-PJ1-PG1-CH04-0001) and Science Research Center Grants from Korea

Science and Engineering Foundation. M.-S.L. is an awardee of the National Research Laboratory Grants from the Korea Institute of Science and Technology Evaluation and Planning (2000-N-NL-01-C-232).

REFERENCES

1. KERR, J.F.R., G.C. GOBE, C.M. WINTERFORD, et al. 1995. Anatomical methods in cell death. Methods Cell Biol. **46:** 1–22.
2. LIU, X., C.N. KIM, J. YANG, et al. 1996. Induction of apoptotic program in cell-free extract: requirement for dATP and cytochrome c. Cell **86:** 147–157.
3. LI, P., D. NIJHAWAN, I. BUDIHARDJO, et al. 1997. Cytochrome c and dATP-dependent formation of Apaf-1/caspase-9 complex initiates an apoptotic protease cascade. Cell **91:** 479–489.
4. KROEMER, G., N. ZAMZAMI & S.A. SUSIN. 1997. Mitochondrial control of apoptosis. Immunol. Today **18:** 44–51.
5. KING, M.P. & G. ATTARDI. 1989. Human cells lacking mtDNA: repopulation with exogenous mitochondria by complementation. Science **246:** 500–503.
6. JACOBSON, M.D., J.F. BURNE, M.P. KING, et al. 1993. Bcl-2 blocks apoptosis in cells lacking mitochondrial DNA. Nature **361:** 365–369.
7. MARCHETTI, P., S.A. SUSIN, D. DECAUDIN, et al. 1996. Apoptosis-associated derangement of mitochondrial function in cells lacking mitochondrial DNA. Cancer Res. **56:** 2033–2038.
8. LIANG, B.C. & E. ULLYATT. 1998. Increased sensitivity to cis-diamminedichloroplatinum induced apoptosis with mitochondrial DNA depletion. Cell Death Differ. **5:** 694–701.
9. JIANG, S.J., J. CAI, D.C. WALLACE, et al. 1999. Cytochrome c-mediated apoptosis in cells lacking mitochondrial DNA. J. Biol. Chem. **274:** 29905–29911.
10. HIGUCHI, M., B.B. AGGARWAL & E.T.H. YEH. 1997. Activation of CPP32-like protease in tumor necrosis factor-induced apoptosis is dependent on mitochondrial function. J. Clin. Invest. **99:** 1751–1758.
11. PARK, K.-S., K.-J. NAM, J.-W. KIM, et al. 2001. Depletion of mitochondrial DNA alters glucose metabolism in SK-Hep1 cells. Am. J. Physiol. **280:** E1007–E1014.
12. KIM, J.Y., Y.H. KIM, I. CHANG, et al. 2002. Resistance of mitochondrial DNA-deficient cells to TRAIL: role of Bax in TRAIL-induced apoptosis. Oncogene **21:** 3139–3148.
13. POLYAK, K., Y. LI, H. ZHU, et al. 1998. Somatic mutations of the mitochondrial genome in human colorectal tumors. Nat. Genet. **20:** 291–293.
14. FLISS, M.S., H. USADEL, O.L. CABELLERO, et al. 2000. Facile detection of mitochondrial DNA mutations in tumors and bodily fluids. Science **287:** 2017–2019.
15. KAYAGAKI, N., N. YAMAGUCHI, M. NAKAYAMA, et al. 1999. Expression and function of TNF-related apoptosis-inducing ligand on murine activated NK cells. J. Immunol. **163:** 1906–1913.
16. GRIFFITH, T.S., S.R. WILEY, M.Z. KUBIN, et al. 1999. Monocyte-mediated tumoricidal activity via the tumor necrosis factor-related cytokine, TRAIL. J. Exp. Med. **189:** 1343–1354.
17. TAKEDA, K., Y. HAYAKAWA, M.J. SMYTH, et al. 2001. Involvement of tumor necrosis factor-related apoptosis-inducing ligand in surveillance of tumor metastasis by liver natural killer cells. Nat. Med. **7:** 94–100.
18. SMYTH, M.J., E. CRETNEY, K. TAKEDA, et al. 2001. Tumor necrosis factor-related apoptosis-inducing ligand (TRAIL) contributes to interferon-γ-dependent natural killer cell protection from tumor metastasis. J. Exp. Med. **193:** 661–670.
19. ARNHEIM, N. & G. CORTOPASSI. 1992. Deleterious mitochondrial DNA mutations accumulate in aging human tissues. J. Biol. Chem. **275:** 157–167.
20. WANG, Y., Y. MICHIKAWA, C. MALLIDIS, et al. 2001. Muscle-specific mutations accumulate with aging in critical human mtDNA control sites for replication. Proc. Natl. Acad. Sci. USA **98:** 4022–4027.
21. BOHR, V.A. & R.M. ANSON. 1995. DNA damage, mutation and fine structure DNA repair in aging. Mutat. Res. **338:** 25–34.

22. SINGH, K.K., J. RUSSELL, B. SIGALA, et al. 1999. Mitochondrial DNA determines the cellular response to cancer therapeutic agents. Oncogene **18:** 6641–6645.
23. SUH, Y., K.-A. LEE, W.-H. KIM, et al. 2002. Aging alters the apoptotic response to genotoxic stress. Nat. Med. **8:** 3–4.
24. VAN ORSOUW, N.J., X. ZHANG, J.Y. WEI, et al. 1998. Mutational scanning of mitochondrial DNA by two-dimensional electrophoresis. Genomics **52:** 27–36.
25. VAN DEN BOSCH, B.J.C., R.F.M. DE COO, H.R. SCHOLTE, et al. 2000. Mutation analysis of the entire mitochondrial genome using denaturing high performance liquid chromatography. Nucleic Acids Res. **28:** e89.

Mitochondrial DNA 4,977-bp Deletion in Paired Oral Cancer and Precancerous Lesions Revealed by Laser Microdissection and Real-Time Quantitative PCR

DAR-BIN SHIEH,[a] WEN-PIN CHOU,[b] YAU-HUEI WEI,[c] TONG-YIU WONG,[a] AND YING-TAI JIN[c]

[a]*Institute of Oral Medicine, National Cheng-Kung University, Tainan, Taiwan, ROC*

[b]*Institute of Molecular Medicine, National Cheng-Kung University, Tainan, Taiwan, ROC*

[c]*Department of Biochemistry, National Yang-Ming University, Tainan, Taiwan, ROC*

ABSTRACT: Oral cancer is the fourth leading cause of cancer deaths among men in Taiwan and is closely associated with areca quid chewing habits. Recent studies showed that mitochondrial DNA (mtDNA) mutations occur in various tumors, including oral cancers, and that the accumulation of mtDNA deletions could be an important contributor to carcinogenesis. Using laser microdissection, we have analyzed mtDNA deletions by pairwise comparisons in oral cancer, precancerous cells, and their adjacent submucosal stoma tissues in 12 patients with areca quid chewing history. Real-time quantitative polymerase chain reaction (RTQPCR) was performed to detect and quantify mtDNA with the 4,977-bp deletion in the histologically defined specified cell groups. Quantitative analysis of 60 samples by RTQPCR revealed that the average proportions of 4,977-bp deleted mtDNA over total mtDNA were 0.137%, 0.367%, and 0.001% in cancer, precancer cells, and lymphocytes of lymph node biopsies, respectively. Pairwise analysis of the proportion of mtDNA deletion in cancer, precancer, and their stroma tissues revealed a consistent trend among these patients. All of the patients (12/12) presented a higher proportion of mtDNA with 4,977-bp deletion in the lesions than in the lymphocytes, with average increases of 198-fold in cancer and 546-fold in precancer cells. A decrease in the proportion of deleted mtDNA was observed in 8 of 12 patients when the disease progressed from precancer to cancer lesions. Interestingly, 7 of 12 cancer tissues and 8 of 12 precancer lesions exhibited an average of 6.3-fold and 17.4-fold increases in the proportion of 4,977-bp deleted mtDNA in the stromal cells than in the lesion cells, respectively. The observation that the proportion of 4,977-bp deleted mtDNA in all oral lesions was higher than normal and consistently decreased during cancer progression from precancer to primary cancer suggests that accumulation and subsequent cytoplasmic segregation of the mutant mtDNA during cell division may play an important role in oral carcinogenesis. This study also demonstrates that laser microdissection combined with RTQPCR is

Address for correspondence: Ying-Tai Jin and Dar-Bin Shieh, Institute of Oral Medicine, National Cheng-Kung University, No. 1 University Rd., Tainan, Taiwan, ROC 701. Voice: 886-6-2353535 ext. 5410, 5376; fax: 886-6-2766626.
yingtai@mail.ncku.edu.tw; dshieh@mail.ncku.edu.tw

Ann. N.Y. Acad. Sci. 1011: 154–167 (2004). © 2004 New York Academy of Sciences.
doi: 10.1196/annals.1293.016

an efficient and sensitive tool to gain insight into the role that mtDNA mutation may play in carcinogenesis.

KEYWORDS: real-time quantitative PCR; laser microdissection; mitochondrial DNA deletion; mouth neoplasm

INTRODUCTION

Oral cancer is the fourth leading cause of male cancer death and in Taiwan attributed to the prevalent areca quid chewing habits.[1] Previous studies have shown that ingredients of areca quid generate significant amounts of reactive oxygen species (ROS) and induce oxidative DNA damage in cultured cells.[2,3] Although a wide spectrum of genomic alterations in nuclear DNA in carcinogenesis has been established, little attention has been paid to the role of mitochondrial DNA (mtDNA) mutations in human cancers. The human mitochondrial genome is composed of a circular double-stranded DNA of 16.6 kb, which encodes 13 subunits of respiratory enzyme complexes, 22 tRNAs, and 2 rRNAs in the organelle. There may be hundreds to thousands of mitochondria in one cell and each may contain 2 to 10 copies of mitochondrial DNA. The ROS-induced mtDNA damage usually is accumulated in somatic tissues and more closely reflects the exposure history than damage to nuclear DNA because mitochondrial genome is more susceptible to environmental insults for being a relatively unprotected DNA structure and having less sophisticated DNA repair mechanisms.[4,5] Moreover, the large number of mtDNA copies and the heteroplasmic nature of the mitochondrial genome enable affected cells to survive with damaged mtDNA, which may quantitatively reflect the extent of exposure to ROS.[6]

Different types of mtDNA mutations have been shown to play important roles in human aging and many metabolic, degenerative, or neuromuscular diseases.[5] Recently, somatic mutations of mtDNA were found to occur in various types of human cancers, including colorectal carcinoma, breast, thyroid, and gastric cancers.[7–12] The observation that most of these mutations are homoplasmic indicates the advantage of the mutant mtDNA toward malignant progression. Although several point mutations in mtDNA have been identified in human tumor tissues, only a few research groups have investigated the presence and accumulation of mtDNA deletions in cancers. Using semiquantitative polymerase chain reaction (PCR) analysis, Lee et al.[13] first demonstrated a significant association between areca quid chewing habit and the occurrence of 4,977-bp mtDNA deletion in paired oral cancer tissues. Interestingly, much higher levels of mtDNA deletion were observed in the stroma tissue than in the tumor tissue in groups with or without areca quid history. Consistent finding that stromal cells harbored a higher proportion of mtDNA with 4,977-bp deletion than the tumor counterparts also was reported by Tan et al.[14] in a study of 18 oral cancer tissues.

Here, we report a new approach toward a more comprehensive quantitative analysis of mtDNA mutations in histologically well-defined cell populations utilizing a combination of laser microdissection (LMD) and real-time quantitative PCR (RTQPCR) analysis. The proportion of mtDNA with 4,977-bp deletion was analyzed by RTQPCR in microdissected peripheral lymph node, precancer, cancer cells, and surrounding stromal cells that were isolated from 12 patients with a history of sequentially developed oral lesions.

MATERIALS AND METHODS

Patient Population and Collection of Tissue Specimens

All paraffin sections of the specimens were collected from the National Areca Quid Related Oral Lesion Bank with consent from the donors and approval by the committee for the use of human tissues in research. A total of 12 oral cancer patients admitted to National Cheng-Kung University Medical Center between 1986 and 1995 were included in this study. All patients were male and had an areca quid chewing habit and previous biopsies had shown proven diagnosis of precancerous lesion. The patients had a mean age of 54.2 years, ranging from 40 to 71 years. Both precancer and cancer paraffin-embedded tissue blocks were obtained for the following analysis.

Laser Microdissection and DNA Extraction

For laser microdissection, the paraffin-embedded tissue samples were cut in 5-μm thick sections on a microtome with a disposable blade and placed on specifically designed LMS slide with PEN foil (2.5 μm thick; Leica Microsystems, Wetzlar, Germany). The slides were deparaffinized in two changes of xylene for 10 min, rehydrated through gradients of ethanol-water mixture, and finally immersed in doubly distilled H_2O. The slides were briefly stained with hematoxylin and eosin for 45 s, rinsed in distilled H_2O for 30 s. The microdissection was performed with the LMD system using a 337-nm nitrogen ultraviolet (UV) laser (Leica Laser Microdissection System; Leica Microsystems). The target cells were dissected from a section and dropped immediately into a microcentrifuge tube cap filled with 30 μL DNA extraction buffer (DNeasy Mini Kit; Qiagen, Valencia, CA). Twenty thousand cells were collected into a 0.5-mL tube, and total genomic DNA was extracted according to the manufacturer's instructions and the final DNA pellet was dissolved in 30 μL of distilled water before PCR analysis. A total of five different tissues were collected in each case. They included lymph node, the precancer cells, the precancer stroma, the cancer cells, and the cancer stroma.

PCR Amplification of Mitochondrial DNA

General PCR analysis was performed in a Perkin-Elmer/Cetus DNA thermal cycler 9700 (Oak Brook, IL). Two mtDNA fragments were amplified by PCR for 45 cycles in a 50-μL reaction mixture with 2 μL of extracted DNA. PCR amplification was conducted under the following conditions: denaturation at 94°C for 5 min; and then 45 cycles of denaturation at 94°C for 45 s; hybridization at 50°C for 30 sec; extension at 72°C for 30 sec; and final extension at 72°C for 5 min. PCR products were separated by electrophoresis in a 1.5% agarose gel at 100 volt for 40 min and were detected by a Canon digital gel documentation system under UV transillumination after ethidium bromide staining. The primer pair P3.1 and P4.1 that we used to amplify a 206-bp DNA of the target mtDNA sequence flanking the 4,977-bp deletion between nucleotide 8342 and 13524 are listed as the following: P3.1: 5'-AACCAA-CACCTCTTTACAGTGAA-3' (8342–8364) and P4.1: 5'-GATGATGTG-GTCTTTGGAGTAGAA-3' (13501–13524).

The primer pair HHC and LHC used for the amplification of a 213-bp reference DNA representing total mtDNA at nucleotides 5527–5739 are as follows: LHC: 5′-ATACAGACCAAGAGCCT-3′ (5527–5543) and HHC: 5′-GCGGGAGAAGTA-GATTGA-3′ (5722–5739).

Real-Time PCR Analysis of 4,977-bp Deleted mtDNA and Total mtDNA

Real-time PCR analysis was performed using the Roche Lightcycler detection system and analysis software (Roche Molecular Biochemicals, Germany). Amplification was conducted in a 20 µL final volume containing 4 mM $MgCl_2$, 0.1 M of each primer, and 2 µL of the Lightcycler-Faststart DNA Master SYBR Green I Mix. The PCR program included initial denaturation at 95°C for 8 min followed by 50 amplification cycles consisting of heating at 20°C/s with a 1-s hold, annealing at 20°C/s to 57°C with a 5-s hold, and extension at 20°C/s to 72°C with a 10-s hold. The fluorescent signal of the amplified product was detected at the last step of each cycle by single acquisition. After amplification, a melting curve was made by heating the product at 20°C/s to 95°C, cooling at 20°C/s to 70°C, and slowly heating the product at 0.1°C/s to 95°C with continuous fluorescence collection. Melting curves were used to determine the specificity of the PCR products. The STD 10 dilution was used as the calibrator to correct the variation of PCR efficiency between different batches. Variations in sample loading or in PCR tube-to-tube efficiency were corrected by performing replicate analysis for each unknown sample.

Standard Curve Design for Quantification of DNA Concentration

The total DNA of each sample with known concentration quantified by fluorescent Hoechst dye H33258 with DyNA Quant 200 (Amersham Biosciences, Uppsala, Sweden) was serially diluted into eight detection concentrations from 50 ng/µL to 5 fg/µL. The standard log-linear regression lines of the signal intensities of the PCR products amplified by use of the two primer sets, HHC-LHC and P3.1-P4.1, were generated by the Lightcycler software. The second derivation of the standard curve was used to calculate the quantity of the target DNA templates in the samples. Every data point from a serially diluted sample in the standard curve was performed in triplicate.

Statistical Analysis

The mean value of the replicate measurements for each sample was used to determine the relative amount of the total mtDNA (HHC-LHC) and the 4,977-bp deleted mtDNA (P3.1–P4.1). The proportion of 4,977-bp deleted mtDNA in the sample was calculated by:

$$\frac{\text{Target (deleted mtDNA)}}{\text{Reference (Total mtDNA)}} = \text{The proportion of 4,977 bp-deleted mtDNA}.$$

The results were analyzed by paired Student's t-test for the significance between alterations in the proportion of 4,977-bp deleted mtDNA in the microdissected cell population and their clinical status of disease progression.

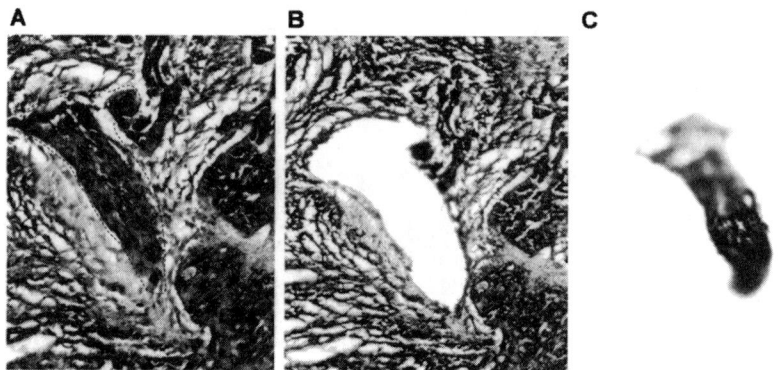

FIGURE 1. Retrieval of histologically defined pure cell population by the Laser microdissection system. (**A**) The tissue section was stained with H&E and examined on the control screen to identify and mark the target cell populations by the software. (**B**) The selected tissue was precisely cut down by an UV laser beam and collected into a designed collecting tube. (**C**) The collected samples were reconfirmed under the "cap-view" mode in the software. The digital images before and after the laser cutting, the collected samples were stored in CDR for future reference, and the cell number isolated then was calculated.

RESULTS

Laser Microdissection and DNA Extraction from Paraffin-Embedded Tissues

To establish a method for quantitative analysis of mtDNA copy number in a histologically defined cell population, we first examined the microdissected cells and the extracted DNA. As shown in FIGURE 1, digital images were taken before and after laser dissection by a pathologist and the collected samples were examined by a collection tube view mode in the LMS software. These procedures ensured that quantitative analysis was performed in histologically defined groups of cells. A total of 20,000 cells were collected for each type of the tissues for the subsequent analysis.

PCR Amplification and Detection of mtDNA Deletion

To investigate the frequency of occurrence of the 4,977-bp mtDNA deletion in paired oral tissues during malignant progression, we amplified DNA extracted from the microdissected tissues using PCR method with the primer pairs HHC/LHC and P3.1/P4.1. The HHC/LHC primer pair amplified a fragment of 213 bp at mtDNA nucleotides 5527–5739. The primer set P3.1/P4.1 was designed to flank the breakpoints of the 4,977-bp deletion. The deletion of nucleotides 8342–13527 of mtDNA enabled the primer pair to amplify a 206-bp product within the given extension time under the designed PCR condition, which discriminate the wild-type sequence. FIGURE 2 shows the PCR products amplified by use of the primer pairs HHC/LHC and P3.1/P4.1 separated on a 2% agarose gel from two representative cases. Both PCR amplifications for HHC/LHC and P3.1/P4.1 specifically amplified single bands of the target DNA sequence of 213 and 206 bp, respectively, in all detectable reactions. HHC/LHC that represented total mtDNA, derived abundant PCR products in all

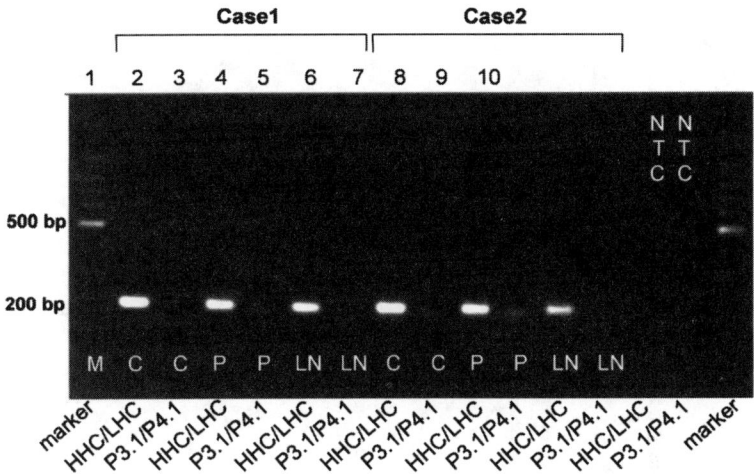

FIGURE 2. The purified microdissected DNA samples from two representative cases were amplified by PCR and separated on a 2% agarose gel. The primer set HHC/LHC amplified a 213-bp fragment representing total mtDNA showed a strong single DNA band. The primer set P3.1/P4.1 amplified a single product of 206-bp DNA fragment for the 4,977-bp deleted mtDNA. The signal intensities of PCR products amplified by P3.1/4.1 were extremely low in most cases and even undetectable in cases such as lanes 2 and *13* on this gel. C, cancer cell; P, precancer cell; LN, lymphocytes in lymph nodes; NTC, no template control.

samples. The DNA band amplified from 4,977-bp deleted mtDNA by P3.1/P4.1, on the other hand, was much weaker and even undetectable in some cases. Only 3 of the 12 patients (cases 1, 5, and 10) had detectable products with extremely weak signal. Because of the saturation and the relatively low copy number for the deleted mtDNA, quantitative determination of the 4,977-bp deleted mtDNA became extremely difficult using the traditional semiquantitative photometric analysis of electrophoretic gel upon UV illumination. Therefore, we performed the real-time quantitative PCR using SYBR Green I detection system in a Lightcycler.

REAL-TIME PCR ANALYSIS OF 4,977-bp DELETED mtDNA AND TOTAL mtDNA

To examine the specificity and the sensitivity of the primer and the designed PCR condition, we performed the primer-shift assay and melting curve analysis. Primer-shift PCR demonstrated the specificity of the primer sets to amplify the target sequence (data not shown). As shown in FIGURE 3A, melting curve analysis performed in all RT-QPCR presented single PCR product for both primer pairs HHC/LHC and P3.1/P4.1. The standard curves for the two primers are presented in FIGURE 3B. Both primer sets were found to show good linearity in the target detection range. The coefficients of variation of replicates in each sample and the standard curve were lower

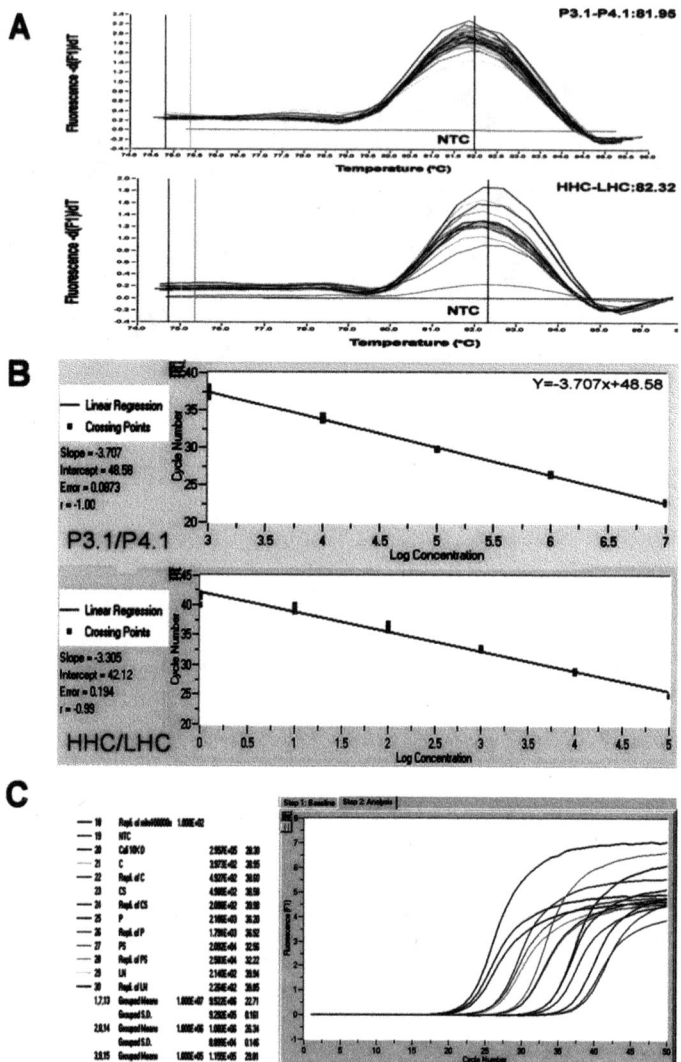

FIGURE 3. Analysis of the proportion of 4,977-bp deleted mtDNA in the laser microdissected formalin-fixed paraffin-embedded oral lesions using real-time quantitative PCR. (**A**) The melting curve analysis of the amplified products showed that only single specific product was generated in the PCR amplification process. (**B**) The logarithm of the input DNA amount ranging from 50 ng/μL to 5 fg/μL was plotted versus the threshold cycle monitored during real-time quantitative PCR analysis by Sybr Green I detection. Both primer sets presented good linearity in the target detection range. (**C**) A representative real-time quantitative PCR curve generated for the 4,977-bp deleted mtDNA in a case with 450 cells each from purified fraction of cancer,[21,22] cancer stroma,[23,24] precancer,[25,26] precancer stroma,[27,28] and the lymph node tissue.[29,30] The original DNA concentration in each sample was calculated accordingly.

TABLE 1. Ratio of 4977-bp deleted mtDNA in microdissected paired human oral tissues and lymph nodes examined by real-time quantitative PCR analysis

Case no.	Age / Sex	Ratio of 4977bp-deleted mtDNA / total mtDNA (fold-increase relative to lymph node)				
		Cancer	Cancer stroma	Precancer	Precancer stroma	Lymph node
1	53 / M	0.109% (90.8)	0.215% (179.2)	0.671% (559.2)	1.4201% (1183.3)	0.001200%
2	57 / M	0.017% (43.3)	0.090% (230.3)	0.007% (15.9)	0.0781% (201.0)	0.000390%
3	71/ M	0.023% (849.1)	0.042% (1600.0)	0.047% (1773.6)	0.0911% (3441.5)	0.000027%
4	41 / M	0.043% (50.2)	0.004% (4.2)	0.108% (127.1)	0.2291% (269.4)	0.000850%
5	58 /M	0.986% (172.5)	0.826% (144.6)	0.011% (1.9)	1.2471% (218.4)	0.005710%
6	48 / M	0.078% (186.8)	0.091% (220.1)	0.244% (587.0)	0.1491% (359.3)	0.000415%
7	61 / M	0.033% (50.9)	0.026% (40.9)	0.176% (272.2)	0.071% (109.1)	0.000647%
8	50 / M	0.0315% (160.8)	0.063% (319.2)	0.245% (1249.2)	0.099% (503.3)	0.000196%
9	54 / M	0.208% (375.3)	0.135% (244.2)	0.246% (444.7)	0.437% (788.8)	0.000554%
10	64 / M	0.055% (20.5)	1.147% (424.9)	2.605% (965.0)	6.492% (2404.3)	0.002700%
11	40 / M	0.056% (249.6)	0.028% (126.2)	0.022% (96.8)	0.012% (52.9)	0.000226%
12	53 / M	0.005% (129.4)	0.050% (1270.9)	0.018% (461.5)	0.020% (515.5)	0.000040%
	mean	0.137% (198.3)	0.226% (400.4)	0.367% (546.2)	0.862% (837.2)	0.001080%

than 3% in all experiments, which indicates high reproducibility of the analysis. The mean value of the calculated initial DNA concentration for each test sample was used to determine the proportion of 4,977-bp deleted mtDNA/total mtDNA. FIGURE 3C is a representative real-time quantitative PCR curve in the quantitative analysis of 4,977-bp deleted mtDNA. The detection was performed in a case with 450 cells each from purified fraction of cancer,[21,22] cancer stroma,[23,24] precancer,[25,26] precancer stroma,[27,28] and lymph node tissues.[29,30] The initial target DNA concentration in each sample was calculated accordingly.

TABLE 1 summarizes the real-time quantitative analysis of the 12 oral cancer patients with sequentially developed precancer lesion and cancer. All patients were male and had areca quid chewing habits. The mean age of the patients was 54.2 years with a range of 40 to 71 years. The proportion of 4,977-bp deleted mtDNA presented a broad spectrum of range from 0.00004% in the lymph node of a 53-year-old male to 6.492% in the precancer stroma of a 64-year-old male. Seven cases (7/12) presented higher proportion of 4,977-bp deleted mtDNA in cancer stroma than in cancer cells. The mean proportion of mtDNA with 4,977-bp deletion in cancer stroma was 0.226% and in the cancer cells it was 0.137%. The precancer stroma showed significantly higher deleted mtDNA (0.862%) than the precancer (0.367%). Eight patients (8/12) had a higher level of 4,977-bp deleted mtDNA in precancer stroma than in precancer cells. Interestingly, eight patients (8/12) had a higher proportion of deleted mtDNA in precancer cells than in cancer cells. The mean proportion of mtDNA with 4,977-bp deletion in precancer cells and cancer cells were 0.367% and 0.137%, re-

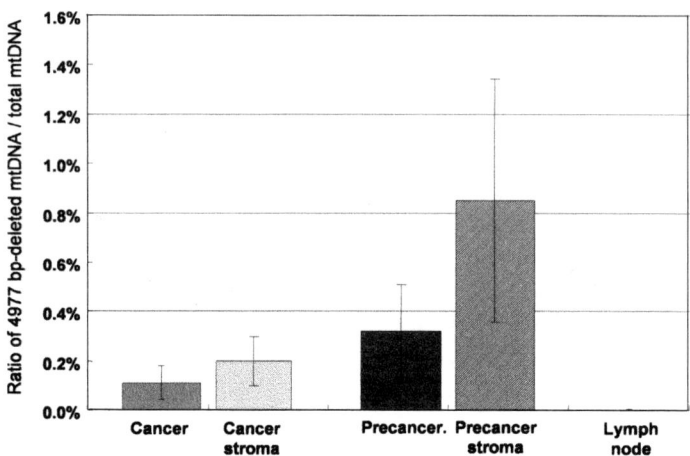

FIGURE 4. Accumulation of 4,977-bp deleted mtDNA in the paired laser microdissected tissues during oral carcinogenesis progression. Indicated tissue cells of the same individual were analyzed using real-time quantitative PCR. As expected, the proportion of the 4,977-bp deleted mtDNA in the lymph node was significantly lower than the DNA isolated from all other types of dissected samples ($P < 0.05$). The mean proportion of the 4,977-bp deleted mtDNA was the highest in the precancer stroma, followed by precancer, cancer stroma, and cancer.

spectively. Moreover, we observed a higher proportion of 4,977-bp deleted mtDNA in precancer stroma than in cancer stroma in nine patients (9/12). The mean proportion of deleted mtDNA was 0.862% for precancer stroma and 0.226% for cancer stroma (FIG. 4).

The proportion of 4,977-bp deleted mtDNA in lymph node tissue may be more susceptible to genetic or systemic factors rather to local environmental exposures. Thus, the measured values were recognized as the baseline level of 4,977-bp deleted mtDNA for each individual case. The proportion of mtDNA with 4,977-bp deletion in each of the other tissues was divided by this baseline level. This presents the fold of increase in the mtDNA deletion that was mostly contributed by regional environmental insults, that is, areca quid chewing. As shown in FIGURE 5, precancer cells presented significantly higher proportion of 4,977-bp deleted mtDNA than did cancer cells ($P < 0.05$). The proportion of 4,977-bp deleted mtDNA in precancer stroma was higher than in cancer stroma ($P < 0.05$), which was higher than that in cancer cells ($P < 0.05$).

DISCUSSION

Areca quid chewing has been implicated as the major risk factor for the rapid increase of oral cancer incidence in Taiwan.[15] Several studies have suggested that ROS and free radicals generated from areca quid chewing are one of the contributory factors to oral carcinogenesis.[2,16] However, most previous studies have focused on ge-

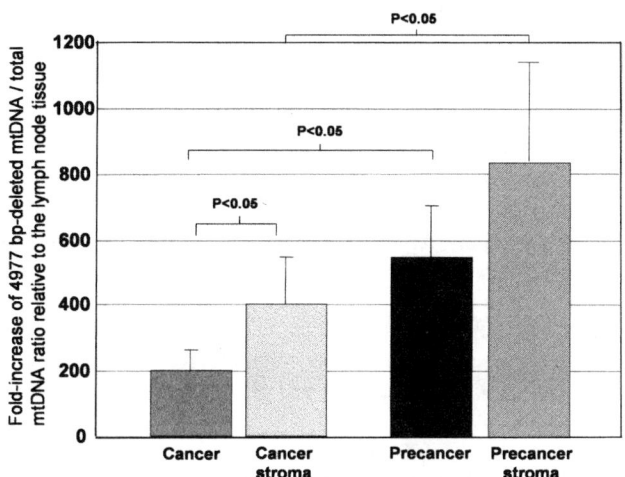

FIGURE 5. Proportions of mtDNA with 4,977-bp deletion in different tissues of patients with oral cancer. The proportion of 4,977-bp deleted mtDNA of each of the collected tissue types was divided by that of each individual's lymph node to obtain the relative fold-increase in the mtDNA deletion associated with regional environmental exposure. The precancer stroma presented significantly increased proportion of 4,977-bp deleted mtDNA than did the cancer stroma ($P < 0.05$). The precancer tissue and the cancer stroma both showed an increased proportion of 4,977-bp deleted mtDNA than did the cancer cells ($P < 0.05$).

netic alterations of nuclear DNA during oral carcinogenesis and their clinical significance. Recent studies have demonstrated that somatic mtDNA mutations may play a role in carcinogenesis including oral cancer.[9,14,17,18] The observations of a wide spectrum of mtDNA alterations in human carcinogenesis suggest that accumulation of mtDNA mutations not only is a contributor to carcinogenesis but also could provide a sensitive molecular marker for disease staging and profiling of human cancers.

It has been shown that mtDNA is 10–20 times more sensitive to ROS-induced DNA damage because of its relatively unprotected genome structure and less efficient DNA repair mechanism.[19] Accumulation of mtDNA deletions has been shown to be an important contributing factor to human aging and degenerative diseases.[19–21] More than 50 different types of mtDNA deletions have been identified in human diseases. Some of which were found only in certain type of tissue, whereas others may occur in many organs or tissues.[18,22,23] The 4,977-bp deletion is the most prevalent and abundant type of mtDNA deletion in humans.[19] Using PCR techniques, Lee and colleagues[13] first analyzed the occurrence and accumulation of 4,977-bp deleted mtDNA in 53 pairs of freshly frozen oral cancer and noncancer oral tissues from patients with or without areca quid chewing history. They found that the proportion of 4,977-bp deleted mtDNA was significantly lower in non-tumor tissue than in cancer tissues irrespective of areca quid chewing history. The proportion of mtDNA with the large-scale deletion was significantly higher in non-tumor tissue of the chewers than nonchewers.[13] More recently, similar results were also reported by Tan et al.[14]

in an analysis of 17 primary oral squamous cell carcinoma tissues by the same detection strategy. However, by semiquantitative analysis of the eletrophoretic gel for the PCR amplified products, they were able to detect 4,977-bp deleted mtDNA only in limited number of cases.[14]

We are the first to our knowledge to introduce a strategy that combines the power of laser microdissection technique and real-time quantitative PCR in the study of mtDNA deletions in oral carcinogenesis. We have been able to demonstrate reproducible, precise quantitative measurements of 4,977-bp deleted mtDNA from histologically defined cell populations of as little as 450 cells derived from formalin-fixed paraffin-embedded tissues. The laser microdissection technique has enabled us to study the response of human cells to different environmental insults.

The importance of acquiring pure cell populations from biopsied tissue in molecular profiling and fine dissection of disease mechanism has been repeatedly emphasized in recent years. There are different types of cells present in a neoplastic lesion, including supporting stromal cells, blood and lymph vessels, infiltrating immune cells, and many others. The proportion of these "contaminating cells" may vary greatly between various lesions and at different time points during disease progression. Molecular profiling using a homogenized total tissue may lead to confusing results. In this study, we differentially collected microdissected cancer cells, precancer cells, adjacent stromal cells, and lymphocytes from peripheral lymph node from each individual patient as the starting material. This allowed a comprehensive study of the longitudinal changes in mtDNA alterations during disease progression. The longitudinal collection of both epithelial and stromal cells enabled us to compare cells with the same genetic and exposure background for each individual. In addition, we have analyzed the baseline level of 4,977-bp deleted mtDNA in each patient's peripheral lymphocytes, which allowed us to translate the proportion of mtDNA with 4,977-bp deletion of tissues into relative fold increase above a baseline that reflects local exposure to environmental ROS. This should normalize the influence of age, genetics, and other personal factors when performing cross-sectional comparative study.

As shown in FIGURE 3B, the real-time quantitative PCR analysis in this study provided a broad spectrum of range for linear quantification of 4,977-bp deleted mtDNA. Compared with the limited detection rate in previous studies, all 60 paired samples from the 12 patients in the current study could be quantitatively analyzed for their initial copy number of total mtDNA and 4,977-bp deleted mtDNA with high reproducibility (coefficient of variation <3%). The proportion of 4,977-bp deleted mtDNA detected covers a wide range, as low as 0.000040% in the lymph node of case 12 and up to 6.493% in the precancer stroma of case 10.

Previous studies on aging-associated mtDNA mutations have suggested that oxidative stress could shift the rate of accumulation of mtDNA mutations from linear to exponential.[24] The observation that all 48 paired oral tissues had proportions of mtDNA with 4,977-bp deletion higher than lymph node tissue in the areca quid chewing patient provides additional evidence that local exposure of ROS may contribute to the accumulation of mtDNA mutations. Although a recent cell fusion study in a colon cancer cell line model demonstrated that a strong selective mechanism existed to maintain certain genotype of altered mtDNA in subsequent cell culture, the role of specific mtDNA alterations in the carcinogenesis process still remained unclear.[25] The 4,977-bp mtDNA fragment that is deleted normally encodes 7 polypeptide components of the mitochondrial respiratory chain and 5 of the 22 tRNAs necessary for mitochondrial protein synthesis. It has been shown that cells containing a high proportion of the 4,977-bp deleted mtDNA were impaired in mitochondrial membrane potential, rate of ATP synthesis, and cellular ATP/ADP ratio, which may further undermine the tumor growth advantage. Our results appear to be consistent with previous studies in which accumulation of the 4,977-bp deleted mtDNA was observed in malignant cells from oral lesions and hepatocellular carcinoma.[13,14,26] All mtDNA deletions in the oral lesions were heteroplasmic and less abundant than in the surrounding stromal cells, as previously reported by Lee et al.[13] and Tan et al.[14] Lee et al.[13] postulated that the dilution of mtDNA molecules with the deletion in the rapidly dividing malignant cells and clonally selective mechanism may explain their observation that tumor cells harbored higher level of mtDNA deletion than did the slow-growing non-tumor stromal cells. In addition to their observations, we also found that precancer lesions harbored a higher proportion of mtDNA with 4,977-bp deletion before malignant progression. Moreover, the precancer stroma exhibited a higher proportion of 4,977-bp deleted mtDNA than did the cancer stroma. Analysis of the proliferation index by PCNA expression in the cancer and precancer lesions further supported the hypothesis. The precancer or cancer cells with higher proliferation index also presented lower level of mtDNA with 4,977-bp deletion (data not shown). However, the abundance of 4,977-bp deleted mtDNA varied greatly between individual cases up to 197-fold in cancer cells and 372-fold in precancer cells. This indicates that other genetic factors also may play some role in the susceptibility of mtDNA to deletion under exposure to environmental ROS and free radicals. In light of the finding that cancer stroma had significantly lower proportion of 4,977-bp deleted mtDNA than did precancer stroma, we propose that cancer cells may stimulate the proliferation of surrounding connective tissue cells and induce neovascularization during the disease progression, which, in turn, may accelerate the turnover of local cell population and further dilute the mtDNA copies with the deletion. The rapidly dividing cells in the reactive stroma may be induced to apoptosis with a high proportion of mtDNA with the deletion caused by elevated energy needs. Besides, pathological examinations of the cancer lesions often observed significant infiltrating cells from peripheral blood or lymphatic origin that may not yet have been exposed to local ROS damage.

Recent studies showed that the deleted DNA fragment may escape from the mitochondria and translocate into the nucleus because insertion of mtDNA sequences into the nuclear DNA has been observed.[27] However, whether the fragments of migrating mtDNA may alter the genetic information or change the expression level of certain nuclear genes and thereby promote the cancer progression remains unclear. Moreover, our recent study also indicated an elevated total mtDNA and mitochondrial mass in cancer cells harboring the deletion. Whether an association exists between the abundance of specific mtDNA deletion and the cellular signaling for mtDNA proliferation remains to be investigated.

In summary, we have demonstrated for the first time to our knowledge tissue type–specific profiling of mtDNA deletion in serially developed oral precancer and cancer lesions by using a combined LMD of formalin-fixed paraffin-embedded tissue sections and real-time PCR analysis. Our results indicate an enhanced accumulation of mtDNA deletion in the areca quid–associated oral precancer and cancer lesions with a consistent trend of decrease in the proportion of 4,977-bp deleted mtDNA from precancer stroma, to precancer, to cancer stroma, and to cancer. The molecular mechanism underlying the observed phenomena remains to be explored. The LMD techniques developed in this study have provided a good quality of mtDNA mutation profiling without the contamination of other surrounding unrelated cell types and will be an important tool in our future study for additional mtDNA mutations or deletions that might be involved in oral carcinogenesis. The RTQPCR system not only allows more accurate analysis of mtDNA mutation or deletion but also offers a rapid and high-throughput profiling of the mitochondrial genomic alterations that may occur during the initiation and progression of carcinogenesis.

ACKNOWLEDGMENT

This study was supported by National Science Council grants NSC 91-2320-B-006-057 and NSC 91-3112-P-006-006-Y.

REFERENCES

1. SHIU, M.N., T.H. CHAN, S.H. CHANG, et al. 2000. Risk factors for leukoplakia and malignant transformation to oral carcinoma: a leukoplakia cohort in Taiwan. Br. J. Cancer **82:** 1871–1874.
2. NAIR, U.J., G. OBE, M. FRIESSEN, et al. 1992. Role of lime in the generation of reactive oxygen species from betel-quid ingredients. Environ. Health Perspect. **98:** 203–205.
3. CHEN, C.L., C.W. CHI & T.Y. LIU. 2002. Hydroxyl radical formation and oxidative DNA damage induced by areca quid in vivo. J. Toxicol. Environ. Health A **65:** 327–336.
4. MANDAVILLI, B.S., J.H. SANTOS & B. VAN HOUTEN. 2002. Mitochondrial DNA repair and aging. Mutat. Res. **509:** 127–151.
5. WEI, Y.H. 1998. Oxidative stress and mitochondrial DNA mutations in human aging. Proc. Soc. Exp. Biol. Med. **217:** 53–63.
6. BIANCHI, N.O., M.S. BIANCHI & S.M. RICHARD. 2001. Mitochondrial genome instability in human cancers. Mutat. Res. **488:** 9–23.
7. MATSUYAMA, W., M. NAKAGAWA, J. WKIMOTO, et al. 2003. Mitochondrial DNA mutation correlates with stage progression and prognosis in non-small cell lung cancer. Hum. Mutat. **21:** 441–443.
8. TONG, B.C., P.K. HA, K. DHIR, et al. 2003. Mitochondrial DNA alterations in thyroid cancer. J. Surg. Oncol. **82:** 170–173.
9. DURHAM, S.E., K.J. KRISHNAN, J. BETTS, et al. 2003. Mitochondrial DNA damage in non-melanoma skin cancer. Br. J. Cancer **88:** 90–95.
10. NAGY, A., M. WILHELM, F. SUKOSD, et al. 2002. Somatic mitochondrial DNA mutations in human chromophobe renal cell carcinomas. Genes Chromosomes Cancer **35:** 256–260.
11. TAMURA, G., S. NISHIZUKA, C. MAESAWA, et al. 1999. Mutations in mitochondrial control region DNA in gastric tumours of Japanese patients. Eur. J. Cancer **35:** 316–319.
12. TAN, D.J., R.K. BAI & L.J. WONG. 2002. Comprehensive scanning of somatic mitochondrial DNA mutations in breast cancer. Cancer Res. **62:** 972–976.
13. LEE, H.C., P.H. YIN, T.N. YU, et al. 2001. Accumulation of mitochondrial DNA deletions in human oral tissues—effects of betel quid chewing and oral cancer. Mutat. Res. **493:** 67–74.
14. TAN, D.J., J. CHANG, W.L. CHEN, et al. 2003. Novel heteroplasmic frameshift and missense somatic mitochondrial DNA mutations in oral cancer of betel quid chewers. Genes Chromosomes Cancer **37:** 186–194.
15. CHEN, Y.K., H.C. HUANG, C.C. LIN, et al. 1999. Primary oral squamous cell carcinoma: an analysis of 703 cases in southern Taiwan. Oral Oncol. **35:** 173–179.

16. STICH, H.F. & F. ANDERS. 1989. The involvement of reactive oxygen species in oral cancers of betel quid/tobacco chewers. Mutat. Res. **214:** 47–61.
17. BHAT, H.K. 2002. Depletion of mitochondrial DNA and enzyme in estrogen-induced hamster kidney tumors: a rodent model of hormonal carcinogenesis. J. Biochem. Mol. Toxicol. **16:** 1–9
18. ROGOUNOVITCH, T.I., VA. SAENKO, Y. SHIMIZU-YOSHIDA, *et al.* 2002. Large deletions in mitochondrial DNA in radiation-associated human thyroid tumors. Cancer Res. **62:** 7031–7041.
19. WEI, Y.H. 1992. Mitochondrial DNA alterations as ageing-associated molecular events. Mutat. Res. **275:** 145–155.
20. BANDY, B. & A.J. DAVISON. 1990. Mitochondrial mutations may increase oxidative stress: implications for carcinogenesis and aging? Free Radical Biol. Med. **8:** 523–539.
21. BOHR, V. & G.L. DIANOV. 1998. Oxidative DNA damage processing and changes with aging. Toxicol. Lett. **102–103:** 47–52.
22. KAWASHIMA, S., S. OHTA, Y. KAGAWA, *et al.* 1994. Widespread tissue distribution of multiple mitochondrial DNA deletions in familial mitochondrial myopathy. Muscle Nerve **17:** 741–746.
23. LEE, H.C., C.Y. PANG, H.S. HSU, *et al.* 1994. Differential accumulations of 4,977 bp deletion in mitochondrial DNA of various tissues in human ageing. Biochim. Biophys. Acta **1226:** 37–43.
24. TURKER, M.S. 2000. Somatic cell mutations: can they provide a link between aging and cancer? Mech. Ageing Dev. **117:** 1–19.
25. POLYAK, K., Y. LI, H. ZHU, *et al.* 1998. Somatic mutations of the mitochondrial genome in human colorectal tumours. Nat. Genet. **20:** 291–293.
26. FUKUSHIMA, S., K. HONDA, M. AWANE, *et al.* 1995. The frequency of 4977 base pair deletion of mitochondrial DNA in various types of liver disease and in normal liver. Hepatology **21:** 1547–1551.
27. TURNER, C., C. KILLORAN, N.S. THOMAS, *et al.* 2003. Human genetic disease caused by de novo mitochondrial-nuclear DNA transfer. Hum. Genet. **112:** 303–309.

Role of Oxidative Stress in Pancreatic β-Cell Dysfunction

YOSHITAKA KAJIMOTO AND HIDEAKI KANETO

Department of Internal Medicine and Therapeutics, Osaka University Graduate School of Medicine, Suita, Osaka 565-0871, Japan

ABSTRACT: Oxidative stress is produced under diabetic conditions and is likely involved in progression of pancreatic β-cell dysfunction found in diabetes. Possibly caused by low levels of antioxidant enzyme expressions, pancreatic β-cells are vulnerable to oxidative stress. When β-cell–derived HIT-T15 cells or isolated rat islets were exposed to oxidative stress, insulin gene expression was markedly decreased. To investigate the significance of oxidative stress in the progression of pancreatic β-cell dysfunction in type 2 diabetes, we evaluated the effects of antioxidants in diabetic C57BL/KsJ-db/db mice. According to an intraperitoneal glucose tolerance test, the treatment with antioxidants retained glucose-stimulated insulin secretion and moderately decreased blood glucose levels. Histological analyses of the pancreata revealed that the β-cell mass was significantly larger in the mice treated with the antioxidants, and the antioxidant treatment suppressed apoptosis in β-cells without changing the rate of β-cell proliferation. The antioxidant treatment also preserved the amounts of insulin content and insulin mRNA, making the extent of insulin degranulation less evident. As possible mechanism underlying the phenomena, expression of pancreatic and duodenal homeobox factor-1 (also known as IDX-1/STF-1/IPF1), an important transcription factor for the insulin gene, was more clearly visible in the nuclei of islet cells after the antioxidant treatment. Under diabetic conditions, JNK is activated by oxidative stress and involved in the suppression of insulin gene expression. This JNK effect appears to be mediated in part by nucleocytoplasmic translocation of PDX-1, which is also downstream of JNK activation. Taken together, oxidative stress and consequent activation of the JNK pathway are involved in progression of β-cell dysfunction found in diabetes. Antioxidants may serve as a novel mechanism-based therapy for type 2 diabetes.

KEYWORDS: glucose toxicity; oxidative stress; insulin gene; PDX-1; JNK; antioxidant

PANCREATIC β-CELL GLUCOSE TOXICITY

The development of type 2 diabetes is usually associated with a combination of pancreatic β-cell dysfunction and insulin resistance. Normal β-cells can compensate for insulin resistance by increasing insulin secretion or β-cell mass, but insufficient compensation leads to the onset of glucose intolerance. Chronic hyperglycemia not only is a marker of poor glycemic control in diabetes but is itself a cause of impairment of insulin biosynthesis and secretion; once hyperglycemia becomes apparent, β-cell function deteriorates,[1] with some of this dysfunction being caused by the adverse effects of chronic hyperglycemia. This process is called β-cell glucose toxicity, which has been demonstrated in various *in vivo*,[2–4] and *in vitro* studies,[5,6] with a key characteristic being the suppression of insulin gene transcription and secretion. Histologically, such a damaged β-cell often reveals extensive degranulation and is clinically associated with development of diabetes in some model animals for type 2 diabetes.

OXIDATIVE STRESS CAUSED BY HYPERGLYCEMIA IN PANCREATIC β-CELLS

Under diabetic conditions, reactive oxygen species (ROS)[7] are produced in various tissues such as nerve cells and vascular cells and are involved in the development of diabetic complications.[8–12] Recently, pancreatic β-cells emerged as a target of oxidative stress–mediated tissue damage.[13–28] Pancreatic β-cells express the high-Km glucose transporter GLUT2 abundantly and thereby display highly efficient glucose uptake when exposed to high glucose concentration. Also, because of the relatively low expression of antioxidant enzymes such as catalase and glutathione peroxidase,[15] pancreatic β-cells may be rather sensitive to ROS attack when they are exposed to oxidative stress. Thus, it is likely that oxidative stress plays a major role in β-cell deterioration in type 2 diabetes.

There are several sources of ROS productions in cells: the nonenzymatic glycosylation reaction,[13,14] the electron transport chain in mitochondria,[11,26] and the hex-

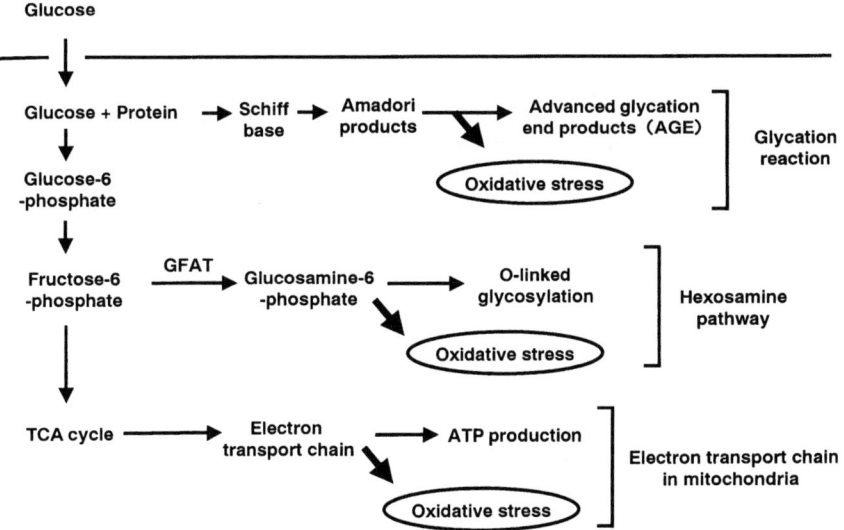

FIGURE 1. Oxidative stress caused by hyperglycemia.

osamine pathway.[23] Among those, the glycation reaction seems to have broad pathological significance in diabetic complications, because under hyperglycemia, the production of various reducing sugars, such as glucose, glucose-6-phosphate, and fructose, increases through glycolysis and the polyol pathway. All of these reducing sugars are known to promote glycation reactions of various proteins. In diabetic animals, glycation is observed extensively in various tissues and organs, and various kinds of glycated proteins such as glycosylated hemoglobin, albumin, and lens crystalline are produced in a nonenzymatical manner through the Maillard reaction. During the reaction which in turn produces Schiff base, Amadori product and advanced glycosylation end products (AGE), ROS are also produced.[13] Also, the electron transport chain in mitochondria is likely to be an important pathway to produce ROS. Indeed, it was suggested that mitochondrial overwork, which causes induction of ROS, is a potential mechanism causing impaired first-phase of glucose-stimulated insulin secretion found in the early stage of diabetes.[26] In addition, ROS are produced also through the hexosamine pathway and that activation of the pathway leads to deterioration of β-cell function by provoking oxidative stress[23] (FIG. 1).

SUPPRESSION OF INSULIN GENE EXPRESSION BY OXIDATIVE STRESS

To understand the molecular background for the β-cell glucose toxicity, we established an *in vitro* model system with HIT-T15 cells. Using D-ribose, which could induce glycation in HIT-T15 cells during a relatively short period (<72 h), we evaluated the potential effects of glycation and the consequent increase of ROS on β-cell function.[14] Among various reducing sugars that potentially induce glycation, D-ribose is outstanding for its very potent activity and thus is often used in *in vitro* studies as an inducer of glycation. In this *in vitro* model system, we showed that induction of glycation suppresses the insulin gene transcription in β-cells. This supports the idea that glycation and ROS-dependent suppression of the insulin gene promoter cause impairment of insulin biosynthesis in HIT-T15 cells. According to the results of reporter gene analyses, the insulin gene promoter was very sensitive to induction of glycation: when the HIT-T15 cells were exposed to D-ribose, insulin gene promoter activity was markedly suppressed, whereas no such changes were observed with the β-actin gene promoter. Because this promoter-suppressing effect was neutralized at least in part with aminoguanidine, an inhibitor of the glycation reaction, or *N*-acetyl-L-cysteine, an antioxidant, the ROS induction should mediate the glycation-dependent suppression of the insulin gene promoter. The glycation-dependent suppression of the insulin gene promoter also caused the reduction of its transcripts, and, in agreement with the observations on promoter activities, aminoguanidine or NAC partly neutralized the effect of D-ribose. In addition, after exposure to D-ribose, insulin content in the cells was also markedly decreased (FIG. 2).

PANCREATIC AND DUODENAL HOMEOBOX FACTOR-1

Pancreatic and duodenal homeobox factor-1 (PDX-1), also known as IDX-1/STF-1/IPF1,[29–31] is a member of the homeodomain-containing transcription factor family. PDX-1 is expressed in the pancreas and duodenum and plays a crucial role in pancreatic development,[32–35] β-cell differentiation,[36–41] and in maintaining normal β-cell function by regulating multiple important β-cell genes.[42–45] At an early stage of embryonic development, PDX-1 is expressed in the gut region when the foregut endoderm becomes committed to common pancreatic precursor cells. During the development of the pancreas, PDX-1 expression is maintained in multipotential precursors that coexpress several hormones and later it becomes restricted to β-cells. Mice homozygous for a targeted mutation in the PDX-1 gene are apancreatic and develop fatal perinatal hyperglycemia.[32] Heterozygous PDX-1–deficient mice, on the other hand, reveal impaired glucose tolerance,[46] also providing support for the crucial role of PDX-1 in pancreas development. Clinically, mutations in PDX-1 are known to cause some cases of maturity-onset diabetes of the young (MODY).[47,48]

REDUCTION OF PDX-1 DNA BINDING ACTIVITY BY OXIDATIVE STRESS

We found that as a possible cause of the reduction in the insulin gene promoter activity by oxidative stress the DNA binding activity of PDX-1 is rather sensitive to glycation and the resulting oxidative stress. When HIT-T15 cells were exposed to D-ribose, marked reduction of PDX-1 binding to the insulin gene was observed.[14] Coexistence of aminoguanidine or NAC in the media prevented this decrease, indicating that D-ribose suppressed the PDX-1 DNA binding activity in a manner dependent on glycation and ROS.

Note, however, that the decrease in the PDX-1 binding activity may not instantly explain the suppression of the insulin gene transcription. According to our previous study,[49] suppression of PDX-1 expression (by ~80% or more) in β-cell–derived MIN6 cells using an antisense oligodeoxynucleotide (ODN) did not lead to a decrease of any β-cell–specific genes investigated such as the insulin gene and glucokinase gene which is also known to be regulated by PDX-1.[45] Therefore, the PDX-1 amount per se may not function as a rate determinant of the insulin gene transcription in pancreatic β-cells at least under normal conditions. Instead, some posttranslational modification may play a role in suppression of its activity under diabetic conditions. We have recently found that oxidative stress alters intracellular distribution of PDX-1 and thereby suppresses its activity in nuclei.[50] Because PDX-1 is involved in the activation of its own gene, this nucleocytoplasmic translocation may cause a progressive and severe reduction of PDX-1 activity. Also, we need to investigate the possible implication of other transcription factors constituting the transcription-controlling machinery to fully understand the molecular basis of the impaired insulin biosynthesis.

PROTECTION OF PANCREATIC β-CELLS AGAINST GLUCOSE TOXICITY BY ANTIOXIDANTS

Because oxidative stress is produced under diabetic conditions and is possibly involved in pancreatic β-cell dysfunction found in diabetes, we evaluated the potential usefulness of antioxidants in treatment for type 2 diabetes.[17] We used diabetic

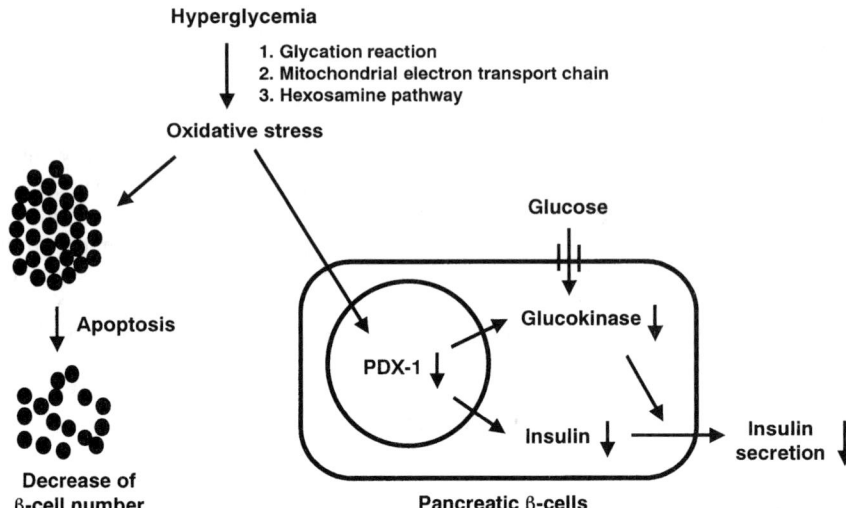

FIGURE 2. A proposed molecular mechanism for β-cell glucose toxicity in diabetes.

C57BL/KsJ-db/db mice, in whom antioxidant treatment was started at 6 weeks of age, and its effects were evaluated at 10 and 16 weeks of age. According to an intraperitoneal glucose tolerance test, the treatment with antioxidants retained glucose-stimulated insulin secretion and moderately decreased blood glucose levels. No such effect on insulin secretion was observed when the same set of antioxidants was given to nondiabetic control mice. Histological analyses of the pancreata revealed that the β-cell mass is significantly larger in the mice treated with the antioxidants. As a possible cause of this, the antioxidant treatment suppressed apoptosis in β-cells without changing the rate of β-cell proliferation, supporting the hypothesis that oxidative stress–induced apoptosis causes reduction of β-cell mass due to chronic hyperglycemia. The antioxidant treatment also preserved the amounts of insulin content and insulin mRNA, making the extent of insulin degranulation less evident. Furthermore, PDX-1 expression was more clearly visible in the nuclei of islet cells after the antioxidant treatment. Taken together, these data indicate that antioxidant treatment can exert beneficial effects for diabetes with preservation of *in vivo* β-cell function. Several other reports also have demonstrated that antioxidant treatment can protect β-cells against glucose toxicity and thus exert beneficial effects for type 2 diabetes.[18,22,25] These data suggest potential usefulness of antioxidants for diabetes and provides further support for the implication of oxidative stress in β-cell dysfunction in diabetes.

MOLECULAR MECHANISM FOR OXIDATIVE STRESS–MEDIATED SUPPRESSION OF INSULIN GENE EXPRESSION

Several signal transduction pathways including c-Jun N-terminal kinase (JNK) (also known as stress-activated protein kinase), p38 mitogen–activated protein kinase (p38 MAPK), and protein kinase C (PKC) are activated by oxidative stress in several cell types including pancreatic β-cells. It has been shown recently that activation of JNK is involved in reduction of insulin gene expression by oxidative stress and that suppression of the JNK pathway can protect β-cells from oxidative stress.[51] Subjecting rat islets to oxidative stress activates c-JNK, p38 MAPK, and PKC, preceding the decrease of insulin gene expression. Adenovirus-mediated overexpression of dominant-negative type (DN) JNK, but not the p38 MAPK inhibitor SB203580 nor the PKC inhibitor GF109203X, protected insulin gene expression and secretion from oxidative stress. Moreover, wild-type (WT) JNK overexpression suppressed both insulin gene expression and secretion. These results were correlated with changes in the binding of the important transcription factor PDX-1 to the insulin promoter; adenoviral overexpression of DN-JNK preserved PDX-1 DNA binding activity in the face of oxidative stress, whereas WT-JNK overexpression decreased PDX-1 DNA binding activity. It is known that β-cell destruction by cytokines such as interleukin-1β (IL-1β)[52–54] can be prevented by inhibition of the JNK pathway,[55–58] implying that JNK plays a role in autoimmune β-cell destruction found in early stage of type 1 diabetes. Also, it has been reported that levels of 8-hydroxy-2′-deoxyguanosine (8-OHdG), a marker for oxidative stress, are increased in the blood of type 2 diabetic patients as well as in islets of type 2 diabetic animal models[8,9,16] and that JNK activation by oxidative stress in islets actually reduces the PDX-1 DNA binding activity and insulin gene transcription.[51] Thus, we assume that JNK is involved in deterioration of β-cell function in both type 2 diabetes and the early stage of type 1 diabetes. Taken together, these results provide new insights into the mechanism through which oxidative stress suppresses insulin gene transcription in pancreatic β-cells, and the finding that this adverse outcome can be prevented by DN-JNK overexpression suggests that the JNK pathway in β-cells could become a new therapeutic target for diabetes.

CONCLUSIONS

Under diabetic conditions, reactive oxygen species are produced through several processes such as the nonenzymatic glycosylation reaction, the electron transport chain in mitochondria, and the hexosamine pathway. Oxidative stress and consequent activation of the JNK pathway are likely involved in progression of pancreatic β-cell dysfunction found in diabetes. Finally, antioxidant treatment can protect β-cells against glucose toxicity and thus exert beneficial effects for type 2 diabetes.

ACKNOWLEDGMENT

The data on the JNK implication described here were obtained in Dr. Gordon Weir's laboratory under his supervision.

REFERENCES

1. WEIR, G.C., D.R. LAYBUTT, H. KANETO, et al. 2001. β-Cell adaptation and decompensation during the progression of diabetes. Diabetes **50:** S154–S159.

2. TOKUYAMA, Y., J. STURIS, A.M. DEPAOLI, *et al.* 1995. Evolution of β-cell dysfunction in the male Zucker diabetic fatty rat. Diabetes **44:** 1447–1457.
3. ZANGEN, D.H., S. BONNER-WEIR, C.H. LEE, *et al.* 1997. Reduced insulin, GLUT2, and IDX-1 in β-cells after partial pancreatectomy. Diabetes **46:** 258–264.
4. JONAS, J.-C., A. SHARMA, W. HASENKAMP, *et al.* 1999. Chronic hyperglycemia triggers loss of pancreatic β-cell differentiation in an animal model of diabetes. J. Biol. Chem. **274:** 14112–14121.
5. ROBERTSON, R.P., H.-J. ZHANG, K.L. PYZDROWSKI & T.F. WALSETH. 1992. Preservation of insulin mRNA levels and insulin secretion in HIT cells by avoidance of chronic exposure to high glucose concentrations. J. Clin. Invest. **90:** 320–325.
6. SHARMA, A., L.K. OLSON, R.P. ROBERTSON & R. STEIN. 1995. The reduction of insulin gene transcription in HIT-T15β chronically exposed to high glucose concentration is associated with loss of RIPE3β1 and STF-1 transcription factor expression. Mol. Endocrinol. **9:** 1127–1134.
7. FINKEL, T. & N.J. HOLBROOK. 2000. Oxidants, oxidative stress and the biology of ageing. Nature **408:** 239–247.
8. DANDONA, P., K. THUSU, S. COOK, *et al.* 1996. Oxidative damage to DNA in diabetes mellitus. Lancet **347:** 444–445.
9. LEINONEN, J., T. LEHTIMAKI, S. TOYOKUNI, *et al.* 1997. New biomarker evidence of oxidative DNA damage in patients with non-insulin-dependent diabetes mellitus. FEBS Lett. **417:** 150–152.
10. BAYNES, J.W. & S.R. THORPE. 1999. Role of oxidative stress in diabetic complications: a new perspective on an old paradigm. Diabetes **48:** 1–9.
11. NISHIKAWA, T., D. EDELSTEIN, X.L. DU, *et al.* 2000. Normalizing mitochondrial superoxide production blocks three pathways of hyperglycaemic damage. Nature **404:** 787–790.
12. BROWNLEE, M. 2001. Biochemistry and molecular cell biology of diabetic complications. Nature **414:** 813–820.
13. KANETO, H., J. FUJII, T. MYINT, *et al.* 1996. Reducing sugars trigger oxidative modification and apoptosis in pancreatic β-cells by provoking oxidative stress through the glycation reaction. Biochem. J. **320:** 855–863.
14. MATSUOKA, T., Y. KAJIMOTO, H. WATADA, *et al.* 1997. Glycation-dependent, reactive oxygen species-mediated suppression of the insulin gene promoter activity in HIT cells. J. Clin. Invest. **99:** 144–150.
15. TIEDGE, M., S. LORTZ, J. DRINKGERN & S. LENZEN. 1997. Relation between antioxidant enzyme gene expression and antioxidative defense status of insulin-producing cells. Diabetes **46:** 1733–1742.
16. IHARA, Y., S. TOYOKUNI, K. UCHIDA, *et al.* 1999. Hyperglycemia causes oxidative stress in pancreatic β-cells of GK rats, a model of type 2 diabetes. Diabetes **48:** 927–932.
17. KANETO, H., Y. KAJIMOTO, J. MIYAGAWA, *et al.* 1999. Beneficial effects of antioxidants for diabetes: possible protection of pancreatic β-cells against glucose toxicity. Diabetes **48:** 2398–2406.
18. TANAKA, Y., C.E. GLEASON, P.O.T. TRAN, *et al.* 1999. Prevention of glucose toxicity in HIT-T15 cells and Zucker diabetic fatty rats by antioxidants. Proc. Natl. Acad. Sci. USA **96:** 10857–10862.
19. KANETO, H., Y. KAJIMOTO, Y. FUJITANI, *et al.* 1999. Oxidative stress induces p21 expression in pancreatic islet cells: possible implication in β-cell dysfunction. Diabetologia **42:** 1093–1097.
20. KAJIMOTO, Y., T. MATSUOKA, H. KANETO, *et al.* 1999. Induction of glycation suppresses glucokinase gene expression in HIT-T15 cells. Diabetologia **42:** 1417–1424.
21. MAECHLER, P., L. JORNOT & C.B. WOLLHEIM. 1999. Hydrogen peroxide alters mitochondrial activation and insulin secretion in pancreatic beta cells. J. Biol. Chem. **274:** 27905–27913.
22. IHARA, Y., Y. YAMADA, S. TOYOKUNI, *et al.* 2000. Antioxidant a-tocopherol ameliorates glycemic control of GK rats, a model of type 2 diabetes. FEBS Lett. **473:** 24–26.
23. KANETO, H., G. XU, K.-H. SONG, *et al.* 2001. Activation of the hexosamine pathway leads to deterioration of pancreatic β-cell function by provoking oxidative stress. J. Biol. Chem. **276:** 31099–31104.

24. TANAKA, Y., P.O. TRAN, J. HARMON & R.P. ROBERTSON. 2002. A role of glutathione peroxidase in protecting pancreatic β cells against oxidative stress in a model of glucose toxicity. Proc. Natl. Acad. Sci. USA **99:** 12363–12368.
25. GOROGAWA, S., Y. KAJIMOTO, Y. UMAYAHARA, et al. 2002. Probucol preserves pancreatic β-cell function through reduction of oxidative stress in type 2 diabetes. Diabetes Res. Clin. Pract. **57:** 1–10.
26. SAKAI, K., K. MATSUMOTO, T. NISHIKAWA, et al. 2003. Mitochondrial reactive oxygen species reduce insulin secretion by pancreatic β-cells. Biochem. Biophys. Res. Commun. **300:** 216–222.
27. ROBERTSON, R.P., J. HARMON, P.O. TRAN, et al. 2003. Glucose toxicity in β-cells: type2 diabetes, good radicals gone bad, and the glutathione connection. Diabetes **52:** 581–587.
28. EVANS, J.L., I.D. GOLDFINE, B.A. MADDUX & G.M. GRODSKY. 2003. Are oxidative stress-activated signaling pathways mediators of insulin resistance and beta-cell dysfunction? Diabetes **52:** 1–8.
29. OHLSSON, H., K. KARLSSON & T. EDLUND. 1993. IPF1, a homeodomain-containing-transactivator of the insulin gene. EMBO J. **12:** 4251–4259.
30. LEONARD, J., B. PEERS, T. JOHNSON, et al. 1993. Characterization of somatostatin transactivating factor-1, a novel homeobox factor that stimulates somatostatin expression in pancreatic islet cells. Mol. Endocrinol. **7:** 1275–1283.
31. MILLER, C.P., R.E. MCGEHEE & J.F. HABENER. 1994. IDX-1: a new homeodomain transcription factor expressed in rat pancreatic islets and duodenum that transactivates the somatostatin gene. EMBO J. **13:** 1145–1156.
32. JONSSON, J., L. CARLSSON, T. EDLUND & H. EDLUND. 1994. Insulin-promoter-factor 1 is required for pancreas development in mice. Nature **37:** 606–609.
33. AHLGREN, U., J. JONSSON & H. EDLUND. 1996. The morphogenesis of the pancreatic mesenchyme is uncoupled from that of the pancreatic epithelium in IPF1/PDX1-deficient mice. Development **122:** 1409–1416.
34. OFFIELD, M.F., T.L. JETTON, P. LABOSKY, et al. 1996. PDX-1 is required for pancreas outgrowth and differentiation of the rostral duodenum. Development **122:** 983–985.
35. KANETO, H., J. MIYAGAWA, Y. KAJIMOTO, et al. 1997. Expression of heparin-binding epidermal growth factor-like growth factor during pancreas development: a potential role of PDX-1 in the transcriptional activation. J. Biol. Chem. **272:** 29137–29143.
36. SHARMA, A., D.H. ZANGEN, P. REITZ, et al. 1999. The homeodomain protein IDX-1 increases after an early burst of proliferation during pancreatic regeneration. Diabetes **48:** 507–513.
37. BONNER-WEIR, S., M. TANEJA, G.C. WEIR, et al. 2000. In vitro cultivation of human islets from expanded ductal tissue. Proc. Natl. Acad. Sci. USA **97:** 7999–8004.
38. WATADA, H., Y. KAJIMOTO, J. MIYAGAWA, et al. 1996. PDX-1 induces insulin and glucokinase gene expression in αTC1 clone 6 cells in the presence of betacellulin. Diabetes **45:** 1826–1831.
39. SERUP, P., J. JENSEN, F.G. ANDERSEN, et al. 1996. Induction of insulin and islet amyloid polypeptide production in pancreatic islet glucagonoma cells by insulin promoter factor 1. Proc. Natl. Acad. Sci. USA **93:** 9015–9020.
40. YOSHIDA, S., Y. KAJIMOTO, T. YASUDA, et al. 2002. PDX-1 induces differentiation of intestinal epithelioid IEC-6 into insulin-producing cells. Diabetes **51:** 2505–2513.
41. KOJIMA, H., T. NAKAMURA, Y. FUJITA, et al. 2002. Combined expression of pancreatic duodenal homeobox 1 and islet factor 1 induces immature enterocytes to produce insulin. Diabetes **51:** 1398–1408.
42. AHLGREN, U., J. JONSSON, L. JONSSON, et al. 1998. β-Cell-specific inactivation of the mouse *Ipf1/Pdx1* gene results in loss of the β-cell phenotype and maturity onset diabetes. Genes Dev. **12:** 1763–1768.
43. WANG, H., P. MAECHLER, B. RITZ-LASER, et al. 2001. Pdx 1 level defines pancreatic gene expression pattern and cell lineage differentiation. J. Biol. Chem. **276:** 25279–25286.
44. WAEBER, G., N. THOMPSON, P. NICOD & C. BONNY. 1996. Transcriptional activation of the GLUT2 gene by the IPF-1/STF-1/IDX-1 homeobox factor. Mol. Endocrinol. **10:** 1327–1334.

45. WATADA, H., Y. KAJIMOTO, Y. UMAYAHARA, *et al.* 1996. The human glucokinase gene β-cell-type promoter: an essential role of insulin promoter factor 1 (IPF1)/PDX-1 in its activation in HIT-T15 cells. Diabetes **45:** 1478–1488.
46. DUTTA, S., S. BONNER-WEIR, M. MONTMINY & C. WRIGHT. 1998. Regulatory factor linked to late-onset diabetes? Nature **392:** 560.
47. STOFFERS, D.A., N.T. ZINKIN, V. STANOJEVIC, *et al.* 1997. Pancreatic agenesis attributable to a single nucleotide deletion in the human IPF1 gene coding sequence. Nat. Genet. **15:** 106–110.
48. STOFFERS, D.A., J. FERRER, W.L. CLARKE & J.F. HABENER. 1997. Early-onset type-II diabetes mellitus (MODY4) linked to IPF1. Nat. Genet. **17:** 138–139.
49. KAJIMOTO, Y., H. WATADA, T. MATSUOKA, *et al.* 1997. Suppression of transcription factor PDX-1/IPF1/STF-1/IDX-1 causes no decrease in insulin mRNA in MIN6 cells. J. Clin. Invest. **100:** 1840–1846.
50. KAWAMORI, D., Y. KAJIMOTO, H. KANETO, *et al.* 2003. Oxidative stress induces nucleocytoplasmic translocation of pancreatic transcription factor PDX-1 through activation of c-Jun NH_2-terminal kinase. Diabetes **52:** 2896–2904.
51. KANETO, H., G. XU, N. FUJII, *et al.* 2002. Involvement of c-Jun N-terminal kinase in oxidative stress-mediated suppression of insulin gene expression. J. Biol. Chem. **277:** 30010–30018.
52. CORBETT, J.A., J.L. WANG, M.A. SWEETLAND, *et al.* 1992. Interleukin 1β induces the formation of nitric oxide by β-cells purified from rodent islets of Langerhans: evidence for the β-cell as a source and site of action of nitric oxide. J. Clin. Invest. **90:** 2384–2391.
53. ANKARCRONA, M., J.M. DYPBUKT, B. BRUNE & P. NICOTERA. 1994. Interleukin-1β-induced nitric oxide production activates apoptosis in pancreatic RINm5F cells. Exp. Cell Res. **213:** 172–177.
54. KANETO, H., J. FUJII, H.G. SEO, *et al.* 1995. Apoptotic cell death triggered by nitric oxide in pancreatic β-cells. Diabetes **44:** 733–738.
55. AMMENDRUP, A., A. MAILLARD, K. NIELSEN, *et al.* 2000. The c-Jun amino-terminal kinase pathway is preferentially activated by interleukin-1 and controls apoptosis in differentiating pancreatic β-cells. Diabetes **49:** 1468–1476.
56. BONNY, C., A. OBERSON, M. STEINMANN, *et al.* 2000. IB1 reduces cytokine-induced apoptosis of insulin-secreting cells. J. Biol. Chem. **275:** 16466–16472.
57. MANDRUP-POULSEN, T. 2001. β-Cell apoptosis: stimuli and signaling. Diabetes **50:** S58–S63.
58. BONNY, C., A. OBERSON, S. NEGRI, *et al.* 2001. Cell-permeable peptide inhibitors of JNK: novel blockers of β-cell death. Diabetes **50:** 77–82.

Initiation of Apoptotic Signal by the Peroxidation of Cardiolipin of Mitochondria

YASUHITO NAKAGAWA

School of Pharmaceutical Sciences, Kitasato University, Minato-ku, Tokyo 108, Japan

ABSTRACT: Overexpression of phospholipid hydroperoxide glutathione peroxidase (PHGPx) in mitochondria of RBL2H3 cells (M15 cells) prevented the release of cytochrome *c* (cyt.c), the activation of caspase-3, and apoptosis caused by 2-deoxyglucose (2DG), whereas cells overexpressing nonmitochondrial PHGPx(L9) and control (S1) cells were induced to apoptosis. Hydroperoxide levels in mitochondria of L9 and S1 cells were significantly enhanced by 2DG-induced apoptosis. In contrast, generation of hydroperoxide in mitochondria was protected in M15 cells, which also showed resistance to apoptosis by etoposide, staurosporine, UV irradiation, cycloheximide, and actinomycin D, stimuli that induce apoptosis by the liberation of cyt.c from mitochondria. Cyt.c preferentially binds to the monolayer of cardiolipin (CL), the specific phospholipid of the inner membrane of mitochondria. The amount of cyt.c bound to the monolayer of cardiolipin hydroperoxide (CL-OOH) was much lower than that bound to CL. Cyt.c bound to liposome containing CL was released by peroxidation with a radical initiator. Adenine nucleotide translocator (ANT), which regulates the opening and closing the permeability transition (PT) pore, potentially was inactivated in apoptosis-induced S1 cells 4 h after the addition of 2DG, coincidentally with cyt.c release from mitochondria. ANT activity was suppressed by the fusion of isolated mitochondria with liposomes containing CL-OOH. ANT activity was expressed in proteoliposomes containing 10% CL, but it was competitively inhibited by the addition of CL-OOH. This study suggests that CL peroxidation might have an initiating role in the liberation of cyt.c from the inner membrane, and in the opening of the PT pore via inactivation of ANT. Mitochondrial PHGPx might play a role as an anti-apoptotic factor by protecting CL and reducing CL-OOH.

KEYWORDS: apoptosis; mitochondria; cardiolipin; lipid hydroperoxide; phospholipid hydroperoxide; glutathione peroxidase

INTRODUCTION

Many experimental findings have suggested a relationship between reactive oxygen species (ROS) and apoptosis, which can be induced by prooxidant agents, such as hydrogen peroxide, diamide, etoposide, and semiquinones.[1,2] Other apoptotic stimuli, such as exposure to tumor necrosis factor (TNF) and ceramide, also elevate intracellular levels of ROS.[3,4] Antioxidants, such as *N*-acetylcysteine (NAC), sup-

---

Address for correspondence: Yasuhito Nakagawa, School of Pharmaceutical Sciences, Kitasato University, 5-9-1 Shirokane, Minato-ku, Tokyo 108, Japan. Voice: 81-3-5791-6235; fax: 81-3-3444-4913.
nakagaway@pharm.kitasato-u.ac.jp

press apoptosis by acting as scavengers of ROS, providing additional evidence that ROS act as signaling molecules to initiate apoptosis.[5]

The production of ROS predominantly occurs within mitochondria, where it is regulated by several antioxidant enzymes, including phospholipid hydroperoxide glutathione peroxidase (PHGPx), classic glutathione peroxidase (cGPx), and Mn-superoxide dismutase (Mn-SOD). The proposal that mitochondrial antioxidant enzymes have a role in preventing apoptosis is based on the observation that Bcl-2, a general inhibitor of apoptosis in mammalian cells predominantly localized in mitochondria, has an apparent antioxidant function;[6] and that apoptosis of MCF-7 cells after exposure to TNF was completely suppressed by the overexpression of Mn-SOD.[7] These reports suggest that antioxidant enzymes localized in mitochondria might be linked to apoptosis and contribute to modulation of apoptotic signals. However, the molecular mechanism of ROS-mediated apoptosis has not been clarified.

PHGPx is a unique intracellular antioxidant enzyme that directly reduces peroxidized lipids produced in cell membranes.[8,9] Although PHGPx might participate in defense systems as an antioxidant enzyme protecting mitochondrial lipids from damage, little is known about the biological significance of mitochondrial PHGPx in apoptosis mediated by mitochondrial pathways. In this study, we demonstrate that mitochondrial PHGPx suppresses apoptotic cell death initiated by the peroxidation of cardiolipin, which is localizes exclusively in mitochondria.

RESULTS AND DISCUSSION

Prevention of 2-Deoxyglucose–Induced Apoptosis by Mitochondrial PHGPx

Three lines of PHGPx-overexpressing rat basophilic 2H3 cells (RBL2H3 cells) have been established[10]: mitochondrial-type PHGPx-overexpressing (M15), nonmitochondrial type PHGPx-overexpressing (L9), and vector-transfected (S1) cells. Apoptosis was induced by hypoglycemia with 2-deoxyglucose (2DG)[11]; and the nature of cell death was studied by LDH release (FIG. 1), and by the pattern of DNA fragmentation by electrophoresis on an agarose gel. The number of dead S1 cells gradually increased and reached approximately 80% of total cells within 12 h. On the other hand, apoptosis was protected by the overexpression of PHGPx in mitochondria M15 cells. However, L9 cells, which overexpressed nonmitochondrial PHGPx, were quite sensitive to apoptosis, as were S1 cells.

Cytochrome c (cyt.c) released from mitochondria was detectable in S1 and L9 cells 4 h after exposure to 2DG, whereas no detectable cyt.c was found in the cytosol of 2DG-treated M15 cells. Activation of caspases was observed in S1 and L9 cells at 6 h, but not in M15 cells.

Our finding that protection from apoptosis was associated with the overexpression of mitochondrial PHGPx suggested that mitochondrial injury caused by 2DG might increase levels of ROS. Marked increases in the levels of intracellular superoxide were, in fact, detectable within 1 h of exposure to 2DG, but there were no differences in the rates of superoxide production among the three lines of cells.[11] The production of hydroperoxide, which accumulated in cells after approximately 2 h, was suppressed by the overexpression of PHGPx, but there was no significant difference in the level of hydroperoxide between L9 and M15 cells. Hydroperoxide levels

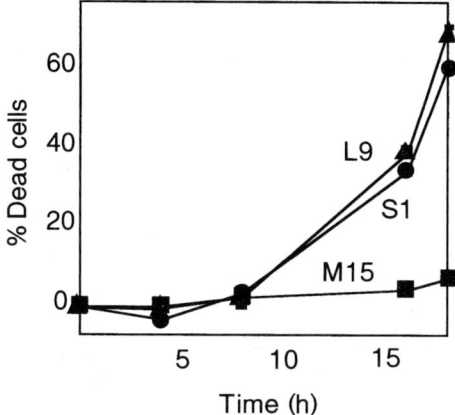

FIGURE 1. Protection of apoptosis by the overexpression of mitochondrial PHGPx. Time course of LDH release from each cell line after treatment with 2DG.

significantly increased in the mitochondria of S1 and L9 cells induced to apoptosis by 2DG. On other hand, there was no significant elevation of hydroperoxide in mitochondria of apoptosis-induced M15 cells. These results suggest that production of hydroperoxide in mitochondria is an important trigger of cyt.c release and apoptosis.

We studied the effects of several kinds of apoptotic agonists on the viability of PHGPx-overexpressing cells. Staurosporine, etoposide, actinomycin D, cycloheximide, and UV irradiation killed S1 and L9 cells, whereas M15 cells were resistant to these apoptotic agents all of which induce the release of cyt.c (the mitochondria-death pathway). In contrast, overexpression of mitochondrial PHGPx failed to protect cells from apoptosis caused by Fas-specific antibodies, which cause apoptosis by direct activation of caspase 8, indicating that M15 cells exhibit resistance to the mitochondria-death pathway. These results suggest that lipid hydroperoxide generated in mitochondria might be involved in the induction of cyt.c release, because lipid peroxide is a preferential substrate for PHGPx.

Two steps could operate in the release of cyt.c from mitochondria to cytosol. It has been shown that cyt.c release is preceded by its dissociation from the inner membrane. As a second step, the opening of pores in the outer membrane is a prerequisite for the release of cyt.c from mitochondria. In 2DG-induced apoptosis, cyt.c could be released through permeability transition (PT) pores, because the release was strongly suppressed by pretreatment of cells with bongkrekic acid, which closes PT pores by binding and inactivating the adenine nucleotide translocator (ANT). ANT has a central role in the regulation of opening and closing of PT pores.[12,13] Cardiolipin (CL), which especially localizes in the inner membrane of mitochondria, is an interesting molecule for the release of cyt.c, because it has a potent affinity to cyt.c and is essential for the activity of ANT.[14,15] Peroxidation of CL was observed in the mitochondria of 2DG apoptosis-induced S1 cells. On the other hand, the formation of cardiolipin hydroperoxide (CL-OOH) in mitochondria was effectively suppressed in 2DG-treated M15 cells. Therefore, we studied the effects of CL-OOH on interactions with cyt.c and on the activities of ANT.

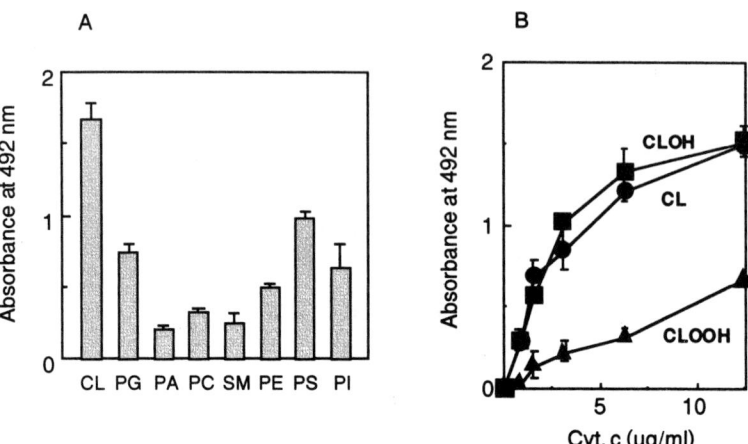

FIGURE 2. Interaction of cyt.c with cardiolipin (CL) and cardiolipin hydroperoxide (CL-OOH). (**A**) Binding of cyt.c to various phospholipids. A solution of cyt.c was placed on each monolayer of phospholipids in microtiter wells and incubated at room temperature for 2 h. Bound cyt.c then was detected with cyt.c antibodies. CL, cardiolipin; PC, phosphatidylcholine; PE, phosphatidylethanolamine; PS, phosphatidylserine; PI, phosphatidylinositol; PA, phosphatidic acid; PG, phosphatidylglycerol; SM, sphingomyelin (**B**) Binding of cyt.c to monolayers of CL (*circles*), CL-OOH (*triangles*), and hydroxycardiolipin (CL-OH; *squares*).

INTERACTIONS OF CYT.C WITH CARDIOLIPIN AND ITS HYDROPEROXIDE

The binding of cyt.c with the monolayers of several standard phospholipids was examined *in vitro* (FIG. 2). Cyt.c associates with the monolayers of CL in preference to other phospholipid classes,[16] and the binding of cyt.c to CL monolayers was drastically decreased after autooxidation of CL monolayer for 48 h. The binding of cyt.c to the monolayers of chemically synthesized CL-OOH was much lower than it was to the monolayers of nonoxidized CL; hydroxyl CL (CL-OH) bound to cyt.c to the same extent as CL.

Liposomes composed of PC and CL in equimolar amounts were incubated with cyt.c. Liposomes bound to cyt.c were peroxidized for 1, 2, and 4 h by 10 mM of the radical initiator AAPH, which caused the liberation of cyt.c from liposomes in a time-dependent manner.

The binding of phospholipids to cyt.c has been studied extensively by nuclear magnetic resonance and electron spin resonance spectroscopy.[17–19] Negatively charged polar head groups of phospholipids interact by hydrogen bonding and/or by electrostatic interactions at the sites of the opening of the cavity of cyt.c.[20,21] The acyl moiety of the phospholipid is inserted into the hydrophobic cavity, and the rest of the acyl chain is embedded in the membrane. These reports suggest that appropriate structures are necessary for the association of cyt.c with CL in bilayers. The loss of molecular interaction between cyt.c and CL-OOH has been demonstrated by nu-

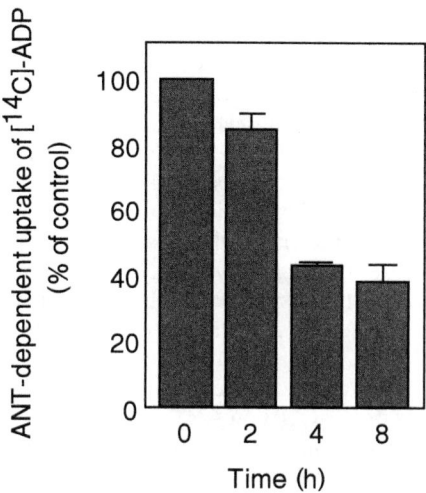

FIGURE 3. Mitochondrial PHGPx suppresses the loss of the translocase activity of ANT and the release of cyt.c and AIF during 2DG-induced apoptosis. S1 cells were treated with 2DG for indicated times, and mitochondria were isolated. ANT activity was measured by ANT-dependent uptake of ^{4}C-ADP into mitochondria.

clear magnetic resonance spectroscopy.[19] Although the reason for the disruption of the interaction between cyt.c and biomembranes that contain CL-OOH is unclear, the interaction of the fatty acid of CL with the hydrophobic cavity of cyt.c might be disturbed by the peroxidation of the acyl chain of CL, which extends into the hydrophobic cavity.[21] Alternatively, it is possible that the failure of CL-OOH to bind cyt.c might be caused by a change in the conformation of cyt.c upon lipid peroxidation. A monoclonal antibody study showed that the conformation of cytosolic cyt.c released from mitochondria was different from that of native cyt.c.[22]

Inactivation of ANT by CL-OOH

ANT activity fell drastically to 50% of control level in S1 cells after 4-h treatment with 2DG (FIG. 3).[23] However, inactivation of ANT was minimal in M15 cells. In parallel, cyt.c was released from mitochondria of S1 cells at 4 h, whereas no similar liberation of proapoptotic factors was observed in M15 cells. These results suggest that inactivation of ANT could induce opening of PT pores and release cyt.c. The opening of PT pores induced by prooxidants can be prevented by the exogenous addition of catalase and by any thiol-specific antioxidant that can directly reduce H_2O_2.[24] Oxidative stress also induces the loss of ANT activity in isolated mitochondria.[25] ANT activity is reduced during apoptosis induced by the withdrawal of interleukin 3 (IL-3), which decreases the rates of glucose metabolism and of ATP production.[26,27] Apoptosis induced by the withdrawal of IL-3 is caused by the production of H_2O_2. These reports suggest that inactivation of ANT by oxidative stress would induce the release of cyt.c through PT pores, although the relationship between the oxidative loss of ANT activity and the release of cyt.c remains to be characterized.

The effect of CL-OOH on ANT activity was estimated by fusing liposomes with isolated mitochondria, basically performed as described by Paradies et al.,[28] to allow the incorporation of various phospholipids into the mitochondrial inner membranes. There was no significant reduction of ANT activity after fusion of liposomes containing nonoxidized phospholipids such as CL, PC, PE, and PS, with mitochondria from S1 cells. On the other hand, ANT activity decreased to 20% of the control level upon fusion of isolated mitochondria with liposomes containing CL-OOH. There was no significant decrease in the activity of ANT when mitochondria were fused with liposomes prepared with phosphatidylcholine hydroperoxide. However, CL-OOH in liposomes failed to reduce ANT activity when liposomes were fused with mitochondria from M15 cells, indicating that mitochondrial PHGPx effectively prevented the decrease in ANT activity by the reduction of CL-OOH.

To evaluate the effects of CL-OOH on ANT activity in a simpler system, we prepared reconstituted liposomes with purified ANT from rat brain.[23] ANT was incorporated into liposomes composed of 90% dioleoylphosphatidylcholine and 10% tested phospholipids, as estimated for the requirements for ANT activities. The translocase activity of ANT was supported by CL, but not by other phospholipids of mitochondria. CL-OOH did not support the expression of ANT activity. CL-OH, a reduced form of CL-OOH, enhanced ANT activity similarly to CL, and ANT activity in proteoliposomes containing CL-OOH became manifest upon reduction of the proteoliposomes with PHGPx. ANT activity was competitively inhibited in a dose-dependent manner by the inclusion of CL-OOH in proteoliposomes containing 10% CL, and ANT activity disappeared in proteoliposomes constituted with the same amounts of CL-OOH (10%) and CL (10%). In contrast, PC-OOH had no similar inhibitory effect on ANT activity, suggesting that CL-OOH might displace CL from specific binding sites in ANT. Although the oxidation of mitochondrial membrane lipids is not specific to CL during apoptotic cell death, ANT was selectively inactivated by CL-OOH, suggesting that ANT might sense levels of CL-OOH in the mitochondrial membrane that provide a signal for apoptotic death.

CL interacts in the inner membrane with many mitochondrial proteins, such as ANT, cyt.c, complexes I, III, and IV, and F_0F_1-ATPase, and is required for the folding and activity of these mitochondrial enzymes.[29,30] Peroxidation of CL causes conformational changes of those proteins and a decrease in their activities. Although the mechanism of the CL-OOH–induced opening of PT pores via modification of ANT remains to be resolved, purified ANT from bovine heart contains six molecules of tightly bound CL per protein dimer,[31] and ^{31}P-NMR analysis revealed that CL is strongly immobilized on the surface of ANT. Furthermore, dissociation of CL from ANT can be achieved only under denaturing conditions and results in the complete loss of ANT function.[32]

This study suggests that the oxidation modification of CL might be a critical event in the regulation of apoptotic signaling. Mitochondrial PHGPx, which is located in the intermembrane space, in particular, at a contact site located in the PT pore complex that includes ANT,[33] suppresses apoptosis that occurs via the mitochondrial death pathway and involves the peroxidation of CL. This study also demonstrates that mitochondrial PHGPx plays an important role in regulating ANT activity and in the opening of PT pores.

ACKNOWLEDGMENTS

I thank Drs. Hong Kyu Lee and Myung-Shik Lee for inviting me to this meeting. I also deeply appreciate the involvement of Dr. H. Imai and my staff in the development of this project. This work was supported in part by Special Coordination Funds for the Promotion of Science Technology and by Grant-in-Aid (No.14572063) from the Ministry of Education, Science and Culture of Japan.

REFERENCES

1. JABS, T. 1999. Reactive oxygen intermediates as mediators of programmed cell death in plants and animals. Biochem. Pharmacol. **57:** 231–245.
2. SKULACHEV, V.P. 1998. Cytochrome c in the apoptotic and antioxidant cascades. FEBS Lett. **423:** 275–280.
3. DONATO, N.J. & M. PEREZ. 1998. Tumor necrosis factor-induced apoptosis stimulates p53 accumulation and p21WAF1 proteolysis in ME-180 cells. J. Biol. Chem. **273:** 5067–5072.
4. QUILLET-MARY, A., J.P. JAFFREZOU, V. MANSAT, et al. 1997. Implication of mitochondrial hydrogen peroxide generation in ceramide-induced apoptosis. J. Biol. Chem. **272:** 21388–21395.
5. ZAMZANI, N., P. MARCHETTI, M. CASTEDO, et al. 1995. Sequential reduction of mitochondrial transmembrane potential and generation of reactive oxygen species in early programmed cell death. J. Exp. Med. **182:** 367–377.
6. HOCKENBERY, D.M., Z.N. OLTVAI, X.M. YIN, et al. 1993. Bcl-2 functions in an antioxidant pathway to prevent apoptosis. Cell **75:** 241–251.
7. MANNA, S.K., H.J. ZHANG, T. YAN, et al. 1998. Overexpression of manganese superoxide dismutase suppresses tumor necrosis factor-induced apoptosis and activation of nuclear transcription factor-B and activated protein-1. J. Biol. Chem. **273:** 13245–13254.
8. URSINI, F., M. MAIORINO, M. VALENTE, et al. 1982. Purification from pig liver of protein which protects liposomes and biomembrane from peroxidative degradation and exhibits glutathione peroxidase activity on phosphatidylcholine hydroperoxide. Biochim. Biophys. Acta **710:** 197–211.
9. URSINI, F., M. MAIORINO & C. GREGOLIN. 1985. The selenoenzyme phospholipid hydroperoxide glutathione peroxidase. Biochim. Biophys. Acta **839:** 62–72.
10. ARAI, M., H. IMAI, T. KOUMURA, et al. 1999. Mitochondrial phospholipid hydroperoxide glutathione peroxidase plays a major role in preventing oxidative injury to cells. J. Biol. Chem. **274:** 4924–4933.
11. NOMURA, K., H. IMAI, T. KOUMURA, et al. 1999. Mitochondrial phospholipid hydroperoxide glutathione peroxidase suppresses apoptosis mediated by a mitochondrial death pathway. J. Biol. Chem. **274:** 29294–29302.
12. JACOTOT, E., P. COSTANTINI, E. LABOUREAU, et al. 1999. Mitochondrial membrane permeabilization during the apoptotic process. Ann. N.Y. Acad. Sci. **887:** 18–30.
13. CROMPTON, M. 1999. The mitochondrial permeability transition pore and its role in cell death. Biochem. J. **341:** 233–249.
14. HOCH, F.L. 1992. Cardiolipins and biomembrane function. Biochim. Biophys. Acta **1113:** 71–133.
15. SCHLAME, M., D. RUA & M.L. GREENBERG. 2000. The biosynthesis and functional role of cardiolipin. Prog. Lipid Res. **39:** 257–288.
16. NOMURA, K., H. IMAI, T. KOUMURA, et al. 2000. Mitochondrial phospholipid hydroperoxide glutathione peroxidase inhibits the release of cytochrome c from mitochondria by suppressing the peroxidation of cardiolipin in hypoglycaemia-induced apoptosis. Biochem. J. **351:** 183–193.
17. BROWN, L.R. & K. WUTHRICH. 1977. NMR and ESR studies of the interactions of cytochrome c with mixed cardiolipin-phosphatidylcholine vesicles. Biochim. Biophys. Acta **468:** 389–410.

18. HEIMBURG, T., P. HILDEBRANDT & D. MARSH. 1991. Cytochrome c-lipid interactions studied by resonance Raman and 31P NMR spectroscopy. Correlation between the conformational changes of the protein and the lipid bilayer. Biochemistry **30:** 9084–9089.
19. SHIDOJI, Y., K. HAYASHI, S. KOMURA, *et al.* 1999. Loss of molecular interaction between cytochrome c and cardiolipin due to lipid peroxidation. Biochem. Biophys. Res. Commun. **264:** 343–347.
20. RYTOMAA, M. & P.K. KINNUNEN. 1995. Reversibility of the binding of cytochrome c to liposomes. Implications for lipid-protein interactions. J. Biol. Chem. **270:** 3197–3202.
21. RYTOMAA, M. & P.K. KINNUNEN. 1994. Evidence for two distinct acidic phospholipid-binding sites in cytochrome c. J. Biol. Chem. **269:** 1770–1774.
22. JEMMERSON, R., J. LIU, D. HAUSAUER, *et al.* 1999. A conformational change in cytochrome c of apoptotic and necrotic cells is detected by monoclonal antibody binding and mimicked by association of the native antigen with synthetic phospholipid vesicles. Biochemistry **38:** 3599–3609.
23. IMAI,H., T. KOUMURA, R. NAKAJIMA, *et al.* 2003. Protection from inactivation of adenine nucleotide translocator during hypo-glycemia-induced apoptosis by mitochondrial phospholipid hydroperoxide glutathione peroxidase. Biochem. J. **371:** 1–11.
24. KOWALTOWSKI, A.J., L.E. NETTO & A.E. VERCESI. 1998. The thiol-specific antioxidant enzyme prevents mitochondrial permeability transition. Evidence for the participation of reactive oxygen species in this mechanism. J. Biol. Chem. **273:** 12766–12769.
25. HALESTRAP, A.P., K.Y. WOODFIELD & C.P. CONNERN. 1997. Oxidative stress, thiol reagents, and membrane potential modulate the mitochondrial permeability transition by affecting nucleotide binding to the adenine nucleotide translocase. J. Biol. Chem. **272:** 3346–3354.
26. VANDER HEIDEN, M.G., N.S. CHANDEL, P.T. SCHUMACKER & C.B. THOMPSON. 1999. Bcl-xL prevents cell death following growth factor withdrawal by facilitating mitochondrial ATP/ADP exchange. Mol. Cell. **3:** 159–167.
27. VANDER HEIDEN, M.G., D.R. PLAS, J.C. RATHMELL, *et al.* 2001. Growth factors can influence cell growth and survival through effects on glucose metabolism. Mol. Cell. Biol. **21:** 5899–5912.
28. PARADIES, G., F.M. RUGGIERO, G. PETROSILLO, *et al.* 1998. Peroxidative damage to cardiac mitochondria: cytochrome oxidase and cardiolipin alterations. FEBS Lett. **424:** 155–158.
29. BOGDANOV, M. & W. DOWHAN. 1999. Lipid-assisted protein folding. J. Biol. Chem. **274:** 36827–36830.
30. LANGE, C., J.H. NETT, B.L. TRUMPOWER, *et al.* 2001. Specific roles of protein-phospholipid interactions in the yeast cytochrome bc1 complex structure. EMBO J. **20:** 6591–6600.
31. BEYER, K. & M. KLINGENBERG. 1985. ADP/ATP carrier protein from beef heart mitochondria has high amounts of tightly bound cardiolipin, as revealed by 31P nuclear magnetic resonance. Biochemistry **24:** 3821–3826.
32. BEYER, K. & B. NUSCHER. 1996. Specific cardiolipin binding interferes with labeling of sulfhydryl residues in the adenosine diphosphate/adenosine triphosphate carrier protein from beef heart mitochondria. Biochemistry **35:** 15784–15790.
33. GODEAS, C., G. SANDRI & E. PANFILI. 1994. Distribution of phospholipid hydroperoxide glutathione peroxidase (PHGPx) in rat testis mitochondria. Biochim. Biophys. Acta **1191:** 147–150.

Diabetes Mellitus with Mitochondrial Gene Mutations in Japan

SUSUMU SUZUKI

Department of Internal Medicine, Division of Molecular Metabolism and Diabetes, Tohoku University Graduate School of Medicine, Sendai 980-8574, Japan

ABSTRACT: Diabetes mellitus due to the mitochondrial DNA 3243(A–G) mutation is reported to represent 0.5–1% of the general diabetic population in Japan. To further elucidate the clinical symptoms and course of diabetes mellitus with the 3243 mutation, we undertook a nationwide cross-sectional case-finding study and observational study of a genetically defined subject group. One hundred sixteen Japanese diabetic patients with the mutation were registered and analyzed. The patients had a higher maternal inheritance of diabetes or deafness, short stature, thin habitus, and early middle-aged onset of diabetes or deafness. Eighty-six percent of the patients required insulin therapy because of progressive insulin secretory defect. Although half of the patients had the phenotype of type 1 diabetes or slowly progressive type 1 diabetes, the patients lacked the presence of autoantibodies to glutamic acid decarboxylase. Diabetes in the mothers was characterized by early middle-aged onset, reduction in the insulin secretory capacity, early requirement of insulin therapy, and increases in the daily insulin dose. The heteroplasmic ratio of the 3243 mutation in leukocytes was low. The patients had mitochondria-related complications such as sensorineural deafness, cardiomyopathy, cardiac conductance disorders, encephalomyopathy, macular pattern dystrophy, and mental disorders. The patients also had advanced microvascular complications. Thus, this study has revealed that (1) diabetes mellitus with the 3243 mutation is a subtype of diabetes mellitus with mitochondria-related complications and (2) insulin secretory ability is more severely impaired in the patients whose mothers were also diabetic.

KEYWORDS: diabetes mellitus; 3243(A–G) mutation; glutamic acid decarboxylase; mitochondria-related complications

INTRODUCTION

The mitochondrial DNA 3243(A–G) mutation in the tRNA$^{\text{Leu (UUR)}}$ gene originally was reported as a pathological mutation of mitochondrial encephalomyopathy, lactic acidosis, and stroke-like episodes (MELAS),[1] but is also known to be a major mitochondrial DNA mutation in diabetes mellitus.[2–4] The 3243 mutation is also associated with neurosensory deafness,[5] cardiomyopathy,[6] Alport-like glomerulopa-

---

Address for correspondence: Susumu Suzuki, Department of Internal Medicine, Division of Molecular Metabolism and Diabetes, Tohoku University Graduate School of Medicine, Sendai 980-8574, Japan. Voice: 81-22-717-7171; fax: 81-22-717-7177.
 ssuzuki@int3.med.tohoku.ac.jp

thy,[7] and cardiac conduction disturbances.[8] Diabetes mellitus with the 3243 mutation is a subtype of diabetes mellitus, referred to as mitochondrial diabetes mellitus[9] or maternal-inherited diabetes mellitus with deafness.[10]

Diabetes mellitus with the 3243 mutation has the following characteristics: (1) it accounts for 0.5–1% of the diabetic population in Japan; (2) it shows maternal inheritance; (3) it presents the clinical phenotypes of type 1 diabetes, slowly progressive type 1 diabetes, or type 2 diabetes; (4) it occurs in nonobese patients; (5) it requires insulin therapy because of the progressive insulin secretory defect; (6) it is not associated with anti–glutamic acid decarboxylase (GAD) antibodies; (7) it is often accompanied by progressive sensorineural hearing loss; (8) it is often complicated by cardiomyopathy, encephalomyopathy, or mental disorders; and (9) it shows rapidly progressive diabetic complications.[3,4,9,10] These clinical characteristics of mitochondrial diabetes were ascertained from several clinical studies with low numbers of the mutant diabetic patients.

To elucidate the symptoms and clinical course of diabetes mellitus with the 3243 mutation in Japan, the Research Committee of the Japan Diabetes Society (JDS) undertook a nationwide case-finding study of diabetic patients with the mutation and then performed a descriptive study using questionnaires.[11] The clinical symptoms, natural history, characteristics of inheritance, mitochondria-related complications, and diabetic complications were analyzed in 116 Japanese mutant diabetic patients with the 3243 mutation.

PATIENTS AND METHODS

The JDS Research Committee undertook a nationwide cross-sectional study of diabetes mellitus caused by the 3243 mutation in Japan.[11] The study protocol was approved by the ethic review board of JDS and by the appropriate institutional review committees. Informed consent was obtained from each subject. The patient survey found 116 diabetic patients in 105 families. Questionnaires were used to ascertain the type of diabetes, treatment regimen, insulin secretory capacities, insulin sensitivities, as well as associated symptoms, such as sensorineural deafness, muscle weakness, ophthalmoplegia, and mental disorders. The insulin secretory capacity was evaluated by measurements of the C-peptide concentration 6 min after intravenous administration of 1 mg glucagon and in 24-h urine samples. Plasma glucose was assayed using the glucose oxidase method. Plasma insulin and C-peptide were assayed using a radioimmunoassay. Sensorineural hearing loss was diagnosed by expert otorhinolaryngologists. Diabetic retinopathy and macular pattern dystrophy was diagnosed by expert ophthalmologists.

RESULTS

The clinical characteristics of the diabetic patients with the 3243 mutation are shown in TABLE 1. These diabetic patients were more lean and had shorter stature. The mean age of the onset of diabetes was 32.8 years. Maternal inheritance of diabetes was significantly higher than paternal inheritance. However, 32% of the mutant diabetic patients did not show maternal inheritance. Anti–GAD autoantibodies were

TABLE 1. Clinical characteristics of diabetic patients with the 3243(A–G) mutation

Characteristic	
Gender	
Male	37
Female	79
Total	116
Age (years)	45.4 ± 11.9
Height (cm)	
Male	161.2 ± 7.4
Female	149.2 ± 5.4
BMI	18.6 ± 2.8
Male	18.6 ± 2.1
Female	18.6 ± 3.1
Onset of diabetes (years)	32.8 ± 12.4
Symptoms or signs at diagnosis	
Hyperglycemic symptoms	37
Health examination	34
Health examination at school	6
Glucosuria at pregnancy	6
Mitochondrial disorders	4
MELAS	3
Cardiomyopathy	1
Other illness	6
Familial survey	2
Duration of diabetes (years)	12.9 ± 8.3
Familial history of diabetes	
Maternal	68%
Paternal	5.4%
DM types	
Type 1	37.4%
SP type 1	12.2%
Type 2	50.4%
GAD antibody (all negative) (n)	78
Mean (after the onset of diabetes)	8.2 years
Insulin secretory capacity	
Insulinogenic index ($n = 27$)	0.072 ± 0.082
24-hour urinary excretion of C-peptide (µg/day, $n = 76$)	26.6 ± 17.8
Plasma C-peptide after breakfast: 2.14 ± 1.89 (ng/mL, $n = 22$)	
Plasma C-peptide at 6 min after glucagon injection (ng/mL, $n = 25$)	2.25 ± 1.34
Insulin sensitivity (hyperinsulinemic euglycemic glucose clamp 7, minimal model 2)	
Reduced	1
Normal	7
Increased	1
Percentile of the 3243 mutation in leukocytes ($n = 69$)	13.1 ± 12.8%

TABLE 1. (*continued*) **Clinical characteristics of diabetic patients with the 3243(A–G) mutation**

Characteristic	
Blood lactate (L) (mg/dL, $n = 88$)	18.8 ± 8.6
Blood pyruvate (P) (mg/dL, $n = 86$)	1.15 ± 1.21
L/P ($n = 86$)	20.5 ± 16.7
HbA1c ($n = 116$)	7.8 ± 1.4%
Treatment	
Diet only	3.5%
Alpha-glucosidase inhibitor	1.7%
Sulfonylurea	7.8%
Insulin	86.1%
Insulin dosage (U/day)	26.3 ± 14.1
Years from diagnosis to insulin therapy	3.03 ± 4.61

negative in all 78 examined patients, although half of the mutant patients had phenotypes of type 1 diabetes or slowly progressive type 1 diabetes. Insulin secretory capacity was reduced in the mutant diabetic patients. The insulin sensitivity assessed by hyperinsulinemic euglycemic glucose clamp or minimal model was normal in a few patients. The percentile of the 3243 mutation in leukocytes was as low as 13%; 86.1% of the mutant diabetic patients required insulin therapy. The interval between diagnosis and start of insulin therapy was approximately 3 years, suggesting the presence of progressive secretory defects.

When mitochondria-associated complications in the diabetic patients were investigated, sensorineural deafness was demonstrated in 92.2% of the diabetic patients (TABLE 2). The age at diagnosis of sensorineural deafness was 33.2 years, similar to that of the onset of diabetes. Maternal inheritance of deafness was significantly higher than paternal inheritance of deafness. Cardiomyopathy was demonstrated in 30.4% of the patients. Cardiac conduction disturbance was demonstrated in 27.8% of the mutant diabetic patients, Wolff-Parkinson-White (WPW) syndrome in 13.0% of patients, and sick sinus syndrome in 4.3%.

Encephalomyopathy was demonstrated in 25.2% of the mutant diabetic patients. MELAS was seen in 14.8% and ophthalmoplegia in 10.4%. Diabetes followed MELAS in three patients, whereas MELAS followed diabetes MELAS in 14 patients. The mean age at onset of MELAS was 37.2 years. Interestingly, the mutant diabetic patients had stroke-like episodes when they lacked carbohydrate intake during sick days. Computed tomography examination demonstrated basal-ganglia calcifications in 71% of mutant diabetic patients, although most patients had no neurological symptoms or signs; 13.4% of mutant diabetic patients had macular pattern dystrophy, and 17.4% had mental disorders.

Diabetic complications also were investigated in these patients. Peripheral neuropathy was found in 49.6% of patients (TABLE 2). Painful peripheral neuropathy was shown in 18.3% of patients and posttreatment neuropathy in 4.3%. Autonomic neuropathy was also demonstrated in 26.1% of patients. Retinopathy was present in 54.8% of patients and nephropathy in 48.7%.

TABLE 2. Mitochondria-related or diabetic complications of diabetic patients with the 3243(A–G) mutation

- *Mitochondria-related complications*
  Sensorineural deafness: 92.2% [age at diagnosis: 33.2 ± 12.0 years]
  Cardiomyopathy: 30.4% [age at diagnosis: 42.2 ± 6.3 years]
    [hypertrophic cardiomyopathy: 7.8%; dilated cardiomyopathy: 18.3%]
  Conduction disorders: 27.8% [WPW: 13.0%; SSS: 4.3%]
  Neuromuscular symptoms: 25.2%
    [MELAS: 14.8%; myopathy and ophthalmoplegia: 10.4%]
    [MELAS to DM: 3 patients; DM to MELAS: 14 patients]
  Basal-ganglia calcifications: 71.4%
  Macular pattern dystrophy: 13.4%
  Optic nerve atrophy: 5.2%
  Neuropsychiatric disturbances: 17.4%
    [mental retardation: 7.1%; dementia: 3.6%; depression: 3.6%]

- *Diabetic complications*
  Peripheral neuropathy: 49.6%
    [painful peripheral neuropathy: 18.3%; posttreatment neuropathy: 4.3%]
  Autonomic neuropathy: 6.1%
    [neurogenic bladder: 8.7%; orthostatic hypotension: 5.2%;
    erectile disorder: 3.5%; diabetic diarrhea: 3.5%; gastricatony: 0.9%]
  Retinopathy: 54.8%
    [proliferative: 12.2%; preproliferative: 8.7%; simple: 33.9%]
  Nephropathy: 48.7%
    [CRF and/or HD: 4.3%; macroalbuminuria: 15.7%; microalbuminuria: 28.7%;
    normoalbuminuria: 50.4%]

When we compared the clinical features in patients with and without maternal inheritance, the onset of diabetes was earlier in patients with diabetic mothers than in patients with mothers with normal glucose tolerance (NGT) (TABLE 3). The onset of deafness was also earlier in patients with diabetic mothers than in patients with NGT mothers. However, there was no difference in the degree of heteroplasmy in leukocytes between patients with and without maternal inheritance. Urinary C-peptide excretion as well as glucagon-induced C-peptide was lower in patients with diabetic mothers than in the patients with NGT mothers. The interval between onset of diabetes and beginning of insulin therapy was shorter in patients with diabetic mothers than in patients with NGT mothers. Daily insulin doses were higher in patients with diabetic mothers than in patients with NGT mothers. Thus, maternal inheritance of diabetes was correlated with early onset of diabetes and deafness, reduced insulin secretory capacity, early requirement of insulin therapy, and the need to increase daily insulin doses.

The degree of heteroplasmy was studied in relation to leukocytes in the presence or absence of mitochondrial complications. Patients with encephalomyopathy had higher heteroplasmic concentrations in leukocytes than patients without encephalomyopathy.[11] Patients with mental disorders had higher heteroplasmic concentrations than patients without mental disorders. There were no differences in blood heteroplasmy between patients with and without deafness, cardiomyopathy, conduction disorders, or ocular disorders.

TABLE 3. Clinical characteristics of mitochondrial diabetic patients with or without maternal inheritance of diabetes [mean ± SD (n)]

Characteristic	Mothers		Significance (DM vs. NGT) (P value)
	Diabetes	NGT	
Onset of diabetes (age)	28.9 ± 11.2 (68)	42.3 ± 12.3 (25)	<0.0001
Onset of deafness (age)	30.9 ± 9.50 (41)	37.9 ± 15.1 (18)	<0.05
Heteroplasmy in leukocytes (%)	14.0 ± 14.1 (40)	6.77 ± 4.58 (15)	NS
Urinary C-peptide excretion (μg/day)	22.6 ± 13.3 (45)	34.2 ± 22.4 (19)	<0.02
Glucagon-induced plasma C-peptide (ng/mL)	1.93 ± 0.98 (36)	3.31 ± 1.90 (12)	<0.002
Years from diagnosis of diabetes to insulin therapy	1.15 ± 2.53 (58)	3.04 ± 4.55 (23)	<0.02
Insulin doses (U/day)	29.0 ± 15.6 (58)	19.1 ± 8.74 (23)	<0.01

DISCUSSION

This study shows the following clinical characteristics of diabetes mellitus due to the 3243 mutation in Japanese patients: (1) maternal inheritance of diabetes and deafness, (2) early middle-aged onset of diabetes and deafness, (3) thin habitus and short stature, (4) absence of anti–GAD autoantibodies, (5) low heteroplasmy in leukocytes, (6) high incidence of neurosensory deafness, and (7) early requirement for insulin therapy due to progressive insulin secretory defect. These results confirm data of previous reports.[2-4,9,10]

One of important findings of this study is the association between diabetes mellitus with the 3243 mutation and mitochondria-related complications, such as cardiomyopathy, cardiac conductance disorders, neuromuscular symptoms, basal-ganglia calcifications, macular pattern dystrophy, and neuropsychiatric disturbance. The 3243 mutation has a wide distribution in tissues, including peripheral leukocytes, skeletal muscle, skin, heart, kidney, and brain, with variable degree of heteroplasmy, which explains the clinical heterogeneity of mitochondria-related complications.

Late-onset MELAS appears to be one of the characteristic complications of diabetes mellitus with the mutation, because 14 mutant diabetic patients developed MELAS after the diagnosis of diabetes. Stroke-like episodes occurred after carbohydrate starvation for 1–3 days in eight mutant diabetic patients. Mitochondria with the 3243 mutation have impaired oxidative phosphorylation, which results in reduced ATP synthesis. The cells with higher heteroplasmic mutations depend to a larger extent on ATP supplied by glycolysis. Therefore, carbohydrate deficiency in mutant diabetic patients during illness might cause ATP deprivation, leading to stroke-like episodes. In this sense, supplementation of carbohydrates during illness seems advisable to prevent stroke-like episodes.

Recent case reports have described cardiomyopathy[6] or WPW syndrome[8] in diabetic patients with the 3243 mutation. Our study demonstrated a relatively high incidence of cardiomyopathy, WPW syndrome, and sick sinus syndrome in Japanese patients. We speculate that cardiac conductance disorders might occur by a rearrangement of the cardiac conductance system induced by the mitochondrial dysfunction.

Macular pattern dystrophy[12] is a recognized complication of diabetes mellitus with the 3243 mutation. Guillausseau et al. demonstrated a high prevalence of macular pattern dystrophy in French patients with MIDD and suggested that macular pattern dystrophy might be used as a marker to select the patients with possible MIDD.[12] The Japanese mutant diabetic patients in this study, however, had a lower incidence of macular pattern dystrophy but a higher incidence of diabetic retinopathy, as compared with the French patients with MIDD.[11] This study confirmed a significant discordance between macular pattern dystrophy and diabetic retinopathy, consistent with the hypothesis that macular pattern dystrophy might protect against the development of diabetic retinopathy through a reduction in retinal metabolism and a decrease in oxygen consumption in the retina.

Several studies have reported the rapid development of diabetic complications in the mutant diabetic patients.[3] Peripheral polyneuropathy was reported to be associated with the 3243 mutation.[13] This study demonstrated a relatively high incidence of peripheral neuropathy and autonomic neuropathy. The relatively high prevalence of painful peripheral neuropathy and autonomic neuropathy in our diabetic patients is noteworthy. Our mutant diabetic patients had a significantly higher prevalence of diabetic retinopathy than age- and duration-matched diabetic patients in the multicenter study in Japan[14]; 4.3% of our patients had end-stage renal disease despite the relatively short duration of diabetes and the low prevalence of hypertension. Iwasaki reported that 5.9% of diabetic patients with end-stage renal disease had the 3243 mutation.[15] Alport-like glomerulopathy[9] and focal and segmental glomerulopathy[16] were associated with the 3243 mutation. The relatively high incidence of end-stage renal disease in the mutant diabetic patients might be explained by these mitochondria-related glomerulopathies.

Another remarkable finding of this study is the association between diabetes of the mothers and insulin secretory defects in the mutant diabetic patients. Maternal inheritance of diabetes was correlated with early middle-aged onset of diabetes, reduction in insulin secretory capacity, early requirement of insulin therapy, and increase the need to the daily insulin doses. Thus, maternal inheritance might be a marker of more severe dysfunction of pancreatic β cells. The heteroplasmic concentrations in leukocytes might account for the effects of maternal inheritance on these clinical features, because there was a significant difference in the degree of blood heteroplasmy between patients with diabetic mothers and the patients with NGT mothers. According to recent epidemiological studies, maternal transmission of diabetes is reported to occur in many ethnic groups.[17] Other maternal genetic factors, such as other mitochondrial DNA mutations, parental imprinting of nuclear genes, or the intrauterine environment, might contribute to the effects of maternal inheritance.

REFERENCES

1. GOTO, Y. et al. 1990. A mutation in the tRNA$^{(Leu)(UUR)}$ gene associated with the MELAS subgroup of mitochondrial encephalomyopathies. Nature **348:** 651–653.
2. VAN DEN OUWELAND, J.M.W. et al. 1992. Mutation in mitochondrial tRNA$^{Leu(UUR)}$ gene in a large pedigree with maternally transmitted type II diabetes mellitus and deafness. Nat. Genet. **1:** 368–371.

3. KADOWAKI, T. *et al.* 1994. A subtype of diabetes mellitus associated with a mutation of mitochondrial DNA. N. Engl. J. Med. **330:** 962–968.
4. SUZUKI, S. *et al.* 1994. Pancreatic beta-cell secretory defect associated with mitochondrial point mutation of the tRNA$^{(LEU(UUR))}$ gene: a study in seven families with mitochondrial encephalomyopathy, lactic acidosis and stroke-like episodes (MELAS). Diabetologia **37:** 818–825.
5. GOLD, M. *et al.* 1994. Non-Mendelian mitochondrial inheritance as a cause of progressive genetic sensorineural hearing loss. Int. J. Pediatr. Otorhinolaryngol. **30:** 91–104.
6. ANAN, R. *et al.* 1995. Cardiac involvement in mitochondrial diseases. A study on 17 patients with documented mitochondrial DNA defects. Circulation **91:** 955–961.
7. CHEONG, H.I. *et al.* 1999. Hereditary glomerulopathy associated with a mitochondrial tRNA$^{(Leu)}$ gene mutation. Pediatr. Nephrol. **13:** 477–480.
8. AGGARWAL, P. *et al.* 2001. Identification of mtDNA mutation in a pedigree with gestational diabetes, deafness, Wolff-Parkinson-White syndrome and placenta accreta. Hum. Hered. **51:** 114–116.
9. GERBITZ, K.D. *et al.* 1995. Mitochondrial diabetes mellitus: a review. Biochim. Biophys. Acta **1271:** 253–260.
10. VAN DEN OUWELAND, J.M. *et al.* 1994. Maternally inherited diabetes and deafness is a distinct subtype of diabetes and associates with a single point mutation in the mitochondrial tRNA$^{(Leu(UUR))}$ gene. Diabetes **43:** 746–751.
11. SUZUKI, S. *et al.* 2003. Clinical features of diabetes mellitus with the mitochondrial DNA 3243 (A-G) mutation in Japanese: maternal inheritance and mitochondria-related complications. Diabetes Res. Clin. Pract. **59:** 207–217.
12. GUILLAUSSEAU, P.J. *et al.* 2001. Maternally inherited diabetes and deafness: a multicenter study. Ann. Intern. Med. **134:** 721–728.
13. FANG, W. 1996. Polyneuropathy in the mtDNA base pair 3243 point mutation. Neurology **46:** 1494–1495.
14. KUZUYA, T. *et al.* 1994. Prevalence of chronic complications in Japanese diabetic patients. Diabetes Res. Clin. Pract. **24** (Suppl.): S159–S164.
15. IWASAKI, N. *et al.* 2001. Prevalence of A-to-G mutation at n ucleotide 3243 of the mitochondrial tRNA$^{(Leu(UUR))}$ gene in Japanese patients with diabetes mellitus and end stage renal disease. J. Hum. Genet. **46:** 330–334.
16. HOTTA, O. *et al.* 2001. Clinical and pathologic features of focal segmental glomerulosclerosis with mitochondrial tRNA$^{Leu(UUR)}$ gene mutation. Kidney Int. **59:** 1236–1243.
17. ALCOLADO, J.C. *et al.* 2002. Maternal transmission of diabetes. Diabetic Med. **19:** 89–98.

Accumulation of Somatic Mutation in Mitochondrial DNA and Atherosclerosis in Diabetic Patients

TAKASHI NOMIYAMA,[a] YASUSHI TANAKA,[a] LIANSHAN PIAO,[b] NOBUTAKA HATTORI,[c] HIROSHI UCHINO,[a] HIROTAKA WATADA,[a] RYUZO KAWAMORI,[a] AND SHIGEO OHTA[d]

[a]*Department of Medicine, Metabolism and Endocrinology, Juntendo University School of Medicine, Tokyo, Japan*

[b]*Department of Endocrinology, The Affiliated Hospital of Yanbian University College of Medicine, Yanji, Jilin, China*

[c]*Department of Neurology, Juntendo University School of Medicine, Tokyo, Japan*

[d]*Department of Biochemistry and Cell Biology, Institute of Gerontology, Nippon Medical School, Kawasaki, Japan*

> ABSTRACT: A point mutation of mitochondrial DNA at nucleotide position 3243 A to G is responsible for the genetic cause of diabetes. Otherwise, this mutation is also reported to occur as a somatic mutation, possibility because of oxidative stress. Because diabetes may cause oxidative stress, we hypothesized that accumulation of the somatic A3243G mutation in mitochondrial DNA may be accelerated by diabetes. DNA was extracted from blood samples of 290 nondiabetic healthy subjects (aged 0–60 years) and from 383 type 2 diabetic patients (aged 18–80 years). Then, the extent of somatic A3243G mutation in total mitochondrial DNA was detected by real-time polymerase chain reaction (PCR) using the TaqMan probe. The genotyping of *ACE I/D* or *p22phox C242T* was done by PCR or PCR–restriction fragment length polymorphism. Although the level of the A3243G mutation was negligible in the newborn group, it increased in healthy subjects aged 20–29 and 41–60 years. In diabetic patients, the mutational rate increased along with age and the duration of diabetes. In the middle-aged group (41–60 years old), the A3243G mutation accumulates fourfold higher in the diabetic patients than in the healthy subjects. Moreover, multiple regression analysis revealed that the most critical factor associated with this mutation in diabetic patients was the duration of diabetes. Furthermore, the genotype of *DD, DI-CC (ACE-p22phox)* has the highest mutational rate and the thickest intima-media thickness of the carotid artery. In conclusion, diabetes accelerates the accumulation of the somatic A3243G mutation in mitochondrial DNA, and this somatic mutation may be a marker for the duration of diabetes and atherosclerosis.
>
> KEYWORDS: diabetes; atherosclerosis; mitochondrial DNA; somatic mutation; oxidative stress

Address for correspondence: Y. Tanaka, Department of Medicine, Metabolism and Endocrinology, School of Medicine, Juntendo University, 2-1-1, Hongo, Bunkyo-ku, Tokyo, 113-8421 Japan. Voice: 81-3-5802-1579.
 y-tanaka@med.juntendo.ac.jp

INTRODUCTION

A point mutation of mitochondrial DNA (mtDNA) at nucleotide position 3243 A to G (A3243G) is a major genetic cause of diabetes.[1] In Japan, approximately 1% of diabetic patients are estimated to have this mutation detectable in blood cells.[2] This point mutation was first identified as being responsible for the mitochondrial myopathy, encephalopathy, lactic acidosis, and stroke-like episodes (MELAS) subgroup of mitochondrial encephalomyopathies.[3] On the other hand, it also has been reported that point mutations and deletions of mtDNA accumulate age-dependently in healthy subjects, possibly because of sustained exposure to reactive oxygen species (ROS) during life in nondividing cells.[4] These acquired mutations are called somatic mutations and are recognized as a marker and a cause of aging.[5] mtDNA has a 5- to 20-fold higher somatic mutational rate compared with nDNA, because mtDNA is located near the respiratory chain, potentially generating ROS.[6] mtDNA 3243 nucleotide position is one of the vulnerable spots for somatic mutation, because this point remains single-stranded for a relatively long time during the duplication of mtDNA. Furthermore, damage to mtDNA by ROS may result in errors in the synthesis of respiratory chain proteins, leading to dysfunction in ATP production and increasing a leak of ROS to cause of further mtDNA mutations. This vicious cycle may decrease cell viability and cause the aging process.

Somatic mtDNA mutations not only accumulate during normal aging[7] but also in various diseases associated with enhancement of ROS. In diabetes, both hyperglycemia and hyperinsulinemia may increase oxidative stress. Accordingly, diabetes has been suggested to enhance DNA damage by ROS,[8] which, in turn, may be a cause of diabetic complications.[9] Thus, we hypothesized that accumulation of the somatic mtDNA A3243G may be accelerated by diabetes.

Atherosclerosis, including coronary artery disease and stroke, is the critical complication of diabetes. We have measured the carotid artery intima-media thickness (IMT) as an important marker of atherosclerosis in diabetic patients.[10] Although the exact mechanism behind the increase of IMT in diabetic patients has not been determined, data obtained from *in vivo* and *in vitro* studies suggest that hyperglycemia-induced oxidative stress may lead to atherogenesis.[11] Furthermore, lots of polymorphisms are reported to associate with the acceleration of atherosclerosis. In particular, the insertion/deletion polymorphism of angiotensin-converting enzyme (*ACE I/D*) is one of the most famous polymorphisms associated with atherosclerosis.[12] Recently, p22phox, the subunit of NAD(P)H oxidase which produces ROS in the downstream of angiotensin II receptor, also has been found to have a polymorphism associated with atherosclerosis.[13]

In this study, we tried to evaluate the potency of the somatic mtDNA mutation as a marker in diabetic patients. Although blood cells may not be expected to accumulate mutant mtDNA because of their rapid turnover, DNA samples are easily collected from peripheral blood. Therefore, it is essential to develop a method to detect small amounts of mutant mtDNA coexisting with wild-type mtDNA. We established a method to measure the mtDNA A3243G at a level of less than 0.001% in mtDNA using the TaqMan probe.

MATERIALS AND METHODS

Patients and Controls

Total DNA was prepared from peripheral blood cells using the QIAamp DNA Blood Mini Kit (Qiagen, Chatsworth, CA) and stored at −20°C. A total of 383 Japanese patients with type 2 diabetes diagnosed by diabetes criteria of the Japan Diabetes Society (247 men and 136 women, 18–80 years old, mean ± SD; 58.9 ± 0.6 years) were recruited from the outpatient clinic of Juntendo University Hospital (Tokyo, Japan). Patients with a diagnosis of MELAS or mitochondrial diabetes were excluded in the study population. All patients gave written informed consent before enrollment in the study, which was approved by the ethics committee of Juntendo University. Blood from the umbilical vein of neonates was obtained with the written informed consent of their parents at the second hospital of Nippon Medical School (Kawasaki, Japan). Healthy adult volunteers also were recruited after giving informed consent. Both the healthy subjects and the diabetic patients were divided into three age groups.

Quantitative Detection of mtDNA with the A3243G Point Mutation

For the detection of a small amount of mtDNA A3243G mutation, the amplification refractory-mutation system method was applied as previously described.[14] A polymerase chain reaction (PCR) primer with a sequence matched to the mutated mtDNA at the 3′ end was used for selective amplification of the mutated mtDNA. The PCR products were quantified with the TaqMan probe using an automatic sequence detection system (Prism 7700 system; Applied Biosystems) at each step of amplification. The primers used were ATTAAAGTCCTACGTGATC (3048–3066) and ATGCGATTACCGGGCC (3258–3243) for detecting the mutant mtDNA, whereas GCCTTCCCCCGTAAATGATAT (3163–3183) and GAAGAGGAATTGAACCTCTGACTG (3298–3275) were used for quantifying total mtDNA. The nucleotide sequence of the TaqMan probe was TGCCATCTTAACAAACCCTGTTCTTGGGTT (3241–3213). The oligonucleotides were highly purified by high-performance liquid chromatography (HPLC) and avoided to repeat freezing and thawing not to contaminate partially hydrolyzed primers. PCR was repeated for 40 cycles at 95°C for 15 s, 51°C for 10 s, and 57°C for 1 min. The reaction mixture (25 mL) contained ABI TaqMan buffer A, plus 0.05% glycerol, 2.5 mM $MgCl_2$, 0.16 mmol/L each of dATP, dGTP, dCTP, and dTTP, 0.6 units of Ampli Taq Gold (ABI), 100 nmol/L TaqMan probe, and 200 nmol/L of each of the forward and reverse primers for quantifying total mtDNA, or 300 nmol/L of the forward primer and 100 nmol/L of the reverse primer for quantifying mutant mtDNA.

Standard curves for quantifying total and mutant mtDNA were obtained using plasmids with the wild-type or the mutant mtDNA fragments, respectively. The correlation obtained over the range of standard samples from 10^3 to 10^{-1} fg. When the correlation coefficient was lower than 0.99, the experiment was reported to ensure intra- and interreproducibility of each assay. The mtDNA fragments with or without mutation were cloned from a MELAS patient and a healthy control, respectively. The total mtDNA concentration initially was determined in each sample as described above, and the concentration was adjusted just before examining the mutant mtDNA.

All the samples were randomly rearranged to avoid any bias. PCR measurements were performed five times for each sample in different PCR wells and the average of the three medium values of the five measurements was calculated to eliminate false values. By taking three mean values out of five measurements, we could restrict the average standard deviation to within ±50%. The total mtDNA content was examined again at the same time as the mutant mtDNA content and a ratio was calculated.

Determination of p22phox C242T and ACE I/D Genotypes

The p22phox genotype was determined by using PCR–restriction fragment length polymorphism, as described previously.[15] In brief, genomic DNA was amplified using a forward primer (5′-TGCTTGTGGGTAAACCAAGGCCCGGTG-3′) and a reverse primer (5′-AACACTGAGGTAAGTGGGGGTGGCTCCTGT-3′). PCR products were digested with *Rsa*I for 3 h at 37°C. Electrophoresis then was done on 2.5% agarose gel followed by staining with ethidium bromide.

ACE I/D polymorphism was analyzed using PCR, as described previously.[12] In brief, genomic DNA was amplified using a forward primer (5′-CTGGAGAC-CACTCCCATCCTTTCT-3′) and a reverse primer (5′-GATGTGGCCATCACAT-TCGTCAGAT-3′), after which the PCR products were subjected to electrophoresis on 2% agarose gel that was stained with ethidium bromide. Insertion and deletion of the allele were detected as a band of 490 and 190 bp, respectively.

Statistical Analysis

Data are expressed as the mean ± SEM. The statistical significance of differences in mean values and frequencies was evaluated by Student's *t*-test or Scheffe's multiple comparison test. To assess the relationship of parameters related to diabetes and the somatic mtDNA A3243G, we performed multiple regression analysis. Then, the stepwise forward selection method was applied for more detailed analysis. Differences were considered statistically significant at *P* value less than 0.05.

RESULTS

Development of a Method for Quantifying a Low Level of the A3243G Mutation in mtDNA

Because only minor accumulation of the mutant mtDNA was expected in blood cells, we first developed a method for quantifying tiny amounts of mutant mtDNA. Under the strict PCR conditions described in MATERIALS AND METHODS, only mutant mtDNA with A3243G was amplified, whereas a 100,000-fold excess of wild-type mtDNA was not detected by agarose electrophoresis (FIG. 1A). The amplified DNA fragment was confirmed to be the corresponding mtDNA region by sequencing. In addition, no band was seen when total DNA derived from HeLa cells lacking mtDNA (so-called ρ^0 cells) was used as a template (data not shown). This finding excluded any influence of nuclear pseudogenes of the mtDNA fragments. For quantitative analysis, the amplified DNA fragment was monitored with the TaqMan probe in DNA sequence detection system 7700 and was estimated using plasmids containing mtDNA fragments with or without the A3243G mutation as the standard. Nuclear

DNA from ρ^0 cells did not disturb the measurement of mutant mtDNA in a mixture of the nuclear DNA and the standard plasmid. FIGURE 1B shows the wide range of linearity for quantification of the mutant mtDNA, whereas even a 10,000-fold excess of wild-type mtDNA could not be detected. In addition, because all the samples were randomly rearranged and measurement was repeated five times, bias could be avoided to ensure reliability. In conclusion, the mutant mtDNA according for 0.001% to 0.1% of total mtDNA could be quantified by this method.

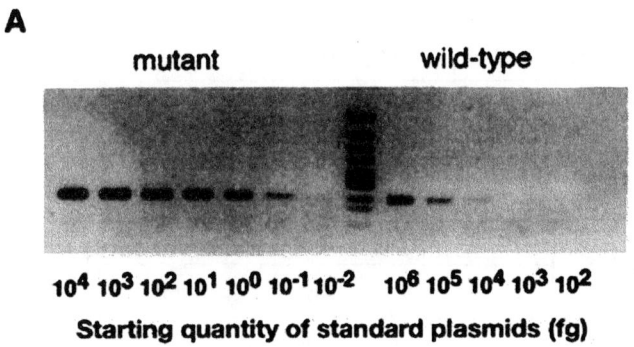

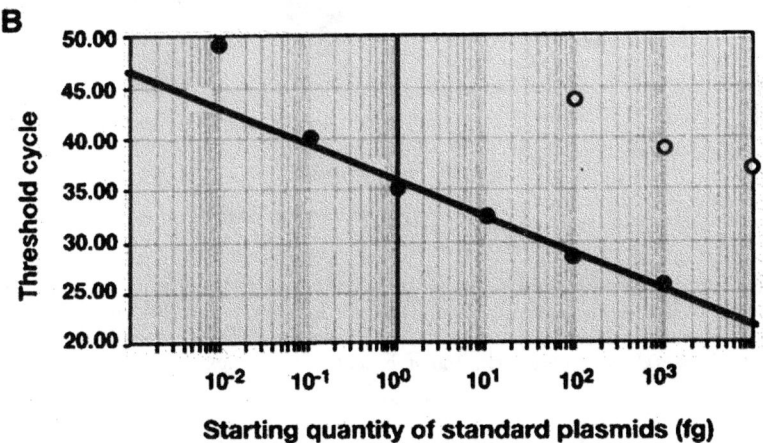

FIGURE 1. A method for quantifying a small amount of A3243G mutant mtDNA. (**A**) PCR was performed for 35 cycles with primers for the mutant mtDNA (*left*) or total mtDNA (*right*), using the indicated quantity of standard plasmids (*left*; A3243G mutant; *right*, wild type) as a template under the conditions described in MATERIALS AND METHODS. PCR products were subjected to agarose gel electrophoresis, followed by ethidium bromide staining. (**B**) A standard curve for quantifying the amount for mutant mtDNA. The PCR products were monitored by fluorescence intensity from the TaqMan probe at each PCR cycle. PCR was performed as described in MATERIALS AND METHODS using the standard plasmids with (*closed circles*) or without (*open circles*) the A3243G mutant.

TABLE 1. List of nondiabetic healthy subjects and diabetic patients

Age (years)	n	Male/female	Duration of diabetes (years)	HbAlc (%)	BMI (kg/m^2)
Healthy subjects					
0	98				
20–29	93				
41–60	99				
Diabetic patients					
18–40	26	15/11	7.4 ± 1.35	7.3 ± 0.26	26.6 ± 1.16
41–60	173	114/59	10.8 ± 0.60^a	7.3 ± 0.11	23.0 ± 0.24^a
61–80	184	118/66	14.5 ± 0.75^{a,b}	7.0 ± 0.11^c	23.0 ± 0.25^a

$^a P < 0.01$ compared with the young-aged group.
$^b P < 0.01$ compared with the middle-aged group.
$^c P < 0.05$ compared with the middle-aged group.

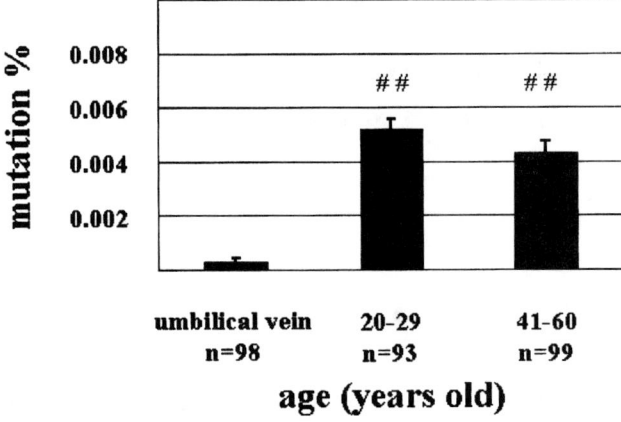

FIGURE 2. Ratio of A3243G mutant mtDNA to total mtDNA in nondiabetic subjects. The ratio of A3243 G mutant mtDNA to wild-type mtDNA was examined according to the method described in MATERIALS AND METHODS. mutation %, percentage of total mtDNA mutated. ##, $P < 0.01$ compared with the umbilical vein samples.

Detection of the A3243G Mutant mtDNA

The characteristics of the subjects are shown in TABLE 1. The stable fractions of glycated hemoglobin (HbA$_{1c}$; references range is <5.8%), a marker of glycemic control for a few months, was slightly lower in the elderly group (61–80 years old) than in the middle-aged group ($P < 0.05$), and body mass index (BMI) was significantly higher in the younger age group (18–40 years old) compared with the other groups ($P < 0.01$). The mtDNA A3243G mutational rates in the healthy subjects are shown in FIGURE 2. The level in the newborn group was negligible (0.0003 ± 0.0001% [mean ± SEM], not detected in 83 of 98 subjects), but both the young group and the middle-aged group showed significantly higher mutational rates (0.0052 ± 0.0003%, not detected in 2 of 93 young subjects, and 0.0043 ± 0.0004%, not detected in 7 of 99 middle-aged subjects, both $P < 0.01$ vs. the newborn group). As shown in FIGURE 3, the mutational rates of mtDNA A3243G in the diabetic patients increased age-dependently, and the mutation was always detectable (0.0115 ± 0.0001% in the young group, 0.0158 ± 0.0001% in the middle-aged group, and 0.0231 ± 0.0040 in the elderly group). There was no relationship between the somatic A3243G mutational rate and the treatment of diabetes (diet only; $n = 55$, % mutation = 0.0233 ± 0.0070, oral hypoglycemic reagents; $n = 187$, % mutation = 0.0161 ± 0.0010, insulin injection therapy; $n = 141$, % mutation = 0.0212 ± 0.0045) or BMI. A comparison between the age-matched healthy subjects and diabetic patients (41–60 years old) is shown in FIGURE 4. The mutational rate was approximately fourfold higher in the diabetic patients than in the healthy subjects ($P < 0.01$).

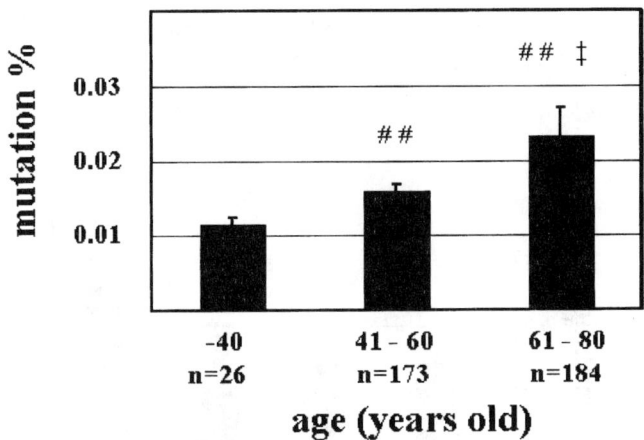

FIGURE 3. Ratio of A3243G mutant mtDNA to total mtDNA in age-divided diabetic patients. ##, $P < 0.01$ compared with the young group. ‡, $P < 0.05$ compared with the middle-aged group.

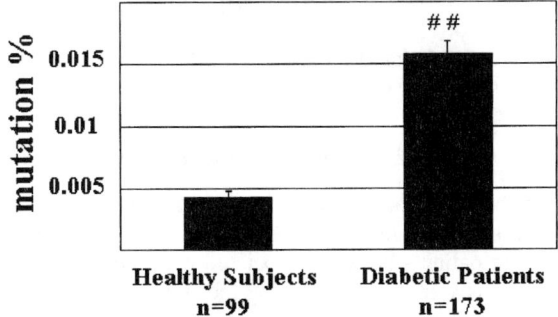

FIGURE 4. Comparison of mtDNA A3243G in 41–60-year-old diabetic patients and nondiabetic subjects. ##, $P < 0.01$ compared with the age-matched nondiabetic subjects.

TABLE 2. Analysis of clinical factors associated with mtDNA A3243G in diabetic patients with multiple regression analysis

	Partial regression coefficient	Partial correlation coefficient	F value	P value
Constant	0.0333		3.9134	0.0486
Age (years)	0.0001	0.0218	0.1790	0.6725
Sex (1 male, 2 females)	−0.0044	−0.0551	1.1518	0.2839
Duration of diabetes (years)	0.0004	0.0903	3.1082	0.0787
HbAlc (%)	−0.0014	−0.0500	0.9474	0.3310
CMV-Hb (pmol CMV/mg Hb)	−0.0003	−0.0678	1.7462	0.1876
Stepwise forward selection analysis				
Constant	0.0212		11.1749	0.0009
Duration of diabetes (years)	0.0004	0.1068	4.3973	0.0367
CMV-Hb (pmol CMV/mg Hb)	−0.0003	−0.0727	2.0216	0.1559

CMV-Hb: hemoglobin with carbonyl methylated valine residue.

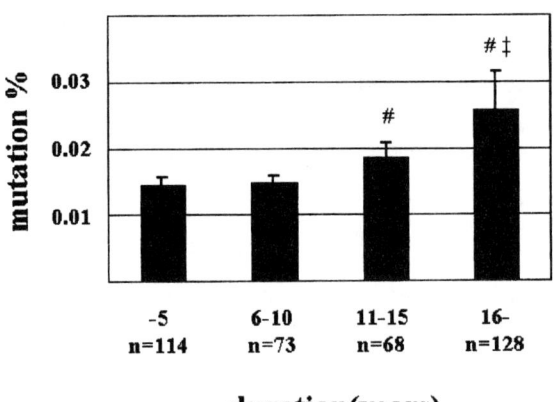

FIGURE 5. mtDNA A3243G in duration-divided diabetic patients. #, $P < 0.05$ compared with 1–5 years; ‡, $P < 0.05$ compared with 6–10 years.

Multiple Regression Analysis of Factors Associated with the A3243G Mutation

The correlations between the A3243G mutational rate in the diabetic patients and various clinical parameters are listed in TABLE 2. The mutational rate was not correlated with the age of the patients but was associated with the estimated duration of diabetes. Then correlations were analyzed using the stepwise forward selection method as shown in TABLE 2. Only the duration of diabetes (but not age) was significantly correlated with the rate of mutation of mtDNA A3243G ($P = 0.0367$). As shown in FIGURE 5, the rate of mtDNA A3243G increased the estimated duration of diabetes dependent-

TABLE 3. Clinical characteristics of patients classified by ACE I/D and p22phox C242T genotype

Characteristic	ACE-p22phox genotype			
	DD, DI-CC ($n = 142$)	II-CC ($n = 78$)	DD, DI-CT, TT ($n = 23$)	II-CT, TT ($n = 19$)
Male/female (n, %)	94 (66.2) / 48 (33.8)	51 (65.4) / 27 (34.6)	18 (78.3) / 5 (21.7)	10 (52.6) / 9 (47.4)
Age (years)	58 ± 0.7	56 ± 0.9	58 ± 1.5	58 ± 2.1
Duration of DM (years)	12 ± 0.7	11 ± 1.8	12 ± 2.2	16 ± 2.2[a]
BMI (kg/m^2)	23.1 ± 0.41	23.7 ± 0.51	22.0 ± 0.72	22.0 ± 0.58
HbA1c (%)	7.12 ± 0.27	7.2 ± 0.21	7.41 ± 0.44	7.27 ± 0.33

[a]$P < 0.05$ vs. other groups.

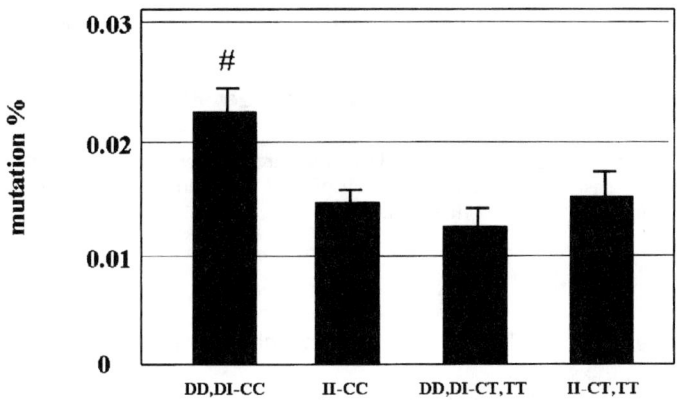

FIGURE 6. Rate of the somatic mutation of mitochondrial DNA, that is, A to G substitution at position 3243, in peripheral blood cells of patients with type 2 diabetes mellitus classified by the combination of *ACE I/D* and *p22phox C242T* genotypes. #, $P < 0.05$ compared with the other groups.

ly and the mutational rate in the longest duration group (≥ 16 years) was about twice higher than that in the shortest duration group (≤ 5 years, $P < 0.05$).

Association of ACE and p22phox Genotypes with Somatic mtDNA A3243G and IMT in Diabetic Patients

We divided total 262 diabetic patients into four groups according to the combination of *ACE I/D* and *p22phox C242T* genotypes. As shown in TABLE 3, the frequencies of *ACE I/D* or *p22phox C242T* genotypes were not significantly different from the previous reports of Japanese diabetic patients.[15,16] The estimated duration of diabetes in the group with *p22phox CT or TT + ACE II* was significantly longer than

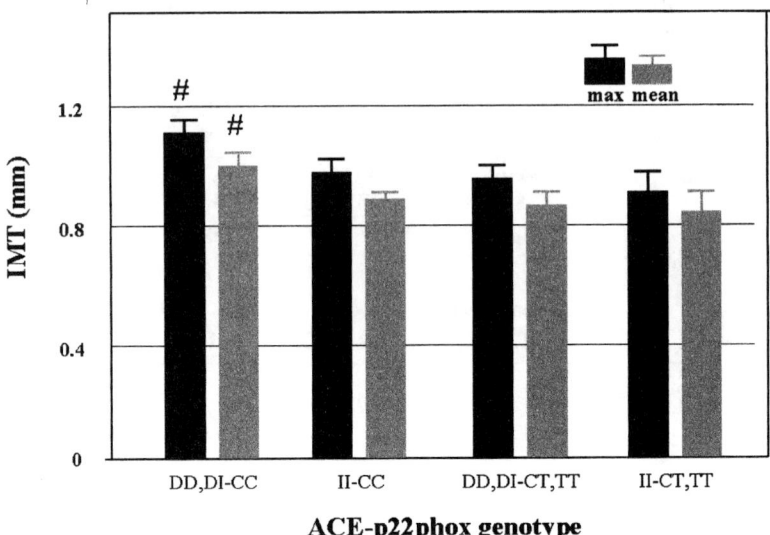

FIGURE 7. Maximum and mean IMT of the carotid artery in patients with type 2 diabetes mellitus classified by the combination of *ACE I/D* and *p22phox C242T* genotypes. #, $P < 0.05$ compared with the other groups.

in the other groups. The HbA_{1c} level and other clinical factors did not differ significantly among the groups.

The mtDNA A3243G mutational rates were shown in FIGURE 6. The *p22phox CC + ACE DD* or *DI* group showed a significantly higher mutational rate than the other groups ($0.0219 \pm 0.0028\%$ vs. $0.0152 \pm 0.0011\%$, $0.0135 \pm 0.0012\%$ and $0.0154 \pm 0.0021\%$, $P < 0.05$). Also, as shown in FIGURE 7, both maximum and mean IMT were significantly higher in this group than in the other groups (maximum: 1.13 ± 0.04 mm vs. 1.00 ± 0.04 mm, 1.02 ± 0.05 mm and 1.01 ± 0.08 mm; mean: 1.01 ± 0.03 mm vs. 0.92 ± 0.04 mm, 0.94 ± 0.04 mm and 0.93 ± 0.06 mm, $P < 0.05$). In contrast, the mtDNA A3243G mutational rate and the IMT were similar among the other three groups.

DISCUSSION

In this study, we recognized the accumulation of the somatic mtDNA A3243G in the healthy subjects after birth, which is consistent with the previous report.[4] Furthermore, no mutation was detected in umbilical venous blood cells. Taken together, these results suggest that A3243G detected by our assay may be acquired as a somatic mutation. However, it is difficult to completely separate somatic mutation from clonal expansion of an inherited mtDNA A3243G; therefore, we should further investigate the A3243G mutation in umbilical blood cells from a large number of neonates. Our second finding was a marked increase of the mutational rate between the newborn and young groups, but no corresponding increase between the young and middle-aged groups. Preliminary evaluation in nondiabetic subjects older than 90 years showed markedly higher mutation rates than those found in the young and middle aged. Thus, the possibility has not been excluded that somatic mutation may increase in an age-dependent manner in healthy subjects.

Our third finding was an age-dependent increase and higher mutational rate in the diabetic subjects compared with age-matched healthy subjects. Multiple regression analysis showed a significant correlation between A3243G and the estimated duration of diabetes, but no correlation with age. In preliminary experiments, we could not detect somatic mutations in human monocytic leukemia-derived THP-1 cells cultured under a high-glucose conditions (50 mM) for 3 months (data not shown), which suggested that somatic mutation did not occur at such a short period. Therefore, considering the life span of peripheral blood cells, mtDNA A3243G in peripheral blood cells may reflect accumulation of somatic mutations in bone marrow stem cells. It was reported that 84% of subjects with impaired glucose tolerance and diabetes shared a 4,977–base pair deletion of mtDNA, whereas only 41% of age-matched healthy subjects had this deletion.[17] Recently, it was also reported that JCR:LA-corpulent rats which were characterized by insulin resistance, compensatory hyperinsulinemia, hypertriglyceridemia, and atherosclerosis showed mtDNA deletion and mutation by both aging and hyperglycemia.[18] These reports evaluated somatic mutations in postmitotic tissues, and the findings suggest that diabetes may accelerate age-induced somatic mutation of mtDNA. Thus, it may be interesting to examine whether or not this common deletion is detectable in peripheral blood cells.

Currently, there is no clinical marker that results persistent hyperglycemia for a longer period than HbA_{1c}. Because the somatic A3243G seems to show a diabetes duration–dependent increase, this may be a potent marker of long-term hyperglycemia. A persistent increase of ROS due to sustained hyperglycemia or insulin resistance–induced compensatory hyperinsulinemia may cause not only somatic mutation, but also the development of diabetic complications.[19] Thus, it is necessary to assess the role of long-term enhancement of ROS in the onset of diabetic complications. Preliminarily, we detected higher mutational rate in the diabetic patients with severe retinopathy or nephropathy showed higher level of the somatic A3243G mutation than in the patients without any complications (data to be published elsewhere). Thus, somatic A3243G also may be a useful marker of persistent oxidative stress and microvascular complications.

Furthermore, we evaluated the relation of the somatic A3243G and the carotid artery IMT of genotypes. *ACE I/D* and *p22phox C242T* were recognized to be associated with atherosclerosis. Because angiotensin II, which is converted from angiotensin I by ACE, produces intracellular ROS through activation of p22phox,[13] it is suggested that the combination of *ACE I/D* and *p22phox C242T* may associate the production of intracellular ROS and the progression of atherosclerosis. As shown in FIGURES 6 and 7, the combination of genotypes, each of which is atherogenic combination according to previous repots, has the higher mutational rate and IMT than the other combinations. This suggests that the somatic mtDNA A3243G may be a marker of atherosclerosis, and that the mutational rate is affected by the genotypes.

In this study, we demonstrated four important findings: (1) the somatic mtDNA A3243G accumulates even in healthy subjects after birth,[20] (2) diabetic patients have a fourfold higher mutational rate than healthy subjects,[20] (3) the somatic mtDNA A3243G may be a marker of the long-term duration of diabetes,[20] and (4) the somatic mtDNA A3243G may be a marker of atherosclerosis.[21] We hope that further study confirms these points and that somatic mtDNA A3243G will be a useful clinical marker in future.

REFERENCES

1. VAN DEN OUWELAND, J.M.W. 1992. Mutations in mitochondrial tRNA$^{Leu(UUR)}$ gene in a large pedigree with maternally transmitted type II diabetes mellitus and deafness. Nat. Genet. **1:** 368–371.
2. KADOWAKI, T. 1994. A subtype of diabetes mellitus associated with a mutation of mitochondrial DNA. N. Engl. J. Med. **330:** 962–968.
3. KOBAYASHI, Y. 1990. A point mutation in the mitochondrial tRNA(Leu)(UUR) gene in MELAS (mitochondrial myopathy, encephalopathy, lactic acidosis and stroke-like episodes). Biochem. Biophys. Res. Commun. **173:** 816–822.
4. MUNSCHER, C. 1993. Human aging is associated with various point mutations in tRNA gene of mitochondrial DNA. Biol. Chem. Hoppe-Seyler **374:** 1099–1104.
5. CORRAL-DEBRINSKI, M. 1992. Mitochondrial DNA deletion in human brain: regional variability and increase with advanced age. Nat. Genet. **2:** 324–329.
6. OZAWA, T. 1995. Mechanism of somatic mitochondrial DNA mutations associated with age and diseases. Biochem. Biophys. Acta **1271:** 177–189.
7. KADENBACH, B. 1995. Human aging is associated with stochastic somatic mutations of mitochondrial DNA. Mutat. Res. **338:** 161–172.
8. DANDONA, P. 1996. Oxidative damage to DNA in diabetes mellitus. Lancet **347:** 444–445.
9. HINOKIO, Y. 1999. Oxidative damage in diabetes mellitus: its association with diabetic complications. Diabetologia **42:** 995–998.
10. KAWAMORI, R. 1995. Asymptomatic hyperglycemia and early atherosclerotic changes. Diabetes Res. Clin. Pract. **40** (Suppl.): S35–S42.
11. NISHIKAWA, T. 2000. Normalizing mitochondrial superoxide production blocks three pathways of hyperglycemic damage. Nature **404:** 787–790.
12. CAMBIEN, F. 1992. Deletion polymorphism in the gene for angiotensin-converting enzyme is a potent risk factor for myocardial infarction. Nature **359:** 641–644.
13. VIEDT, C. 2000. Differential activation of mitogen-activated protein kinases in smooth muscle cells by angiotensin II: involvement of P22phox and reactive oxygen species. Arterioscler. Thromb. Vasc. Biol. **20:** 949–948.
14. HAYASHI, J. 1994. Nuclear but not mitochondrial genome involvement in human age-related mitochondrial dysfunction. Functional integrity of mitochondrial DNA from aged subjects. J. Biol. Chem. **269:** 6878–6883.
15. INOUE, N. 1998. Polymorphism of the NADH/NADPH oxidase p22phox gene in patients with coronary artery disease. Circulation **97:** 135–137.
16. KOGAWA, K. 1997. Effect of polymorphism of apoplipoprotein E and angiotensis-converting enzyme genes on arterial wall thickness. Diabetes **46:** 682–687.
17. LIANG, P. 1997. Increased prevalence of mitochondrial DNA deletions in skeletal muscle of older individuals with impaired glucose tolerance: possible marker of glycemic stress. Diabetes **46:** 920–923.
18. FUKAGAWA, N.K. 1999. Aging and high concentrations of glucose potentiate injury to mitochondrial DNA. Free Radic. Biol. Med. **27:** 1437–1443.
19. SUZUKI, S. 1999. Oxidative damage to mitochondrial DNA and its relationship to diabetic complications. Diabetes Res. Clin. Pract. **45:** 161–168.
20. NOMIYAMA, T. 2002. Accumulation of somatic mutation in mitochondrial DNA extracted from peripheral blood cells in diabetic patients. Diabetologia **45:** 1577–1583.
21. PIAO, L.S. 2002. Combined genotypes of ACE and NADPH oxidase p22phox associated with somatic mutation of mtDNA and carotid intima-media thickness in Japanese patients with type 2 diabetes mellitus. Curr. Ther. Res. **12:** 842–852.

Changes of Mitochondrial DNA Content in the Male Offspring of Protein-Malnourished Rats

HYEONG KYU PARK,[a] CHENG JI JIN,[b] YOUNG MIN CHO,[b,c] DO JOON PARK,[b,c] CHAN SOO SHIN,[b,c] KYONG SOO PARK,[b,c] SEONG YEON KIM,[b,c] BO YOUN CHO,[b,c] AND HONG KYU LEE[b,c]

[a]*Department of Internal Medicine, SoonChunHyang University College of Medicine, Seoul, Korea*

[b]*Metabolism and Hormone Research Center, Clinical Research Institute Seoul National University Hospital, Seoul, Korea*

[c]*Department of Internal Medicine, Seoul National University College of Medicine, Seoul, Korea*

ABSTRACT: Nutritional deprivation of the fetus and infant is associated with susceptibility to the development of impaired glucose tolerance or type 2 diabetes in adult life. Quantitative changes in mitochondrial DNA (mtDNA) seem to be associated with type 2 diabetes, but the effect of protein malnutrition on mtDNA content is not known. This study investigated the effects of protein malnutrition in fetus and early life on mtDNA content and glucose-insulin metabolism in adult life. Male offspring of dams fed a low-protein (LP) diet (8% casein) during pregnancy and lactation were weaned onto either a control (18% casein) diet (recuperated group, R) or a LP diet, and they were compared with the control group (C). The mtDNA content in the liver was lower in the R and LP groups than in the C group at 5 weeks of age, but higher in the R and LP groups than in the C group at 15 weeks of age. The mtDNA content in skeletal muscle and pancreas was significantly lower in the R and LP groups than in the C group at 25 weeks of age. Fetal-malnourished rats showed decreased pancreatic β-cell mass and reduced insulin secretory responses to glucose load, but no differences in glucose tolerance or insulin sensitivity. Our findings imply that protein malnutrition *in utero* causes changes in mtDNA content, impaired β-cell development, and insulin secretion, which may contribute to the development of type 2 diabetes in later life.

KEYWORDS: protein malnutrition; mitochondrial DNA (mtDNA); *in utero*; type 2 diabetes

INTRODUCTION

Numerous reports have demonstrated a relation between low birth weight and impaired glucose tolerance or type 2 diabetes in later life.[1–4] To explain the association

Address for correspondence: Hong Kyu Lee, M.D., Department of Internal Medicine, Seoul National University College of Medicine, 28 Yongon-dong, Chongno-gu, Seoul, 110-744, South Korea. Voice: +82-2-760-2266; fax: +82-2-765-7966.
  hkleemd@plaza.snu.ac.kr

between reduced fetal growth and glucose intolerance in adult life, Hales and colleagues[5] first proposed the "thrifty phenotype" hypothesis. The thrifty phenotype implicates fetal malnutrition in the development of type 2 diabetes; nutritional deprivation leads to long-term changes in the structure or function of certain organs and tissues, that is, programming, which predisposes individuals to type 2 diabetes in adult life. It has been confirmed subsequently by epidemiologic studies.[6,7]

In addition to epidemiologic research, animal studies have demonstrated that the endocrine pancreas development and insulin secretory response to glucose are impaired in fetal-malnourished rats.[8,9] In recent experiments, the activities of insulin-sensitive enzymes in liver and insulin sensitivity of adipocytes were altered in the offspring of protein-malnourished rats.[10,11] However, the mechanisms underlying the relation between low birth weight and impaired glucose tolerance or type 2 diabetes are largely unclear.

The mitochondrion is the major site of intracellular respiration and energy metabolism. It contains its own genome, mitochondrial DNA (mtDNA), encoding some proteins of the respiratory chain, responsible for electron transport and ATP synthesis.[12] Mutations and deletions in mtDNA have been implicated in the pathogenesis of diabetes mellitus.[13-15] In addition to qualitative changes, quantitative changes in mtDNA seem to be associated with type 2 diabetes. The mtDNA copy number was decreased in the skeletal muscle and peripheral blood leukocytes of type 2 diabetic patients.[16,17] The mtDNA is known to be exclusively maternally transmitted. Therefore, mtDNA content could be affected by protein malnutrition *in utero*. In the current study, we examined the effects of protein malnutrition during fetal and early life on mtDNA content. We also examined insulin secretory responses to glucose load and whole body insulin sensitivity in a model of fetal malnutrition to explore whether mtDNA content may be related to the development of type 2 diabetes in later life.

METHODS

Animals and Material

Male and female Sprague-Dawley rats were obtained from Daehan Laboratory Animal Research Ltd. (Seoul, Korea) and housed in wire-meshed cages in a temperature- (21 ± 2°C) and humidity-controlled (55%) room on a standard 12:12-hour light-dark cycle (light from 7:00 AM). The animal facilities were free of specific pathogens. All rats were freely given a standard rat diet (Samyang, Wonju, Korea) and tap water. The special diet was purchased from Harlan-Teklad Ltd. (Madison, WI). Control diet and low protein diet were isocaloric but only differed in protein content (TABLE 1). All experimental procedures were approved by the ethnic committee for animal experiments at Seoul National University Hospital.

Study Design

Eight-week-old female rats were divided into two groups and fed either a control diet (casein 180 g/kg diet) or a low-protein diet (casein 80 g/kg diet) from 15 days before mating to the end of the weaning period. They were mated with 10-week-old

TABLE 1. Composition of control and low-protein diets

Ingredient	Control diet 18% casein diet (g/kg)	Low-protein diet 8% casein diet (g/kg)
Casein	180.0	80.0
dl-Methionine	0.0	2.0
Corn starch	677.85	774.99
Peanut oil	80.0	80.0
Cellulose	20.0	20.0
Vitamin mix	10.0	10.0
Mineral mix	13.4	13.4
Calcium phosphate	16.5	19.51
Calcium carbonate	2.25	0.1

male Sprague-Dawley rats fed a control diet. Male offspring born from low-protein-fed mothers were randomly divided into two groups at 4 weeks after birth and weaned onto either a low protein (low-protein group) or a control diet (recuperated group). As a control group, male offspring born from control diet-fed mothers were weaned onto a control diet. Animals in each group were randomly divided again into three groups and sacrificed at 5, 15, and 25 weeks of age, respectively ($n = 7$ per group). Body weight was measured before the animals were sacrificed. Liver, quadriceps muscle, and pancreas were quickly collected under ether anesthesia, immediately frozen in liquid nitrogen, and stored at $-70°C$ until measurement. At 25 weeks of age, all rats in each group underwent an intravenous glucose tolerance test or measurement of *in vivo* insulin action by the glucose-insulin clamp technique.

Quantitation of mtDNA

Frozen specimens were homogenized and DNA was totally extracted using a DNeasy Tissue Kit (Qiagen, Hilden, Germany), and the total concentration of DNA was measured with a spectrophotometer (Beckman, Fullerton, CA). To facilitate the quantitation of mtDNA from a small amount of samples, we developed quantitative polymerase chain reaction (PCR). The internal standard was designed to use the same primer set as the target gene but to yield a different sized PCR product (694 versus 770 bp), and was prepared by PCR using specially designed primers (TABLE 1). Two independent PCR amplifications using sets of MT3: 5'-AGGACTTAACCA-GACCCAAACACG-3' (4395-4418) and Dr1: 5'-TGTTGGTTGG//GTTGAT-AGGGTTGAGCAGTT-3' (4721-4730, 4807-4826), and MT4: 5'-CCTCTTTTCTGATAGGCGGG-3' (5145-5164) and Df1: 5'-CCCTATCAAC//CCAACCAACAACAACTCCAA-3' (4711-4730, 4807-4816) produced 346 and 369 bp, respectively. Secondary PCR amplification using the aforementioned products and primers, MT3 and MT4, produced a 694-bp fragment containing sequences from mtDNA positions 4395-4730 and 4807-5164, with deletion of the intervening 76 bp (from position 4731 to 4806). A known amount of serially diluted standard plasmid was added to 50 ng of total cellular DNA and subjected to PCR using a set of primers (MT3 and MT4). The PCR mixture contained 5 pmol of each primer, 175 µM of each dNTP, 1 unit of Taq polymerase, 10 mM Tris-HCl, 1.5 mM $MgCl_2$, 50 mM KCl, 0.05% Tween 20, and 0.001% gelatin. Reaction took place under the following condition: one cycle of 3 min at 94°C, 30 s at 60°C, 40 s at 72°C, and 29 cycles of 30 s at 94°C, 30 s at 60°C, 40 s at 72°C, and final extension of 5 min at 72°C. The PCR product was analyzed on 1.5% agarose gel by electrophoresis. Gels were stained with ethidium bromide and photographed under UV light, and intensities of the target DNA band (770 bp) and competitor band (694 bp) were quantitated using TINA 2.0 Image software of BAS 2500 (Fujifilm, Tokyo). The ratio of each target DNA/internal standard product was plotted against the number of copies of internal standard added to yield the equivalence point between internal standard and target DNA. The r value of the standard curve was between 0.95 and 1.00. The interassay variance of mtDNA measurement was 12.2%.

In Vivo *Glucose-Induced Insulin Secretion Tests*

Each rat was fitted with a catheter (PE-10, Intramedic, Clay Adams, Parsippany, NJ). Food was removed at 8:00 AM, and intravenous glucose tolerance tests (IVGTT) were performed in awake, unstressed 25-week-old rats 6 h after food withdrawal. A single injection of glucose (0.5 g glucose/kg body wt) was administered via a tail vein. The catheter was then flushed with saline. Blood samples (200 µL) were collected sequentially from the tail artery before (0 min) and at 2, 6, 10, 20, and 50 min and were stored on ice. They were then centrifuged, and the plasma was separated. Plasma glucose concentration was immediately determined, and the remaining plasma was stored at –70 °C until assayed for insulin.

Euglycemic-Hyperinsulinemic Clamp Studies

Each rat underwent placement of catheters into two tail veins and a tail artery for infusion and blood sampling, respectively. Food was removed at 8:00 AM, and the animal was transferred to a quiet isolated room and weighed. Whole body glucose kinetics were estimated in awake, unstressed 25-week-old rats 6 h after food withdrawal. Animals were allowed to rest for 40 min before the first blood sample (200 µL). Patency of the arterial line was maintained by a slow (0.015 mL/min) infusion of saline. A continuous intravenous infusion of purified human insulin (Novolin R; Novo Nordisk, Bagsvaerd, Denmark) was then started at a rate of 72 pmol/kg/min and continued for 120 min with a syringe pump (Medfusion 2010i, Medexinc, GA) through a tail vein. Insulin was dissolved in 0.9% NaCl containing 0.2% bovine serum albumin (Sigma, St. Louis, MO). Blood samples (50 µL) were taken from the tail artery and immediately centrifuged for glucose measurement at 10-min intervals. Blood samples of 200 µL were collected 60 and 120 min after insulin infusion for determination of plasma insulin concentrations. Steady-state plasma glucose levels were reached after 40–50 min. Dextrose solution (25%) was infused through the other venous line at variable rates to maintain plasma glucose at basal levels.

β-*Cell Immunohistochemistry and Morphometry*

After excision, whole pancreases were immediately weighed, fixed in 10% formalin solution overnight, and embedded in paraffin for immunostaining of insulin. Each pancreas was subsequently sectioned (5 µm thick) throughout its length, and about 10 sections taken at regular intervals were immunostained for insulin with a technique adapted from the peroxidase indirect labeling method, as previously described.[18] Labeling was performed using a peroxidase-conjugated rabbit anti-guinea pig IgG. Activity was revealed with a peroxidase substrate kit. The area of insulin-positive cells and the total area of pancreatic cells were evaluated in each section. The total β-cell mass per pancreas was derived by multiplying the β-cell volume by the total pancreatic weight.

Analysis and Statistics

Plasma glucose levels were analyzed using a glucose oxidase method (YSI 2300; Yellow Springs, OH), and plasma insulin levels were measured by radioimmunoassay using kits for rat insulin (IVGTT and clamp basal insulin) and human insulin (clamp insulin; Linco, St. Charles, MO). Results are given as means ± SE. Statistical analyses were performed using one-way ANOVA for comparisons of unpaired data between groups. Individual between-group comparisons were performed using a least significance difference test after ANOVA. Differences were considered significant at $P < 0.05$.

RESULTS

Body Weight Gain

The average body weights were lower in the low-protein group than in the control group at all ages examined (at 15 weeks: low-protein 315 ± 18 g; control 471 ± 6 g, $P < 0.001$; FIG. 1). In the recuperated group, the average body weight was similar to that of the low-protein group at 5 weeks of age, and thereafter, the average body weights were greater than those in the low-protein group. However, the recuperated

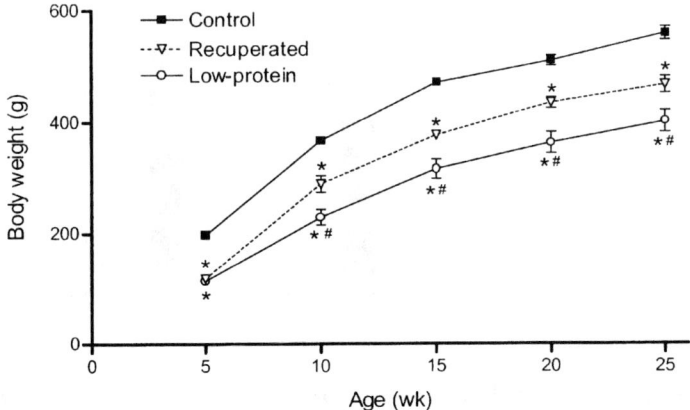

FIGURE 1. Body weight curves of rats. Body weights were lower in the recuperated and low-protein groups than in the control group at all time points studied (*$P < 0.005$ vs. control group). Postnatal protein malnutrition resulted in further lower weight gain (#$P < 0.05$ vs. recuperated group).

group never caught up in weight and remained smaller than the control group throughout the experiment (at 25 weeks: control 559 ± 12 g, recuperated 467 ± 15 g, $P < 0.001$ vs. control, low-protein 401 ± 19 g, $P < 0.05$ vs. control and recuperated; FIG. 1).

Effect of Fetal Protein Malnutrition on mtDNA Content

The mtDNA content increased with age in skeletal muscle, but in liver it decreased slightly at 15 weeks of age and increased afterward (FIG. 2). The mtDNA content in liver was lower in the recuperated and low-protein groups than in the control group at 5 weeks of age (control 84.9 ± 7.7; recuperated 39.6 ± 7.7, $P < 0.001$ vs. control; low-protein 34.3 ± 3.8 amol/100 ng DNA, $P < 0.001$ vs. control; FIG. 2A),

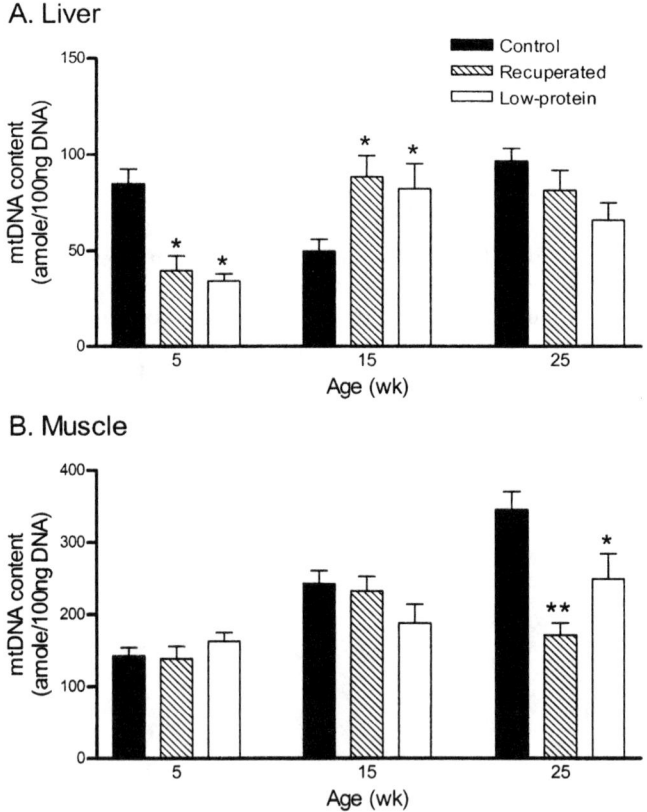

FIGURE 2. Mitochondrial DNA content in liver (**A**) and skeletal muscle (**B**) of rats. In liver, mtDNA content was lower in recuperated and low-protein group than control group at 5 weeks of age, but higher in recuperated and low-protein group than control group at 15 weeks of age (*$P < 0.05$ vs. control group). In skeletal muscle, mtDNA content was significantly lower in recuperated and low-protein groups than control group (*$P < 0.05$, **$P < 0.01$ vs. control group).

TABLE 2. Results of IVGTT and euglycemic hyperinsulinemic clamp in rats at 25 weeks of age

	Control	Recuperated	Low-protein
Fasting plasma glucose (mmol/L)	6.47 ± 0.20	6.11 ± 0.14	6.29 ± 0.11
Fasting plasma insulin (nmol/L)	0.32 ± 0.03	0.24 ± 0.02	0.30 ± 0.02
AUC glucose (mmol/L/min)	9.53 ± 0.11	9.26 ± 0.18	9.32 ± 0.16
AUC insulin (nmol/L/min)	0.54 ± 0.03	0.39 ± 0.03**	0.43 ± 0.02*
Insulin sensitivity index	10.7 ± 0.6	9.2 ± 0.7	11.2 ± 0.7

Values are means ± SE; AUC, areas under the curve. Insulin sensitivity index was calculated from glucose infusion rates divided by steady-state plasma insulin concentrations and expressed with arbitrary unit. AUC insulin were significantly decreased in recuperated and low-protein group than control group (*$P < 0.05$, **$P < 0.005$ vs. control group).

but it was higher in the recuperated and low-protein groups than in the control group at 15 weeks of age (control 49.8 ± 6.2, recuperated 88.5 ± 11.1, $P <0.05$ vs. control, low-protein 82.3 ± 13.2 amol/100 ng DNA, $P <0.05$ vs. control). In skeletal muscle, the mtDNA content did not show any differences at 5 and 15 weeks of age. At 25 weeks of age, however, the mtDNA content was significantly lower in the recuperated and low-protein groups than in the control group (control 345.4 ± 25.2, recuperated 171.6 ± 16.7; $P <0.001$ vs. control; low-protein 249.8 ± 34.8 amol/100 ng DNA, $P <0.05$ vs. control; FIG. 2B). As in skeletal muscle, the mtDNA content in pancreatic tissue was decreased in both the recuperated and the low-protein groups compared with the control group at 25 weeks of age (control 43.7 ± 2.6, recuperated 20.7 ± 2.4, $P <0.001$ vs. control; low-protein 33.0 ± 3.2 amol/250 ng DNA, $P <0.05$ vs. control; TABLE 3).

Insulin Secretion, Insulin Sensitivity, and Pancreatic β-Cell Development

Fasting plasma glucose and insulin concentrations were not significantly different among the three groups (TABLE 2). Plasma insulin concentrations 20 min after intravenous glucose administration were significantly lower in the recuperated and low-protein groups than in the control group (control 0.43 ± 0.05, recuperated 0.24 ± 0.03, $P <0.005$ vs. control; low-protein 0.28 ± 0.02 nmol/L, $P <0.05$ vs. control; FIG. 3). In response to intravenous glucose load, the mean insulin areas under the curve (AUC) in the recuperated and low-protein groups were decreased compared with those in the control group. Whole body insulin sensitivity was measured during hyperinsulinemic clamp studies. Insulin sensitivity index was calculated from glucose infusion rates divided by steady-state plasma insulin concentrations. No significant difference in the insulin sensitivity index was noted among the three groups. The 25-week-old fetal protein-malnourished rats showed a pancreas weight significantly lower than that of the control group and the difference persisted when values were referred to body weight (TABLE 3). The β-cell mass per body weight was significantly lower in the low-protein group than in the control group.

TABLE 3. Pancreas and β-cell development in rats at 25 weeks of age

	Control Group	Recuperated Group	Low-protein Group
Pancreas weight (mg)	873 ± 37	546 ± 27**	558 ± 40**
Pancreas wt / body wt (arbitrary unit)	0.17 ± 0.01	0.12 ± 0.01**	0.14 ± 0.01*
Relative β-cell mass (μg/g body wt)	34.8 ± 5.6	22.4 ± 3.9	13.2 ± 3.2*
Pancreas mtDNA content	43.7 ± 2.6	20.7 ± 2.4*	33.0 ± 3.2*

Values are means ± SE. Pancreas mtDNA was quantitated in 250 ng DNA of pancreatic tissue. Recuperated and low-protein groups showed a pancreas weight significantly lower than that of the control group, and the difference still persisted when values were expressed as the value per body weight. The relative β-cell mass was significantly lower in the low-protein group than in the control group. Pancreatic mtDNA content was significantly lower in the recuperated and low-protein groups than in the control group ($*P < 0.05$, $**P < 0.001$ vs. control group).

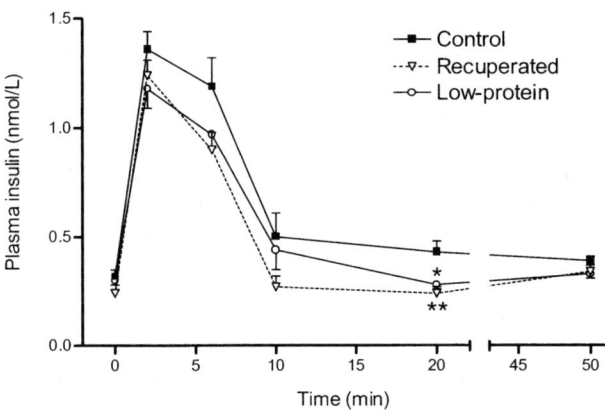

FIGURE 3. Changes in plasma insulin concentrations to intravenous glucose tolerance tests (IVGTT) in rats at 25 weeks of age. Plasma insulin concentrations in the recuperated and low-protein groups were significantly lower than in the control group at 20 min after an IV glucose load ($*P < 0.05$, $**P < 0.005$ vs. control group).

DISCUSSION

Previous epidemiologic studies and experimental animal research have demonstrated that fetal malnutrition is associated with reduced insulin secretion or impaired glucose tolerance in adult life. In the curent study, we found that protein malnutrition *in utero* causes changes in mtDNA content, and decreased β-cell mass and impaired insulin secretion in adulthood. However, the offspring of protein-malnourished rats showed no differences in glucose tolerance or insulin sensitivity in adulthood compared with controls.

The mtDNA content changed differently according to tissue and age. In liver, mtDNA content was significantly lower in offspring of protein-malnourished rats at 5 weeks, but higher than controls at 15 weeks of age. On the contrary, mtDNA content in skeletal muscle was significantly lower in the offspring of protein-malnourished rats than in controls at 25 weeks of age. How the mtDNA content in animal cells is controlled is not yet clear. However, the mtDNA copy number seems to be affected by cellular metabolic demand and aging. The mtDNA increased proportionately with the increase in oxidative capacity in skeletal muscle.[19,20] An age-related reduction of mtDNA copy number proportional to tissue oxidative capacities was demonstrated in skeletal muscle and liver.[21] Thus, the alterations of mtDNA content in liver may imply altered metabolic demands during postnatal growth in the offspring of protein-malnourished rats.

The decrease in mtDNA content seems to be associated with type 2 diabetes. The mtDNA copy number is decreased in skeletal muscle and peripheral blood of type 2 diabetic patients, and lower mtDNA levels in peripheral blood precede the development of diabetes.[16,17] Moreover, peripheral blood mtDNA content was related to insulin sensitivity in offspring of type 2 diabetic patients.[22] In this study, insulin sensitivity did not differ among the three rat groups in adulthood, although mtDNA content in skeletal muscle was significantly decreased in offspring of protein-malnourished rats. Therefore, it does not seem that mtDNA content in skeletal muscle is directly related to insulin sensitivity.

Protein malnutrition is associated with reduced antioxidative enzyme activities, and mtDNA is known to be vulnerable to oxidative stress.[23,24] Thus, protein malnutrition may, in part, contribute to the decrease in mtDNA content. It is of interest that the decrease in mtDNA content in skeletal muscle was greater in fetal-malnourished rats weaned onto a normal diet than in rats persistently fed a low-protein diet. This may imply that mtDNA content is affected by changes in nutritional environment as well as by poor intrauterine nutrition. This finding is in line with the thrifty phenotype hypothesis that metabolic adaptations by the undernourished fetus may lead to adverse consequences later in life under conditions of both adequate nutrition and overnutrition. It has been proposed that mitochondrial gene expression is mainly dependent on mtDNA copy number.[19] Thus, the changes in mtDNA content in liver and skeletal muscle observed in offspring of protein-malnourished rats might be related to alterations in tissue oxidative capacities and impairment of insulin action, although we did not measure mitochondrial encoded transcript levels in this study. Recently, a defect in mitochondria with impaired ATP synthesis was demonstrated in insulin-resistant skeletal muscle of intrauterine growth-retarded rats.[25] Therefore, it is possible that intrauterine malnutrition causes changes of mtDNA content that may affect ATP synthesis and thus glucose-insulin metabolism in later life.

In the current study, offspring of protein-malnourished rats did not show any glucose intolerance or insulin resistance until 25 weeks of age. In an animal model of undernourishment *in utero*, glucose tolerance was impaired in a 15-month-old offspring.[26] In this offspring, several abnormalities were found. Hepatic glucokinase activity and phosphoenolpyruvate carboxykinase (PEPCK) activity were altered in a direction expected to result in an insulin-resistant liver.[10] Impaired phosphatidylinositol 3-kinase activation was also observed in adipocytes.[11] Recently, it was shown that protein malnutrition *in utero* resulted in reduced insulin response in first-generation rats, but in a markedly elevated insulin response following a glucose chal-

The mtDNA content changed differently according to tissue and age. In liver, mtDNA content was significantly lower in offspring of protein-malnourished rats at 5 weeks, but higher than controls at 15 weeks of age. On the contrary, mtDNA content in skeletal muscle was significantly lower in the offspring of protein-malnourished rats than in controls at 25 weeks of age. How the mtDNA content in animal cells is controlled is not yet clear. However, the mtDNA copy number seems to be affected by cellular metabolic demand and aging. The mtDNA increased proportionately with the increase in oxidative capacity in skeletal muscle.[19,20] An age-related reduction of mtDNA copy number proportional to tissue oxidative capacities was demonstrated in skeletal muscle and liver.[21] Thus, the alterations of mtDNA content in liver may imply altered metabolic demands during postnatal growth in the offspring of protein-malnourished rats.

The decrease in mtDNA content seems to be associated with type 2 diabetes. The mtDNA copy number is decreased in skeletal muscle and peripheral blood of type 2 diabetic patients, and lower mtDNA levels in peripheral blood precede the development of diabetes.[16,17] Moreover, peripheral blood mtDNA content was related to insulin sensitivity in offspring of type 2 diabetic patients.[22] In this study, insulin sensitivity did not differ among the three rat groups in adulthood, although mtDNA content in skeletal muscle was significantly decreased in offspring of protein-malnourished rats. Therefore, it does not seem that mtDNA content in skeletal muscle is directly related to insulin sensitivity.

Protein malnutrition is associated with reduced antioxidative enzyme activities, and mtDNA is known to be vulnerable to oxidative stress.[23,24] Thus, protein malnutrition may, in part, contribute to the decrease in mtDNA content. It is of interest that the decrease in mtDNA content in skeletal muscle was greater in fetal-malnourished rats weaned onto a normal diet than in rats persistently fed a low-protein diet. This may imply that mtDNA content is affected by changes in nutritional environment as well as by poor intrauterine nutrition. This finding is in line with the thrifty phenotype hypothesis that metabolic adaptations by the undernourished fetus may lead to adverse consequences later in life under conditions of both adequate nutrition and overnutrition. It has been proposed that mitochondrial gene expression is mainly dependent on mtDNA copy number.[19] Thus, the changes in mtDNA content in liver and skeletal muscle observed in offspring of protein-malnourished rats might be related to alterations in tissue oxidative capacities and impairment of insulin action, although we did not measure mitochondrial encoded transcript levels in this study. Recently, a defect in mitochondria with impaired ATP synthesis was demonstrated in insulin-resistant skeletal muscle of intrauterine growth-retarded rats.[25] Therefore, it is possible that intrauterine malnutrition causes changes of mtDNA content that may affect ATP synthesis and thus glucose-insulin metabolism in later life.

In the current study, offspring of protein-malnourished rats did not show any glucose intolerance or insulin resistance until 25 weeks of age. In an animal model of undernourishment *in utero*, glucose tolerance was impaired in a 15-month-old offspring.[26] In this offspring, several abnormalities were found. Hepatic glucokinase activity and phosphoenolpyruvate carboxykinase (PEPCK) activity were altered in a direction expected to result in an insulin-resistant liver.[10] Impaired phosphatidylinositol 3-kinase activation was also observed in adipocytes.[11] Recently, it was shown that protein malnutrition *in utero* resulted in reduced insulin response in first-generation rats, but in a markedly elevated insulin response following a glucose chal-

23. HUANG, C.J. & M.L. FWU. 1993. Degree of protein deficiency affects the extent of the depression of the antioxidant enzyme activities and the enhancement of tissue lipid peroxidation in rats. J. Nutr. **123:** 803–810.
24. RICHTER, C. 1992. Reactive oxygen and DNA damage in mitochondria. Mutat. Res. **275:** 249–255.
25. SELAK, M.A., B.T. STOREY, I. PETERSIDE, et al. 2003. Impaired oxidative phosphorylation in skeletal muscle of intrauterine growth retarded rats. Am. J. Physiol. Endocrinol. Metab. **285:** E130–137.
26. HALES, C.N., M. DESAI, S.E. OZANNE, et al. 1996. Fishing in the stream of diabetes: from measuring insulin to the control of fetal organogenesis. Biochem. Soc. Trans. **24:** 341–350.
27. MARTIN, J.F., C.S. JOHNSTON, C.-T. HAN, et al. 2000. Nutritional origins of insulin resistance: a rat model for diabetes-prone human populations. J. Nutr. **130:** 741–744.
28. SWENNE, I., C.J. CRACE & R.D.G. MILNER. 1987. Persistent impairment of insulin secretory response to glucose in adult rats after limited period of protein-calorie malnutrition early in life. Diabetes **36:** 454–458.
29. SIMMONS, R.A., L.J. TEMPLETON & S.J. GERTZ. 2001. Intrauterine growth retardation leads to the development of type 2 diabetes in the rat. Diabetes **50:** 2279–2296.
30. GAROFANO, A., P. CZERNICHOW & B. BREANT. 1997. In utero undernutrition impairs rat beta-cell development. Diabetologia **40:** 1231–1234.
31. BERTIN, E., M.N. GANGNERAU, D. BAILBE, et al. 1999. Glucose metabolism and beta-cell mass in adult offspring of rats protein and/or energy restricted during the last week of pregnancy. Am. J. Physiol. **277**(EM 40): E11–17.
32. PETRIK, J., E. REUSSENS, E. ARANY, et al. 1999. A low protein diet alters the balance of islet cell replication and apoptosis in the fetal and neonatal rat and is associated with a reduced pancreatic expression of insulin-like growth factor II. Endocrinology **140:** 4861–4873.
33. SOEJIMA, A., K. INOUE, D. TAKAI, et al. 1996. Mitochondrial DNA is required for regulation of glucose-stimulated insulin secretion in a mouse pancreatic beta cell line, MIN6. J. Biol. Chem. **271:** 26194–26199.
34. KENNEDY, E.D., P. MAECHLER & C.B. WOLLHEIM. 1998. Effects of depletion of mitochondrial DNA on metabolism-secretion coupling in INS-1 cells. Diabetes **47:** 374–380.
35. SERRADAS, P., M.-H. GIROIX, C. SAULNIER, et al. 1995. Mitochondrial deoxyribonucleic acid content is specifically decreased in adult, but not fetal, pancreatic islets of the Goto-Kakizaki rat, a genetic model of noninsulin-dependent diabetes. Endocrinology **136:** 5623–5631.
36. SILVA, J.P., M. KOHLER, C. GRAFF, et al. 2000. Impaired insulin secretion and beta-cell loss in tissue-specific knock-out mice with mitochondrial diabetes. Nat. Genet. **26:** 336–340.
37. RASSCHAERT, J., B. REUSENS, S. DAHRI, et al. 1995 Impaired activity of rat pancreatic islet mitochondrial glycerophosphate dehydrogenase in protein malnutrition. Endocrinology **136:** 2631–2634.

REFERENCES

1. HALES, C.N., D.J.P. BARKER, P.M.S. CLARK, et al. 1991. Fetal and infant growth and impaired glucose tolerance at age 64. Br. Med. J. **313:** 1019–1022.
2. ROBINSON, S., R.J. WALTON, P.M. CLARK, et al. 1992. The relation of fetal growth to plasma glucose in young man. Diabetologia **35:** 444–446.
3. MCCANCE, D.R., D.J. PETTITT, R.L. HANSON, et al. 1994. Birth weight and non-insulin dependent diabetes: thrifty genotype, thrifty phenotype, or surviving small baby genotype? Br. Med. J. **308:** 942–945.
4. LITHELL, H.O., P.M. MCKEIGUE, L. BERGLUND, et al. 1996. Relation of size at birth to non-insulin dependent diabetes and insulin concentrations in man aged 50-60 years. Br. Med. J. **312:** 406–410.
5. HALES, C.N. & D.J.P. BARKER. 1992. Type 2 diabetes mellitus: the thrifty phenotype hypothesis. Diabetologia **35:** 595–601.
6. RICH-EDWARDS, J.W., G.A. COLDITZ, M.J. STAMPER, et al. 1999. Birth weight and the risk for type-2 diabetes mellitus in adult women. Ann. Int. Med. **130:** 278–284.
7. LI, C., M.S. JOHNSON & M.I. GORAN. 2001. Effects of low birth weight on insulin resistance syndrome in Caucasian and African-American children. Diabetes Care **24:** 2035–2042.
8. SNOECK, A., C. REMACLE, B. REUSENS, et al. 1990. Effect of a low protein diet during pregnancy on the fetal rat endocrine pancreas. Biol. Neonate **57:** 107–118.
9. DAHRI, S., A. SNOECK, B. REUSENS-BILLEN, et al. 1991. Islet function in offspring of mothers on low-protein diet during gestation. Diabetes **40:** 115–120.
10. DESAI, M., C.D. BYRNE, J. ZHANG, et al. 1997. Programming of hepatic insulin-sensitive enzymes in offspring of rats dams fed a protein-restricted diet. Am. J. Physiol. **35:** G1083–1090.
11. OZANNE, S.E., M.W. DORLING, C.L. WANG, et al. 2001. Impaired PI 3-kinase activation in adipocytes from early growth-restricted male rats. Am. J. Physiol. Endocrinol. Metab. **280:** E534–E539.
12. WALLACE, D.C. 1992. Mitochondrial genetics: a paradigm for aging and degenerative disease. Science **256:** 628–632.
13. VAN DEN OUWELAND, J.M.W., H.H.P.J. LEMKES, W. RUITENBEEK, et al. 1992. Mutation in mitochondrial tRNALeu(UUR) gene in a large pedigree with maternally transmitted type II diabetes mellitus and deafness. Nat. Genet. **1:** 368–371.
14. BALLINGER, S.W., J.M. SHOFFNER, E.V. HEDAYA, et al. 1992. Maternally transmitted diabetes and deafness associated with a 10.4 kb mitochondrial DNA deletion. Nat. Genet. **1:** 11–15.
15. REARDON, W., R.J.M. ROSS, M.G. SWEENEY, et al. 1992. Diabetes mellitus associated with a pathogenic point mutation in mitochondrial DNA. Lancet **340:** 1376–1379.
16. ANTONETTI, D.A., C. REYNET & C.R. KAHN. 1995. Increased expression of mitochondrial-encoded genes in skeletal muscle of humans with diabetes mellitus. J. Clin. Invest. **95:** 1383–1388.
17. LEE, H.K., J.H. SONG, C.S. SHIN, et al. 1998. Decreased mitochondrial DNA content in peripheral blood precedes the development of non-insulin-dependent diabetes mellitus. Diabetes Res. Clin. Pract. **42:** 161–167.
18. MICHEL, C., J. CHARIOT, M. SOUCHARD, et al. 1982. Modifications of the endocrine pancreas in rats after ethionine destruction of acini. Cell Mol. Biol. **28:** 135–148.
19. WILLIAMS, R.S. 1986. Mitochondrial gene expression in mammalian striated muscle: evidence that variation in gene dosage is the major regulatory event. J. Biol. Chem. **261:** 12390–12394.
20. PUNTSCHART, A., H. CLAASSEN, K. JOSTARNDT, et al. 1995. mRNA of enzymes involved in energy metabolism and mtDNA are increased in endurance-trained athletes. Am. J. Physiol. **269** (Cell Physiol. 38): C619–C625.
21. BARAZZONI, R., K.R. SHORT & K.S. NAIR. 2000. Effects of aging on mitochondrial DNA copy number and cytochrome c oxidase gene expression in rat skeletal muscle, liver, and heart. J. Biol. Chem. **275:** 3343–3347.
22. SONG, J.H., Y.M. KIM PAK, J.Y. OH, et al. 2001. Peripheral blood mitochondrial DNA content is related to the insulin sensitivity in offspring of type 2 diabetic patients. Diabetes Care **24:** 865–869.

Mitochondrial Encephalomyopathies

Diagnostic Approach

SALVATORE DIMAURO, STACEY TAY, AND MICHELANGELO MANCUSO

Department of Neurology, Columbia University College of Physicians & Surgeons, New York, New York 10032, USA

ABSTRACT: Mitochondrial diseases have extremely heterogeneous clinical presentations due to the ubiquitous nature of mitochondria and the dual genetic control of the respiratory chain. Thus, mitochondrial disorders can be multisystemic (mitochondrial encephalomyopathies) or confined to a single tissue, and they can be sporadic or transmitted by mendelian or maternal inheritance. Mendelian disorders are usually inherited as autosomal recessive traits, tend to present earlier in life, and usually "breed true" in each family. By contrast, mitochondrial DNA-related diseases usually start later and vary in their presentation within members of the same family. Precise diagnosis is often a challenge; we go through the traditional steps of the diagnostic process, trying to highlight clues to mitochondrial dysfunction in the family history, physical and neurological examinations, routine and special laboratory tests, and histochemical and biochemical results of the muscle biopsy. The ultimate goal is to reach, whenever possible, a definitive molecular diagnosis, which permits rational genetic counseling and a prenatal diagnosis.

KEYWORDS: mitochondria; encephalomyopathies

INTRODUCTION

Mitochondrial diseases (or mitochondrial encephalomyopathies), even when defined restrictively as disorders due to defects in the mitochondrial respiratory chain (excluding diseases caused by defects in other mitochondrial metabolic pathways, such as pyruvate oxidation, beta-oxidation, or the Krebs cycle), are relatively common, as several epidemiologic studies from Northern Europe have documented.[1-4] This stresses the importance of detection and accurate diagnosis, not an easy task for clinicians faced with a mind-boggling variety of symptoms and signs, hereditary patterns, and molecular etiologies. The protean nature of mitochondrial diseases is due to the ubiquitous presence of mitochondria, the dual genetic control of the respiratory chain, and the fact that mitochondrial genetics follows the rules of population genetics rather than the "all or none" rules of mendelian genetics.[5]

Address for correspondence: Salvatore DiMauro, M.D., 4-420 College of Physicians & Surgeons, 630 West 168th Street, New York, NY 10032. Voice: 212-305-1662; fax: 212-305-3986.
sd12@columbia.edu

As a brief reminder, the mitochondrial respiratory chain is the "business end" of oxidative metabolism, where most cellular ATP is generated through a series of biochemical and biophysical (chemiosmotic) events (FIG. 1). High-energy electrons derived from the oxidation of pyruvate or fatty acids are transported along the four complexes of the electron transport chain (complexes I to IV) and reoxidized to form water; the resulting energy is harnessed to pump protons from the inside to the outside of the inner mitochondrial membrane, thus generating an electrochemical proton gradient. As protons flow back into the mitochondrial matrix through a tiny rotary engine, ATP synthase (complex V), physical energy is reconverted to chemical energy in the form of ATP.

Thirteen proteins of the respiratory chain are encoded by mitochondrial DNA (mtDNA) and synthesized within the mitochondria: seven are subunits of complex I (ND1-ND4, ND4L, ND5, ND6), one is the cytochrome *b* (cyt *b*) subunit of complex III, three are subunits of complex IV (cytochrome *c* oxidase, COX I–III), and two are subunits of complex V (ATPase 6 and ATPase 8). All other subunits of complexes I, III, IV, and V, all of complex II, and the two mobile electron carriers, coenzyme Q10 (CoQ10) and cytochrome *c* (cyt *c*), are encoded by nuclear DNA (nDNA), synthesized and targeted to mitochondria in the cytoplasm, from which they are carried into the organelle through a complex importation machinery.[6]

The diagnostic process is no different from that employed for other diseases and includes patient and family history, physical and neurologic examination, routine and special laboratory tests, exercise physiology, muscle biopsy for morphology and biochemistry, and molecular genetic screening. We will review these tools, trying to stress how each can help unravel mitochondrial dysfunction and lead to the correct diagnosis. In so doing, we will update a previous similar review.[7]

PATIENT AND FAMILY HISTORY

Family history has to be recorded with great care because of the dual genetic control of the respiratory chain. Thus, clear evidence of maternal inheritance directs attention to mtDNA mutations, which, if the clinical presentation is sufficiently characteristic, may allow us to zero in on molecular analysis of blood or urine, bypassing a number of laboratory tests. On the other hand, mendelian inheritance indicates mutations in nDNA-encoded proteins; these can be subunits of respiratory chain complexes, ancillary proteins needed for the correct assembly of one or more complexes, components of the protein transport machinery, factors required for mtDNA replication and maintenance (intergenomic signaling), factors needed for mitochondrial fission and mitochondrial motility, and factors controlling the lipid milieu of the inner mitochondrial membrane.

The sporadic occurrence of a mitochondrial disease suggests that a single mtDNA deletion is implicated in the etiology and is characteristic of the three conditions associated with single mtDNA deletions: Kearns-Sayre syndrome (KSS), Pearson syndrome (PS), and chronic progressive external ophthalmoplegia (CPEO).

Maternal inheritance may be difficult to recognize because of the quantitative nature of mitochondrial genetics, illustrated by polyplasmy, heteroplasmy, and mitotic segregation. *Polyplasmy* indicates that each cell contains multiple copies of mtDNA, which in normal individuals are identical to one another (*homoplasmy*).

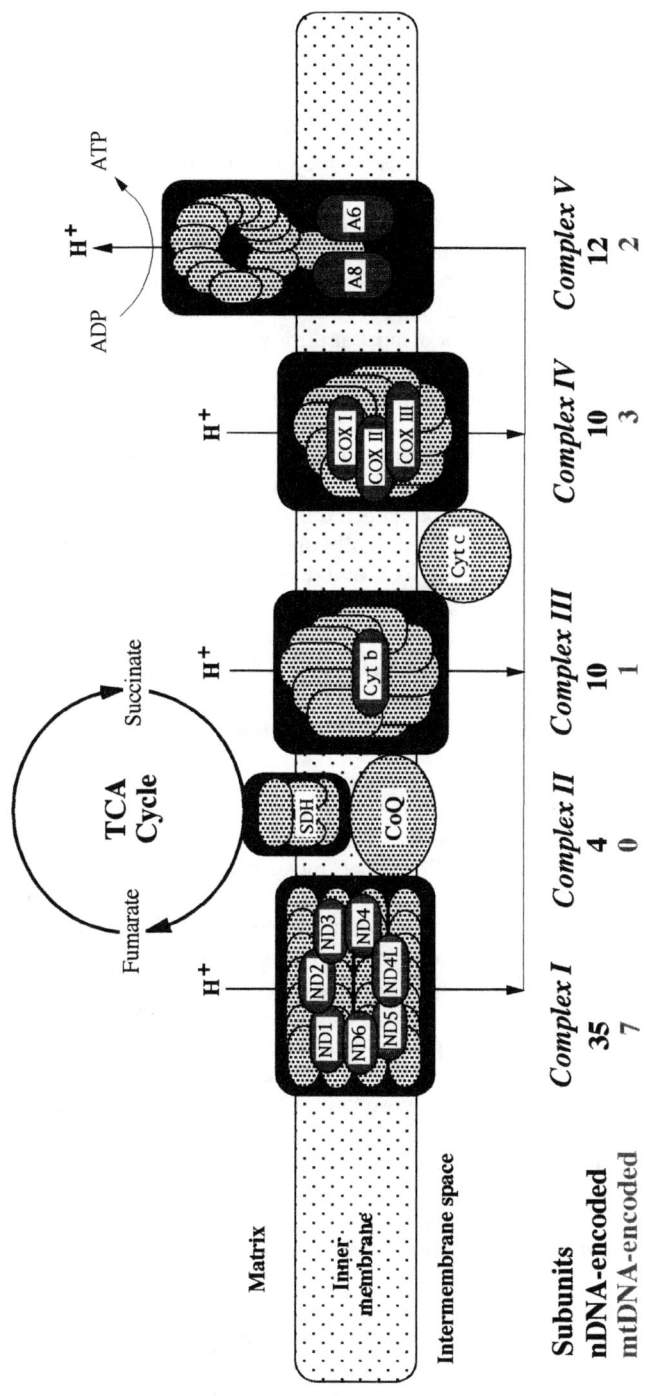

FIGURE 1. The mitochondrial respiratory chain, showing nDNA-encoded (stippled) and mtDNA-encoded (light gray) subunits. Ptotons (H$^+$) are first pumped from the matrix to the intermembrane space through Complexes I, III, and IV. They are then pumped back into the matrix through Complex V to produce ATP. Coenzyme Q (CoQ) and cytochrome c (Cyt c) are electron tranfer carriers encoded by nDNA.

Heteroplasmy refers to the coexistence of two populations of mtDNA, normal and mutated, in tissues from patients. A corollary of heteroplasmy is the *threshold effect*: mutated mtDNAs in a given tissue have to reach a minimum critical number before oxidative metabolism is impaired severely enough to cause dysfunction, that is, symptoms. Therefore, the pathogenic threshold will vary from tissue to tissue according to the relative dependence of each tissue on oxidative metabolism. This explains the selective vulnerability of brain, skeletal, and cardiac muscle, the retina, the renal tubule, the organ of Corti, and other "high maintenance" tissues. Another important feature of mitochondrial genetics is *mitotic segregation*; the mutation load can change from one cell generation to the next and, with time, it can either surpass or fall below the pathogenic threshold. This explains how certain manifestations of mtDNA-related disorders can appear or disappear during the course of the illness, an unusual phenomenon and a useful diagnostic clue.

The maternal inheritance of a pathogenic mtDNA point mutation may be blurred by heteroplasmy; for example, in families with MELAS and the A3243G mutation, there is often only one index case with the full-fledged syndrome, whereas maternal relatives carrying the same mutation can be totally asymptomatic or oligosymptomatic. Recognition of oligosymptomatic patients requires the use of a fine-tooth comb in eliciting the family history, asking specifically about "soft signs," such as short stature, migraine-like headache, hearing loss, and diabetes.

PHYSICAL AND NEUROLOGICAL EXAMINATIONS

Disorders due to mtDNA Mutations

TABLE 1 summarizes the variety of symptoms and signs associated with mtDNA mutations. As mitochondria are ubiquitous, it is hardly surprising that mtDNA-related disorders can affect every tissue in the body and can involve every subspecialty of "mitochondrial medicine."[8] In fact, most, but not all, mtDNA-related disorders are multisystemic. As highlighted in TABLE 1, certain constellations of symptoms and signs are so characteristic as to make the diagnosis self-evident. Until proven otherwise, a patient with progressive external ophthalmoplegia (PEO), pigmentary retinopathy, and cardiac conduction block has Kearns-Sayre syndrome (KSS); a patient with seizures, ataxia, and myoclonus has myoclonus epilepsy and ragged-red fibers (MERRFs); a young patient with migraine headache, seizures, and stroke has mitochondrial encephalomyopathy, lactic acidosis, and stroke-like syndrome (MELAS); and a patient with neuropathy, ataxia, and retinitis pigmentosa has NARP by definition.

In general, the apparently unrelated involvement of two or more tissues should suggest the possibility of mitochondrial disease, especially when symptoms include one or more of the following "red flags": short stature, neurosensory hearing loss, PEO, axonal neuropathy, diabetes mellitus, hypertrophic cardiomyopathy, or renotubular acidosis.

However, tissue-specific disorders are not incompatible with mtDNA mutations, due to at least two mechanisms, somatic mutations or skewed heteroplasmy. *Somatic mtDNA mutations* are spontaneous *de novo* mutations in the mitochondrial genome occurring in the oocyte or embryo, but affecting exclusively the precursor cells of

TABLE 1. Clinical features of mitochondrial diseases associated with mtDNA mutations

TISSUE	SYMPTOM / SIGN	Δ-mtDNA		tRNA		ATPase	
		KSS	PEARSON	MERRF	MELAS	NARP	MILS
CNS	Seizures	-	-	[+]	+	-	+
	Ataxia	+	-	[+]	+	[+]	±
	Myoclonus	-	-	[+]	±	-	-
	Psychomotor retardation	-	-	-	-	-	+
	Psychomotor regression	+	-	±	+	-	-
	Hemiparesis/hemianopia	-	-	-	[+]	-	-
	Cortical blindness	-	-	-	[+]	-	-
	Migraine-like headaches	-	-	-	[+]	-	-
	Dystonia	-	-	-	+	-	+
PNS	Peripheral neuropathy	±	-	±	±	[+]	-
Muscle	Weakness/exercise intolerance	+	-	+	+	+	+
	Ophthalmoplegia	+	±	-	-	-	-
	Ptosis	[+]	-	-	-	-	-
Eye	Pigmentary retinopathy	[+]	-	-	-	[+]	±
	Optic atrophy	-	-	-	-	±	±
	Cataracts	-	-	-	-	-	-
Blood	Sideroblastic anemia	±	[+]	-	-	-	-
Endocrine	Diabetes mellitus	±	-	-	±	-	-
	Short stature	+	-	+	+	-	-
	Hypoparathyroidism	±	-	-	-	-	-
Heart	Conduction block	[+]	-	-	±	-	-
	Cardiomyopathy	±	-	-	±	-	±
GI	Exocrine pancreatic dysfunction	±	[+]	-	-	-	-
	Intestinal pseudo-obstruction	-	-	-	-	-	-
ENT	Sensorineural hearing loss	-	-	+	+	±	-
Kidney	Fanconi syndrome	±	±	-	±	-	-
Laboratory	Lactic acidosis	+	+	+	+	-	±
	Muscle biopsy: RRF	+	±	+	+	-	-
Inheritance	Maternal	-	-	+	+	+	+
	Sporadic	+	+	-	-	-	-

Boxes highlight typical features of different syndromes, except for maternally inherited Leigh syndrome (MILS), which is defined by neuroradiologic or neuropathologic criteria. Δ-mtDNA, deleted mtDNA; RNA, ribnucleic acid; KSS, Kearns-Sayre syndrome; MERRF, myoclonus epilepsy with ragged-red fibers; MELAS, mitochondrial encephalomyopathy, lactic acidosis, and strokelike episodes; NARP, neuropathy, ataxia, retinitis pigmentosa; MILS, maternally inherited Leigh syndrone.

+, present; −, absent; ±, possible.

one tissue after germ-layer differentiation. Examples include PEO with or without proximal myopathy due to single large-scale deletions of mtDNA[9] or exercise intolerance due to mutations in protein-coding genes (commonly the cytochrome *b* gene).[9,10] *Skewed heteroplasmy* describes the situation in which a ubiquitous and often maternally inherited mtDNA mutation is abundant enough to surpass the pathogenic threshold only in one tissue, thus resulting in phenotypic tissue specificity. Again, myopathies offer good examples of skewed heteroplasmy.[9]

Disorders due to nDNA Mutations

By and large, disorders due to mutations in nDNA are more stereotypical in their presentations due to the "all-or-none" nature of mendelian genetics (TABLE 2). With few exceptions, mutations affecting the respiratory chain are autosomal recessive; as a rule, heterozygous carriers are asymptomatic, and homozygous patients in the same family have comparable phenotypes. In other words, the disease breeds true in the family.

Like mtDNA-related disorders, mendelian mitochondrial diseases also tend to be multisystemic, but tissue involvement is less variable and onset is earlier. A mitochondrial disease manifesting at or soon after birth is more likely to be associated with nDNA than with mtDNA mutations. For example, most causes of Leigh syndrome (LS) involve mutations in genes encoding subunits or assembly proteins of respiratory chain complexes,[11] and both the myopathic and the hepatopathic forms of congenital mtDNA depletion are inherited as autosomal recessive traits.[12]

Tissue specificity in mendelian mitochondrial diseases could be due to mutations in genes encoding tissue-specific respiratory chain subunits or assembly protein. Although, in fact, no such pathogenic mechanism has yet been documented, it has been postulated in several disorders, including "fatal infantile myopathy"[13] and "benign infantile myopathy"[14] with COX deficiency.

ROUTINE LABORATORY TESTS

The most useful diagnostic test is the measurement of serum lactic acid. To avoid spurious rises due to tourniquet placement and active resistance in children, the use of arterious blood is recommended. A disturbance of the intramitochondrial oxidation of pyruvate at the level of the pyruvate dehydrogenase complex (PDHC), the Krebs cycle, or the electron transport chain will cause a compensatory increase of glycolysis with overproduction of pyruvate. Excess pyruvate can be transaminated to alanine or reduced to lactate; in fact, all three of these compounds are increased in the blood (and urine) of patients with mitochondrial diseases. Depending on the oxidation-reduction potential in tissues, the lactate:pyruvate ratio can be preserved or abnormally elevated. Inherited defects of the respiratory chain (as well as hypoxia) cause the intramitochondrial NAD:NADH ratio to fall, which in turn causes the lactate:pyruvate ratio to increase markedly (reaching values of 50:1 to 250:1, the normal ratio being 25:1). By contrast, defects of PDHC do not alter significantly the redox state of the cell and, therefore, do not affect the lactate:pyruvate ratio. This difference is a useful clue to the differential diagnosis.

Lactate should also be measured in the cerebrospinal fluid (CSF) because some patients with mitochondrial encephalopathy have increased lactate values (and lactate:pyruvate ratios) in the CSF in the presence of normal or only mildly increased values in blood. An alternative, and noninvasive, way to measure lactate in the CSF and in the brain is magnetic proton resonance spectroscopy imaging in voxels containing mainly lateral ventricular fluid or brain cortex.[15] Important as it is, lactic acidosis should not be considered an absolute requisite for diagnosis; there are mitochondrial diseases, including, for example, NARP and its infantile variant, maternally inherited LS (MILS), in which blood lactic acid is often normal or only mildly increased.

Serum creatine kinase values are usually moderately elevated in patients with mitochondrial diseases. One notable exception is the myopathic variant of the mtDNA depletion syndrome (MDS), where creatine kinase values in the thousands are often seen.[16] In the childhood form of MDS, this renders more difficult the differential diagnosis from muscular dystrophies.

In patients with clinical features suggestive of mitochondrial neurogastrointestinal encephalomyopathy, a useful diagnostic test is the measurement of thymidine in blood; because of the specific deficiency of thymidine phosphorylase, these patients have extremely elevated values of thymidine.[17]

SPECIAL LABORATORY TESTS

Neuroradiology

Patients with different mitochondrial syndromes have characteristic MRI patterns, which are very useful for diagnosis. For example, the diagnosis of LS cannot be established without neuroradiological evidence of bilateral signal hyperintensities in the basal ganglia and brain stem. Stroke-like lesions in the posterior cerebral hemispheres, especially the occipital lobe, are typical of MELAS. Diffuse signal abnormalities of the central white matter are characteristic of KSS, and basal ganglia calcifications are seen in both KSS and MELAS.

As just mentioned, proton magnetic resonance spectroscopy can reveal lactate accumulation in the ventricular CSF and in specific areas of the brain. This technique also allows us to compare the concentrations of lactate with those of choline, creatine, and N-acetyl-L-aspartate, which is a good indicator of cell viability.

Exercise Physiology

Oxygen extraction by exercising muscles can be detected by near infrared spectroscopy, a noninvasive technique that measures the degree of deoxygenation of hemoglobin. This is characteristically decreased or abolished in patients with mitochondrial diseases.[18]

The energetics of muscle exercise can be studied noninvasively by ^{31}P-magnetic resonance spectroscopy, following the ratio of phosphocreatine (PCr) to inorganic phosphate (Pi) at rest, during exercise, and during recovery. Typically, in patients with mitochondrial diseases, PCr/Pi ratios are lower than normal at rest, decline excessively during exercise, and show a slow return to baseline.[19]

These exercise physiology tests are very useful, but cannot be applied to infants or small children because they require the patient's active cooperation.

MUSCLE BIOPSY

Histochemistry

A good histochemical examination (which requires appropriate freezing of the specimen and a battery of stains) provides invaluable diagnostic clues. The modified

Gomori trichrome stain reveals massive mitochondrial proliferation as irregular reddish patches under the sarcolemma and in the midst of what have been labeled "ragged-red fibers" (RRFs).[20] A more sensitive "detector" of mitochondrial proliferation is the stain for succinate dehydrogenase (SDH): not only do RRFs stain intensely (ragged-blue fibers?), but more subtle proliferation of mitochondria can also be detected in the form of blue "rims" at the periphery of fibers. Because the SDH stain is **not** affected by mtDNA mutations (complex II is exclusively encoded by nDNA), it is an excellent marker of mitochondrial "mass" and a good counterpart of the histochemical stain for COX, which is decreased or abolished by mtDNA mutations affecting mitochondrial protein synthesis (mtDNA deletions, tRNA or rRNA mutations, and mtDNA depletion) or by mutations in the mtDNA COX genes. Thus, the finding of scattered SDH-intense COX-negative fibers is a sure indication that the patient harbors an mtDNA mutation affecting protein synthesis. On the other hand, the presence of scattered SDH-intense COX-positive fibers suggests an mtDNA mutation in one of the protein-coding genes, excluding, of course, the three COX genes. The only exception to this rule is MELAS; patients with typical MELAS presentation have SDH-intense and COX-positive (although not intensely positive) fibers. In fact, the SDH and COX stains can be usefully superimposed in the same tissue section to reveal partial COX deficiency. In normal fibers, the brown COX stain prevails; only fibers lacking or deficient in COX will stand out as blue.[21]

A generalized, more or less severe defect of COX stain can be the only morphological abnormality in patients with nDNA-related COX deficiencies (e.g., LS due to *SURF1* or *SCO2* mutations).[22]

Mitochondria proliferate not only in skeletal muscle but also in smooth muscle and enthothelial cells of the intramuscular vessels; "strongly SDH-reactive vessels" (SSVs) are especially evident in MELAS,[23] but can occur in other mitochondrial diseases.

Biochemistry

A long-standing controversy in mitochondrial circles is whether fresh tissue is needed or frozen muscle is adequate to obtain reliable biochemical results. Fresh tissue is needed if functionally intact mitochondria are to be isolated and used in polarographic analyses. However, we think that frozen tissue can be used reliably for analysis of individual enzymes (or segments) of the respiratory chain, and these studies are diagnostically adequate. We perform routinely the following assays: NADH-cytochrome *c* reductase (complexes I + III); NADH-CoQ reductase (complex I); NADH dehydrogenase (complex I); succinate-cytochrome *c* reductase (complexes II + III); cytochrome *c* oxidase (complex IV); succinate dehydrogenase (complex II); and citrate synthase, a matrix enzyme of the Krebs cycle. Citrate synthase is a good marker of mitochondrial abundance, and we refer the activities of respiratory chain enzymes to those of citrate synthase to correct for increased (more rarely decreased) numbers of mitochondria.[24]

Certain respiratory chain enzyme profiles give useful diagnostic clues. Thus, isolated defects of complex I, complex III, or complex IV activity suggest either mutations in mtDNA genes encoding for individual subunits of the defective complex or mutations in nuclear genes, which may encode subunits of the defective complex or ancillary proteins needed for its assembly. Examples of the three situations are (i) mutations in the cytochrome *b* gene associated with complex III deficiency;[10] (ii) mutations in the *NDUFS7* gene encoding the homonymous subunit of complex I and causing LS associated with complex I deficiency;[25] and (iii) mutations in the *SCO2* gene, encoding a protein needed for the insertion of copper into the COX holoenzyme and causing a rapidly fatal infantile cardioencephalo-myopathy with COX deficiency.[26]

Isolated defects of complex II must be due to mutations in nuclear genes, some of which are associated with LS.[27] Combined defects of complexes I, III, and IV, but not of complex II, suggest mtDNA mutations affecting mitochondrial protein synthesis *in toto*, such as large-scale mtDNA deletions, point mutations in tRNA or rRNA genes, or mtDNA depletion. Because of heteroplasmy, these combined enzyme defects are often partial, and sometimes they affect one complex more than the others: for example, complex I activity is especially vulnerable to the MELAS A3243G mutation, whereas COX activity is more susceptible to the MERRF A8344G mutation.[28]

Combined defects of complex I, II, and III, but not IV, suggest a problem with iron-sulfur (Fe-S) clusters, which are indispensable components of the first three respiratory chain complexes but not of COX. The suspicion is reinforced when the activity of aconitase, a Krebs cycle enzyme containing Fe-S clusters, is also decreased.[29–32]

MOLECULAR GENETIC SCREENING

The information gathered from clinical phenotype, family history, laboratory tests, muscle histochemistry, and muscle biochemistry should orient the astute clinician as to which molecular defects to screen for or, at least, on the priority of different genetic tests.

In the ideal situation, the clinical syndrome is typical (let us say MERRF), maternal inheritance is evident, there is lactic acidosis, and the diagnosis of MERRF has only to be confirmed by molecular genetic analysis. Even before muscle biopsy, the most common MERRF mutation, A8344G, can be sought in blood. While the MERRF mutation is almost always detectable in blood, other tRNA mutations, notably the most common A3243G MELAS mutation, are sometimes present in very low amounts in blood. Other easily accessible tissues in which to look for the mutation include urinary sediment, buccal mucosa, hair follicles, or cultured skin fibroblasts.

When there is evidence of maternal inheritance and histochemical and biochemical data suggest a problem with mitochondrial protein synthesis, a first screening of the 22 mtDNA tRNA genes can be accomplished by single strand conformational polymorphism analysis: abnormal segments can then be sequenced directly to reveal mutations.

In patients with PEO, family history and clinical presentation are crucial in choosing molecular strategies. If a sporadic patient has a multisystem disorder with all or some of the features of KSS, Southern blot of muscle mtDNA is the test of choice. If muscle is not available, blood or urinary sediment can be used with good probability of showing a deletion. However, if a sporadic patient has a pure myopathy, Southern blot has to be performed in muscle DNA, because other tissues are spared both clinically and genetically. Patients with maternally inherited PEO (often associated with other symptoms, such as hearing loss, seizures, retinopathy, or endocrinopathy), should be screened for the A3243G MELAS mutation first[33] and for rarer mutations in tRNA genes later.[34] Patients with autosomal dominant PEO (adPEO), which is usually associated with other symptoms (parkinsonism, bipolar disorder, and peripheral neuropathy), should be screened for multiple mtDNA deletions by Southern blot of muscle DNA (TABLE 2). If these are found, nDNA from any tissue should be screened for mutations in three candidate genes: *ANT1*, encoding the adenine nucleotide translocator protein; *twinkle*, a mitochondrial helicase; and *POLG*, encoding the mitochondrial polymerase γ.[34] Patients with autosomal recessive PEO (arPEO), which is usually part of a multisystemic illness, also show multiple mtDNA deletions in Southern blots of muscle DNA, and their nDNA should be screened for mutations in *POLG*.[34] It is important to keep in mind that mutations in *ANT1*, *twinkle*, and *POLG* do not explain all cases of mendelian PEO with multiple mtDNA deletions in muscle; other mutant nuclear genes remain to be identified.[34] One special form of arPEO is a multisystemic syndrome known as mitochondrial neurogastrointestinal encephalomyopathy (MNGIE), a unique clinical presentation in which PEO is accompanied by intestinal dysfunction, peripheral neuropathy, and leukoencephalopathy.[35] The finding of greatly increased blood levels of thymidine[17] and of decreased activity of thymidine phosphorylase (TP) in leukocytes[34] establishes the diagnosis, which is confirmed by documenting mutations in the *TP* nuclear gene.[35,36]

The clinical suspicion of mtDNA depletion syndrome (MDS) nowadays encompasses more than the two severe infantile clinical presentations described initially, congenital mitochondrial myopathy (with or without renal dysfunction) or fatal hepatopathy,[37,38] to include childhood myopathy,[39] spinal muscular atrophy phenotypes,[40–42] and variable involvement of the CNS.[16] When MDS is suspected, the diagnosis has to be confirmed by comparison of the relative amounts of the two genomes through Southern blot of total DNA in affected tissues.[43] The next step is to screen nDNA for mutations in two genes, *dGK*, encoding deoxyguanosine kinase, and *TK2*, encoding thymidine kinase 2; many, but not all, patients with hepatopathic MDS have mutations in *dGK*,[44,45] and many, but not all, patients with myopathic MDS have mutations in *TK2*.[41,46]

Leigh syndrome (LS) poses a special problem because of the striking biochemical and genetic heterogeneity.[47] Some clinical features may offer clues to the diagnosis; for example, in COX-deficiency LS, onset is relatively late, seizures are infrequent,[48,49] and the presence of retinitis pigmentosa is virtually pathognomonic of the maternally inherited LS (MILS) due to the T8993G mutation in the ATPase 6 gene of mtDNA.[49] Family history is also helpful; the occurrence of LS or NARP (neuropathy, ataxia, and retinitis pigmentosa) in maternal relatives of a child with LS makes sequencing of the ATPase 6 gene imperative,[50–52] whereas evidence of X-linked transmission invites sequencing the gene for the E1α subunit of the PDHC. However, in most LS patients inheritance is autosomal recessive, and the choice of nuclear genes to be sequenced comes from biochemical studies of muscle (or cultured fibroblasts). Mutations in genes encoding several complex I subunits have been associated with LS or childhood leukodystrophy.[53] Mutations in genes encoding complex II subunits can cause LS[27,54] or adult-onset encephalopathy.[55] Defects in complex III and complex IV have not been associated with genes encoding subunits

TABLE 2. Nuclear gene defects in mitochondrial diseases[a]

Disease	Gene	mtDNA	RC Biochemistry
adPEO (hearing loss, bipolar disease, weakness)	ANT1	multiple Δs	RC complexes I+III+IV ⇓
adPEO (weakness, psychiatric problems)	Twinkle	multiple Δs	RC complexes I+III+IV ⇓
adPEO (neuropathy,bipolar disease, cataracts, hypogonadism, parkinsonism)	POLG	multiple Δs	RC complexes I+III+IV ⇓
adPEO	?	multiple Δs	RC complexes I+III+IV ⇓
MNGIE (PEO, intestinal dismotility, neuropathy, leukoencephalopathy)	TP	Δs+depletion	RC complexes I+III+IV ⇓
ARCO (PEO, cardiopathy)	?	multiple Δs	RC complexes I+III+IV ⇓
arPEO	POLG	multiple Δs	RC complexes I+III+IV ⇓
arPEO	?	multiple Δs	RC complexes I+III+IV ⇓
Fatal infantile myopathy; juvenile myopathy; tubular acidosis	TK2	depletion	RC complexes I+III+IV ⇓
Fatal infantile hepatopathy ± encephalopathy	dGK	depletion	RC complexes I+III+IV ⇓
LS, infantile cardiomyopathy, Leigh-like leukoencephalopathy	NDUSF2		complex I deficiency
LS + cardiomyopathy	NDUSF4		complex I deficiency
LS	NDUSF7		complex I deficiency
LS + cardiomyopathy	NDUSF8		complex I deficiency
Infantile encephalopathy; myoclonic epilepsy	NDUFV1		complex I deficiency
LS	SDHA		complex II deficiency
Late-onset encephalopathy	SDHA		complex II deficiency
Hereditary paraganglioma	SDHD		complex II deficiency
Neonatal tubulopathy, hepatopathy, encephalopathy	BCS1L		complex III deficiency
GRACILE syndrome	BCS1L		complex III deficiency
LS	SURF1		complex IV deficiency
Infantile cardioencephalomyopathy	SCO2		complex IV deficiency
Infantile cardioencephalomyopathy	COX15		complex IV deficiency
Infantile hepatoencephalomyopathy	SCO1		complex IV deficiency
Infantile nephroencephalomyopathy	COX10		complex IV deficiency

See following page for footnote to TABLE 2.

of these complexes, but with genes encoding assembly proteins, an indirect pathogenetic mechanism. Thus, the most common form of COX-deficient LS is due to over 40 different mutations in the *SURF1* gene, which encodes a protein essential for the stepwise assembly of the COX holoenzyme.[56,58,59] Mutations in other COX-assembly genes (*SCO2, SCO1, COX10,* and *COX15*) have been associated with generalized disorders of infancy or childhood, in which brain involvement is accompanied or overshadowed by dysfunction of the heart,[22,26,57] liver,[60] or kidney[61] (TABLE 2).

ACKNOWLEDGMENTS

Part of the work described here was supported by National Institutes of Health Grants HD32062 and NS11766, by a grant from the Muscular Dystrophy Association, and by a generous gift from the Medical Illness Counseling Center, Chevy Chase, Maryland.

REFERENCES

1. MAJAMAA, K., J.S. MOILANEN, S. UIMONEN, *et al.* 1998. Epidemiology of A3243G, the mutation for mitochondrial encephalomyopathy, lactic acidosis, and strokelike episodes: prevalence of the mutation in an adult population. Am. J. Hum. Genet. **63:** 447–454.
2. DARIN, N., A. OLDFORS, A.-R. MOSLEMI, *et al.* 2001. The incidence of mitochondrial encephalomyopathies in childhood: clinical features and morphological, biochemical, and DNA abnormalities. Ann. Neurol. **49:** 377–383.
3. CHINNERY, P.F., T.M. WARDELL, R. SINGH-KLER, *et al.* 2000. The epidemiology of pathogenic mitochondrial DNA mutations. Ann. Neurol. **48:** 188–193.
4. CHINNERY, P.F. & D.M. TURNBULL. 2001. Epidemiology and treatment of mitochondrial disorders. Am. J. Med. Genet. **106:** 94–101.
5. SCHON, E.A., M. HIRANO & S. DIMAURO. 2002. Molecular genetic basis of the mitochondrial encephalomyopathies. *In* Mitochondrial Disorders in Neurology 2. A.H.V. Schapira & S. DiMauro, Eds.: 69–113. Butterworth-Heinemann. Boston.
6. PASCHEN, S.A. & W. NEUPERT. 2001. Protein import into mitochondria. IUBMB Life **52:** 101–112.
7. DIMAURO, S., E. BONILLA & D.C. DE VIVO. 1999. Does the patient have a mitochondrial encephalomyopathy? J. Child Neurol. **14:** S23–S35.
8. LUFT, R. 1994. The development of mitochondrial medicine. Proc. Natl. Acad. Sci. USA **91:** 8731–8738.
9. DIMAURO, S., E. BONILLA, M. MANCUSO, *et al.* 2003. Mitochondrial myopathies. Basic Appl. Myol. **13:** 145–155.

---

[a]Diseases asociated with defects of mtDNA are also called "defects of intergenomic signaling."

ABBREVIATIONS: ANT1, adenine nucleotide translocator 1; ARCO, autosomal recessive cardiopathy and ophthalmoplegia; COX, cytochrome *c* oxidase; dGK, deoxyguanosine kinase; GRACILE, growth retardation, aminoaciduria, cholestasis, iron overloaded, lactoacidosis, and early death; LS, Leigh syndrome; MNGIE, mitochondrial neurogastrointestinal encephalomyopathy; PEO, progressive external ophthalmoplegia; POLG, polymerase γ; SDH, succinate dehydrogenase; TK2, thymidine kinase 2; TP, thymidine phosphorylase. Δ signifies deletion.

10. ANDREU, A.L., M.G. HANNA, H. REICHMANN, et al. 1999. Exercise intolerance due to mutations in the cytochrome *b* gene of mitochondrial DNA. N. Engl. J. Med. **341**: 1037–1044.
11. ZEVIANI, M. 2001. The expanding spectrum of nuclear gene mutations in mitochondrial disorders. Cell Dev. Biol. **12**: 407–416.
12. HIRANO, M., R. MARTI, C. FERREIRA-BARROS, et al. 2001. Defects of intergenomic communication: autosomal disorders that cause multiple deletions and depletion of mitochondrial DNA. Cell Dev. Biol. **12**: 417–427.
13. DIMAURO, S., J.R. MENDELL, Z. SAHENK, et al. 1980. Fatal infantile mitochondrial myopathy and renal dysfunction due to cytochrome-*c*-oxidase deficiency. Neurology **30**: 795–804.
14. DIMAURO, S., J.F. NICHOLSON, A.P. HAYS, et al. 1983. Benign infantile mitochondrial myopathy due to reversible cytochrome *c* oxidase deficiency. Ann. Neurol. **14**: 226–234.
15. SHUNGU, D.C., M. SANO, W.S. MILLAR, et al. 1999. Metabolic, structural and neuropsychological deficits in mitochondrial encephalomyopathies assessed by 1H MRSI, MRI and neuropsychological testing. Proc. Int. Soc. Mag. Reson. Med. **7**: 49.
16. VU, T.H., M. SCIACCO, K. TANJI, et al. 1998. Clinical manifestations of mitochondrial DNA depletion. Neurology **50**: 1783–1790.
17. SPINAZZOLa, A., R. MARTI, I. NISHINO, et al. 2002. Altered thymidine metabolism due to defects of thymidine phosphorylase. J. Biol. Chem. **277**: 4128–4132.
18. BANK, W., J. PARK, G. LECH, et al. 1998. Near-infrared spectroscopy in the diagnosis of mitochondrial disorders. BioFactors **7**: 243–245.
19. ARGOV, Z. & W.J. BANK. 1991. Phosphorus magnetic resonance spectroscopy (31P MRS) in neuromuscular disorders. Ann. Neurol. **30**: 90–97.
20. ENGEL, W.K. & G. CUNNIGHAM. 1963. Rapid examination of muscle tissue: an improved trichome stain method for fresh-frozen biopsy sections. Neurology **13**: 919–923.
21. SCIACCO, M. & E. BONILLA. 1996. Cytochemistry and immunocytochemistry of mitochondria in tissue sections. Meth. Enzymol. **264**: 509–521.
22. SUE, C.M., C. KARADIMAS, N. CHECCARELLI, et al. 2000. Differential features of patients with mutations in two COX assembly genes, *SURF-1* and *SCO2*. Ann. Neurol. **47**: 589–595.
23. HASEGAWA, H., T. MATSUOKA, I. GOTO, et al. 1991. Strongly succinate dehydrogenase-reactive blood vessels in muscles from patients with mitochondrial myopathy, encephalopathy, lactic acidosis, and stroke-like episodes. Ann. Neurol. **29**: 601–605.
24. DIMAURO, S., S. SHANSKE, A. NAINI, et al. 2002. Biochemical evaluation of metabolic myopathies. *In* Clinical Evaluation and Diagnostic Tests for Neuromuscular Disorders. T.E. Bertorini, Ed.: 535–564. Butterworth-Heinemann. Amsterdam.
25. TRIEPELS, R.H., L.P. VAN DEN HEUVEL, J.L.C.M. LOEFFEN, et al. 1999. Leigh syndrome associated with a mutation in the NDUFS7 (PSST) nuclear encoded subunit of complex I. Ann. Neurol. **45**: 787–790.
26. PAPADOPOULOU, L.C., C.M. SUE, M.M. DAVIDSON, et al. 1999. Fatal infantile cardioencephalomyopathy with COX deficiency and mutations in *SCO2*, a COX assembly gene. Nat. Genet. **23**: 333–337.
27. BOURGERON, T., P. RUSTIN, D. CHRETIEN, et al. 1995. Mutation of a nuclear succinate dehydrogenase gene results in mitochondrial respiratory chain deficiency. Nat. Genet. **11**: 144–149.
28. LOMBES, A., N.B. ROMERO, G. TOUATI, et al. 1996. Clinical and molecular heterogeneity of cytochrome *c* oxidase deficiency in the newborn. J. Inherited Metab. Dis. **19**: 286–295.
29. SCHAPIRA, A., J.M. COOPER, J.A. MORGAN-HUGHES, et al. 1990. Mitochondrial myopathy with a defect of mitochondrial protein transport. N. Engl. J. Med. **323**: 37–42.
30. HALLER, R.G., K.G. HENRIKSSON, L. JORFELDT, et al. 1991. Deficiency of skeletal muscle succinate dehydrogenase and aconitase. J. Clin. Invest. **88**: 1197–1206.
31. HALL, R.E., K.G. HENRIKSSON, S.F. LEWIS, et al. 1993. Mitochondrial myopathy with succinate dehydrogenase and aconitase deficiency. Abnormalities of several iron-sulfur proteins. J. Clin. Invest. **92**: 2660–2666.

32. MANCUSO, M., A. NAINI, M. FILOSTO, et al. 2003. A defect of mitochondrial iron-sulfur clusters transport. Basic Appl. Myol. **13:** 22 (abstr.).
33. MORAES, C.T., F. CIACCI, G. SILVESTRI, et al. 1993. Atypical clinical presentations associated with the MELAS mutation at position 2343 of human mitochondrial DNA. Neuromusc. Disord. **3:** 43–50.
34. HIRANO, M. & S. DIMAURO. 2001. *ANT1, Twinkle, POLG*, and *TP*: new genes open our eyes to ophthalmoplegia. Neurology **57:** 2163–2165.
35. NISHINO, I., A. SPINAZZOLA, A. PAPADIMITRIOU, et al. 2000. Mitochondrial neurogastrointestinal encephalomyopathy: an autosomal recessive disorder due to thymidine phosphorylase mutations. Ann. Neurol. **47:** 792–800.
36. NISHINO, I., A. SPINAZZOLA & M. HIRANO. 1999. Thymidine phosphorylase gene mutations in MNGIE, a human mitochondrial disorder. Science **283:** 689–692.
37. BOUSTANY, R.N., J.R. APRILLE, J. HALPERIN, et al. 1983. Mitochondrial cytochrome deficiency presenting as a myopathy with hypotonia, external ophthalmoplegia, and lactic acidosis in an infant and as fatal hepatopathy in a second cousin. Ann. Neurol. **14:** 462–470.
38. MORAES, C.T., S. SHANSKE, H.J. TRITSCHLER, et al. 1991. MtDNA depletion with variable tissue expression: a novel genetic abnormality in mitochondrial diseases. Am. J. Hum. Genet. **48:** 492–501.
39. TRITSCHLER, H.J., F. ANDREETTA, C.T. MORAES, et al. 1992. Mitochondrial myopathy of childhood associated with depletion of mitochondrial DNA. Neurology **42:** 209–217.
40. PONS, R., F. ANDREETTA, C.H. WANG, et al. 1996. Mitochondrial myopathy simulating spinal muscular atrophy. Pediatr. Neurol. **15:** 153–158.
41. MANCUSO, M., L. SALVIATI, S. SACCONI, et al. 2002. Mitochondrial DNA depletion. Mutations in thymidine kinase gene with myopathy and SMA. Neurology **59:** 1197–1202.
42. MANCUSO, M., M. FILOSTO, M. HIRANO, et al. 2003. Spinal muscular atrophy and mtDNA depletion. Acta Neuropathol. **105:** 621–622.
43. MORAES, C.T., E. RICCI, E. ARNAUDO, et al. 1993. Quantitative defects of mitochondrial DNA. *In* Mitochondrial DNA in Human Pathology. S. DiMauro & D.C. Wallace, Eds.: 97–108. Raven Press. New York.
44. MANDEL, H., R. SZARGEL, V. LABAY, et al. 2001. The deoxyguanosine kinase gene is mutated in individuals with depleted hepatocerebral mitochondrial DNA. Nat. Genet. **29:** 337–341.
45. SALVIATI, L., S. SACCONI, M. MANCUSO, et al. 2002. Mitochondrial DNA depletion and *dGK* gene mutations. Ann. Neurol. **52:** 311–317.
46. SAADA, A., A. SHAAG, H. MANDEL, et al. 2001. Mutant mitochondrial thymidine kinase in mitochondrial DNA depletion myopathy. Nat. Genet. **29:** 342–344.
47. DIMAURO, S. & D.C. DE VIVO. 1996. Genetic heterogeneity in Leigh syndrome. Ann. Neurol. **40:** 5–7.
48. VAN COSTER, R., A. LOMBES, D.C. DEVIVO, et al. 1991. Cytochrome c oxidase-associated Leigh syndrome: phenotypic features and pathogenetic speculations. J. Neurol. Sci. **104:** 97–111.
49. SANTORELLI, F.M., S. SHANSKE, A. MACAYA, et al. 1993. The mutation at nt 8993 of mitochondrial DNA is a common cause of Leigh syndrome. Ann. Neurol. **34:** 827–834.
50. HOLT, I.J., A.E. HARDING, R.K. PETTY, et al. 1990. A new mitochondrial disease associated with mitochondrial DNA heteroplasmy. Am. J. Hum. Genet. **46:** 428–433.
51. TATUCH, Y., J. CHRISTODOULOU, A. FEIGENBAUM, et al. 1992. Heteroplasmic mtDNA mutation (T>G) at 8993 can cause Leigh disease when the percentage of abnormal mtDNA is high. Am. J. Hum. Genet. **50:** 852–858.
52. CIAFALONI, E., F. SANTORELLI, S. SHANSKE, et al. 1993. Maternally inherited Leigh syndrome. J. Pediatr. **122:** 419–422.
53. SMEITINK, J. & L. VAN DEN HEUVEL. 1999. Protein biosynthesis '99: Human mitochondrial complex I in health and disease. Am. J. Hum. Genet. **64:** 1505–1510.
54. PARFAIT, B., D. CHRETIEN, A. ROTIG, et al. 2000. Compound heterozygous mutation in the flavoprotein gene of the respiratory chain complex II in a patient with Leigh syndrome. Hum. Genet. **106:** 236–243.

55. TAYLOR, R.W., M.A. BIRCH-MACHIN, J. SCHAEFER, et al. 1996. Deficiency of complex II of the mitochondrial respiratory chain in late-onset optic atrophy and ataxia. Ann. Neurol. **39:** 224–232.
56. PEQUIGNOT, M.O., R. DEY, M. ZEVIANI, et al. 2001. Mutations in the *SURF1* gene associated with Leigh syndrome and cytochrome *c* oxidase deficiency. Hum. Mut. **17:** 374–381.
57. ANTONICKA, H., A. MATTMAN, C.G. CARLSON, et al. 2003. Mutations in COX15 produce a defect in the mitochondrial heme biosynthetic pathway, causing early-onset fatal hypertrophic cardiomyopathy. Am. J. Hum. Genet. **72:** 101–114.
58. TAYLOR, R.W., M.A. BIRCH-MACHIN, J. SCHAEFER, et al. 1996. Deficiency of complex II of the mitochondrial respiratory chain in late-onset optic atrophy and ataxia. Ann. Neurol. **39:** 224–232.
59. ZHU, Z., J. YAO, T. JOHNS, et al. 1998. *SURF1*, encoding a factor involved in the biogenesis of cytochrome *c* oxidase, is mutated in Leigh syndrome. Nat. Genet. **20:** 337–343.
60. TIRANTI, V., K. HOERTNAGEL, R. CARROZZO, et al. 1998. Mutations of SURF-1 in Leigh disease associated with cytochrome *c* oxidase deficiency. Am. J. Hum. Genet. **63:** 1609–1621.
61. VALNOT, I., J.-C. von KLEIST-RETZOW, A. BARRIENTOS, et al. 2000. A mutation in the human heme-A:farnesyltransferase gene (COX 10) causes cytochrome *c* oxidase deficiency. Hum. Mol. Genet. **9:** 1245–1249.
62. VALNOT, I., S. OSMOND, N. GIGAREL, et al. 2000. Mutations of the *SCO1* gene in mitochondrial cytochrome *c* oxidase deficiency with neonatal-onset hepatic failure and encephalopathy. Am. J. Hum. Genet. **67:** 1104–1109.

Mitochondrial Encephalomyopathies

Therapeutic Approach

SALVATORE DiMAURO, MICHELANGELO MANCUSO, AND ALI NAINI

Department of Neurology, Columbia University College of Physicians & Surgeons, New York, New York 10032, USA

ABSTRACT: Therapy for mitochondrial diseases is woefully inadequate. However, lack of cure does not equate with lack of treatment. In this review, we consider sequentially several different therapeutic approaches. Palliative therapy is dictated by good medical practice and includes anticonvulsant medication, control of endocrine dysfunction, and surgical procedures. Removal of noxious metabolites is centered on combating lactic acidosis, but it extends to other metabolites, such as thymidine in patients with the mitochondrial neuro-gastrointestinal encephalomyopathy syndrome. Attempts to bypass blocks in the respiratory chain by administration of artificial electron acceptors have not been successful, but this concept may be amenable to genetic engineering. Administration of metabolites and cofactors is the mainstay of real-life therapy and includes both components of the respiratory chain and other natural compounds. There is increasing interest in the administration of reactive oxygen species scavengers both in primary mitochondrial diseases and in neurodegenerative diseases directly or indirectly related to mitochondrial dysfunction. Aerobic exercise and physical therapy prevent or correct deconditioning and improve exercise tolerance in patients with mitochondrial myopathies due to mtDNA mutations. Gene therapy is a challenge because of polyplasmy and heteroplasmy, but interesting experimental approaches are being pursued and include, for example, decreasing the ratio of mutant to wild-type mitochondrial genomes (gene shifting), converting mutated mtDNA genes into normal nDNA genes (allotropic expression), importing cognate genes from other species, or correcting mtDNA mutations with specific restriction endonucleases. Germline therapy raises ethical problems but is being seriously considered to prevent maternal transmission of mtDNA mutations. Preventive therapy through genetic counseling and prenatal diagnosis is still limited for mtDNA-related disorders but is becoming increasingly important for nDNA-related disorders.

KEYWORDS: encephalomyopathies; mitochondria

INTRODUCTION

In contrast to the spectacular progress in our understanding of the biochemical and molecular bases of the mitochondrial encephalomyopathies (defined restrictively as disorders due to defects in the mitochondrial respiratory chain; see DiMauro

---

Address for correspondence: Salvatore DiMauro, M.D., 4-420 College of Physicians & Surgeons, 630 West 168th Street, New York, NY 10032. Voice: 212-305-1662; fax: 212-305-3986.
sd12@Columbia.edu

et al., this volume), we are still extremely limited in our ability to treat these conditions.

Following the organization of a previous review of the subject,[1] we will separately consider: palliative therapy; pharmacological or physical approaches aimed at removing toxic metabolites; pharmacological attempts to bypass blocks in the respiratory chain by means of artificial electron acceptors; supplementation with vitamins and cofactors to "strengthen" the respiratory chain; scavenging excessive oxygen radicals (or reactive oxygen species, ROS); aerobic exercise and physical therapy; gene therapy for mutations in mitochondrial DNA (mtDNA) or mutations in nuclear DNA (NDNA); and preventive measures such as genetic counseling and prenatal diagnosis.

Any discussion of therapy requires some knowledge of the structure and function of the respiratory chain and of the basic concepts of mitochondrial genetics. A sche-

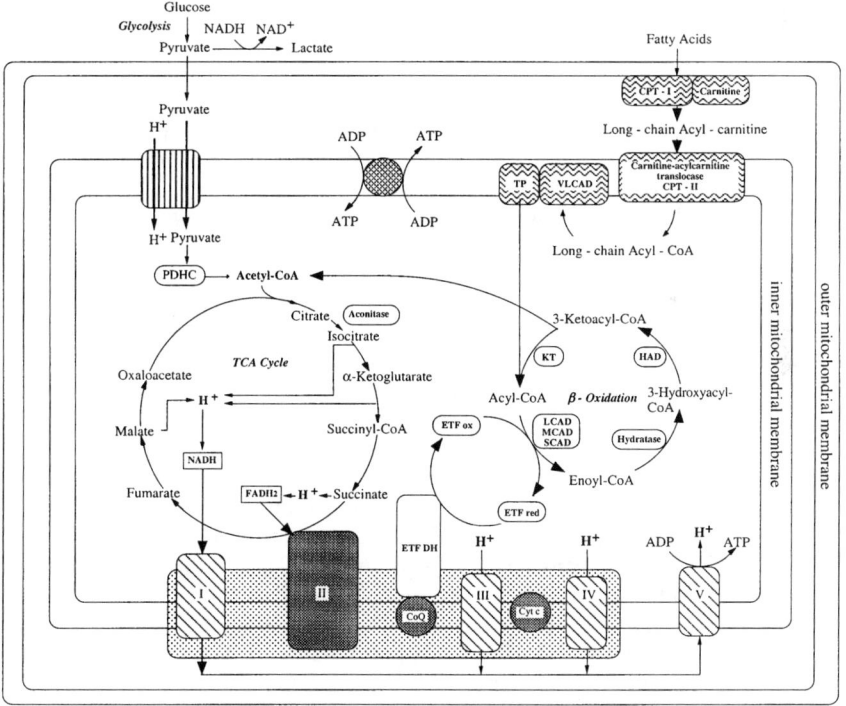

FIGURE 1. Mitochondrial metabolism. Respiratory chain components or complexes encoded exclusively by nDNA are *solid*; complexes containing some subunits encoded by nDNA and others encoded by mtDNA are *cross-hatched*. *Abbreviations*: PDHC, pyruvate dehydrogenase complex; CPT, carnitine palmitoyltransferase; VLCAD, very long-chain acyl-CoA dehydrogenase; TP, trifunctional protein; LCAD, long chain acyl-CoA dehydrogenase; MCAD, medium chain acyl-CoA dehydrogenase; SCAD, short chain acyl-CoA dehydrogenase; HAD, 3-hydroxy-CoA dehydrogenase; KT, 3-ketothiolase; ETFox, oxidized form of electron transfer favoprotein; ETFred, reduced form of electron transfer flavoprotein; ETF-DH, ETF-coenzyme Q oxidoreductase.

matic view of the respiratory chain in the general context of mitochondrial metabolism (FIG. 1) indicates the points of entry of carbohydrate and lipid "fuels." Pyruvate, the terminal product of anaerobic glycolysis, is carried across the inner mitochondrial membrane (IMM) through a symport system "in the wake" of hydrogen ions flowing inward down their electrochemical gradient. Transport of free fatty acids requires a more complex system, which includes two enzymes (carnitine palmitoyltransferase I [CPT I] and carnitine palmitoyltransferase II [CPT II]), a carrier molecule (L-carnitine), and a translocase (carnitine acylcarnitine translocase). After pyruvate has been oxidized by the pyruvate dehydrogenase complex and fatty acyl-CoAs by the reactions of β-oxidation "spirals," the common resulting metabolite, acetyl-CoA, is further oxidized in the Krebs cycle. The reducing equivalents produced by the Krebs cycle and by the β-oxidation spirals are carried along a chain of proteins embedded in the IMM (electron transport chain). The electron transport chain consists of four multimeric complexes (complexes I to IV) and two small electron carriers (coenzyme Q or ubiquinone and cytochrome c). The energy generated by these reactions pumps protons from the mitochondrial matrix (space within the IMM) to the space between the IMM and the outer mitochondrial membrane (OMM) at three sites (complex I, complex III, and complex IV). This builds up an electrochemical proton gradient across the IMM. When protons flow back into the intermembrane space through a fifth multimeric complex (complex V or ATP synthase) that functions as a rotary motor, the physical energy of the proton gradient is converted into chemical energy in the form of ATP.

The *respiratory chain*, which comprises the electron transport chain and ATP synthase, is the only cellular pathway under dual genetic control. Thirteen subunits, seven in complex I (ND1–ND4, ND4L, ND5, and ND6), one in complex III (cytochrome b), three in complex IV (COX I–COX III), and two in complex V (ATPase6 and ATPase8) are encoded by mtDNA, and all others by nDNA (FIG. 1; see also *Mitochondrial Encephalomyopathies: Diagnostic Approach,* this volume). The rules of mitochondrial genetics are discussed elsewhere in this volume: they include maternal inheritance, heteroplasmy and threshold effect, and mitotic segregation. Briefly, maternal inheritance refers to the fact that all mtDNA comes to the zygote from the oocyte: thus, as a rule, pathogenic mutations of mtDNA (and related diseases) are transmitted from a woman to all her children, but only her daughters will pass it on to their progeny, with no evidence of male-to-child transmission. *Heteroplasmy* refers to the coexistence of mutant and wild-type mtDNAs in the same cell, tissue, and individual, and is predicated on the notion that mtDNA is present in hundreds or thousands of copies in each cell (polyplasmy). A corollary of heteroplasmy is the threshold effect: a certain minimum number of mutant mtDNAs is needed to impair oxidative phosphorylation and cause symptoms, and the threshold will be lower in tissues that are highly dependent on oxidative metabolism.

PALLIATIVE THERAPY

Lack of a cure does not equate with lack of therapy. Symptomatic therapy can be very effective in patients with mitochondrial disorders. Let us consider some of the most common manifestations of mitochondrial encephalomyopathies.

Seizures usually respond to conventional anticonvulsants. However, valproic acid should be used with caution and in association with L-carnitine, because it inhibits carnitine uptake.[2]

In patients with progressive external ophthalmoplegia, severe ptosis can be ameliorated by surgery, at least transiently. Congenital cataracts are also treated surgically.

Diabetes mellitus, whether insulin-dependent or not, responds to dietary or pharmacological therapy. The use of growth hormone in children with growth retardation is controversial because the increased metabolic demands may be ill tolerated by an already metabolically challenged patient. In children with impaired growth due to feeding problems, recurrent vomiting, or severe gastroesophageal reflux, useful surgical procedures include percutaneous endoscopic gastrostomy or fundoplication.

Timely placement of a pacemaker can be life-saving in patients with Kearns-Sayre syndrome (KSS) and blocks of cardiac conduction. Heart transplantation is controversial in patients with cardiomyopathy and multisystemic disorders. However, when cardiac involvement is the predominant or exclusive problem, cardiac transplantation is justified.[3-5]

Neurosensory hearing loss can be alleviated by cochlear implants.[6]

Liver failure, often associated with the mtDNA depletion syndrome, may benefit from liver transplantation, especially if other organs seem to be spared.[7,8]

Recurrent myoglobinuria is seen in patients with primary coenzyme Q10 (CoQ10) deficiency[9-11] or with mutations in mtDNA protein-coding gene.[12-14] Patients with CoQ10 deficiency benefit from CoQ10 supplementation. During acute episodes, all patients should be vigorously rehydrated and subjected to renal dialysis when myoglobinuria is complicated by renal failure.

REMOVAL OF NOXIOUS METABOLITES

As FIGURE 1 illustrates, any severe block in the respiratory chain will result in accumulations of substrates upstream, all the way to pyruvate, which is reduced to lactate and transaminated to alanine. In fact, all three compounds are usually increased in the blood, cerebrospinal fluid, and urine of patients with mitochondrial encephalomyopathies (see DiMauro et al., this volume).

As excessive concentrations of lactic acid are neurotoxic, it is reasonable to control lactic acidosis; unfortunately, this is usually done through the buffering effect of bicarbonate, which is transient and may actually exacerbate cerebral symptoms.[15] A more specific tool to combat lactic acidosis is dichloroacetate, a well-studied inhibitor of PDH kinase. Keeping PDH in the dephosphorylated, active form favors pyruvate metabolism and decreases lactate concentration.[16] There is anecdotal evidence of clinical improvement in children with mitochondrial encephalomyopathy, lactic acidosis, and stroke-like episodes (MELAS),[17-21] but the first double-blind, placebo-controlled trial in a large cohort of MELAS patients is still underway in our center. One serious disadvantage of dichloroacetate is that it often causes or exacerbates peripheral neuropathy, even when it is administered together with thiamine (8.6 mg/kg).

A severe autosomal recessive syndrome due to a defect of intergenomic signaling is mitochondrial neurogastrointestinal encephalomyopathy (MNGIE).[22] Patients with this multisystemic disorder have mutations in the *TP* gene, encoding the enzyme thymidine kinase, and accumulate excessive amounts of thymidine in blood.[23] Although the pathogenic role of thymidine accumulation is not known, restoring normal levels of circulating thymidine is a logical first approach to therapy, which is being pursued by dialysis and pharmacological means (Hirano, personal communication).

ADMINISTRATION OF ARTIFICIAL ELECTRON ACCEPTORS

An ingenious attempt to bypass a block in complex III of the respiratory chain in a young woman with mitochondrial myopathy and severe exercise intolerance[24,25] used two artificial electron acceptors (menadiol diphosphate, 40 mg daily, and vitamin C, 4 g daily) whose redox potentials fit the gap created by the cytochrome *b* dysfunction.[26] The patient initially improved dramatically, as documented by ^{31}P nuclear magnetic resonance spectroscopy (NMR) of muscle,[27] but the improvement was not sustained. Other myopathic patients with similar biochemical and molecular defects (complex III deficiency due to mutations in the cytochrome *b* gene of mtDNA) have not responded to this treatment.

Replacement of defective mammalian respiratory chain subunits with corresponding subunits from yeast has been achieved *in vitro* through molecular engineering: these experiments are discussed below under "gene therapy."

ADMINISTRATION OF METABOLITES AND COFACTORS

Various "cocktails" of vitamins and cofactors have been used–and are still used– in patients with mitochondrial encephalomyopathies, including riboflavin (vitamin B_2), thiamine (vitamin B_1), folic acid, CoQ10, L-carnitine, and creatine. All of these are natural compounds and presumably harmless at the doses used. Some, such as riboflavin and CoQ10, are components of the respiratory chain, but there is no evidence that they are decreased in primary mitochondrial diseases.[28] Others appear to be decreased in certain conditions; for example, folic acid was lower than normal in the blood and CSF of patients with KSS.[29] Still others are decreased secondarily; free carnitine tends to be lower than normal in the blood of patients with respiratory chain defects, whereas esterified carnitine tends to be increased. This shift may reflect a partial impairment of β-oxidation, whose reducing equivalents enter the respiratory chain at the level of CoQ10 through the action of the electron transfer flavoprotein (FIG. 1). The "cocktail" of our choice is a combination of L-carnitine (1,000 mg three times a day) and CoQ10 (100 mg three times a day), with the rationale of restoring free carnitine levels and exploiting the oxygen radical scavenger properties of CoQ10.

The rationale for using some of these compounds is powerful when the compound in question is specifically and markedly decreased, either because of defective transport or because of defective synthesis. This is illustrated by primary carnitine deficiency and primary CoQ10 deficiency.

Although primary carnitine deficiency is not a defect of the respiratory chain, we briefly describe it here because of the life-saving effect of replacement therapy. Primary systemic carnitine deficiency is an autosomal recessive disorder due to genetic defects of the plasma membrane carnitine transporter.[30,31] The most common presentation is childhood cardiomyopathy, which is progressive. Echocardiography shows dilated cardiomyopathy, and electrocardiography shows peaked T waves and signs of ventricular hypertrophy. Endomyocardial biopsies or postmortem cardiac specimens show massive lipid storage and, when measured, carnitine concentration was less than 5% of normal. There is a dramatic response to carnitine supplementation, and indices of cardiac function return to normal within a few months;[32,33] hence the importance of measuring blood carnitine concentration in all children with unexplained cardiomyopathy.

Primary CoQ10 deficiency is a defect of the respiratory chain and, therefore, a *bona fide* mitochondrial encephalomyopathy. There are three main clinical presentations. The first is characterized by the following triad: (1) myopathy with recurrent myoglobinuria; (2) ragged-red fibers (RRF) and lipid storage in the muscle biopsy; and (3) CNS involvement, with seizures, ataxia, or mental retardation.[9-11] The second variant is a devastating multisystem disease of infancy, with encephalopathy, hepatopathy, and nephropathy.[34] The third presentation is dominated by ataxia and cerebellar atrophy, often associated with weakness, pyramidal signs, seizures, or mental retardation.[35,36] All patients respond to CoQ10 administration, although patients with the ataxic form require higher doses (up to 1,000 mg daily) and respond less dramatically, probably because of irreversible cerebellar damage.

Creatine monohydrate has been tried in six patients with MELAS and one with undefined mitochondrial disease in a controlled study; there was improvement of high-intensity activities but not of lower intensity aerobic exercise.[37]

In vitro studies have revealed a potentially useful therapeutic approach to a fatal infantile form of encephalocardiomyopathy associated with COX deficiency and due to mutations in the *SCO2* gene, which encodes a COX-assembly protein needed for the insertion of copper into the holoenzyme.[38,39] When copper was added to the medium of cultured COX-deficient myoblasts harboring *SCO2* mutations, COX activity was restored.[40,41] This suggests that copper supplementation should be tried in desperately ill infants with cardiopathy and *SCO2* mutations, especially because copper administration has already been used in infants with Menkes disease without untoward effects.[42]

ADMINISTRATION OF OXYGEN RADICAL SCAVENGERS

Defects of the respiratory chain have detrimental effects that go beyond impairing ATP production and include altered intracellular calcium buffering,[43] excessive production of ROS,[44] and promoting apoptosis.[45,46] Increased production of ROS damages cell membranes through lipid peroxidation and further accelerates the high mutation rate of mtDNA, creating a vicious cycle. Evidence of oxidative stress has been provided not only in primary mitochondrial diseases but also in vast numbers of neurodegenerative disorders, in which nDNA mutations affect mitochondrial or nonmitochondrial proteins. These include Friedreich's ataxia, Wilson's disease, some forms of hereditary spastic paraplegia, Huntington's disease, amyotrophic lateral sclerosis, and Parkinson's disease.[44,46,47]

In an attempt to quench the effects of ROS, several oxygen radical scavengers have been utilized in most of the disorders just listed, including vitamin E, CoQ10, idebenone, and dihydrolipoate.

CoQ10 has been widely used for primary mitochondrial diseases, and the multitude of generally positive anecdotal data[48–55] together with the lack of negative side effects has contributed to its widespread use in these patients. However, there is a need for controlled trials in large cohorts of patients.

CoQ10 and idebenone have been increasingly employed in therapeutic trials of neurodegenerative disorders whose pathogenetic mechanisms are thought to involve excessive production of ROS. Both CoQ10 and idebenone improved cardiac dysfunction in patients with Friedreich's ataxia but not their CNS symptoms.[56–60] A recent multicenter, randomized, parallel–group, double-blind, placebo-controlled study of 80 patients with early Parkinson's disease treated with three doses of CoQ10 (300, 600, or 1200 mg daily) showed not only that CoQ10 was well tolerated, but also that less disability developed in subjects taking CoQ10 and the benefit was greatest in patients on the highest dosage.[61] A small open-labeled pilot study of CoQ10 (600 mg/day) in amyotrophic lateral sclerosis showed some positive trends,[62] but larger controlled studies are underway and will yield more conclusive results.

EXERCISE AND PHYSICAL THERAPY

The role of exercise–whether, in fact, exercise and physical therapy had to be considered at all–in patients with mitochondrial diseases remained controversial until recently. Detailed functional studies have defined the exercise limitations in a large cohort of patients with mitochondrial myopathies due to diverse mtDNA mutations.[63] Mean work capacity and oxygen uptake were decreased, as oxidative capacity was limited by reduced ability of muscle to extract oxygen from blood; ventilation and cardiac output relative to VO_2 were exaggerated; and there was an inverse relation between the proportion of mutant mtDNAs in muscle and peak oxygen extraction during exercise.

Inactivity in patients with mitochondrial diseases and exercise intolerance is often condoned for fear of causing muscle damage, but it only increases deconditioning. Instead, aerobic training in these patients has remarkable effects, increasing work and oxidation capacity.[64] After 14 weeks of bicycle exercise training, 10 patients with exercise intolerance and different mtDNA mutations showed increased peak work and oxidative capacities, increased systemic arteriovenous O_2 differences, and improved metabolic indices by ^{31}P magnetic resonance spectroscopy (^{31}P-MRS). There was mitochondrial proliferation and the activities of some, but not all, defective respiratory chain enzymes increased. These results suggested that mitochondrial proliferation was accompanied by an increased ratio of wild-type to mutant mtDNAs; somewhat paradoxically, this was true in only three of nine patients, whereas in six of nine patients this ratio actually decreased. Although this apparent contradiction remains to be explained, there is little doubt that patients with mitochondrial myopathies due to mtDNA mutations benefit from aerobic training.

The possible role of isometric exercise as an intentional muscle injury intended to favor regeneration and "gene shifting" [65] is discussed under "gene therapy."

GENE THERAPY

A simple form of gene therapy was used in the second patient with nonthyroidal hypermetabolism (Luft's disease); to reduce the excessive number of muscle mitochondria, the patient was given chloramphenicol, an inhibitor of mitochondrial protein synthesis. There was a mild reduction of metabolic rate and some subjective improvement, but the trial was short-lived because of drug toxicity,[66] and a second trial was unsuccessful.[67]

For mitochondrial diseases due to mutations in nuclear genes, the problems are no different from those vexing gene therapy for other mendelian disorders, including choice of appropriate viral or nonviral vectors, delivery to the affected tissues, and potential immunological reactions.

The problems are even more complex for mtDNA-related diseases because of polyplasmy and heteroplasmy and because nobody has yet been able to transfect DNA into mitochondria in a heritable manner.

The most promising approach is to influence heteroplasmy, reducing the ratio of mutant to wild-type mitochondrial genomes, or "gene shifting."[65] This can be achieved in various manners. A pharmacological approach employed oligomycin *in vitro*, which, in appropriate culture conditions, allowed for the rapid selection of wild-type over mutant fibroblasts harboring heteroplasmic levels of the T8993G NARP/MILS (neuropathy, ataxia, retinitis pigmentosa/maternally inherited Leigh's syndrome) mutation.[68] With all due caution because oligomycin is a mitochondrial poison, this system could be adapted to human therapy.

A molecular approach used peptide nucleic acids homologous to specific mutant mtDNAs, which selectively bind to complementary molecules and inhibit their replication. Some success was obtained in decreasing the ratio of A8344G MERRF (myoclonus epilepsy with RRF) mutants *in vitro*,[69] but there were problems with the delivery of peptide nucleic acids to human mitochondria.[70]

A pharmacological and a functional approach to "gene shifting" were based on the same observation, that satellite cells and myoblasts contain lesser amounts of mutant mtDNAs than do mature muscle fibers.[65,71–73] The pharmacological approach used a myotoxic agent, bupivocaine, to cause limited muscle necrosis, which would be "repaired" by tissue harboring less mutant mtDNA. Unfortunately, unilateral injection of bupivocaine in levator palpebrae muscles of five patients with progressive external ophthalmoplegia or KSS did not cause any improvement.[74] The functional approach exploits the notion that isometric exercise leads to "microtraumas" and limited necrosis of exercising muscles.[65] Indirect support for this concept comes from our studies of one patient with recurrent myoglobinuria due to a nonsense mutation (G5920A) in the COX I gene of mtDNA.[14] The few COX-positive fibers in the muscle biopsy of this patient were totally devoid of mutant mtDNA, suggesting that these were indeed regenerated fibers. This putative, and apparently paradoxical, beneficial effect of massive muscle necrosis suggests the potential therapeutic value of induced attacks of myoglobinuria in patients with severe exercise intolerance or weakness due to muscle-specific mtDNA mutations.

Genetic engineering provides multiple interesting approaches to therapy. One of these is to convert an mtDNA-related disease into an nDNA-related disease. For example, the ATPase 6 gene of mtDNA can be converted from the mitochondrial into the nuclear genetic code. To be sure that the novel nuclear protein encoded by the converted gene is recognized by, and transported into, mitochondria, it has to be provided with a leader peptide, whose genetic sequence can be "borrowed" from another mtDNA-encoded protein. Once this genetic "Trojan horse" has been carried into the nucleus (which would present the same problems existing for gene therapy of any mendelian disorder), its translation product in the cytoplasm would be transported into the mitochondria, freed of the leader peptide, and assembled into the F_0 component of complex V together with its ATPase 8 counterpart synthesized within the mitochondria. Circuitous as it is, this approach (known as "allotopic expression") has been realized *in vitro* to correct the biochemical defect in cybrid cells harboring the T8993G NARP/MILS mutation[75] and in cybrids harboring the G11778A LHON (Leber's hereditary optic neuropathy) mutation.[76]

A more direct molecular approach is to import specific restriction endonucleases as "magic bullets" to selectively destroy mutant mtDNAs, a system proven effective in cybrids carrying the T8993G NARP/MILS mutation.[77]

Still another molecular "trick" is to correct a respiratory chain defect due to an mtDNA mutation by transfecting affected mammalian cells with either mitochondrial or nuclear genes from other organisms but encoding cognate proteins.[78-80]

Finally, normal yeast tRNAs can be imported from the cytoplasm to compensate for mutant mitochondrial tRNAs, and human mitochondria can internalize yeast tRNA derivatives in the presence of a specific yeast transport factor.[81] In yeast, the deleterious effects of mutations that are counterparts of human MELAS mutations can be compensated by the overexpression of a mitochondrial elongation factor, EF-TU.[82] This yeast model opens new therapeutic vistas for human mitochondrial diseases, based on the exploitation of genes that modify or suppress pathological phenotypes.

GERMLINE THERAPY

Theoretically, a woman carrying an mtDNA mutation could have her oocytes "cleansed" *in vitro* of the cytoplasm, with all mitochondria and all mtDNAs. The naked nucleus could then be transferred to a normal enucleated host oocyte, which could be fertilized *in vitro* and implanted in the woman's uterus. If successful, this approach would guarantee a normal progeny with all the nuclear, and physiognomic, traits of both parents.[83] Partial replacement (5–10%) of the cytoplasm of aged oocytes is actually used to "rejuvenate" them and improve the success rate of *in vitro* fertilization.[84] This approach, which is being actively investigated in the UK, raises ethical questions regarding germline genetic manipulation.[85]

GENETIC COUNSELING

Prenatal diagnosis for tRNA point mutations, including the more common ones associated with MELAS and MERRF, is made practically impossible by two concerns. First, the mutation load in amniocytes or chorionic villi does not necessarily correspond to that of other fetal tissues. Second, mutation load measured in prenatal samples may shift *in utero* or after birth due to mitotic segregation.

At the other end of the spectrum, large-scale deletions of mtDNA as a rule are neither inherited nor transmitted and either arise *de novo* in oogenesis or in early embryogenesis, or, when they are present in a fertilized oocyte, are unlikely to slip through the bottleneck between ovum and embryo that allows only a small minority of maternal mtDNAs to populate the fetus.[86] However, in counseling a woman who harbors a large-scale mtDNA deletion, the possibility of transmission should not be excluded completely, as three such events are recorded in the literature.[87–89]

There is good evidence that mutations in ATPase 6 associated with NARP/MILS do not show tissue- or age-related variations,[90] thus making prenatal diagnosis feasible for parents who have lost a child to maternally inherited Leigh's syndrome.[91]

The rapid progress in our molecular knowledge of nDNA-related defects of the respiratory chain is improving genetic counseling and making prenatal diagnosis an option for families with fatal infantile conditions, such as Leigh's syndrome or mtDNA depletion syndromes.

ACKNOWLEDGMENTS

Part of the work described herein was supported by National Institutes of Health Grants HD32062 and NS11766, by a grant from the Muscular Dystrophy Association, and by a generous gift from the Medical Illness Counseling Center, Chevy Chase, Maryland.

REFERENCES

1. DiMauro, S. 2002. Treatment of mitochondrial diseases. *In* Mitochondrial Disorders: From Pathophysiology to Acquired Defects. C. Desnuelle & S. DiMauro, Eds.: 307-325. Springer-Verlag. Paris.
2. Tein, I., S. DiMauro, Z.-W. Xie, *et al.* 1993. Valproic acid impairs carnitine uptake in cultured human skin fibroblasts. An *in vitro* model for pathogenesis of valproic acid-associated carnitine deficiency. Pediatr. Res. **34:** 281–287.
3. Tranchant, C., B. Mousson, M. Mohr, *et al.* 1993. Cardiac transplantation in an incomplete Kearns-Sayre syndrome with mitochondrial DNA deletion. Neuromusc. Disord. **3:** 561–566.
4. Bohlega, S., K. Tanji, F.M. Santorelli, *et al.* 1996. Multiple mitochondrial DNA deletions associated with autosomal recessive ophthalmoplegia and severe cardiomyopathy. Neurology **46:** 1329–1334.
5. Santorelli, F.M., M.G. Gagliardi, C. Dionisi-Vici, *et al.* 2002. Hypertrophic cardiomyopathy and mtDNA depletion. Successful treatment with heart transplantation. Neuromusc. Disord. **12:** 56–59.
6. Sue, C.M., L.J. Lipsett, D.S. Crimmins, *et al.* 1998. Cochlear origin of hearing loss in MELAS syndrome. Ann. Neurol. **43:** 350–359.
7. Dubern, B., P. Broue, C. Dubuisson, *et al.* 2001. Orthotopic liver transplantation for mitochondrial respiratory chain disorders: a study of 5 children. Transplantation **71:** 633–637.
8. Salviati, L., S. Sacconi, M. Mancuso, *et al.* 2002. Mitochondrial DNA depletion and *dGK* gene mutations. Ann. Neurol. **52:** 311–317.
9. Ogasahara, S., A.G. Engel, D. Frens, *et al.* 1989. Muscle coenzyme Q deficiency in familial mitochondrial encephalomyopathy. Proc. Natl. Acad. Sci. USA **86:** 2379–2382.

10. SOBREIRA, C., M. HIRANO, S. SHANSKE, et al. 1997. Mitochondrial encephalomyopathy with coenzyme Q10 deficiency. Neurology **48:** 1238–1243.
11. DI GIOVANNI, S., M. MIRABELLA, A. SPINAZZOLA, et al. 2001. Coenzyme Q10 reverses pathological phenotype and reduces apoptosis in familial CoQ10 deficiency. Neurology **57:** 515–518.
12. KEIGHTLEY, J.A., K.C. HOFFBUHR, M.D. BURTON, et al. 1996. A microdeletion in cytochrome c oxidase (COX) subunit III associated with COX deficiency and recurrent myoglobinuria. Nat. Genet. **12:** 410–415.
13. ANDREU, A.L., M.G. HANNA, H. REICHMANN, et al. 1999. Exercise intolerance due to mutations in the cytochrome b gene of mitochondrial DNA. N. Engl. J. Med. **341:** 1037–1044.
14. KARADIMAS, C.L., P. GREENSTEIN, C.M. SUE, et al. 2000. Recurrent myoglobinuria due to a nonsense mutation in the COX I gene of mtDNA. Neurology **55:** 644–649.
15. DE VIVO, D.C. & S. DIMAURO. 1999. Mitochondrial diseases. In Pediatric Neurology: Principles & Practice. Vol. 1. K.F. Swaiman & S. Ashwal, Eds. :494–509. Mosby. St. Louis, MO.
16. STACPOOLE, P.W. 1989. The pharmacology of dichloroacetate. Metabolism **38:** 1124–1144.
17. DE VIVO, D.C., A. JACKSON, C. WADE, et al. 1990. Dichloroacetate treatment of MELAS-associated lactic acidosis. Ann. Neurol. **28:** 437.
18. DE STEFANO, N., P.M. MATTHEWS, B. FORD, et al. 1995. Short-term dichloroacetate treatment improves indices of cerebral metabolism in patients with mitochondrial disorders. Neurology **45:** 1193–1198.
19. TAIVASSALO, T., P.M. MATTHEWS, N. DE STEFANO, et al. 1996. Combined aerobic training and dichloroacetate improve exercise capacity and indices of aerobic metabolism in muscle cytochrome oxidase deficiency. Neurology **47:** 529–534.
20. KURODA, Y., M. ITO, E. NAITO, et al. 1997. Concomitant administration of sodium dichloroacetate and vitamin B1 for lactic acidemia in children with MELAS syndrome. J. Pediatr. **131:** 450–452.
21. SAITOH, S., M.Y. MOMOI, T. YAMAGATA, et al. 1998. Effects of dichloroacetate in three patients with MELAS. Neurology **50:** 531–534.
22. NISHINO, I., A. SPINAZZOLA, A. PAPADIMITRIOU, et al. 2000. Mitochondrial neurogastrointestinal encephalomyopathy: an autosomal recessive disorder due to thymidine phosphorylase mutations. Ann. Neurol. **47:** 792–800.
23. SPINAZZOLA, A., R. MARTI, I. NISHINO, et al. 2002. Altered thymidine metabolism due to defects of thymidine phosphorylase. J. Biol. Chem. **277:** 4128–4132.
24. KENNAWAY, N.G., N.R. BUIST, V.M. DARLEY, et al. 1984. Lactic acidosis and mitochondrial myopathy associated with deficiency of several components of complex III of the respiratory chain. Pediatr. Res. **18:** 991–999.
25. KEIGHTLEY, J.A., R. ANITORI, M.D. BURTON, et al. 2000. Mitochondrial encephalomyopathy and complex III deficiency associated with a stop-codon mutation in the cytochrome b gene. Am. J. Hum. Genet. **67:** 1400–1410.
26. ELEFF, S., N.G. KENNAWAY, N.R. BUIST, et al. 1984. 31P-NMR study of improvement in oxidative phosphorylation by vitamins K3 and C in a patient with a defect in electron transport at complex III in skeletal muscle. Proc. Natl. Acad. Sci. USA **81:** 3529–3533.
27. ARGOV, Z., W.J. BANK, J. MARIS, et al. 1986. Treatment of mitochondrial myopathy due to complex III deficiency with vitamins K3 and C: A 31P-NMR follow-up study. Ann. Neurol. **19:** 598–602.
28. MATSUOKA, T., H. MAEDA, Y. GOTO, et al. 1992. Muscle coenzyme Q10 in mitochondrial encephalomyopathies. Neuromusc. Disord. **1:** 443–447.
29. ALLEN, R.J., S. DIMAURO, D.L. COULTER, et al. 1983. Kearns-Sayre syndrome with reduced plasma and cerebrospinal fluid folate. Ann. Neurol. **13:** 679–682.
30. LAMHONWAH, A.-M. & I. TEIN. 1998. Carnitine uptake defect: frameshift mutations in the human plasmalemmal carnitine transporter gene. Biochem. Biophys. Res. Commun. **252:** 396–401.
31. NEZU, J.-I., I. TAMAI, A. OKU, et al. 1999. Primary systemic carnitine deficiency is caused by mutations in a gene encoding sodium ion-dependent carnitine transporter. Nat. Genet. **21:** 91–94.

32. STANLEY, C.A., S. DE LEEUW, P.M. COATES, et al. 1991. Chronic cardiomyopathy and weakness or acute coma in children with a defect in carnitine uptake. Ann. Neurol. **30:** 709–716.
33. TEIN, I., D.C. DEVIVO, F. BIERMAN, et al. 1990. Impaired skin fibroblast carnitine uptake in primary systemic carnitine deficiency manifested by childhood carnitine-responsive cardiomyopathy. Pediatr. Res. **28:** 247–255.
34. ROTIG, A., E.-L. APPELKVIST, V. GEROMEL, et al. 2000. Quinone-responsive multiple respiratory-chain dysfunction due to widespread coenzyme Q10 deficiency. Lancet **356:** 391–395.
35. MUSUMECI, O., A. NAINI, A.E. SLONIM, et al. 2001. Familial cerebellar ataxia with muscle coenzyme Q10 deficiency. Neurology **56:** 849–855.
36. LAMPERTI, C., A. NAINI, M. HIRANO, et al. 2003. Cerebellar ataxia and coenzyme Q10 deficiency. Neurology **60:** 1206–1208.
37. TARNOPOLSKY, M.A., B.D. ROY & J.R. MACDONALD. 1997. A randomized, controlled trial of creatine monohydrate in patients with mitochondrial cytopathies. Muscle Nerve **20:** 1502–1509.
38. PAPADOPOULOU, L.C., C.M. SUE, M.M. DAVIDSON, et al. 1999. Fatal infantile cardio-encephalomyopathy with COX deficiency and mutations in *SCO2*, a COX assembly gene. Nat. Genet. **23:** 333–337.
39. SUE, CM. & E.A. SCHON. 2000. Mitochondrial respiratory chain diseases and mutations in nuclear DNA: a promising start? Brain Pathol. **10:** 442–450.
40. SALVIATI, L., E. HERNANDEZ-ROSA, W.F. WALKER, et al. 2002. Copper supplementation restores cytochrome *c* oxidase activity in cultured cells from patients with *SCO2* mutations. Biochem. J. **363:** 321–327.
41. JAKSCH, M., C. PARET, R. STUCKA, et al. 2001. Cytochrome *c* oxidase deficiency due to mutations in *SCO2*, encoding a mitochondrial copper-binding protein, is rescued by copper in human myoblasts. Hum. Mol. Genet. **10:** 3025–3035.
42. SHERWOOD, G., B. SARKAR & A.S. KORTSAK. 1989. Copper histidinate therapy in Menkes' disease: prevention of progressive neurodegeneration. J. Inherit. Metab. Dis. **12** (Suppl. 2): 393–396.
43. BRINI, M., P. PINTON, M.P. KING, et al. 1999. A calcium signaling defect in the pathogenesis of a mitochondrial DNA inherited oxidative phosphorylation deficiency. Nature Med. **5:** 951–954.
44. BEAL, M.F. 2002. Mitochondria in neurodegeneration. *In* Mitochondrial Disorders: From Pathophysiology to Acquired Defects. C. Desnuelle & S. DiMauro, Eds.: 17–35. Springer-Verlag. Paris, France.
45. SERVIDEI, S., S. DI GIOVANNI, A. BROCCOLINI, et al. 2002. Apoptosis and oxidative stress in mitochondrial disorders. *In* Mitochondrial Disorders: From Pathophysiology to Acquired Defects. C. Desnuelle & S. DiMauro, Eds.: 37–43. Springer. Paris.
46. SCHON, E.A. & G. MANFREDI. 2003. Neuronal degeneration and mitochondrial dysfunction. J. Clin. Invest. **111:** 303–312.
47. TABRIZI, S.J. & A.H.V. SCHAPIRA. 2002. Mitochondrial abnormalities in neurodegenerative disorders. *In* Mitochondrial Disorders in Neurology 2. A.H.V. Schapira & S. DiMauro, Eds.: 143–174. Butterworth-Heinemann. Boston.
48. OGASAHARA, S., Y. NISHIKAWA, S. YORIFUJI, et al. 1986. Treatment of Kearns-Sayre syndrome with coenzyme Q10. Neurology **36:** 45–53.
49. YAMAMOTO, M., T. SATO, M. ANNO, et al. 1987. Mitochondrial myopathy, encephalopathy, lactic acidosis, and strokelike episodes with recurrent abdominal symptoms and coenzyme Q10 administration. J. Neurol. Neurosurg. Psychiatry **50:** 1475–1481.
50. BRESOLIN, N., L. BET, A. BINDA, et al. 1988. Clinical and biochemical correlations in mitochondrial myopathies treated with coenzyme Q10. Neurology **38:** 892–899.
51. IHARA, Y., R. NAMBA, S. KURODA, et al. 1989. Mitochondrial encephalomyopathy (MELAS): pathological study and successful therapy with coenzyme Q10 and idebenone. J. Neurol. Sci. **90:** 263–271.
52. IKEJIRI, Y., E. MORI, K. ISHII, et al. 1996. Idebenone improves cerebral mitochondrial oxidative metabolism in a patient with MELAS. Neurology **47:** 583–585.
53. CHAN, A., H. REICHMANN, A. KOGEL, et al. 1998. Metabolic changes in patients with mitochondrial myopathies and effects of coenzyme Q10 therapy. J. Neurol. **245:** 681–685.

54. BENDAHAN, D., C. DESNUELLE, D. VANUXEM, *et al.* 1992. ^{31}P NMR spectroscopy and ergometer exercise test as evidence for muscle oxidative performance improvement with coenzyme Q in mitochondrial myopathies. Neurology **42:** 1203–1208.
55. ABE, K., Y. MATSUO, J. KADEKAWA, *et al.* 1999. Effect of coenzyme Q10 in patients with mitochondrial myopathy, encephalopathy, lactic acidosis, and stroke-like episodes (MELAS): Evaluation by noninvasive tissue oximetry. J. Neurol. Sci. **162:** 65–68.
56. LODI, R., P.E. HART, B. RAJAGOPALAN, *et al.* 2001. Antioxidant treatment improves *in vivo* cardiac and skeletal muscle bioenergetics in patients with Friedreich's ataxia. Ann. Neurol. **49:** 590–596.
57. RUSTIN, P., J.-C. VON KLEIST-RETZOW, K. CHANTREL-GROUSSARD, *et al.* 1999. Effect of idebenone on cardiomyopathy in Friedreich's ataxia: a preliminary study. Lancet **354:** 477–479.
58. FILLA, A. & A.J. MOSS. 2003. Idebenone for treatment of Friedreich's ataxia? Neurology **60:** 1569–1570.
59. MARIOTTI, C., A. SOLARI, D. TORTA, *et al.* 2003. Idebenone treatment in Friedreich patients: one-year-long randomized placebo-controlled trial. Neurology **60:** 1676–1679.
60. BUYSE, G., L. MERTENS, G. DI SALVO, *et al.* 2003. Idebenone treatment in Friedreich's ataxia. Neurology **60:** 1679–1681.
61. SHULTS, C.W., D. OAKES, K. KIEBURTZ, *et al.* 2002. Effects of coenzyme Q10 in early Parkinson disease. Arch. Neurol. **59:** 1541–1550.
62. HAYES, S., M. DEL BENE, W. TROJABORG, *et al.* 2000. Therapeutic trial of coenzyme Q10 (CoQ10) in amyotrophic lateral sclerosis (ALS/MND). Amyotrophic Lateral Sclerosis Suppl. **3:** 119.
63. TAIVASSALO, T., T. DYSGAARD JENSEN, N.J. KENNAWAY, *et al.* 2003. The spectrum of exercise tolerance in mitochondrial myopathies: a study of 40 patients. Brain **126:** 413–423.
64. TAIVASSALO, T., E.A. SHOUBRIDGE, J. CHEN, *et al.* 2001. Aerobic conditioning in patients with mitochondrial myopathies: physiological, biochemical, and genetic effects. Ann. Neurol. **50:** 133–141.
65. TAIVASSALO, T., K. FU, T. JOHNS, *et al.* 1999. Gene shifting: a novel therapy for mitochondrial myopathy. Hum. Mol. Genet. **8:** 1047–1052.
66. HAYDAR, N.A., H.L. CONN, A. AFIFI, *et al.* 1971. Severe hypermetabolism with primary abnormality of skeletal muscle mitochondria. Ann. Int. Med. **74:** 548–558.
67. DIMAURO, S., E. BONILLA, C.P. LEE, *et al.* 1976. Luft's disease. Further biochemical and ultrastructural studies of skeletal muscle in the second case. J. Neurol. Sci. **27:** 217–232.
68. MANFREDI, G., N. GUPTA, M.E. VAZQUEZ-MEMIJE, *et al.* 1999. Oligomycin induces a decrease in the cellular content of a pathogenic mutation in the human mitochondrial ATPase 6 gene. J. Biol. Chem. **274:** 9386–9391.
69. TAYLOR, R.W., P.F. CHINNERY, D.M. TURNBULL, *et al.* 1997. Selective inhibition of mutant human mitochondrial DNA replication *in vitro* by peptide nucleic acids. Nat. Genet. **15:** 212–215.
70. CHINNERY, P.F., R.W. TAYLOR, K. DIEKERT, *et al.* 1999. Peptic nucleic acid delivery to human mitochondria. Gene Therapy **6:** 1919–1928.
71. Clark, K.M., L.A. Bindoff, R.N. Lightowlers, *et al.* 1997. Reversal of a mitochondrial DNA defect in human skeletal muscle. Nat. Genet. **16:** 222–224.
72. FU, K., R. HARTLEN, T. JOHNS, *et al.* 1996. A novel heteroplasmic tRNAleu(CUN) mtDNA point mutation in a sporadic patient with mitochondrial encephalomyopathy segregates rapidly in skeletal muscle and suggests an approach to therapy. Hum. Mol. Genet. **5:** 1835–1840.
73. SHOUBRIDGE, E.A., T. JOHNS & G. KARPATI. 1997. Complete restoration of a wild-type mtDNA genotype in regenerating muscle fibers in a patient with a tRNA point mutation and mitochondrial encephalomyopathy. Hum. Mol. Genet. **6:** 2239–2242.
74. ANDREWS, RM., P.G. GRIFFITHS, P.F. CHINNERY, *et al.* 1999. Evaluation of bupivacaine-induced muscle regeneration in the treatment of ptosis in patients with chronic progressive external ophthalmoplegia and Kearns-Sayre syndrome. Eye **13:** 769–772.

75. MANFREDI, G., J. FU, J. OJAIMI, et al. 2002. Rescue of an ATP synthesis deficiency in mtDNA-mutant human cells by transfer of *MTATP6*, a mtDNA-encoded gene, to the nucleus. Nat. Genet. **30:** 394–399.
76. GUY, J., X. QI, F. PALLOTTI, et al. 2002. Rescue of a mitochondrial deficiency causing Leber hereditary optic neuropathy. Ann. Neurol. **52:** 534–542.
77. TANAKA, M., H.J. BORGELD, J. ZHANG, et al. 2002. Gene therapy for mitochondrial disease by delivering restriction endonuclease SmaI into mitochondria. J. Biomed. Sci. **9:** 534–541.
78. SEO, B.B., T. KITAJIMA-IHARA, E.K.L. CHAN, et al. 1998. Molecular remedy of complex I defects: rotenone-insensitive internal NADH-quinone oxidoreductase of *Saccharomyces cerevisiae* mitochondria restores the NADH oxidase activity of complex I-deficient mammalian cells. Proc. Natl. Acad. Sci. USA **95:** 9167–9171.
79. BAI, Y., P. HAJEK, A. CHOMYN, et al. 2001. Lack of complex I activity in human cells carrying a mutation in MtDNA-encoded ND4 subunit is corrected by the *Saccharomyces cerevisiae* NADH-quinone oxidoreductase (NDI1) gene. J. Biol. Chem. **276:** 38808–38813.
80. OJAIMI, J., J. PAN, S. SANTRA, et al. 2002. An algal nucleus-encoded subunit of mitochondrial ATP synthase rescues a defect in the analogous human mitochondrial-encoded subunit. Mol. Biol. Cell. **13:** 3836–3844.
81. KOLESNIKOVA, O.A., N.S. ENTELIS, H. MIREAU, et al. 2000. Suppression of mutations in mitochondrial DNA by tRNAs imported from the cytoplasm. Science **289:** 1931–1933.
82. FEUERMANN, M., S. FRANCISCI, T. RINALDI, et al. 2003. The yeast counterparts of human "MELAS" mutations cause mitochondrial dysfunction that can be rescued by overexpression of the mitochondrial translation factor EF-TU. EMBO Rep. **4:** 53–58.
83. RUBENSTEIN, D.S., D.C. THOMASMA, E.A. SCHON, et al. 1995. Germ-line therapy to cure mitochondrial disease: protocol and ethics of *in vitro* ovum nuclear transplantation. Cambridge Q. Healthcare Eth. **4:** 316–339.
84. BARRITT, J.A., C.A. BRENNER, H.E. MALTER, et al. 2001. Mitochondria in human offspring derived from ooplasmic transplantation. Hum. Reprod. **16:** 513–516.
85. THORBURN, D.R., H.-H.M. DAHL & K.K. SINGH. 2001. The pros and cons of mitochondrial manipulation in the human germ line. Mitochondrion **1:** 123–127.
86. DIMAURO, S. & E.A. SCHON. 2003. Mitochondrial respiratory-chain diseases. N. Engl. J. Med. **348:** 2656–2668.
87. BERNES, S.M., C. BACINO, T.R. PREZANT, et al. 1993. Identical mitochondrial DNA deletion in mother with progressive external ophthalmoplegia and son with Pearson marrow-pancreas syndrome. J. Pediatr. **123:** 598–602.
88. SHANSKE, S., Y. TANG, M. HIRANO, et al. 2002. Identical mitochondrial DNA deletion in a woman with ocular myopathy and in her son with Pearson syndrome. Am. J. Hum. Genet. **71:** 679–683.
89. PUOTI, G., F. CARRARA, S. SAMPAOLO, et al. 2003. Identical large-scale rearrangement of mitochondrial DNA causes Kearns-Sayre syndrome in a mother and son. Am. J. Med. Genet. **40:** 858–863.
90. WHITE, S.L., S. SHANSKE, J.J. MCGILL, et al. 1999. Mitochondrial DNA mutations at nucleotide 8993 show a lack of tissue- or age-related variation. J. Inher. Metab. Dis. **22:** 899–914.
91. WHITE, S.L., S. SHANSKE, I. BIROS, et al. 1999. Two cases of prenatal analysis for the pathogenic T to G substitution at nucleotide 8993 in mitochondrial DNA. Prenatal Diagnosis **19:** 1165–1168.

Comprehensive Molecular Diagnosis of Mitochondrial Disorders

Qualitative and Quantitative Approach

LEE-JUN C. WONG

Institute for Molecular and Human Genetics, Georgetown University Medical Center, Washington, DC 20007, USA

ABSTRACT: Mitochondrial disorders can be caused by mutations in nuclear or mitochondrial encoded genes. Point mutations and large deletions in mitochondrial DNA (mtDNA) are responsible for a small portion of the molecular defects in the mitochondrial oxidative phosphorylation system. A significant number of molecular defects of respiratory chain disorders are probably due to mutations in nuclear genes. Molecular diagnosis of mitochondrial disorders has been difficult because of broad genetic and clinical heterogeneity. Mutational analysis of common point mutations of mtDNA such as A3243G, A8344G, and T8993G/C is routinely performed. However, many patients who clearly have clinical manifestations and muscle pathology consistent with oxidative phosphorylation deficiency do not have detectable common mtDNA point mutations. A more comprehensive mutation screening method, temporal temperature gradient gel electrophoresis, was used to scan for unknown mutations in the entire mitochondrial genome. Novel mutations have been discovered but only account for a small portion of patients with suspected mitochondrial RC disorders. Real-time quantitative PCR analysis was used to measure cellular mtDNA content. Abnormal levels of mtDNA were found in many patients with respiratory chain disorders. Molecular analysis revealed that mutations in the thymidine phosphorylase gene are not seen in young patients with severe mtDNA depletion who do not demonstrate clinical features of mitochondrial neurogastrointestinal encephalomyopathy. It was also noted that an increase in the size or number of mitochondria was not necessarily associated with an increase in mtDNA content. On the contrary, in some cases the mtDNA was depleted. Respiratory activity in patients with a defective mitochondrial genome due to either point mutations or deletions may be compensated by amplification of mtDNA. Therefore, a comprehensive molecular analysis of mitochondrial respiratory chain disorders should include qualitative identification of the mutation and quantitative measurement of both the degree of mutant heteroplasmy and the total amount of mtDNA.

KEYWORDS: mtDNA mutations; molecular diagnosis; mitochondrial disorder; mtDNA deletion; mitochondria

Address for correspondence: Lee-Jun C. Wong, Ph.D., Institute for Molecular and Human Genetics, Georgetown University Medical Center, M4000, 3800 Reservoir Road, NW, Washington, DC 20007. Voice: 202-444-0760; fax: 202-444-1770.
wonglj@georgetown.edu

INTRODUCTION

Mitochondrial respiratory chain disorders are a group of clinically and genetically heterogeneous diseases.[1–3] Previous mutational screening of most common mtDNA mutations in 2,000 patients suspected of having respiratory chain disorders by the use of Southern blot and multiplex PCR/allele-specific oligonucleotide (ASO) dot blot hybridization revealed that only about 6% of the patients had identifiable mtDNA mutations.[4] In view of the low detection rate, a more effective and sensitive mutation detection method was developed to screen for unknown mutations in the entire mitochondrial genome.[5] The results demonstrated that the number of mtDNA mutations found are not sufficient to account for the molecular defects in the number of patients with suspected respiratory chain disorders.[5,6] Quantitative rather than qualitative changes in mtDNA and mutations in nuclear genes encoding oxidative phosphorylation enzyme complex subunits should be investigated. In this report, we present advances in the detection of mtDNA mutations and discuss the challenge of a comprehensive molecular diagnosis.

MITOCHONDRIAL GENETICS

The human mitochondrial genome is composed of a circular double-stranded DNA of 16,569 bp which encodes 13 polypeptides essential for oxidative phosphorylation (OXPHOS) enzyme complexes, 2 rRNAs, and 22 tRNA genes required for mitochondrial protein synthesis.[2] Most cells contain hundreds to thousands of mitochondria, and each mitochondrion contains 1–10 copies of the mitochondrial genome. The OXPHOS system consists of five inner membrane bound multi-subunit complexes containing approximately 74 nuclear encoded proteins and 13 mitochondrial encoded polypeptides. Most of the mitochondrial proteins including those involved in mtDNA replication, transcription, and translation are encoded by the nuclear genome, synthesized in the cytoplasm, and imported into the mitochondria. Thus, intergenomic communication between the mitochondria and the nucleus is necessary for mitochondrial biogenesis and function. Mutations in either genome can cause mitochondrial dysfunction.

MITOCHONDRIAL DNA MUTATIONS

Mitochondrial DNA features a high mutation rate due to close proximity to the electron transport system where reactive oxygen species are continuously produced, the limited DNA repair capabilities, and the lack of protective histone proteins. Germline mtDNA mutations cause maternally inherited neuromuscular disease with a broad spectrum of clinical manifestations.[1] The most common point mutations are A3243G in tRNA^{leu} and A8344G in tRNA^{lys}, causing mitochondrial encephalopathy, lactic acidosis, and stroke-like episodes (MELAS) and myoclonic epilepsy and ragged-red fiber (MERRF) syndrome, respectively.[1] Other common point mutations in protein coding regions are T8993G/C in ATPase subunit 6 for neuropathy, ataxia, and retinitis pigmentosa (NARP) and Leigh syndrome, and G11778A in ND4 subunit of complex I for Leber hereditary optic neuropathy

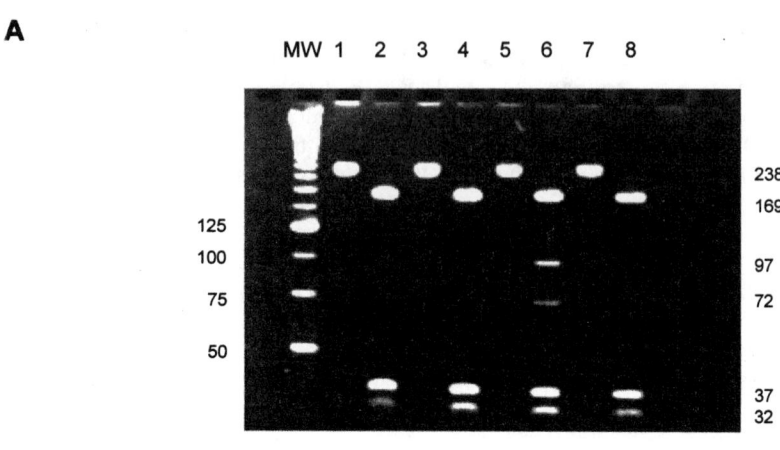

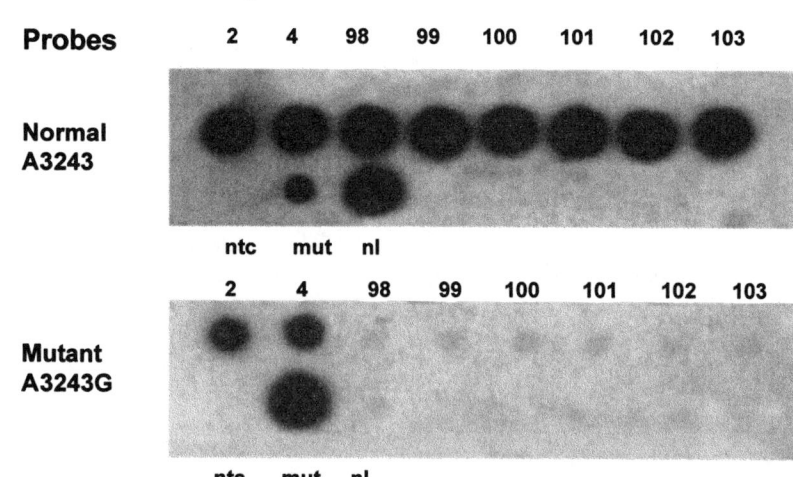

FIGURE 1. Mutational analysis of mtDNA point mutations. RFLP analysis of A3243G mutations. Odd number lanes: uncut. *Lanes 2* and *4*: 5% A3243G mutant DNA; *lane 6*: 17% A3243G mutant DNA; *lane 8*: normal control. The PCR product from mtF3116 and mtR3353 amplification was digested with *Hae*III followed by PAGE gel analysis. ASO dot blot detection of low heteroplasmy. PCR products of *lanes 1* and *3* from FIGURE 1A, and normal control specimens 98–103 were spotted on a membrane and hybridized with either the normal A3243 or the mutant A3243G probe. ntc, no template control; mut, mutant positive control; nl, normal control.

(LHON).[1,2,4,7] Kearns-Sayre syndrome (KSS) and Pearson syndrome are due to large mtDNA deletions,[8] and mitochondrial neurogastrointestinal encephalomyopathy (MNGIE) is due to mtDNA depletion and multiple deletions.[9]

DETECTION OF COMMON POINT MUTATIONS AND THE CHALLENGE OF A COMPREHENSIVE MUTATION PANEL

The complications of molecular diagnosis of mtDNA mutations arise from the genetic and clinical heterogeneity of the disease. For example, different mutations such as A3243G, T3271C, and A8344G may result in a similar clinical phenotype. Conversely, mitochondrial OXPHOS disease is characterized by phenotypic pleiotropy whereby the identical point mutation may lead to different disease phenotypes. For example, the A3243G MELAS mutation is also found in some patients with maternally inherited diabetes/deafness syndrome.[10] The mainstay of mtDNA point mutation, analysis has included PCR amplification of the mtDNA region containing the point mutation followed by restriction fragment length polymorphism (RFLP) analysis, as shown in FIGURE 1A. Ordinarily, this is performed step by step. First, we analyze the most common A3243G mutation followed by T3271C or A8344G mutations. If negative, the less frequent mutations are then analyzed. Because of the overlap in clinical manifestations, the NARP mutations may also be analyzed if the

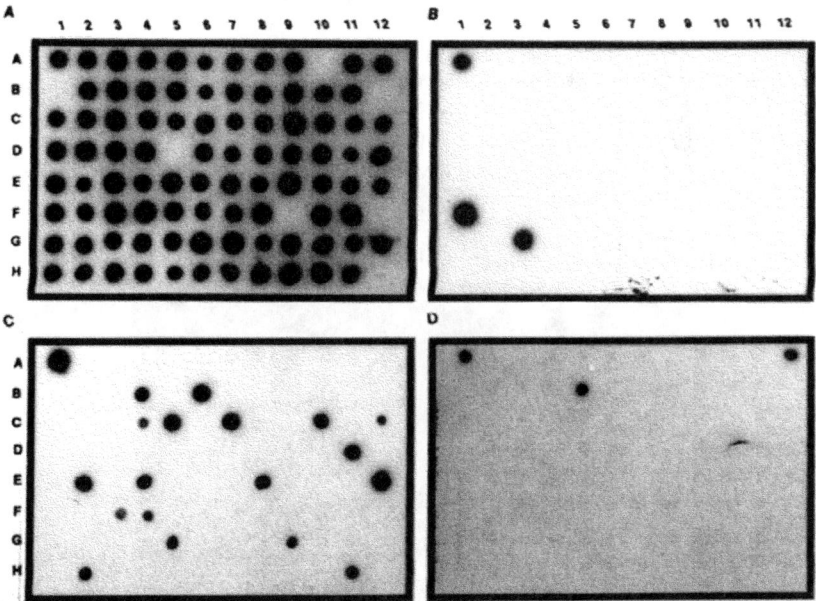

FIGURE 2. Multiplex PCR/ASO dot blot analysis. The PCR products from the amplification with mtF3130 and mtR3758 were spotted on a Biodyne B+ membrane. Blots A, B, C, and D were hybridized with an ASO probe of normal A3243, mutant T3271, mutant A3243G, and mutant T8356C, respectively. The A1 position is either the normal or the positive mutant control.

MELAS and MERRF mutations are found to be negative. There is a set of 4–5 primary mutations and several secondary mutations to be analyzed for patients with LHON. Given its complexity and heterogeneity, testing the mutations one by one may not be the most efficient approach. Furthermore, RFLP is not sensitive enough to detect mutations at a low percentage of heteroplasmy. For example, lanes 2 and 4 in FIGURE 1A had nondetectable levels of mutant load by RFLP. In fact, these two specimens had approximately 5% of A3243G mutant mtDNA, as shown in FIGURE 1B. Thus, an effective and high throughput multiplex PCR/ASO dot blot analysis was developed to simultaneously screen for multiple point mutations in mtDNA.[4,11] Our comprehensive mutation panel includes two MELAS mutations (A3243G and T3271C), two MERRF mutations (A8344G and T8356C), one cardiomyopathy mutation G8363A, two NARP mutations (T8993G and T8993C), and four LHON mutations (G3460A, G11778A, G14459A, and T14484C). This method uses three pairs of primers to amplify mutation hot-spot regions followed by detection of mutation with ASO probes (FIG. 2).[4,11] The multiplex PCR/ASO method is cost-effective and time-efficient because of multiplexing, and it is sensitive enough to detect a low percentage (2%) of mutant heteroplasmy. Currently, we are developing a single step, real-time PCR method using TaqMan probes. This method will allow the detection and quantification of a heteroplasmic mutation at the same time.

SCREENING THE ENTIRE MITOCHONDRIAL GENOME FOR UNKNOWN MUTATIONS BY TTGE

Southern analysis and ASO screening of the 12 common mtDNA point mutations known to cause mitochondrial OXPHOS deficiency, only detected mutations in 5.9% of the 2,000 patients suspected of having mitochondrial disorders. Evidently, other screening methods are necessary to detect unknown mutations in the entire mitochondrial genome. mtDNA is highly polymorphic. There are numerous benign

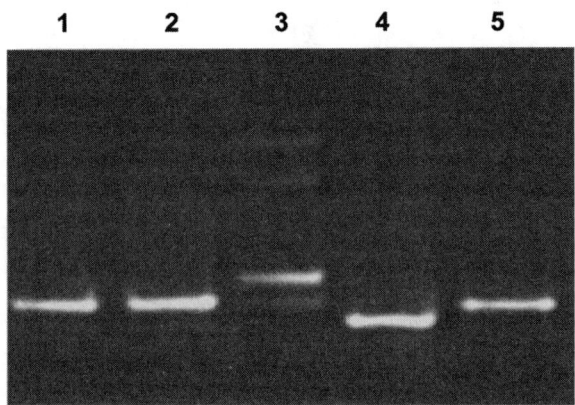

FIGURE 3. TTGE analysis of the tRMA^{val} region of mtDNA. *Lanes 1, 2,* and *5* are wild-type. *Lane 3*: heteroplasmic mutation shows multiple bands. *Lane 4*: homoplasmic mutation shows band-shift.

polymorphisms distinguishing two normal individuals. Thus, a unique requirement for an mtDNA mutation screening method is the ability to distinguish pathogenic heteroplasmic mutations from benign homoplasmic variations. The temporal temperature gradient gel electrophoresis (TTGE) method fits this purpose.[5,6] It is superior to the commonly used DGGE, SSCP, and heteroduplex analyses (HA).[12] TTGE detects homoplasmic mutations as band-shifts and heteroplasmic mutations as multiple bands, as shown in lanes 4 and 3, respectively, in FIGURE 3.[5,6] DNA fragments showing abnormal banding patterns are then sequenced for identification of the mutations. Thirty-two overlapping primer pairs are used to amplify the entire mitochondrial genome for TTGE analysis.[5] TABLE 1 lists some of the novel mutations found.[5]

SIGNIFICANCE OF NOVEL mtDNA MUTATIONS

One of the most difficult tasks in the molecular diagnosis of mtDNA disorders is the determination of the biochemical and clinical significance of sequence alterations. Not all homoplasmic mtDNA mutations are benign and not all heteroplasmic mutations are pathogenic. The mtDNA mutations must fit the following criteria in order to be classified as disease-causing deleterious mutations. First, the mutation should co-segregate with the disease in the affected family. The disease severity of the affected individual should roughly correlate with the mutant loads in the affected tissues, if the mutation is heteroplasmic. Second, the mutation occurs in a structurally/functionally important and evolutionarily conserved region. If a mutation in the mRNA results in a frame shift, nonsense mutation, or a nonconserved amino acid

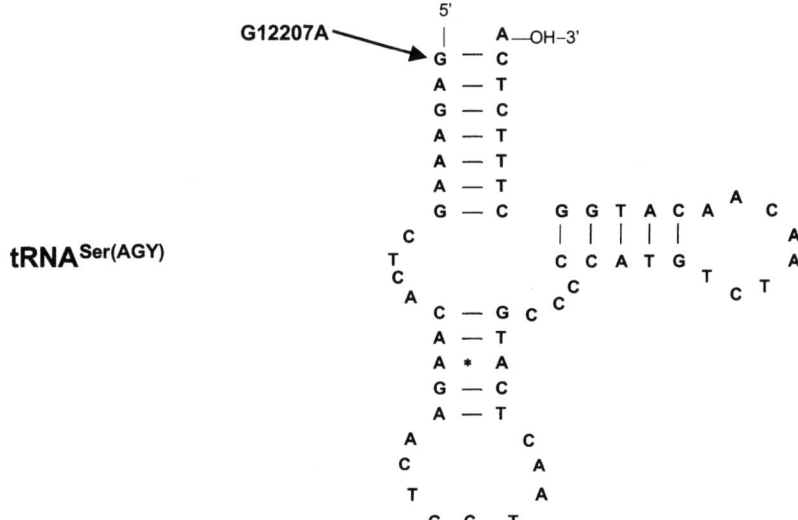

FIGURE 4. Secondary structure of tRNA$^{ser(AGY)}$. *Arrow* points to the mutation G12207A, which disrupts the first base pair of the stem region.

TABLE 1. Novel mtDNA mutations found by TTGE and sequencing

Mutation	Gene	Codon	Amino acid change	Homo-/hetero-plasmy	Significance
286-291 del AA	D loop			Homo	mtTF1 binding site
A723C	12S rRNA			Hetero	
A745G	12S rRNA			Hetero	
A1386T	12S rRNA			Hetero	
G1442A	12S rRNA			Homo	
C1721T	16S rRNA			Hetero	
G2098A	16S rRNA	CTC-CTT	L9L	Homo	
C3333T	ND1	ATC-ACC	I68T	Homo	
T3509C	ND1	ATC-ATT	I241I	Homo	
C4029T	ND1	GAC-AAC	D248N	Homo	
G4048A	ND1	GCA-GCG	A299A	Hetero	
A4203G	ND1			Hetero	
T4363C	Q			Homo	
T4454C	M	CAA-AAA	Q6K	Hetero	Conserved
C4485A	ND2	CCC-CCG	P7P	Hetero	
C4490G	ND2	GCT-GCC	A39A	Hetero	
T4586C	ND2	ACT-AAT	T89N	Homo	
C4735A	ND2	GCC-GCT	A331A	Homo	
C5462T	ND2	ATA-GTA	M343V	Homo	
A5496G	ND2	CTA-TTA	L344L	Hetero	
C5499T	ND2			Hetero	
A5539G	W			Homo	
T5561C	W			Homo	
T5567C	W			Homo	
T5580C	Non-coding			Hetero	W+4(A-7)
A5584G	Non-coding			Homo	W+8(A-3)
G5585A	Non-coding			Homo	W+9(A-2)
C5840T	Y	ACG-ACA	T181T	Hetero	
G6446A	COI	ATC-ATT	I190I	Homo	
C6473T	COI	TAC-TAT	Y270Y	Homo	
C6713T	COI	CTA-CTG	L283L	Homo	
A6752G	COI	TAT-CAT	Y496H	Homo	
T7389C	COI	ACA-ACG	T48T	Homo	
A7729G	COII	ATA-TTA	M152L	Homo	
A8039T	COII			Hetero	Highly conserved
A8326G	K	CAA-CAG	Q67Q	Homo	
A8566G	ATPase 8	ATC-ATT	I43I	Homo	
C8655T	ATPase 6	ATG-GTG	M104V	Homo	
A8836G	ATPase 6	CAC-CAT	H172H	Homo	
C9042T	ATPase 6	CTA-CTG	L47L	Homo	
A9347G	COIII	GAG-GAA	E183E	Homo	

TABLE 1. (*continued*) Novel mtDNA mutations found by TTGE and sequencing

Mutation	Gene	Codon	Amino acid change	Homo-/hetero-plasmy	Significance
G9755A	COIII	CAC-CAT	H204H	Homo	
C9818T	COIII			Homo	
T10007C	G			Homo	
T10031C	G	GTC-GTT	V65V	Homo	
C10664T	ND4L	GTG-TTG	V73L	Homo	
G10686T	ND4L	GTG-GTA	V73V	Hetero	
G10688A	ND4L	GGC-GTC	G74V	Homo	
G10690T	ND4L	CTT-CTC	L18L	Hetero	
T10810C	ND4	CTA-CTG	L39L	Homo	
A10876G	ND4	TTT-CTT	F50L	Homo	
T10907C	ND4	AAA-AAG	K93K	Homo	
A11038G	ND4	TTA-GTA	L325V	Homo	
T11732G	ND4	CTA-CTC	L375L	Hetero	
A11884C	ND4			Hetero	
G12207A	S			Hetero	First base of tRNA absent in mother
C12239T	S			Homo/hetero	
A12273C	L(CUN)			Homo	
A12280G	L(CUN)			Homo	
T12338C	ND5	ATA-ACA	M1T	Hetero	Initiation codon of ND5
A13276G	ND5	ATG-GTA	M314V	Homo	
C13506T	ND5	TAC-TAT	Y390Y	Homo	
C14284T	ND6	GAG-GAT	E111E	Homo	
G15596A	CytB	GTC-ATC	V284I	Homo	
A15848G	CytB	ACT-GCT	T368A	Hetero	Highly conserved
C15849T	CytB	ACT-ATT	T368I	Hetero	Highly conserved
T15852C	CytB	ATC-ACC	I39T	Hetero	Mildly conserved
G15884C	CytB	GCC-CCC	A380P	Hetero	
G15995A	P			Hetero	Absent in mother

change in a conserved region, the mutation is likely to be deleterious. FIGURE 4 shows a nucleotide substitution, G12207A, located at the first base pair of the tRNA$^{ser(AGY)}$ sequence, which disrupts the stem region and may affect the charging of the amino acid to the tRNA molecule.

MITOCHONDRIAL DNA DELETION SYNDROME

Large mtDNA deletions can be detected by Southern RFLP analysis.[8] Usually, two restriction enzymes are used. *Eag*I has one cut, which linearizes the circular

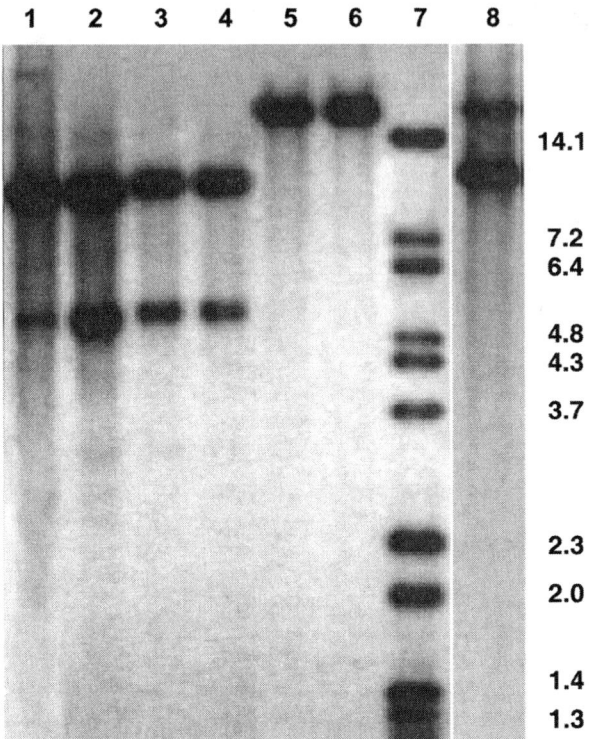

FIGURE 5. Southern analysis of mtDNA deletion. *Lanes 1–4*: *Hind*III digest; *lanes 5, 6,* and *8*: *Eag*I digest; *lane 7*: molecular weight markers. *Lanes 2–4* and *5* and *6* are normal, *lanes 1* and *8* have mtDNA deletion of about 5.5 kb. In *lane 1* the deleted mtDNA overlaps with the 10.5 kb *Hind*III fragment, which shows abnormal ratio of band intensity.

mtDNA (lanes 5 and 6 in FIG. 5). *Hind*III has three cuts (lanes 2–4 in FIG. 5, the smallest band not shown). Lane 8 shows a heteroplasmic 5.5-kb deletion that on *Hind*III digestion coincides with the 10.5-kb band (lane 1), but the signal intensity ratio of the two bands differs from that of the normal band (lanes 2–4). Thus, it is necessary to always use at least two different restriction enzymes. Multiple deletions should show multiple bands or smear on Southern analysis, if each of the deleted mtDNA species is present at undetectable levels. To evaluate this, three primer sets, mtF7234/mtR16133, mtF8295/mtR14499, and mtF5681/mtR14686, covering the commonly deleted regions are used to detect any deletions undetectable by Southern blot analysis in several muscle specimens. As shown in FIGURE 6, multiple bands are revealed in some samples, indicating multiple deletions in these regions. Although mtDNA deletions present at levels undetectable by Southern blot may often be considered clinically insignificant, the percentage and the total amount of wild-type mtDNA determines the actual levels of OXPHOS activity. Multiple deletions may also reflect the consequence of nuclear gene defects that affect the deoxynucleotide pools used for mtDNA synthesis or mtDNA replication; for example, thymidine

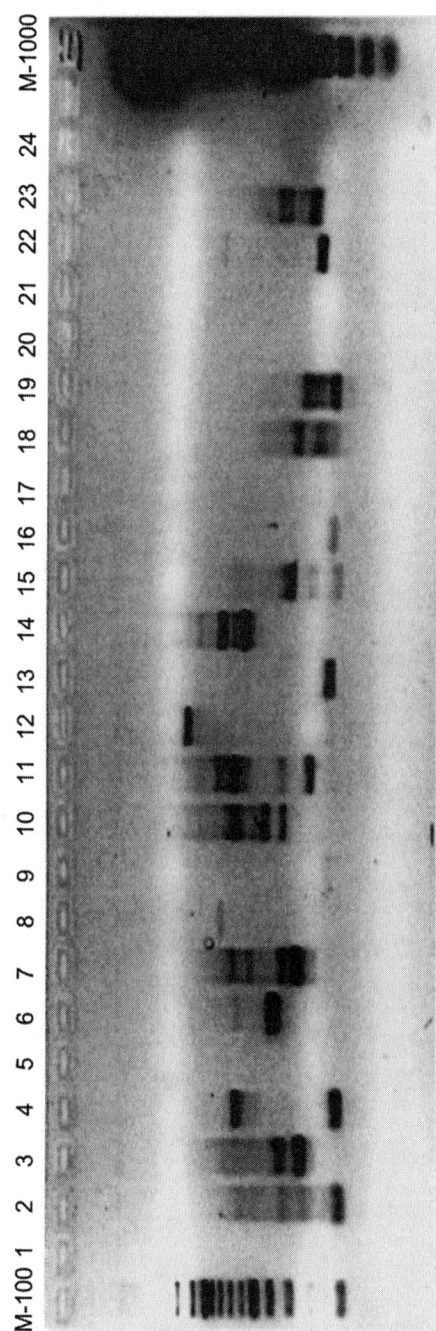

FIGURE 6. Detection of multiple deletions by PCR. Muscle samples were PCR amplified by either mtF7234/mtR16133, mtF8295/mtR14499, or mt5681/mt14686 primer pairs (*lanes 1–8, 9–16,* and *17–24,* respectively) and analyzed on agarose gel. Multiple bands indicate the presence of multiple deletions in these regions.

phosphorylase (TP), DNA polymerase gamma, or DNA helicase, each plays a role in maintaining these processes.[9,13–19]

mtDNA CONTENT

Multiple deletions may be associated with mtDNA depletion. On the other hand, mitochondrial proliferation is a means to compensate for respiratory chain deficiency. Therefore, an abnormal level of mtDNA content may be an indication of respiratory disease. Real-time quantitative PCR (RT Q-PCR) was used to evaluate the mtDNA content in 300 muscle specimens.[20] The primers were mtF3212/mtR3319 for mtDNA and 18S1546F/18S1650R for nuclear DNA (nDNA), 18S rRNA gene.[20] The TaqMan probes, 6FAM-5′TTACCGGGCTCTGCCATCT3′-TAMRA and VIC-5′AGCAATAACAGGTCTGTGATG3′-TAMRA, were labeled at the 5′ end with the fluorescent reporters, 6FAM or VIC, for the mtDNA and the nDNA 18S rRNA gene, respectively, whereas the 3′ ends were labeled with a quencher TAMRA. Real-time quantitative PCR analysis was performed using the Sequence Detector System ABI-Prism 7700.[20] The copy number of the mtDNA and the nDNA 18S rRNA gene was calculated from the threshold cycle number, C_T, and the standard curve.[20] The ratio of copy number between mtDNA and nDNA is a measurement of mtDNA content. It was found that the mtDNA content in muscle increases from birth to about 5 years of age, and remains pretty constant after that. Twenty samples with mtDNA levels below 20% of age-matched mean were analyzed for mutations in the TP gene by sequencing. Only one 20-year-old patient who had clinical manifestations of MNGIE had mutations in the TP gene.[20] Mutations in the TP gene were not detected in the remaining patients who had markedly reduced mtDNA without MNGIE. These patients were very young (<10 years old). Mutations in other nuclear genes, such as DNA polymerase gamma, DNA helicase, thymidine kinase, and adenine nucleotide translocase, are currently under investigation.

CORRELATION BETWEEN mtDNA LEVELS AND STRUCTURE/FUNCTION OF MITOCHONDRIA

Muscle biopsy from a 4-month-old male infant with only 2.6% mtDNA compared to the age-matched mean had increased size and number of mitochondria. A 2-year-old girl showed an increased number of mitochondria on electron microscopic examination, but the mtDNA level was only 14% of the mean. A 4-year-old female child had limb-girdle muscular dystrophy and severe mtDNA depletion (1.9% of age-matched mean). The muscle biopsy of a 4-year old male child revealed a drastic respiratory enzyme complex deficiency, correlated with an abnormally low level of mtDNA content (7.3% of mean). These observations indicate that neuromuscular disorders and low respiratory chain activity may be associated with mtDNA depletion, but morphologic changes in size and number of mitochondria are not necessarily correlated with the change in mtDNA content. This suggests that disproportional expression of mitochondrial proteins, either mitochondrial or nuclear encoded or both, may cause mitochondrial proliferation, but these mitochondria may be devoid of a mitochondrial genome if the mechanism for mtDNA synthesis is defective. Two adult patients had mtDNA depletion. One had mutations in the TP gene, as just mentioned. The other one had acquired mitochondrial myopathy due to prolonged antiretroviral treatment for HIV.

About 20 patients had elevated mtDNA content above twofold of the age-matched mean. The mtDNA amplification in these patients probably reflects a mechanism to compensate for deficient mitochondrial function. This is supported by the finding of two novel mtDNA mutations, G13928C (S531T) in ND5 and C4312T in tRNA^{Ile}, and numerous mtDNA variations in a patient with a sevenfold increase in mtDNA. A woman with 92% heteroplasmy for mtDNA deletion also had a 10-fold increase in total mtDNA content. These findings suggest that mutations may cause mtDNA amplification.

CONCLUSION

A comprehensive molecular diagnosis of mitochondrial respiratory chain disorders should consist of both qualitative mutational analysis and quantitative measurements of both mutant and total mtDNA content. Diagnosis can be performed stepwise. First, the routine mutation panel should be analyzed to rule out common mutations for MELAS, MERRF, NARP, LHON, and mtDNA deletion syndrome. If these are negative, more extensive studies including mutation screening of the entire mitochondrial genome and quantitative evaluation of the mtDNA content can follow. Based on biochemical, histochemical, and ultrastructural results, mutational analysis of specific genes encoding the OXPHOS enzyme complex subunits, complex assembly proteins, and enzymes involved in mtDNA biogenesis and gene expression should all be considered.

REFERENCES

1. SMEITINK, J., L. VAN DEN HEUVEL & S. DIMAURO. 2001. The genetics and pathology of oxidative phosphorylation. Nature Rev. Genet. **2**: 342–352.
2. SHOFFNER, J.M. & D.C. WALLACE. 1995. Oxidative phosphorylation diseases. *In* The Metabolic and Molecular Bases of Inherited Disease, 7th edit. C.R. Scriver, W.S. Sly & D. Valle, Eds.: 1535–1629. McGraw-Hill. New York.
3. WALLACE, D.C. 1999. Mitochondrial disease in man and mouse. Science **283**: 1482–1488.
4. LIANG, M.H. & L.-J.C. WONG. 1998. Yield of mtDNA mutation analysis in 2000 patients. Am. J. Med. Genet. **77**: 385–400.
5. WONG, L.-J.C., M.-H. LIANG, H. KWON, *et al.* 2002. Comprehensive scanning of the whole mitochondrial genome for mutations. Clin. Chem. **48**: 1901–1912.
6. CHEN, T.J., R. BOLES & L.-J.C. WONG. 1998. Detection of mitochondrial DNA mutations by temporal temperature gradient gel electrophoresis. Clin. Chem. **45**: 1162–1167.
7. ZEVIANI, M., C.T. MORAES, S. DIMAURO, *et al.* 1988. Deletions of mitochondrial DNA in Kearns-Sayre syndrome. Neurology **38**: 1339–1346.
8. WONG, L.-J.C. 2001. Recognition of mitochondrial DNA deletion syndrome with non-neuromuscular multisystemic manifestation. Genet. Med. **3**: 399–404.
9. NISHINO, I., A. SPINAZZOLA, A. PAPADIMITRIOU, *et al.* 2000. Mitochondrial neurogastrointestinal encephalomyopathy: an autosomal recessive disorder due to thymidine phosphorylase mutations. Ann. Neurol. **47**: 792–800.

10. REARDON, W., R.J. ROSS, M.G. SWEENEY, et al. 1992. Diabetes mellitus associated with a pathogenic point mutation in mitochondrial DNA. Lancet **340:** 1376–1379.
11. WONG, L.-J.C. & D. SENADHEERA. 1997 Direct detection of multiple point mutations in mitochondrial DNA. Clin. Chem. **43:** 1857–1861.
12. NOLLAU, P. & C. WAGENER. 1997. Methods for detection of point mutations: performance and quality assessment. The IFCC Scientific Division, Committee on Molecular Biology Techniques. J. Int. Fed. Clin. Chem. **9:** 162–170.
13. VAN GOETHEM, G., B. DERMAUT, A. LOFGREN, et al. 2001. Mutation of *POLG* is associated with progressive external ophthalmoplegia characterized by mtDNA deletions. Nat. Genet. **28:** 211–212.
14. VAN GOETHEM, G., J.J. MARTIN & C. VAN BROECKHOVEN. 2002. Progressive external ophthalmoplegia and multiple mitochondrial DNA deletions. Acta Neurol. Belg. **102:** 39–42.
15. SPELBRINK, J.N., F.Y. LI, V. TIRANTI, et al. 2001. Human mitochondrial DNA deletions associated with mutations in the gene encoding Twinkle, a phage T7 gene 4-like protein localized in mitochondria. Nat. Genet. **28:** 223–231.
16. NISHINO, I., A. SPINAZZOLA & M. HIRANO. 1999. Thymidine phosphorylase gene mutations in MNGIE, a human mitochondrial disorder. Science **283:** 689–692.
17. MORAES, C.T. 2001 A helicase is born. Nat. Genet. **28:** 200–201.
18. HIRANO, M., R. MARTI, C. FERREIRO-BARROS, et al. 2001. Defects of intergenomic communication: autosomal disorders that cause multiple deletions and depletion of mitochondrial DNA. Semin. Cell. Dev. Biol. **12:** 417–427.
19. KAUKONEN, J., J.K. JUSELIUS, V. TIRANTI, et al. 2002. Role of adenine nucleotide translocator 1 in mtDNA maintenance. Science **289:** 782–785.
20. WONG, L.-J.C. & R. BAI. 2002. Real time quantitative PCR analysis of mitochodnrial DNA in patients with mitochondrial disease. Am. J. Hum. Genet. **71:** 501.

Molecular Pathogenetic Mechanism of Maternally Inherited Deafness

MIN-XIN GUAN

Division and Program in Human Genetics and Center for Hearing and Deafness Research, Cincinnati Children's Hospital Medical Center, and Department of Pediatrics, University of Cincinnati College of Medicine, Cincinnati, Ohio 45229, USA

ABSTRACT: Mutations in the mitochondrial DNA (mtDNA) have been shown to be one important cause of deafness. In particular, mutations in the mtDNA have been associated with both syndromic and nonsyndromic forms of sensorineural hearing loss. The deafness-linked mutations often occur in the mitochondrial 12S rRNA gene and in tRNA genes. Mutations in the 12S rRNA gene account for most of the cases of aminoglycoside ototoxicity. The other hot spot for mutations associated with hearing impairment is the tRNA$^{Ser(UCN)}$ gene, as five deafness-linked mutations have been identified in this gene. Nonsyndromic deafness-linked mtDNA mutations are often homoplasmic or at high levels of heteroplasmy, indicating a high threshold for pathogenicity. Phenotypic expression of these mtDNA mutations requires the contribution of other factors such as nuclear modifier gene(s), environmental factor(s), or mitochondrial haplotype(s).

KEYWORDS: deafness; mitochondrial DNA; mutations; maternally inherited deafness

INTRODUCTION

Hearing loss is a common congenital disorder, affecting 1 in 1,000 newborns.[1,2] More than 50% of cases of deafness in the pediatric population have a genetic etiology or predisposition with autosomal dominant, autosomal recessive, X-linked, or mitochondrial patterns of inheritance.[3,4] Deafness can result from a mutation in a single gene or from a combination of mutations in different genes. Hearing loss can also be caused by environmental factors, including perinatal infection, acoustic or cerebral trauma affecting the cochlea, or ototoxic drugs such as aminoglycoside antibiotics, or by interactions between genetic and environmental factors.[3,4] Mutations in mitochondrial DNA (mtDNA) have been associated with both syndromic and nonsyndromic deafness in various families of different ethnic backgrounds.[5,6] In this review, the main features of the mitochondrial genetic system are summarized

Address for correspondence: Min-Xin Guan, Ph.D., Division and Program in Human Genetics, Cincinnati Children's Hospital Medical Center, 3333 Burnet Avenue, Cincinnati, Ohio 45229-3039. Voice: 513-636-3337; fax: 513-636-3486.
min-xin.guan@chmcc.org

and the identification and characterization of deafness-linked mtDNA mutations are reviewed.

MITOCHONDRIAL GENOME AND HUMAN DISEASES

A variety of mtDNA mutations have been associated with many clinical abnormalities, including various forms of hearing loss, neuropathy, myopathy, cardiomyopathies, diabetes, as well as Alzheimer's and Parkinson's disease.[7,8] These mutations include deletions or insertions, point mutations, and nuclear driven mtDNA mutations. Point mutations occur in the genes encoding proteins or genes for components of the protein-synthesizing apparatus (tRNAs and rRNAs). The mitochondrial genome is just 16,569 bp in length and comprises only about 1–2% of the total DNA in mammalian cells. The mitochondrial genome encodes 13 essential polypeptides of oxidative phosphorylation as well as the two rRNAs and 22 tRNAs required for mitochondrial protein synthesis.[9] Mitochondria are essential not only for the generation of cellular energy in the form of ATP by the process of oxidative phosphorylation (OXPHO),[7,8] but also for the control of apoptosis, and they are major producers of reactive oxygen species.[7] They possess their own organelle-specific DNA replication, transcription, and translation systems.[10] However, most mitochondrial proteins, encoded by the nuclear genome, are synthesized in the cytosol and then imported into the organelles.[10] Thus, deleterious mutations in mtDNA adversely affect cellular bioenergetics through the reduction of ATP and consequently cause a wide spectrum of diseases. Mutations in nuclear genes can also affect OXPHOS, often resulting in Mendelian diseases with phenotypes similar to those caused by mtDNA mutations.

The unique features of mitochondrial genetics result from cytoplasmic location, high copy number, high mutation rate, random segregation in postmitotic tissues, and maternal inheritance.[11] Due to maternal inheritance, disease manifestations are concentrated along the maternal lineage.[12] Because mitochondria have a low activity DNA repair system and no histones, and are continuously exposed to oxygen radicals that are leaked from the mitochondrial electron-transfer chain, somatic mutations in mtDNA are common. Because each cell contains multiple mitochondria, and each mitochondrion contains multiple (10–100) DNA molecules, any mutation that arises produces heteroplasmy, a mixture of normal and mutant mtDNA in a cell. New mutations, however, segregate rapidly in the female germline early in oogenesis,[13] and as a result, most individuals are homoplasmic for a single species of mtDNA.[14]

Phenotypic heterogeneity is a hallmark of mitochondrial disorders. Families harboring deleterious mtDNA mutations (either heteroplasmic or homoplasmic) typically present a wide range of clinical symptoms among family members, from asymptomatic to death in childhood, although all matrilineal relatives inherited the same mtDNA.[7,11] In the case of heteroplasmy, different proportions of wild-type and mutant mtDNA accumulate in different tissues, and because different organs rely on mitochondrial OXPHOS to a different extent, the variable clinical phenotypes may be due to organ-specific energetic vulnerability.[15] In the case of homoplasmy, clinical heterogeneity presumably results from different nuclear backgrounds or mitochondrial haplotypes.

The correlation of genotype and phenotype in mitochondrial disorders is complex and often poorly understood. Some mtDNA mutations can produce clinical manifestations that are systemic, as in Kearns-Sayre syndrome (KSS),[16] myoclonus epilepsy and ragged red fibers (MERRF),[17] mitochondrial encephalomyelopathy, lactic acidosis, stroke-like symptoms (MELAS),[18] and Leigh disease.[19] Other mtDNA mutations, however, produce clinical manifestations that are confined to specific tissues, such as the optic nerve in Leber's hereditary optic neuropathy (LHON), the β cells of the pancreas in diabetes mellitus, and the cardiac muscle in hypertrophic cardiomyopathy.[20] In some cases, different mtDNA mutations cause the same disorder, as seen in LHON, which is associated with eighteen mtDNA mutations, four of which are considered to be causative in and by themselves.[21] Conversely, in some cases, the same mutation can cause different disorders in different individuals. The most striking example of this phenotypic heterogeneity is the A3243G mutation in the tRNA$^{Leu(UUR)}$ gene. This mutation can give rise to different conditions including MELAS,[18] diabetes mellitus and neurosensory deafness (MIDD),[22,23] pregnancy-induced hypertension,[24] or progressive external ophthalmoplegia (PEO).[7,8]

mtDNA MUTATIONS ASSOCIATED WITH SYNDROMIC AND NONSYNDROMIC DEAFNESS

Hearing loss often occurs as one of several symptoms in syndromic diseases caused by mitochondrial defects. Nonsyndromic hearing loss can also be caused by mtDNA mutations, as demonstrated by mtDNA mutations in families with maternally inherited hearing loss that is nonsyndromic in many patients. Recently, a number of mtDNA mutations have been associated with both syndromic and nonsyndromic forms of sensorineural deafness, as summarized in TABLE 1.

TABLE 1. mtDNA mutations associated with deafness

Hearing Impairment	Mutations identified	Inherited	Acquired	Homoplasmy	Heteroplasmy
Syndromic					
Syst. Neuromuscular	Del. A3243G-tRNA$^{Leu(UUR)}$	Rare	Usually	-	+
Diabete+Deafness	A3243G-tRNA$^{Leu(UUR)}$	+	Possible	-	+
	A8296G-tRNA^{Lys}	+	N/A	-	+
	T14709C-tRNA^{Glu}	+	not observed	-	+
PPK+Deafness	A7445G- tRNA$^{Ser(UCN)}$	+	not observed	+	-
KSS	Large deletions	+	not observed	-	+
Deafness+other	Large Deletions/Duplication	+	not observed	-	+
Non-Syndromic					
	A1555G-12S rRNA	+	not observed	+	-
	C1494T-12S rRNA	+	not observed	+	-
	A7445G-COI/- tRNA$^{Ser(UCN)}$	+	not observed	+	-
	7442 insG- - tRNA$^{Ser(UCN)}$	+	not observed	Nearly	+
	T7510C- - tRNA$^{Ser(UCN)}$	+	not observed	+	-
	T7511C- - tRNA$^{Ser(UCN)}$	+	not observed	Nearly	+
Ototoxic	A1555G-12S rRNA	+	not observed	+	-
	961 ins/DelC-12S rRNA	+	Possible	+	-
	C1494T-12S rRNA	+	not observed	+	-
Presbyacusis	Random	N/A	Yes	-	+

The most frequent forms of mitochondrial syndromic hearing impairments are part of classic mitochondrial disorders, such as Kearns-Sayre syndrome,[25] MELAS,[18,23] and MERRF.[20] Excluding large rearrangements involving several genes, all mitochondrial mutations leading to syndromic hearing impairment are point mutations in tRNA genes. In these cases, the heteroplasmic mtDNA mutation(s), either arrangements or point mutations, can generally be found with the highest abundance in muscle. Maternally inherited diabetes mellitus and sensorineural hearing loss in several families were found to harbor the heteroplasmic A3243G mutation in the tRNA$^{Leu(UUR)}$ gene[22,23] or a large deletion.[26] The hearing loss is sensorineural, and it usually develops only after the onset of diabetes. Other heteroplasmic mtDNA mutations associated with maternally transmitted syndromic deafness include: T14709C in the tRNA^{Glu} gene,[27] A8296G in the tRNA^{Lys} gene,[28] C12258A in the tRNA$^{Ser(AGY)}$ gene,[29] G5540A in the tRNA^{Trp} gene,[30] a heteroplasmic C-nucleotide insertion at position 7472,[31] and the T7512C[32] in the tRNA$^{Ser(UCN)}$ gene.

In some families, hearing loss is the only symptom of mitochondrial disease, suggesting that hearing is strongly dependent on mitochondrial function. To date, five mtDNA mutations have been shown to give rise to nonsyndromic hearing impairment in some patients. These mutations are the A7445G in the precursor of tRNA$^{Ser(UCN)}$ gene[33–35] and 7472insG,[32,36] the T7510C,[37] and the T7511C mutations[38–40] in the same gene, and the A1555G mutation in the 12S rRNA gene.[41–50] Unlike the other pathogenic mtDNA mutations causing syndromic deafness, such as the A3243G mutation in the tRNA$^{Leu(UUR)}$ gene[18] and the A8344G mutation in the tRNA^{Lys} gene,[20] these nonsyndromic deafness-linked mtDNA mutations are often homoplasmic, or nearly so, indicating a high threshold for pathogenicity. These mtDNA mutations produce variable clinical severity, with age of onset ranging from birth to adulthood. However, for three of these five mutations, patients have additional symptoms besides hearing impairment.

The A1555G mutation in the 12S rRNA gene was first discovered in a large Arab-Israeli family[41] and was subsequently found in many families of different ethnic backgrounds.[41–50] In the absence of exposure to aminoglycosides, the A1555G mutation produces a clinical phenotype that varies considerably among family members and ranges from severe congenital deafness, to moderate progressive hearing loss of later onset,[41,44] to completely normal hearing.[41,44] Functional characterization in an Arab-Israeli family demonstrated that more severe biochemical defects were observed in mutant lymphoblastoid cell lines derived from symptomatic individuals than in cell lines derived from asymptomatic individuals.[51] However, under a constant nuclear background, a nearly identical degree of mitochondrial dysfunction was observed in cybrid cell lines derived from symptomatic or asymptomatic individuals from this family.[52] These genetic and biochemical data strongly point out that the A1555G mutation is a primary factor underlying the development of deafness and that a nuclear modifier gene(s) plays a role in modulating the phenotypic expression of the hearing loss associated with the A1555G mutation.[41,51–54]

The A7445G mutation in the tRNA$^{Ser(UCN)}$ gene (FIG. 1A) and COI gene was found in genetically unrelated pedigrees from Scotland,[34] New Zealand, and Japan.[33,35] In the latter two pedigrees, palmoplantar keratoderma also segregated in the maternal line in some individuals.[35] Interestingly, in the Scottish pedigree the penetrance of this mutation for hearing loss is low, whereas in the New Zealand and

Japanese pedigrees the penetrance is very high. The A7445G mutation is a silent change of both the last nucleotide of the COI gene on the heavy strand and of the nucleotide immediately adjacent to the 3' end of the tRNA$^{Ser(UCN)}$ gene on the light strand. Mechanistically, the mutation affected normal processing of the light-strand polycistronic RNA and led to significant decreases in the amount of both tRNA$^{Ser(UCN)}$[55,56] and cotranscripted ND6 mRNA.[56] As a result, it impaired the rate of both mitochondrial protein synthesis and respiration in the mutant cells.[56] In this situation, the difference in penetrance appears to be due to a difference in mitochondrial haplotype. In the New Zealand pedigree, complete sequencing of the mtDNA revealed three additional sequence changes in complex I protein genes, two of which have also been labeled as secondary LHON mutations,[56–58] but these are not present in the Scottish pedigree.[59]

The mitochondrial 7472insG mutation in the tRNA$^{Ser(UCN)}$ gene was first reported in a Sicilian family[31]: the insertion of an extra G into a run of six G residues located within the T-arm of the tRNA$^{Ser(UCN)}$ (FIG. 1B) is associated with a syndromic disorder that includes hearing loss, ataxia, and myoclonus. Most family members carrying the mutation had progressive hearing loss, either in isolation or in combination with ataxia, dysarthria, and, more rarely, focal myoclonus. This mutation was also found in a large Dutch family with maternally inherited progressive hearing loss.[36] The hearing loss was nonsyndromic in all but one family member, a single person, who also suffered from ataxia and myoclonus. The homoplasmic 7472insG mutation has also been reported in some maternal members of several small families.[60,61] In addition to hearing loss, they had myoclonic epilepsy, ataxia, and cognitive impairment. However, most individuals in nt 7472 families, even with very high levels of mutant mtDNA, suffer only from deafness, with other neurologic features appearing only late in life, if at all.[36] Biochemical characterization revealed that cell lines carrying the 7472inG mutation exhibited a reduction in the level of this tRNA comparable to the effect of the A7445G mutation. Protein synthesis was not affected or only mildly so.[62]

A novel homoplasmic T-to-C mutation in the tRNA$^{Ser(UCN)}$ gene at position 7510 (T7510C) was identified in a family with maternally inherited progressive non-syndromic deafness. Furthermore, Friedman *et al.*[63] studied a large African American family with maternally inherited hearing impairment. The family history revealed no muscle disease, visual problems, or neurologic disorders. The hearing loss was progressive with variable age at onset. Mutation analysis of mtDNA revealed the point mutation T7511C in the tRNA$^{Ser(UCN)}$ gene.[38] The T7511C mutation exists in almost homoplasmic levels in most members of this pedigree, in heteroplasmic form in some members, but not in controls.[38] As shown in FIGURE 2B, the 7511C mutation disrupts a highly conserved base-pair in the acceptor stem of tRNA$^{Ser(UCN)}$, which is important for tRNA identity and for interaction with the mitochondrial RNA processing enzyme, RNase P.[64] This mutation has also been found in several genetically unrelated pedigrees with nonsyndromic deafness, including French and Japanese families.[39,40] In addition, the homoplasmic mutations T3308C in the ND1 gene and T5655C in the tRNA^{Ala} gene were found in all maternal members of this African American pedigree and also in some controls.[38] A functional study in cybrid cell lines derived from affected individuals in this family[64] has shown that the T7511C mutation leads to a decrease (~75%) in the amount of tRNA$^{Ser(UCN)}$ but not of the cotranscripted ND6 mRNA. The T5655C mutation also reduces (~50%) the tRNA^{Ala}

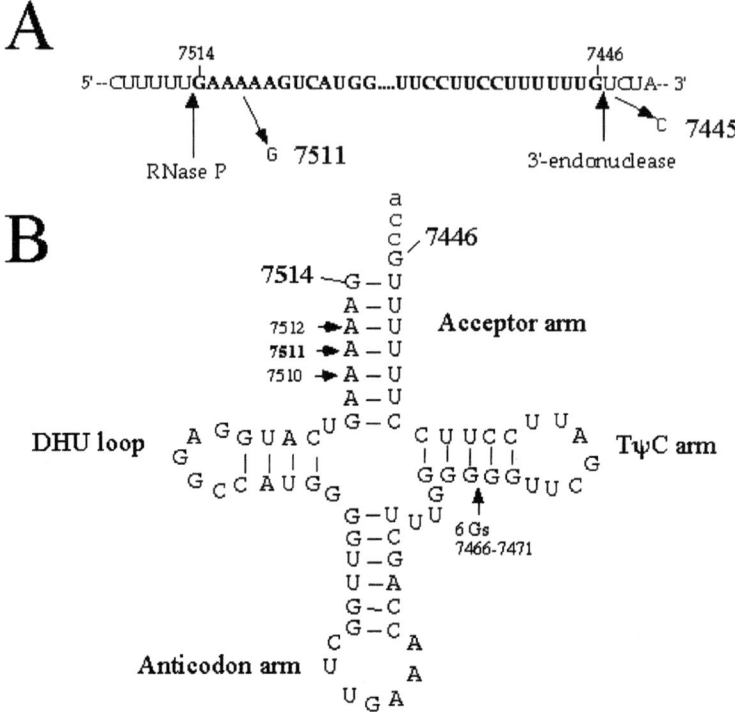

FIGURE 1. Secondary structure and deafness-linked mutations in the mitochondrial tRNA$^{Ser(UCN)}$. (A) Processing site in the mitochondrial tRNA$^{Ser(UCN)}$ precursor, experimentally determined for RNase P and predicted from the sequencing data for the 3' endonuclease.[56] (B) Location of the deafness-linked mutations in the secondary structure of human mitochondrial tRNA$^{Ser(UCN)}$.

level in mutant cells. Strikingly, the T3308C mutation causes a significant decrease in the amount of both the ND1 mRNA and the adjacent tRNA$^{Leu(UUR)}$ in mutant cell lines. A combination of these alterations leads to significant biochemical defects in the rate of mitochondrial protein synthesis, substrate-dependent respiration, and the rate of growth in medium containing galactose in mutant cell lines. These observations provide the first direct biochemical evidence that two mtDNA mutations in different genes contribute to the high penetrance of nonsyndromic deafness in the African American family carrying the T7511C mutation.[64]

mtDNA MUTATIONS ASSOCIATED WITH AMINOGLYCOSIDE OTOTOXICITY

Aminoglycoside antibiotics, such as gentamicin, streptomycin, and tobramycin, are clinically important drugs. They are mainly used in the treatment of hospitalized patients with aerobic gram-negative bacterial infections, particularly chronic infections such as tuberculosis or infections associated with cystic fibrosis.[65,66] These drugs are known to exert their antibacterial effects by binding to 16S ribosomal RNA (rRNA) in the 30S subunit of the bacterial ribosome, causing mistranslation or premature termination of protein synthesis.[67–69] Use of these drugs can frequently lead to toxicity, which involves the renal, auditory, and vestibular systems,[65,66] because the drugs are concentrated in renal tubular cells and in the perilymph and endolymph of the inner ear.[70,71] The renal damage is usually reversible, but the auditory and vestibular ototoxicity frequently is not. Although all aminoglycosides can affect cochlear and vestibular functions, some (streptomycin and gentamicin) produce predominately vestibular damage, whereas others (neomycin and kanamycin) cause mainly cochlear damage. Tobramycin affects both equally.[66]

Mitochondrial ribosomes share more similarities with bacterial ribosomes than do cytosolic ribosomes.[43] Therefore, it is thought that the ototoxic site of action of aminoglycoside antibiotics is the mitochondrial ribosome. The fact that aminoglycoside hypersensitivity is often maternally transmitted[5] suggests that mtDNA mutation(s) are involved in aminoglycoside ototoxicity. In particular, Hu et al.[72] described 36 Chinese families with maternally transmitted predisposition to aminoglycoside ototoxicity, whereas Higashi[73] reported that 26 of 28 families with streptomycin-induced deafness had maternally inherited transmission. Sequence analyses of the mitochondrial genome in patients with aminoglycoside ototoxicity have led to the identification of several ototoxic mtDNA mutations in the 12S rRNA gene: insertion or deletion at position 961, A1555G, and C1494T mutations.

The mutations at position 961 occur in genetically unrelated families, including Chinese,[50,74] Japanese,[75] and Italian[76] pedigrees, affected by aminoglycoside-induced deafness, clearly indicating that this mutation is involved in the pathogenesis of the disorder. The 961 mutation localizes at the C-cluster of a region between loop 21 and 22 of 12S rRNA.[77] This region is not very evolutionarily conserved and its function is not well defined, specifically for its possible interaction with aminoglycosides in bacterial homologues. It is possible that alteration of the tertiary or quaternary structure of this rRNA by the 961 mutation may indirectly affect the binding of aminoglycosides. Alternatively, this alternation may result in a mitochondrial translational defect.[50]

The A1555G mutation is one of the most common causes of aminoglycoside-induced deafness.[41–44,78] This mutation was first discovered in a large Arab-Israeli family[41] and subsequently found in various ethnic groups from Europe,[44,45] Asia,[43,46,47,50] and Africa.[48,49] In certain areas, such as Spain, the A1555G mutation appears to be more frequent.[44] This mutation is homoplasmic in the affected or at-risk individual. The A1555G mutation is located at a highly conserved region of the 12S rRNA that is an essential part of the decoding site of the small ribosomal subunit[79–81] and important for the action of aminoglycosides.[82–84] In particular, the nucleotide at position 1555 in the human 12S rRNA (equivalent to position 1491 in the *Escherichia coli* 16S rRNA) in wild-type cells is A, which, when mutated to a G, as in the Arab-Israeli family, would pair with the C at position 1494 (FIG. 2). This transition makes the secondary structure of the RNA resemble more closely the corresponding region of *E. coli* 16S rRNA (FIG. 2) and consequently leads to defects in mitochondrial translation.[51,85] This new G-C pair in 12S rRNA is also expected to create a binding site for aminoglycosides, which facilitates interaction with these drugs.[86] Thus, exposure to aminoglycosides causes hearing loss in individuals carrying the A1555G mutation.

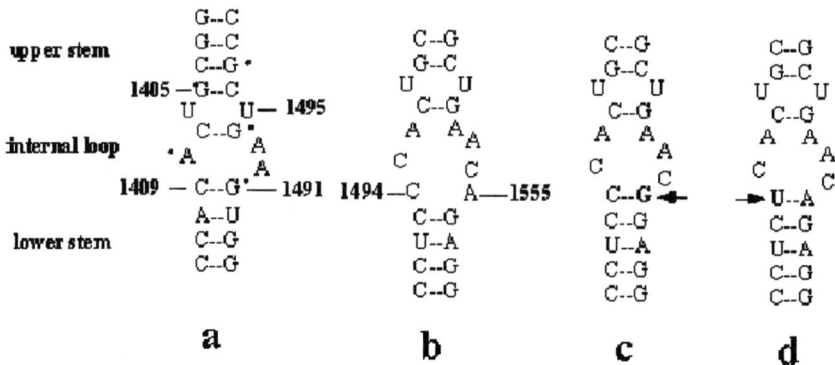

FIGURE 2. The site of the A1555G and C1494T mutations in the decoding region of mitochondrial 12S rRNA. The A-site of *E. coli* 16S rRNA oligonucleotide showing the DMS footprints, observed in the presence of the aminoglycosides neomycin and paromomycin,[82–84] is marked with a dot (**a**). The corresponding region of the human mitochondrial 12S rRNA is shown as the wild-type version[53,54] (**b**) and in the version containing the A1555G mutation (**c**) and C1494T mutation (**d**), respectively. The sites for the A1555G or C1494T mutation are indicated by *arrows*.

Recently, Nye *et al.*[87] found that symptoms of progressive matrilineal hearing loss, premature graying, depigmented patches, and digital anomalies in a large Filipino-American family were also associated with the A1555G mutation in both affected and unaffected maternal relatives. Functional studies demonstrated a decrease in growth rates and in the rate of mitochondrial protein synthesis of lymphoblastoid cells derived from symptomatic and asymptomtic members of an Arabic Israeli pedigree in the presence of high concentration of neomycin or paromomycin.[51,85]

Recently, a homoplasmic C-to-T transition at position 1494 (C1494T) in the 12S rRNA gene was found in a large Chinese family with maternally transmitted aminoglycoside-induced and nonsyndromic deafness.[88] Maternal members of this family showed variable severity and age of onset for the hearing impairment. The C1494T mutation is expected to form a novel 1494U-A1555 base-pair, which is in the same position as the 1494C-G1555 pair created by the deafness-linked A1555G mutation, at the highly conserved A-site of 12S rRNA. This site has been implicated as the main target of aminoglycoside toxicity. Thus, it is anticipated that this alteration in the tertiary structure of 12S rRNA may lead to sensitivity to aminoglycosides.

Sensitivity to the aminoglycosides (paromomycin or neomycin) has been analyzed in lymphoblastoid cell lines derived from four deaf individuals and two normal individuals from this Chinese family and four unrelated controls. In the presence of high concentrations of paromomycin or neomycin, the C1494T mutation-carrying cell lines exhibited a significant increase in doubling time when compared to control cell lines. Furthermore, this mutation was absent in 364 unrelated controls, suggesting specific segregation with the disorder.

These data suggest that mutations in mtDNA are frequently associated with ototoxicity. The following aspects of mitochondrial oxidative phosphorylation are certainly important in the pathogenesis of aminoglycoside ototoxicity. (1) Mitochondria generate cellular energy in the form of ATP by the process of oxidative phosphorylation, which is essential for cellular function, including auditory and vestibular hair cells. (2) Oxidative phosphorylation is the major endogenous source of the reactive oxygen species (ROS) (O2–, H2O2, and OH–), which are toxic byproducts of respiration. Chronic ROS exposure can result in oxidative damage to mitochondrial and cellular protein, lipids, and nuclear acids. (3) Mitochondria also provide a major switch for the initiation of apoptosis.

Based on this genetic and biochemical evidence, we propose the following mechanism for aminoglycoside ototoxicity. We hypothesize that the site of susceptibility to aminoglycosides is the mitochondrial ribosomes. Aminoglycosides accumulate in the cochlear and vestibular mitochondria, where they inhibit mitochondrial protein synthesis by interacting with the 12S rRNA, especially that carrying these A1555G or C1494T mutations. These mitochondrial translational defects result in a decline in ATP production in the cochlear and vestibular cells. At same time, these defects in oxidative phosphorylation lead to increased generation of ROS, thereby damaging mitochondrial and cellular proteins, lipids, and nucleic acids. Consequently, the mitochondrial permeability transition pore opens and activates apoptosis. This causes a loss of cochlear and vestibular cell function or cell death and gives rise to hearing impairment.

ACKNOWLEDGMENTS

These investigations were supported by National Institutes of Health Grants DC04958 and DC05230 from the National Institute on Deafness and Other Communication Disorders and research grant awards from the United Mitochondrial Disease Foundation and Deafness Research Foundation.

REFERENCES

1. NANCE, W.E. & A. SWEENEY. 1975. Symposium on sensorineural hearing loss in children: early detection and intervention: genetic factors in deafness in early life. Otolaryngol. Clin. North Am. **8:** 9–48.
2. MORTON, N.E. 1991. Genetic epidemiology of hearing impairment. Ann. N.Y. Acad. Sci. **630:** 16–31.
3. KALATZIS, V. & C. PETIT. 1998. The fundamental and medical impacts of recent progress in research on hereditary hearing loss. Hum. Mol. Genet. **7:** 1589–1597.
4. MORTON, C.C. 2002. Genetics, genomics and gene discovery in the auditory system. Hum. Mol. Genet. **11:** 1229–1240.
5. FISCHEL-GHODSIAN, N. 1999. Mitochondrial deafness mutations reviewed. Hum. Mut. **13:** 261–270.
6. VAN CAMP, G. & R.J. SMITH. 2000. Maternally inherited hearing impairment. Clin. Genet. **57:** 409–414.
7. WALLACE, D.C. 1999. Mitochondrial diseases in man and mouse. Science **283:** 1482–1488.
8. SCHON, E.A., E. BONILLA & S. DIMAURO. 1997. Mitochondrial DNA mutations and pathogenesis. J. Bioenerg. Biomembr. **29:** 131–149.

9. ANDERSON, S., A.T. BANKIER, B.G. BARRELL, et al. 1981. Sequence and organization of the human mitochondrial genome. Nature **290:** 457–465.
10. ATTARDI, G. & G. SCHATZ. 1988. Biogenesis of mitochondria. Annu. Rev. Cell. Biol. **4:** 289–333.16
11. WALLACE, D.C. 1992. Diseases of the mitochondrial DNA. Annu. Rev. Biochem. **61:** 1175–1212.
12. GILES, R.E., H. BLANC, H.M. CANN & D.C. WALLACE. 1980. Maternal inheritance of human mitochondrial DNA. Proc. Natl. Acad. Sci. USA **77:** 6715–6719.
13. JENUTH, J.P., A.C. PATERSON, K. FU & E.A. SHOUBRIDGE. 1997. Random genetic shift in the female germline explains the rapid segregation of mammalian mitochondrial DNA. Nat. Genet. **14:** 146–151.
14. MONNAT, R. & L. LOEB. 1985 Nucleotide sequence preservation of human mitochondrial DNA. Proc. Natl. Acad. Sci. USA **82:** 2895–2899.
15. SHOFFNER, J.M., & D.C. WALLACE. 1990. Oxidative phosphorylation diseases. Disorders of two genomes. Adv. Hum. Genet. **19:** 267–330.
16. ZEVIANI, M., C.T. MORAES, S. DIMAURO, et al. 1988 Deletions of mitochondrial DNA in Kearns-Sayre syndrome. Neurology **51:** 1525–1346.
17. SHOFFNER, J.M., M.T. LOTT, A.M. LEZZA, et al. 1990. Myoclonic epilepsy and ragged-red fiber disease (MERRF) is associated with a mitochondrial DNA tRNA^{Lys} mutation. Cell **61:** 931–937.
18. GOTO Y., I. NONAKA & S. HORAI. 1990. A mutation in the tRNA$^{Leu(UUR)}$ gene associated with the MELAS subgroup of mitochondrial encephalomyopathies. Nature **348:** 651–653.
19. SANTORELLI, F.M., S. SHANSKE, A. MACAYA, et al. 1993. The mutation at nt 8993 of mitochondrial DNA is a common cause of Leigh's syndrome. Ann. Neurol. **34:** 827–834.
20. SHOFFNER, J.M. & D.C. WALLACE. 1992. Mitochondrial genetics: principles and practice. Am. J. Hum. Genet. **51:** 1179–1186.
21. HOWELL, N. 1997. Leber hereditary optic neuropathy: mitochondrial mutations and degeneration of the optic nerve. Vision Res. **37:** 3495–3507.
22. REARDON, W., R.J. ROSS, M.G. SWEENEY, et al. 1992 Diabetes mellitus associated with a pathogenic point mutation in mitochondrial DNA. Lancet **340:** 1376–1379.
23. VAN DEN OUWELAND, J.M.W., R.C. TREMBATH, R. ROSS, et al. 1994. Maternally inherited diabetes and deafness is a distinct subtype of diabetes and associates with a single point mutation in the mitochondrial tRNA$^{Leu(UUR)}$ gene. Diabetes **43:** 746–751.
24. FOLGERO, T., N. STORBAKK, T. TORBERGSEN & P. OIAN. 1996. Mutations in mitochondrial transfer ribonucleic acid genes in preeclampsia. Am. J. Obstet. Gynecol. **174:** 1626–1630.
25. MORAES, C.T., S. DIMAURO, M. ZEVIANI, et al. 1989. Mitochondrial DNA deletions in progressive external ophthalmoplegia and Kearns-Sayre syndrome. N. Engl. J. Med. **320:** 1293–1299.
26. BALLINGER, S.W., J.M. SHOFFNER, E.V. HEDAYA, et al. 1992 Maternally transmitted diabetes and deafness associated with a 10.4kb mitochondrial DNA deletion. Nat. Genet. **1:** 11–15.
27. VIALETTES, B.H., V. PAQUIS-FLUCKLINGER, J.F. PELISSIER, et al. 1997 Phenotypic expression of diabetes secondary to a T14709C mutation of mitochondrial DNA. Comparison with MIDD syndrome (A3243G mutation): a case report. Diabetes Care **20:** 1731–1737.
28. KAMEOKA, K., H. ISOTANI, K. TANAKA, et al. 1998. Novel mitochondrial DNA mutation in tRNA(Lys) (8296A→G) associated with diabetes. Biochem. Biophys. Res. Commun. **17:** 523–527.
29. MANSERGH, F.C., S. MILLINGTON-WARD, A. KENNAN, et al. 1999. Retinitis pigmentosa and progressive sensorineural hearing loss caused by a C12258A mutation in the mitochondrial MTTS2 gene. Am. J. Hum. Genet. **64:** 971–985.
30. SILVESTRI, G., T. MONGINI, F. ODOARDI, et al. 2000. A new mtDNA mutation associated with a progressive encephalopathy and cytochrome c oxidase deficiency. Neurology **54:** 1693–1696.

31. TIRANTI, V., P. CHARIOT, F. CARELLA, et al. 1995. Maternally inherited hearing loss, ataxia and myoclonus associated with a novel point mutation in mitochondrial tRNA$^{Ser(UCN)}$ gene. Hum. Mol. Genet. **4:** 1421–1427.
32. JAKSCH, M., T. KLOPSTOCK, G. KURLEMANN, et al. 1998. Progressive myoclonus epilepsy and mitochondrial myopathy associated with mutations in the tRNA$^{Ser(UCN)}$ gene. Ann. Neurol. **44:** 635–460.
33. FISCHEL-GHODSIAN, N., T.R. PREZANT, P. FOURNIER, et al. 1995. Mitochondrial mutation associated with non-syndromic deafness. Am. J. Otolaryngol. **16:** 403–408.
34. REID, F.M., G.A. VERNHAM & H.T. JACOBS. 1994. A novel mitochondrial point mutation in a maternal pedigree with sensorineural deafness. Hum. Mut. **3:** 243–247.
35. SEVIOR, K.B., A. HATAMOCHI, I.A. STEWART, et al. 1998. Mitochondrial A7445G mutation in two pedigrees with palmoplantar keratoderma and deafness. Am. J. Med. Genet. **75:** 179–185.
36. VERHOEVEN, K., R.J. ENSINK, V. TIRANTI, et al. 1999. Hearing impairment and neurological dysfunction associated with a mutation in the mitochondrial tRNA$^{Ser(UCN)}$ gene. Eur. J. Hum. Genet. **7:** 45–51.
37. HUTCHIN, T.P., M.J. PARKER, I.D. YOUNG, et al. 2000. A novel mutation in the mitochondrial tRNA$^{Ser(UCN)}$ gene in a family with non-syndromic sensorineural hearing impairment. J. Med. Genet. **37:** 692–694.
38. SUE, C.M., K. TANJI, G. HADJIGEORGIOUS, et al. 1999. Maternally inherited hearing loss in a large kindred with a novel T7511C mutation in the mitochondrial DNA tRNA$^{Ser(UCN)}$ gene. Neurology **52:** 1905–1908.
39. ISHIKAWA, K., Y. TAMAGAWA, K. TAKAHASHI, et al. 2002. Nonsyndromic hearing loss caused by a mitochondrial T7511C mutation. Laryngoscope **112:** 1494–1499.
40. CHAPIRO, E., D. FELDMANN, F. DENOYELLE, et al. 2002. Two large French pedigrees with non syndromic sensorineural deafness and the mitochondrial DNA T7511C mutation: evidence for a modulatory factor. Eur. J. Hum. Genet. **10:** 851–856.
41. PREZANT, T.R., J.V. AGAPIAN, M.C. BOHLMAN, et al. 1993. Mitochondrial ribosomal RNA mutation associated with both antibiotic-induced and nonsyndromic deafness. Nat. Genet. **4:** 289–294.
42. FISCHEL-GHODSIAN, N., T.R. PREZANT, X. BU, et al. 1993. Mitochondrial ribosomal RNA gene mutation in a patient with sporadic aminoglycoside ototoxicity. Am. J. Otolaryngol. **4:** 399–403.
43. HUTCHIN, T., I. HAWORTH, K. HIGASHI, et al. 1993. A molecular basis for human hypersensitivity to aminoglycoside antibiotics. Nucleic Acids Res. **21:** 4174–4179.
44. ESTIVILL, X., N. GOVEA, A. BARCELO, et al. 1998. Familial progressive sensorineural deafness is mainly due to the mtDNA A1555G mutation and is enhanced by treatment with aminoglycosides. Am. J. Hum. Genet. **62:** 27–35.
45. CASANO, R.A.M.S., D.F. JOHNSON, M. HAMON, et al. 1998. Hearing loss due to the mitochondrial A1555G mutation in Italian families. Am. J. Med. Genet. **79:** 388–391.
46. PANDYA, A., X. XIA, J. RADNAABAZAR, et al. 1997. Mutation in the mitochondrial 12S ribosomal-RNA gene in 2 families from Mongolia with matrilineal aminoglycoside ototoxicity. J. Med. Genet. **34:** 169–172.
47. INOUE, K., D. TAKAI, A. SOEJIMA, et al. 1996. Mutant mtDNA at 1555 A to G in 12S rRNA gene and hypersusceptibility of mitochondrial translation to streptomycin can be co-transferred to rhoo HeLa cells. Biochem. Biophys. Res. Commun. **223:** 496–501.
48. MATTHIJS, G., S. CLAES, B. LONGO-BBENZA, et al. 1996 Non-syndromic deafness associated with a mutation and a polymorphism in the mitochondrial 12S ribosomal RNA gene in a large Zairean pedigree. Eur. J. Hum. Genet. **4:** 46–51.
49. GARDNER, J.C., R. GOLIATH, D. VILJOEN, et al. 1997. Familial streptomycin ototoxicity in a South African family: a mitochondrial disorder. J. Med. Genet. **34:** 904–906.
50. LI, R., G. XING, M. YAN, et al. 2004. Cosegregation of Cinsertion at position 961 with A1555G mutation of mitochondrial 12S rRNA gene in a large Chinese family with maternally inherited hearing loss. Am. J. Med. Genet. **124A:** 113–117.
51. GUAN, M.X., N. FISCHEL-GHODSIAN & G. ATTARDI. 1996. Biochemical evidence for nuclear gene involvement in phenotype of non-syndromic deafness associated with mitochondrial 12S rRNA mutation. Hum. Mol. Genet. **6:** 963–972.

52. GUAN, M.X., N. FISCHEL-GHODSIAN & G. ATTARDI. 2001. Nuclear background determines biochemical phenotype in the deafness-associated mitochondrial 12S rRNA mutation. Hum. Mol. Genet. **10:** 573–580.
53. LI, X. & M.X. GUAN. 2002. A human mitochondrial GTP binding protein related to tRNA modification may modulate the phenotypic expression of the deafness-associated mitochondrial 12S rRNA mutation. Mol. Cell. Biol. **22:** 7701–7711.
54. LI, X., R. LI, X. LIN & M.X. GUAN. 2002. Isolation and characterization of the putative nuclear modifier gene MTO1 involved in the pathogenesis of deafness-associated mitochondrial 12S rRNA A1555G mutation. J. Biol. Chem. **277:** 27256–27264.
55. REID, F.M., A. ROVIO, I.J. HOLT, et al. 1997. Molecular phenotype of a human lymphoblastoid cell-line homoplasmic for the np 7445 deafness-associated mitochondrial mutation. Hum. Mol. Genet. **6:** 443–449.
56. GUAN, M.X., J.A. ENRIQUEZ, N. FISCHEL-GHODSIAN, et al. 1998. The deafness-associated mtDNA 7445 mutation, which affects tRNA$^{Ser(UCN)}$ precursor processing, has long-range effects on NADH dehydrogenase ND6 subunit gene expression. Mol. Cell. Biol. **18:** 5868–5879.
57. HOWELL, N., L.A. BINDOFF, D.A., MCCULLOUGH, et al. 1991. Leber hereditary optic neuropathy: identification of the same mitochondrial ND1 mutation in six pedigrees. Am. J. Hum. Genet. **49:** 939–950.
58. HOWELL, N., I. KUBACKA, M. XU, et al. 1991. Leber hereditary optic neuropathy: involvement of the mitochondrial ND1 gene and evidence for an intragenic suppression mutation. Am. J. Hum. Genet. **48:** 935–942.
59. REID, F.M., G.A. VERNHAM & H.T. JACOBS. 1994. Complete mtDNA sequence of a patient in a maternal predigree with sensorineural deafness. Hum. Mol. Genet. **3:** 1435–1436.
60. JAKSCH, M., T. KLOPSTOCK, G. KURLEMANN, et al. 1998. Progressive myoclonus epilepsy and mitochondrial myopathy associated with mutations in the tRNA$^{Ser(UCN)}$ gene. Ann. Neurol. **44:** 635–640.
61. SCHUELKE, M., M. BAKKER, G. STOLTENBURG, et al. 1998. Epilepsia partialis continua associated with a homoplasmic mitochondrial tRNA$^{Ser(UCN)}$ mutation. Ann. Neurol. **44:** 700–704.
62. TOOMPUU, M., V. TIRANTI, M. ZEVIANI, et al. 1999. Molecular phenotype of the np 7472 deafness-associated mitochondrial mutation in osteosarcoma cell cybrids. Hum. Mol. Genet. **8:** 2275–2283.
63. FRIEDMAN, R.A., Y. BYKHOVSKAYA, C.M. SUE, et al. 1999. Maternally inherited nonsyndr omic hearing loss. Am. J. Med. Genet. **84:** 369–372.
64. LI, X., N. FISCHEL-GHODSIAN, F. SCHWARTZ, et al. 2004. Biochemical characterization of the mitochondrial tRNA$^{Ser(UCN)}$ T7511C mutation associated with nonsyndromic deafness. Nucleic Acids Res. **32:** 867–877.
65. LORTHOLARY, O., M. TOD, Y. COHEN, et al. 1995. Aminoglycosides. Med. Clin. North Am. **79:** 761–798.
66. SANDE, M.A. & G.L. MANDELL. 1990. Antimicrobial agents. In Goodman and Golman's The Pharmacological Basis of Therapeutics, 8th edit. A.G. Gilman, T.W. Rall, A.S. Nies & P. Taylor, Eds.: 1098–1116. Pergamon Press. Elmsford, NY.
67. CHAMBER, H.F. & M.A. SANDE. 1996. The aminoglycosides. In The Pharmacological Basis of Therapeutics, 9th Ed. J.G. Hardman, L.E. Limbird, P.B. Molinoff, R.W. Ruddon & A. Gilman, Eds.: 1103–1221. McGraw-Hill. New York.
68. DAVIS, J. & B.D. DAVIS. 1968. Misreading of ribonucleic acid code words induced by aminoglycoside antibiotics. J. Biol. Chem. **243:** 3312–3316.
69. NOLLER, H.F. 1991. Ribosomal RNA and translation. Annu. Rev. Biochem. **60:** 191–227.
70. HENLEY, C.M. & J. SCHACHT. 1988 Pharmacokinetics of aminoglycoside antibiotics in blood, inner ear fluids and their relationship to ototoxicity. Audiology **27:** 137–146.
71. VRABEC, D.P., D.T. CODY & J.A. ULRICH. 1965. A study of the relative concentrations of antibiotics in the blood, spinal fluid and perilmphy in animals. Ann. Otol. Rhino. Laryngol. **74:** 689–678.
72. HU, D.N., W.Q. QUI, B.T. WU, et al. 1991. Genetic aspects of antibiotic induced deafness: mitochondrial inheritance. J. Med. Genet. **28:** 79–83.

73. HIGASHI, K. 1989. Unique inheritance of streptomycin-induced deafness. Clin. Genet. **35:** 433–436.
74. BACINO, C., T.R. PREZANT, W. BU, et al. 1995. Susceptibility mutations in the mitochondrial small ribosomal RNA gene in aminoglycoside induced deafness. Pharmacogenetics **5:** 165–172.
75. YOSHIDA, M., T. SHINTANI, M. HIRAO, et al. 2002. Aminoglycoside-induced hearing loss in a patient with the 961 mutation in mitochondrial DNA. ORL J. Otorhinolaryngol. Relat. Spec. **64:** 219–222.
76. CASANO, R.A., D.F. JOHNSON, Y. BYKHOVSKAYA, et al. 1999. Inherited susceptibility to aminoglycoside ototoxicity: genetic heterogeneity and clinical implications. Am. J. Otolaryngol. **20:** 151–156.
77. NEEFS, J.M., Y. VAN DE PEER, P. DE RIJIK, et al. 1991. Compilation of small ribosomal subunit RNA sequences. Nucleic Acids Res. **19:** 1987–2018.
78. FISCHEL-Ghodsian, N. 1999. Genetic factors in aminoglycoside toxicity. Ann. N.Y. Acad. Sci. **884:** 99–109.
79. ZIMMERMANN, R.A., C.L. THOMAS & J. WOWER. 1990. Structure and function of rRNA in the decoding domain and at the peptidyltransferase center. *In* The Ribosome: Structure, Function and Evolution. W.E. Hill, P.B. Moore, A. Dahlberg, D. Schlessinger, R.A. Garrett & J.R. Warner, Eds.: 331–347. American Society for Microbiology. Washington, DC.
80. GREGORY, S.T. & A.E. DAHLBERG. 1995. Nonsense suppressor and antisuppressor mutations at the 1409-1491 base pair in the decoding region of *Escherichia coli* 16S rRNA. Nucleic Acids Res. **23:** 4234–4238.
81. CHERNOFF, Y.O., A. VINCENT & S.W. LIEBMAN. 1994. Mutations in eukaryotic 18S ribosomal RNA affect translational fidelity and resistance to aminoglycoside antibiotics. EMBO J. **13:** 906–913.
82. MOAZED, D. & H.F. NOLLER. 1986. Transfer RNA shields specific nucleotides in 16S ribosomal RNA from attack by chemical probes. Cell **47:** 389–394.
83. RECHT, M.I., D. FOURMY, S.C. BLANCHARD, et al. 1996. RNA sequence determinants for aminoglycoside bind to an A-site rRNA model oligonucleotide. J. Mol. Biol. **262:** 421–436.
84. FOURMY, D., M.I. RECHT, S.C. BLANCHARD, et al. 1996. Structure of the A-site of *Escherichia coli* 16S ribosomal RNA complexed with an aminoglycoside antibiotic. Science **274:** 1367–1371.
85. GUAN, M.X., N. FISCHEL-Ghodsian & G. ATTARDI. 2000. A biochemical basis for the inherited susceptibility to aminoglycoside ototoxicity. Hum. Mol. Genet. **9:** 1787–1793.
86. HAMASAKI, K. & R.R. RANDO. 1997. Specific binding of aminoglycosides to a human rRNA construct based on a DNA polymorphism, which causes aminoglycoside-induced deafness. Biochemistry **36:** 12323–12328.
87. NYE, J.S., E.A. HAYES, M. AMENDOLA, et al. 2000. Myelocystocele-cloacal exstrophy in a pedigree with a mitochondrial 12S rRNA mutation, aminoglycoside-induced deafness, pigmentary disturbances, and spinal anomalies. Teratology **61:** 165–171.
88. ZHAO, H., R. LI, Q. WANG, et al. 2003. Maternally inherited aminoglycoside-induced and non-syndromic deafness is associated with the novel C1494T mutation in the mitochondrial 12S rRNA gene in a large Chinese family. Am. J. Hum. Genet. **74:** 139–152.

Genetic and Functional Analysis of Mitochondrial DNA–Encoded Complex I Genes

YIDONG BAI,[a] PEIQING HU,[a] JEONG SOON PARK,[a] JIAN-HONG DENG,[a] XIUFENG SONG,[a] ANNE CHOMYN,[b] TAKAO YAGI,[c] AND GIUSEPPE ATTARDI[b]

[a]*Department of Cellular and Structural Biology, University of Texas Health Science Center at San Antonio, San Antonio, Texas 78229, USA*

[b]*Division of Biology, California Institute of Technology, Pasadena, California 91125, USA*

[c]*Division of Biochemistry, Department of Molecular and Experimental Medicine, The Scripps Research Institute, La Jolla, California 92037, USA*

ABSTRACT: Mammalian mitochondrial NADH dehydrogenase (complex I) is a multimeric complex consisting of at least 45 subunits, 7 of which are encoded by mitochondrial DNA (mtDNA). The function of these subunits is largely unknown. We have established an efficient method to isolate and characterize cells carrying mutations in various mtDNA-encoded complex I genes. With this method, 15 mouse cell lines with deficiencies in complex I-dependent respiration were obtained, and two near-homoplasmic mutations in mouse ND5 and ND6 genes were isolated. Furthermore, by generating a series of cell lines with the same nuclear background but different content of an mtDNA nonsense mutation, we analyzed the genetic and functional thresholds in mouse mitochondria. We found that in wild-type cells, about 40% of ND5 mRNA is in excess of that required to support a normal rate of ND5 subunit synthesis. However, there is no indication of compensatory upsurge in either transcription or translation with the increase in the proportion of mutant ND5 genes. Interestingly, the highest ND5 protein synthesis rate was just sufficient to support the maximum complex I–dependent respiration rate, suggesting a tight regulation at the translational level. In another line of research, we showed that the mitochondrial NADH-quinone oxidoreductase of *Saccharomyces cerevisiae* (NDI1), although consisting of a single subunit, can completely restore respiratory NADH dehydrogenase activity in mutant human cells that lack the essential mtDNA-encoded subunit ND4. In particular, in these transfected cells, the yeast enzyme becomes integrated into the human respiratory chain and fully restores the capacity of the cells to grow in galactose medium.

KEYWORDS: complex I; mtDNA mutation; threshold; NDI1

Address for correspondence: Dr. Yidong Bai, Department of Cellular and Structural Biology, University of Texas Health Science Center at San Antonio, San Antonio, TX 78229. Voice: 210-567-0561; fax: 210-567-3800.
baiy@uthscsa.edu

Ann. N.Y. Acad. Sci. 1011: 272–283 (2004). © 2004 New York Academy of Sciences.
doi: 10.1196/annals.1293.026

INTRODUCTION

Mammalian respiratory NADH-ubiquinone oxidoreductase (complex I) is the largest and least understood component of the mitochondrial oxidative phosphorylation system, consisting of at least 45 subunits.[1,2] This enzyme, like its counterpart in *Neurospora crassa*,[3] has an overall L-shaped structure, with one arm, which contains all the subunits encoded by the mtDNA, buried in the mitochondrial inner membrane, and the other arm, which contains the catalytic center, protruding into the mitochondrial matrix.[4] Although the sequence of mtDNA-encoded subunits was determined a long time ago,[5,6] their function is not clear, partly because not many mtDNA mutations in this complex are available.

Mitochondrial genetics has several distinguishing features besides maternal transmission. First, the mitochondrial genome has a high copy number; most mammalian cells have hundreds or thousands of mitochondria, and each mitochondrion has on average of five mtDNA molecules.[7] Because mitochondria are a major source of reactive oxygen species (ROS) and have a relatively less sophisticated DNA protecting system, mtDNA has a relatively high mutation rate.[8] As a result, mtDNA within a cell can be a mixture of both wild-type and mutant species, a status called heteroplasmy. We can see how when mutant mtDNA reaches a certain level, normal mitochondrial function can no longer be supported.

Impaired complex I activity has been implicated in several neurodegenerative diseases and in the normal aging process.[9,10] Mutations in mtDNA-encoded subunits of complex I are associated with Leber's hereditary optical neuropathy (LHON).[11] Restoring respiratory function in complex I–defective cells holds promise for the treatment of mitochondrial diseases such as LHON.

In contrast to the multisubunit enzyme of mammalian cells, the mitochondrial NADH-Q oxidoreductase of *Saccharomyces cerevisiae*, NDI1, is a simple subunit of 513 amino acid residues, including the NH_2-terminal 26-residue signal sequence, which guides its import into the mitochondria.[12,13] Furthermore, NDI1, in contrast to mammalian complex I, contains no proton-translocating site and is rotenone insensitive. The NDI1 enzyme was shown to rescue the deficiency of NADH dehydrogenase in mammalian cells when the NDI1 gene was transfected into Chinese hamster cells carrying a mutation in an essential nuclear-encoded complex I subunit.[14] Subsequently, it was reported that the NDI1 gene can function in human embryonic kidney 293 cells[15] and in nonproliferating human cells.[16]

The work presented here centers on mtDNA-encoded complex I gene mutations. We have established an efficient approach to isolate mtDNA-encoded complex I subunit mutations from cultured mouse cells, based on selection against high concentrations of the specific complex I inhibitor rotenone. Ethidium bromide (EB), a specific inhibitor of mtDNA replication and transcription, was also used to obtain cells carrying near-homoplasmic mutations.[17,18]

Special efforts were made to construct a set of transmitochondrial cell lines carrying, in a constant nuclear background, various levels of mutant ND5 gene, from 0 to near 100%. In these cell lines, analysis of mRNA (transcription), protein synthesis (translation), and complex I-dependent respiration has revealed stringent regulation of ND5 gene expression and respiration.[18]

In a further attempt to explore the potential of the NDI1 gene in gene therapy of mitochondrial diseases, the yeast NDI1 gene was introduced into a human cell line carrying a homoplasmic frameshift mutation in the mitochondrial ND4 gene. Two transformants with different expression levels of the NDI1 gene were isolated. Respiration rates in both intact and permeabilized cells, P:O ratio, and growth capacity analysis revealed that the yeast NDI1 enzyme can modulate the oxidative phosphorylation of host cells.[19]

ISOLATION OF MOUSE CELLS CARRYING COMPLEX I mtDNA MUTATIONS

The approach previously used to isolate human cell lines of mtDNA mutants defective in complex I, which was based on resistance to high concentrations of rotenone,[20,21] was applied to mouse fibroblast cells.[17,18] A preliminary experiment indicated that the growth capacity of the wild-type cells was significantly affected when the rotenone concentration reached 0.6 µM, with almost complete inhibition by 1.2 µM retenone. The procedure was as follows: about 5 million wild-type cells in each dish were treated with 0.8 mM of rotenone for about 3 weeks, individual rotenone-resistant colonies were picked, and these clones were subjected to stepwise increases in the concentration of rotenone, from 1.0 to 1.2 µM. Those cell lines that adapted to grow in the presence of 1.2 µM were candidates for having mutations in mtDNA.

Complex I function analysis was achieved by measuring the respiratory capacity of the rotenone-resistant cell lines. Both the overall respiration in intact cells and specific substrate-driven respiration in permeabilized cells were measured. In partic-

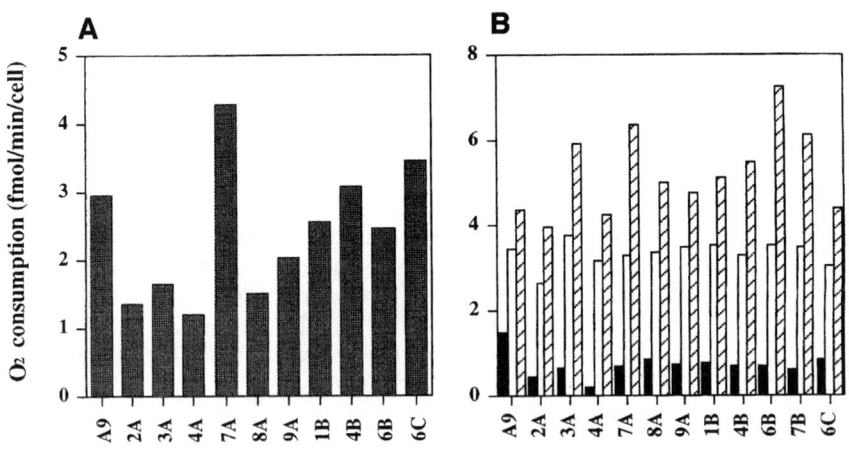

FIGURE 1. Total respiration rate (**A**) and complex-dependent respiration rate (**B**) in the original rotenone-resistant clones and in the parent cell line A9. (**A**) Total respiration rate was measured on $\sim 5 \times 10^6$ cells. (**B**) The complex-dependent respiration was determined on $\sim 5 \times 10^6$ cells as respiration dependent on malate/glutamate (*filled bars*, for complex I), succinate/G-3-P (*open bars*, for complex III), and TMPD/ascorbate (*hatched bars*, for complex IV). Adapted, with permission, from EMBO J. **17**: 4849.

ular, a combination of malate and glutamate was used to measure complex I activity. As controls, the succinate/glycerol-3-phosphate (G-3-P)–driven respiration, which usually reflects the activity of complex III, and the N,N,N',N'-tetramethyl-*p*-phenylenediamine (TMPD)/ascorbate-driven respiration, which reflects the activity of complex IV, were also measured. FIGURE 1 shows the results of respiration measurements in 12 rotenone-resistant cell lines. All rotenone-resistant cells exhibited different extents of decrease in complex I-dependent respiration, and several of them also revealed a decrease in overall oxygen consumption.

To have some indication of the possible sites responsible for the complex I defects, mitochondrial proteins were labeled with [^{35}S] methionine in the presence of emetine, to inhibit cytoplasmic protein synthesis. The labeled products were electrophoresed through SDS-polyacrylamide gel (15–20% exponential gradient). The electrophoretic patterns between wild-type and mutant cell lines were carefully compared. We observed that ND6 protein was abolished in cell line 4A cell. Subsequent DNA sequence analysis revealed a frameshift mutation at position 13,885, which created a stop codon downstream, resulting in a truncated polypeptide.[17]

For those cell lines that showed no obvious deficiency in mitochondrial protein synthesis, treatment with a low concentration of EB was carried out.[22] The goal was to reduce the mtDNA level to an average of one molecule per cell, a state in which most cells would be expected to be homoplasmic for mutant or wild-type mtDNA. Withdrawal of EB then allowed the residual mtDNA molecules to repopulate the cell.[23] The resulting clones were cultured in both normal medium and in a medium containing galactose instead of glucose. The latter medium was shown to severely curtail the growth of cells deficient in oxidative phosphorylation. Cells that failed to grow in galactose medium were isolated, and again respiration rate and mitochondrial protein synthesis were analyzed. An ND5 mutation was isolated in this way.[18] Using this method, we recently also obtained subclones of four of the original rotenone-resistant lines that exhibited 20–50% decreases in respiration rates (Hu, Park, and Bai, in preparation).

Because mitochondria are under dual genetic control, nuclear and mitochondrial, mutations in either genome could potentially cause enzyme defects. To verify that a mutation in mtDNA was solely responsible for the respiration deficiency observed, an mtDNA–less (ρ^0) cell repopulation approach was utilized.[24] Mitochondria carrying a mutation in the ND5 or ND6 gene were transferred to mouse ρ^0 cells by cytoplast-cell fusion, thereby placing those mutant genomes in a new nuclear background. The defects in respiration capacities in the transmitochondrial cells confirmed the essential role of ND5 and ND6 in the function of NADH dehydrogenase.[17,18]

THRESHOLD ANALYSIS IN ND5 GENE EXPRESSION IN MOUSE MITOCHONDRIA

A set of cells containing the same nuclear background but with different content of functional ND5 genes, ranging from about 4–100% of the wild-type level, was constructed by the procedures of mitochondria-mediated transformation and EB treatment. The mutation was identified by DNA sequencing. As shown in FIGURE 2, the C to A mutation at position 12,081 creates a nonsense mutation, resulting in a truncated protein.

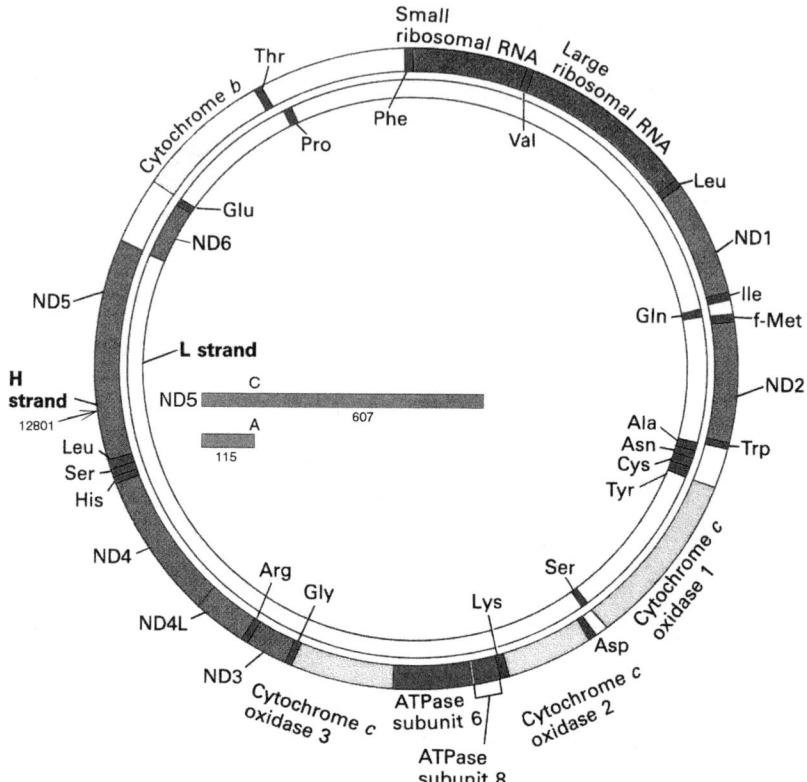

FIGURE 2. Position of the ND5 mutation. A C-to-A transversion was found at position 12081, which changes an arginine codon CGA to the mitochondrial stop codon AGA. The mouse ND5 gene encodes a 607-amino-acid polypeptide, and this mutation resulted in a 115 amino acid truncated peptide.

After verifying that the mutant cells were deficient in both overall and complex I-dependent respirations, further characterization of ND5 mutant cells was carried out. The NADH oxidoreductase activity was determined with a water-soluble ubiquinone analogue (Q1), and the NADH-$K_3Fe(CN)_6$ oxidoreductase activity, which is catalyzed by the nuclear-encoded flavoprotein fragment from the same isolated mitochondrial membrane, was also measured as control. A severe deficiency in the former enzyme activity was observed, in agreement with the results from the respiration measurements.

The availability of this series of cell lines enabled us to conduct a comprehensive analysis of the genetic and functional thresholds operating in the expression of the ND5 gene. In particular, we investigated the effects of the dosage of ND5 DNA on the ND5 mRNA level, the ND5 protein synthesis rate, and the assembly of a functional NADH dehydrogenase.

The total mRNA level was examined by RNA transfer hybridization experiments. In all cell lines, except for two that were near-homoplasmic for mutant mtDNA, the

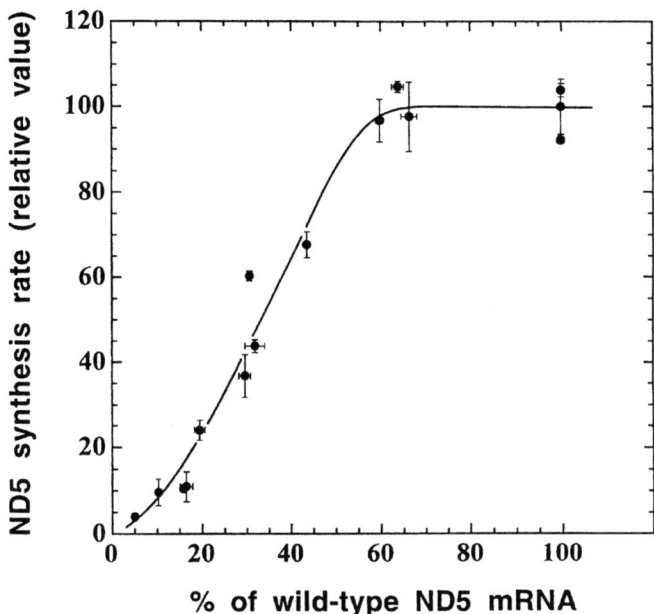

FIGURE 3. Threshold at translation level. Relation between the ND5 synthesis rate, expressed relative to the rate in the wild-type cell, and the proportion of wild-type ND5 mRNA. Individual values for the rate of ND5 protein labeling, as determined by laser densitometry of appropriately exposed fluorograms, were normalized to the overall protein labeling. Adapted, with permission, from Mol. Cell Biol. **20**: 811.

ND5 mRNA level was approximately constant, independent of the proportion of mutant genes.[18] There seemed to be no compensatory upregulation at the transcriptional level with the decrease in wild-type ND5 gene content.

How the rate of ND5 protein synthesis varied in different cell lines was investigated by plotting the rate of protein synthesis versus the percentage of wild-type ND5 mRNA (FIG. 3). It appears that there is a threshold at 60% of wild-type ND5 mRNA. This means that 60% of the normal level of functional ND5 mRNA is adequate to support a normal rate of ND5 protein synthesis. In other words, 40% ND5 mRNA is in excess in wild-type cells. Below this threshold, ND5 protein synthesis declined progressively, indicating that there is no upregulation at the translational level with the decreasing percentage of functional mRNA.

To study the role of ND5 subunit synthesis on the assembly of a functional complex I, the malate/glutamate-dependent respiration rate was plotted against the corresponding ND5 protein synthesis rates. Interestingly, the malate/glutamate-dependent respiration, which is determined by complex I activity, decreased nearly in parallel with the decreasing protein synthesis rate (FIG. 4a). This result clearly indicated that there is very little excess of ND5 protein synthesis capacity over that required to maintain assembly of a functional complex I.

To further understand the control that the ND5 protein synthesis rate exerts on complex I activity in intact cells, the rotenone-sensitive endogenous respiration rate, which reflects the complex I contribution to total respiration, was plotted against the corresponding rates of ND5 protein synthesis (FIG. 4b). We found that rotenone-sensitive respiration remained fairly constant with decreasing ND5 protein synthesis until it reached about 80% of the control value, then it declined progressively to near zero.

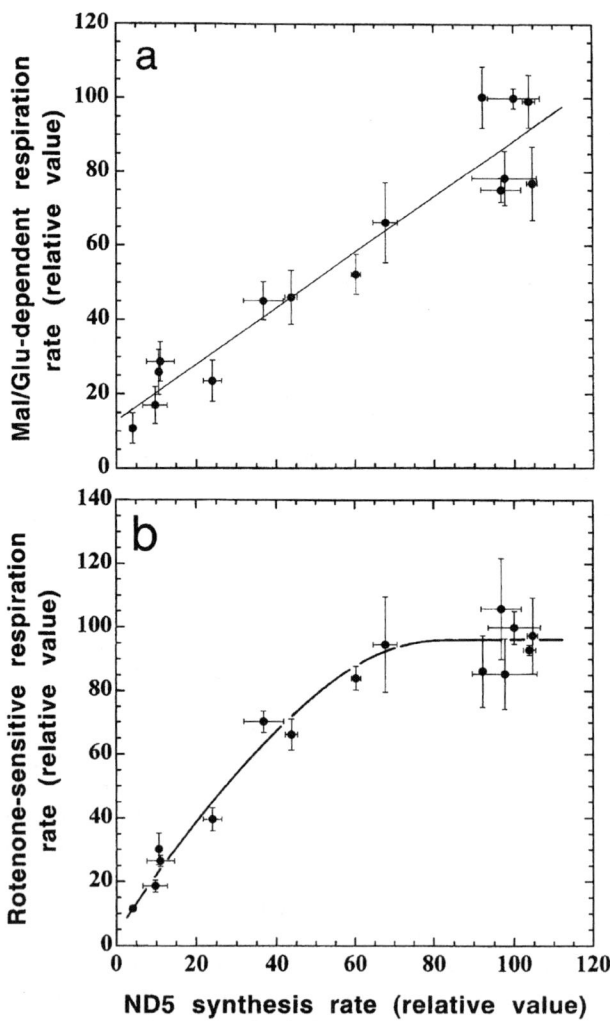

FIGURE 4. Tight regulation of complex I respiration by ND5 protein synthesis. Relation between rate of malate-glutamate-dependent respiration (**a**) or rate of rotenone-sensitive respiration (**b**), as expressed relative to the rate in the wild-type cell, and relative ND5 synthesis rate. Adapted, with permission, from Mol. Cell Biol. **20:** 812.

Interestingly, ND5 gene expression has been reported to be regulated at the translational level in isolated rat brain synaptosomes,[25] and the ND5 subunit was shown to play an important role in proton or sodium ion translocation.[26] The control of ND5 expression could have remarkable biological significance.

RESTORATION OF MAMMALIAN COMPLEX I DEFICIENCY WITH YEAST NDI1 GENE

C4T is a human cell line with a homoplasmic frameshift mutation in the mitochondrial ND4 gene.[20] The abolished synthesis of the ND4 subunit caused by the mutation disrupted the assembly of mtDNA-encoded complex I subunits and consequently abolished its respiratory function and enzyme activity.[20] The yeast NDI1

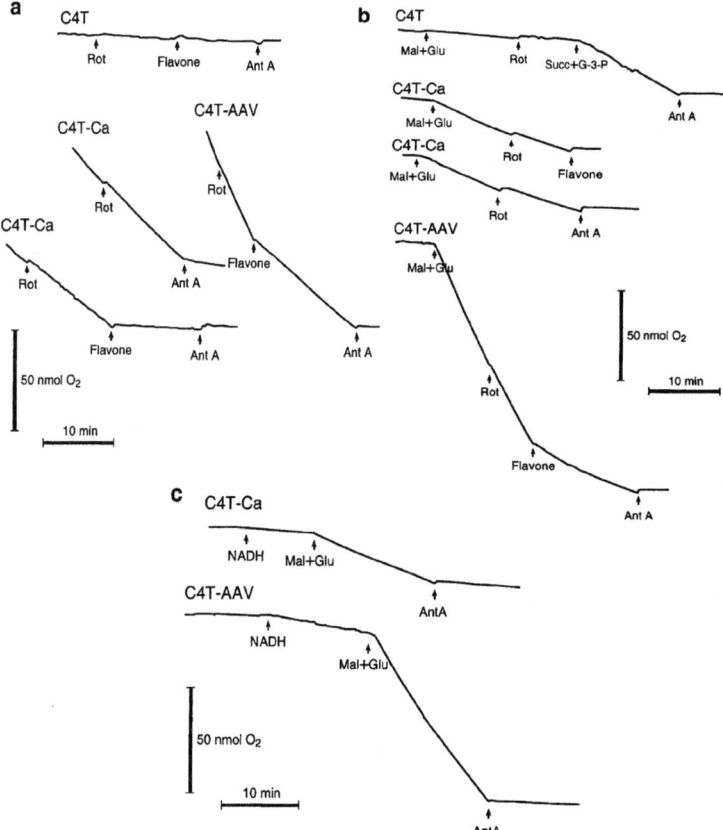

FIGURE 5. Recovery of complex I activity in NDI1 transfected C4T cells. (**a**) Endogenous respiration in intact C4T, C4T-Ca, and C4T-AAV cells; (**b**) substrate-dependent respiration in cells permeabilized with digitonin; (**c**) low response to NADH (0.5 mM) of C4T-Ca and C4T-AAV cells. Adapted, with permission, from J. Biol. Chem. **276:** 38811.

gene was successfully introduced to C4T cells, and two transformants, C4T-Ca and C4T-AAV, were isolated. It was also shown that the NDI1 protein localized to the host mitochondria in both transformants.

To assess the activity of yeast NDI1 in human cells, the respiration capacities of the mutant C4T cells and of the two NDI1 transfectants, C4T-Ca and C4T-AAV cells, were analyzed in detail. A significant difference in the sensitivity to the complex I inhibitor, rotenone, was used to distinguish the contribution from the mammalian complex I or the yeast NDI1 enzyme.

The endogenous respiration rates in intact C4T cells and in their NDI1 transformants are shown in FIGURE 5a. C4T exhibited a severe defect in endogenous respiration. NDI1 restored respiration in the transformants, the extent of which restoration corresponded to the different expression levels of the NDI1 gene. The restored activities in the NDI1 transformants were insensitive to the complex I inhibitor rotenone, but sensitive to flavone, a known inhibitor of the NDI1 enzyme. NDI1-dependent respiration was also shown to be sensitive to the mammalian complex III inhibitor, antimycin, indicating that the yeast NDI1 had integrated the endogenous respiratory chain. To further investigate the respiratory activities in transformed cells, the malate/glutamate-dependent respiration in permeabilized cells was also measured. As shown in FIGURE 5b, similar degrees of restoration were found in two transformants. NADH stimulated only slightly the respiration of digitonin-permeabilized transformants (FIG. 5c). These observations indicated that the NADH-binding site of the expressed NDI1 faces the matrix compartment, as in yeast mitochondria.

The lack of a proton-translocating site in the yeast NDI1 predicted that if NDI1 in human cells did transfer electrons from NADH to ubiquinone, the P:O ratio (defined as nanomoles of ATP produced per nanoatoms of oxygen consumed during ATP-stimulated respiration, coupled to NADH oxidation), measured with respiratory substrates malate/glutamate, would be lower in the transformants than in the wild-type control 143B cells, whereas the P:O ratio coupled to succinate oxidation would be similar in the transformants and the controls. As shown in TABLE 1, these predictions proved to be true. These experiments further established that the respiration restored by NDI1 in the defective C4T cells was coupled with ATP synthesis.

Mammalian cells rely on both mitochondrial oxidative phosphorylation and glycolysis for ATP production. However, they are unable to utilize galactose in the glycolytic pathway.[27,28] In the medium containing galactose instead of glucose, mammalian cells would be forced to rely exclusively on functional mitochondria to generate ATP. As shown in FIGURE 6, whereas wild-type 143B cells grow well in

TABLE 1. Comparison of P:O ratios in 143B cells and NDI1-transfected C4T cells[a]

Cell lines	Malate/glutamate	Cell lines
143B	2.10 ± 0.07	143B
C4T-Ca	1.16	1.31 ± 0.15
C4T-AAV	1.34 ± 0.04	1.46 ± 0.01

[a]Malate/glutamate and succinate were used as respiratory substrates, respectively. Each P:O is the average ± SE of two or three independent measurements. Adapted, with permission, from JBC **276:** 38811.

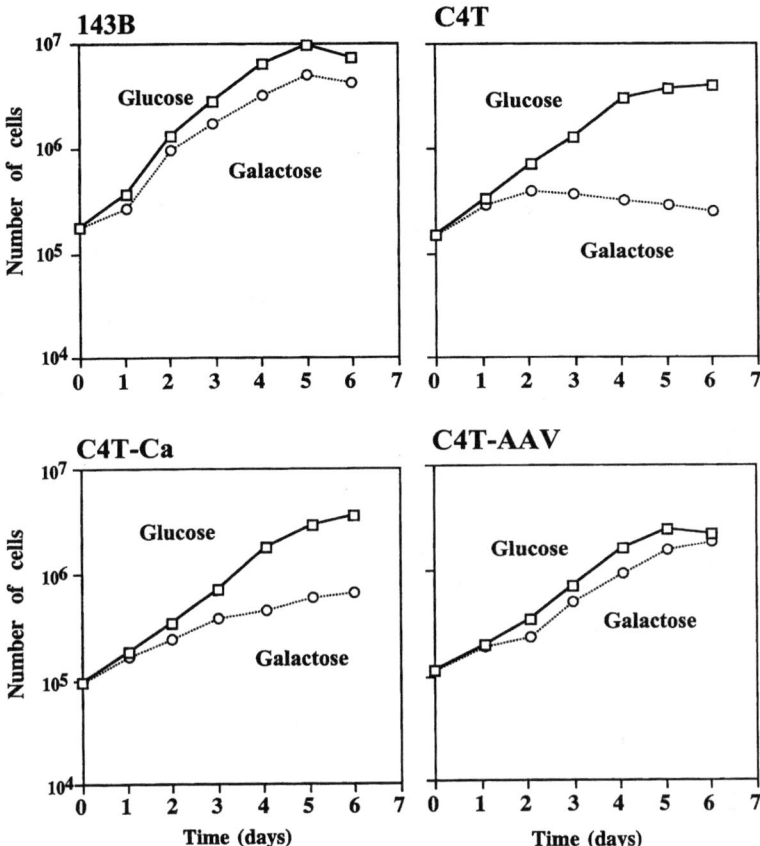

FIGURE 6. Recovery of growth capacity of NDI1 transfected C4T cells in galactose medium. Cells were plated on multiple 10-cm plates at 10^5 per plate and counted on a daily basis for 7 days. Adapted, with permission, from J. Biol. Chem. **276:** 38812.

both glucose- and galactose-containing medium, mutant C4T cells grow well in glucose medium but fail to grow in galactose medium. The growth capacity in the galactose medium was restored to different extents in two NDI1 transformants, corresponding to their mitochondrial respiration abilities.

Recently, NDI1 was also demonstrated to be actively expressed in differentiated and nondifferentiated dopaminergic cell lines.[29] The yeast NDI1 gene as a therapeutic tool for complex I defect is promising.

SUMMARY

To lay a foundation for systematic analysis of the mammalian mitochondrial genome, we developed an efficient approach to isolate and characterize cells carrying mutations in mtDNA-encoded complex I genes. Several cell lines with defective complex I-dependent respiration were obtained, and two cell lines with near-homoplasmic mutations in ND5 and ND6 gene were identified. With a series of cell lines containing the same nuclear background but a different percentage of functional mtDNA, we carried out a comprehensive study of the genetic and functional thresholds operating in the expression of the ND5 gene. We found no upregulation on the transcription level with the increase in the mutant ND5 gene content. In wild-type cells, about 40% of ND5 mRNA was in excess of that required for ND5 protein synthesis. A very high ND5 protein synthesis rate was required to maintain the maximum complex I-dependent respiration rate, with no upregulation of translation occurring with decreasing wild-type mRNA levels. Overall respiration is tightly regulated by ND5 gene expression. Finally, single subunit yeast NDI1 gene was shown to sufficiently compensate for the human mitochondrial defects in a cell line carrying the ND4 mutation.

ACKNOWLEDGMENTS

This work was supported by National Institutes of Health Grants GM-11726 (to G.A.) and DK53244 (to T.Y.) and an Ellison Medical Foundation New Scholar in Aging Award AG-NS-0183-02 (to Y.B).

REFERENCES

1. CARROLL, J., R.J. SHANNON, I.M. FEARNLEY, et al. 2002. Definition of the nuclear encoded protein composition of bovine heart mitochondrial complex I. Identification of two new subunits. J. Biol. Chem. **277:** 50311–50317.
2. WALKER, J.E. 1992. The NADH:ubiquinone oxidoreductase (complex I) of respiratory chains. Q. Rev. Biophys. **25:** 253–324.
3. HOFHAUS, G., H. WEISS & K. LEONARD. 1991. Electron microscopic analysis of the peripheral and membrane parts of mitochondrial NADH dehydrogenase (complex I). J. Mol. Biol. **221:** 1027–1043.
4. GRIGORIEFF, N. 1998. Three-dimensional structure of bovine NADH:ubiquinone oxidoreductase (complex I) at 22 A in ice. J. Mol. Biol. **277:** 1033–1046.
5. CHOMYN, A. 1985. Six unidentified reading frames of human mitochondrial DNA encode components of the respiratory-chain NADH dehydrogenase. Nature **314:** 592–597.
6. CHOMYN, A., P. MARIOTTINI, M.W. CLEETER, et al. 1986. URF6, last unidentified reading frame of human mtDNA, codes for an NADH dehydrogenase subunit. Science **234:** 614–618.
7. GROSSMAN, L.I. & E.A. SHOUBRIDGE. 1996. Mitochondrial genetics and human disease. Bioessays **18:** 983–991.
8. MANDAVILLI, B.S., J.H. SANTOS & B. VAN HOUTEN. 2002. Mitochondrial DNA repair and aging. Mutat. Res. **509:** 127–151.
9. TRIEPELS, R.H., L.P. VAN DEN HEUVEL, J.M. TRIJBELS, et al. 2001. Respiratory chain complex I deficiency. Am. J. Med. Genet. **106:** 37–45.
10. LENAZ, G., C. BOVINA, M. D'AURELIO, et al. 2002. Role of mitochondria in oxidative stress and aging. Ann. N.Y. Acad. Sci. **959:** 199–213.
11. MAN, P.Y., D.M. TURNBULL & P.F. CHINNERY. 2002. Leber hereditary optic neuropathy. J. Med. Genet. **39:** 162–169.
12. DE VRIES, S. & L.A. GRIVELL. 1998. Purification and characterization of a rotenone-insensitive NADH:Q6 oxidoreductase from mitochondria of *Saccharomyces cerevisiae*. Eur. J. Biochem. **176:** 377–384.

13. DE VRIES, S., R. VAN WITZENBURG, L.A. GRIVELL, et al. 1992. Primary structure and import pathway of the rotenone-insensitive NADH-ubiquinone oxidoreductase of mitochondria from *Saccharomyces cerevisiae*. Eur. J. Biochem. **203:** 587–592.
14. SEO, B.B. 1998. Molecular remedy of complex I defects: rotenone-insensitive internal NADH-quinone oxidoreductase of *Saccharomyces cerevisiae* mitochondria restores the NADH oxidase activity of complex I-deficient mammalian cells. Proc. Natl. Acad. Sci. USA **95:** 9167–9171.
15. SEO, B.B. A. MATSUNO-YAGI & T. YAGI. 1999. Modulation of oxidative phosphorylation of human kidney 293 cells by transfection with the internal rotenone-insensitive NADH-quinone oxidoreductase (NDI1) gene of *Saccharomyces cerevisiae*. Biochim. Biophys. Acta **1412:** 56–65.
16. SEO, B.B., J. WANG, T.R. FLOTTE, et al. 2000. Use of the NADH-quinone oxidoreductase (NDI1) gene of *Saccharomyces cerevisiae* as a possible cure for complex I defects in human cells. J. Biol. Chem. **275:** 37774–37778.
17. BAI, Y. & G. ATTARDI. 1998. The mtDNA-encoded ND6 subunit of mitochondrial NADH dehydrogenase is essential for the assembly of the membrane arm and the respiratory function of the enzyme. EMBO J. **17:** 4848–4858.
18. BAI, Y., R.M. SHAKELEY & G. ATTARDI. 2000. Tight control of respiration by NADH dehydrogenase ND5 subunit gene expression in mouse mitochondria. Mol. Cell. Biol. **20:** 805–815.
19. BAI, Y., P HAJEK, A. CHOMYN, et al. 2001. Lack of complex I activity in human cells carrying a mutation in MtDNA-encoded ND4 subunit is corrected by the *Saccharomyces cerevisiae* NADH-quinone oxidoreductase (NDI1) gene. J. Biol. Chem. **276:** 38808–38813.
20. HOFHAUS, G. & G. ATTARDI. 1993. Lack of assembly of mitochondrial DNA-encoded subunits of respiratory NADH dehydrogenase and loss of enzyme activity in a human cell mutant lacking the mitochondrial ND4 gene product. EMBO J. **12:** 3043–3048.
21. HOFHAUS, G. & G. ATTARDI. 1995. Efficient selection and characterization of mutants of a human cell line which are defective in mitochondrial DNA-encoded subunits of respiratory NADH dehydrogenase. Mol. Cell Biol. **15:** 964-974.
22. KING, M.P. 1996. Use of ethidium bromide to manipulate ratio of mutated and wild-type mitochondrial DNA in cultured cells. Methods Enzymol. **264:** 339–334.
23. WISEMAN, A. & G. ATTARDI. 1978. Reversible tenfold reduction in mitochondria DNA content of human cells treated with ethidium bromide. Mol. Gen. Genet. **167:** 51–63.
24. KING, M.P. & G. ATTARDI. 1989. Human cells lacking mtDNA: repopulation with exogenous mitochondria by complementation. Science **246:** 500–503.
25. POLOSA, P.L. & G. ATTARDI. 1991. Distinctive pattern and translational control of mitochondrial protein synthesis in rat brain synaptic endings. J. Biol. Chem. **266:** 10011–10017.
26. NAKAMARU-OGISO, E. 2003. The ND5 subunit was labeled by a photoaffinity analogue of fenpyroximate in bovine mitochondrial complex I. Biochemistry **42:** 746–754.
27. HAYASHI, J., S. OHTA, A. KIKUCHI, et al. 1991. Introduction of disease-related mitochondrial DNA deletions into HeLa cells lacking mitochondrial DNA results in mitochondrial dysfunction. Proc. Natl. Acad. Sci. USA **88:** 10614–10618.
28. ROBINSON, B.H. 1992. Nonviability of cells with oxidative defects in galactose medium: a screening test for affected patient fibroblasts. Biochem. Med. Metab. Biol. **48:** 122–126.
29. SEO, B.B., E. NAKAMARU-OGISO, T.R. FLOTTE, et al. 2002. A single-subunit NADH-quinone oxidoreductase renders resistance to mammalian nerve cells against complex I inhibition. Mol. Ther. **6:** 336–341.

Genome-Wide Analysis of Signal Transducers and Regulators of Mitochondrial Dysfunction in *Saccharomyces cerevisiae*

KESHAV K. SINGH,[a] ANNE KARIN RASMUSSEN,[b] AND LENE JUEL RASMUSSEN[b]

[a]*Department of Cancer Genetics, Roswell Park Cancer Institute, Buffalo, New York 14263, USA*

[b]*Department of Life Sciences and Chemistry, Roskilde University, 4000 Roskilde, Denmark*

ABSTRACT: Mitochondrial dysfunction is a hallmark of cancer cells. However, genetic response to mitochondrial dysfunction during carcinogenesis is unknown. To elucidate genetic response to mitochondrial dysfunction we used *Saccharomyces cerevisiae* as a model system. We analyzed genome-wide expression of nuclear genes involved in signal transduction and transcriptional regulation in a wild-type yeast and a yeast strain lacking the mitochondrial genome (rho^0). Our analysis revealed that the gene encoding cAMP-dependent protein kinase subunit 3 (PKA3) was upregulated. However, the gene encoding cAMP-dependent protein kinase subunit 2 (PKA2) and the VTC1, PTK2, TFS1, CMK1, and CMK2 genes, involved in signal transduction, were downregulated. Among the known transcriptional factors, OPI1, MIG2, INO2, and ROX1 belonged to the upregulated genes, whereas MSN4, MBR1, ZMS1, ZAP1, TFC3, GAT1, ADR1, CAT8, and YAP4 including RFA1 were downregulated. RFA1 regulates DNA repair genes at the transcriptional level. RFA is also involved directly in DNA recombination, DNA replication, and DNA base excision repair. Downregulation of RFA1 in rho^0 cells is consistent with our finding that mitochondrial dysfunction leads to instability of the nuclear genome. Together, our data suggest that gene(s) involved in mitochondria-to-nucleus communication play a role in mutagenesis and may be implicated in carcinogenesis.

KEYWORDS: mitochondria; mutagenesis; repair; yeast; cancer

INTRODUCTION

Mitochondrial dysfunction is an important feature of cancer cells. As early as the last century, studies by Warburg and others[1–4] suggested that mitochondrial defect may result in dedifferentiation and neoplastic transformation that may cause cancer. More recent research extends and provides new insight into the role of mitochondria

Address for correspondence: Keshav K. Singh, Ph.D., Department of Cancer Genetics, Cell and Virus Building, Room 247, Roswell Park Cancer Institute, Elm and Carlton Streets, Buffalo, NY 14263. Voice: 716-845-8017; fax: 716-845-1047.
keshav.singh@roswellpark.org

Ann. N.Y. Acad. Sci. 1011: 284–298 (2004). © 2004 New York Academy of Sciences.
doi: 10.1196/annals.1293.027

in cancer.[5–8] Cancer cells have altered metabolism, including increased glycolysis, glucose transport, gluconeogenesis, lactic acid production, glutaminolytic activity, glycerol and fatty acid turnover, pentose phosphate, and decreased pyruvate oxidation and fatty acid oxidation.[9] Interestingly, there are differences between the mitochondria of normal and transformed cells, such as the preference for substrates oxidized, the magnitude of the acceptor control ratio, the rates of electron and anion transport, the capacity to accumulate and retain calcium, the amounts and forms of DNA, and the rate of protein synthesis and organelle turnover.[9]

Mitochondrial dysfunction is also involved in the pathogenesis of many other diseases. Mitochondrial dysfunction is found in diseases as diverse as infertility, diabetes, heart disease, blindness, deafness, kidney disease, liver disease, stroke, and migraine.[5,8,10] Mitochondrial dysfunction is also involved in aging and neurodegenerative diseases such as Parkinson's and Alzheimer's dementia. Mitochondrial diseases can affect any organ in the body at any age. Mitochondrial diseases are severely debilitating and characteristically complex in nature. They can be transmitted by mendelian or maternal inheritance, but they can also be sporadic or induced by the environment.

Mitochondria contribute to several aspects of the cell biology, including energy production, redox status, and programmed cell death. Mitochondria rely on the nucleus for their biogenesis, synthesis, and function. These organelles respond to intra- and extracellular signals independently, and a highly coordinated "cross-talk" exists between mitochondrial and nuclear signals that can greatly influence cell behavior.[11–15] Mutation in mitochondrial DNA (mtDNA) has been found in almost all cancer examined to date, suggesting that mutation in mtDNA may be involved in carcinogenesis.[8] Because human mtDNA lacks introns, it has been suggested that most mutations will occur in coding sequences, and subsequent accumulation of mutations may lead to tumor formation. The mitochondrial genome depends on the nuclear genome for transcription, translation, replication, and repair, but the precise mechanisms for how the two genomes interact, integrate, and affect each other's genetic stability are poorly understood.

MATERIAL AND METHODS

Media and Strains. Growth media were prepared as described in Ref. 16. The *Saccharomyces cerevisiae* strains used in this study are DL1 (D273-10B)[17] and DL1 rho^0 (this study). The DL1 rho^0 strain lacking mitochondrial DNA was generated by the treatment of DL1 with ethidium bromide.[16] Briefly, ethidium bromide was added to a concentration 10 mg/mL, and the cells were incubated at room temperature, with agitation, for approximately 24 h in YPD. Following a second and third treatment with 10 mg/mL ethidium bromide for 24 h, the cells were diluted (1:100). Following incubation, the cells were diluted in water and plated on YPD for single colonies. The rho^0 cells were selected as cells unable to form colonies on yeast extract-peptone-glycerol (YPG) plates. The loss of mtDNA was verified by DAPI (4, 6-diamidino-2-phenylindole) staining.

Gene Array Analysis. Total RNA was isolated from exponentially growing *S. cerevisiae* cells according to the manufacturer's guidelines using RNeasy (QIAGEN). Total RNA (5 µg) was converted into double-stranded cDNA by GIBCO

BRL's SuperScript Choice system for cDNA synthesis (Life Technologies) and a T7-$(dT)_{24}$ oligomer provided by Research Genetics (Huntsville, Alabama). Double-stranded cDNA was purified by phenol/chloroform extraction and ethanol precipitation. *In vitro* transcription was performed with T7 RNA polymerase following the instructions from the BioArray high-yield RNA transcript labeling kit from Enzo (distributed by Affymetrix). The biotin-labeled cRNA was purified on an affinity resin (RNeasy mini cleanup, QIAGEN), and the amount of labeled cRNA was determined by measuring absorbance at 260 nm and using the conversion that 1 OD at 260 nm corresponds to 40 μg/mL RNA. The yeast Genome S98 Array (Affymetrix) containing approximately 6400 open reading frames (ORFs) of *S. cerevisiae* genome was used for gene expression analysis. Hybridization, reading, and analysis were performed by Research Genetics. The classification of genes into functional groups was done as described in the Munich Information Center for Protein Sequences (MIPS) database (http://mips.gsf.de).

RESULTS

Response of Genes Involved in Signal Transduction from Mitochondria-to-Nucleus (TABLE 1)

cAMP-Dependent Protein Kinase (PKA). The yeast PKA has three catalytic subunits that are encoded by *TPK* genes (*TPK1*, *TPK2*, and *TPK3*).[18] Transcription of mitochondrial genes in *S. cerevisiae* is controlled in part by cAMP levels, and such transcriptional control is mediated by cAMP-dependent protein kinase activity (FIG. 1).[19] PKA controls Msn2 and Msn4 localization and is thereby negatively regulating STRE-dependent transcription, and it was suggested that stress and cAMP-regulated intracellular localization of Msn2 is a key step in STRE-dependent transcription and in the general stress response.[20] The *TPK2* gene negatively regulates genes involved in iron uptake and positively regulates genes involved in trehalose degradation and water homeostasis.[21] We find that *TPK2* is downregulated in rho^0 cells (TABLE 2), suggesting that dysfunctional mitochondria undergo increased iron uptake. This could result in intracellular accumulation of iron, and in the presence of iron the tryptophan metabolite quinolinate causes intense lipid peroxidation. Interestingly, we found that mtDNA depletion of HeLa cells caused increased lipid peroxidation and oxidative damage to DNA.[21a] In addition to DNA damage, iron also may play a crucial role in telomere repair, by activating telomerase. Therefore, by inhibiting apoptosis and enhancing chromosome repair, iron may confer immortality on the cancer cell. Iron is one of the triggers of mitosis, and increased iron levels may be essential for the rapid growth characteristic of many malignancies.[22] Along these lines, it has been shown that a human neuroblastoma rho^0 cell line, highly sensitive to oxidative stress, accumulates iron.[23]

TABLE 1A. Genes involved in signal transduction that are upregulated in rho^0 cells

ORF	Gene name	Description	Fold change
YKL166C	TPK3, PKA3	cAMP-dependent protein kinase 3	2.1

TABLE 1B. Genes involved in signal transduction that are downregulated in rho^0 cells

ORF	Gene name	Description	Fold change
YER072W	VTC1, NRF1	Homolog of *S. pombe* Nrf1, Negative regulator of Cdc42	4.3
YPL203W	TPK2, YKR1, PKA2, PKA3	cAMP-dependent protein kinase	2.2
YJR059W	PTK2, STK2	Serine/threonine protein kinase	2.6
YLR178C	TFS1, DKA1, NSP1	Cdc25-dependent nutrient and ammonia-response cell cycle regulator	3.1
YFL014W	HSP12, GLP1	Heat shock protein	9.4
YFR014C	CMK1	Ca^{2+}/calmodulin-dependent ser/thr protein kinase type I	3.3
YOL016C	CMK2	Ca^{2+}/calmodulin-dependent ser/thr protein kinase type II	2.3
YDR461W	MFA1	Mating pheromone a-factor	3.4

VTC1. The VTC1 protein is a negative regulator of Cdc42, which belongs to the Rho family of small GTPases and is essential for budding initiation in *S. cerevisiae*. Activation of Cdc42 by the guanine-nucleotide exchange factor Cdc24 triggers polarization of the actin cytoskeleton at bud emergence as well as in response to mating pheromones. Studies indicate that mitochondria enter the bud immediately after bud emergence, interact with the actin cytoskeleton for linear, polarized movement of mitochondria from mother to bud, but are equally distributed among mother and daughter cells. We find that *VTC1* is downregulated in rho^0 cells, suggesting that budding formation, polarization of cytoskeleton, and movement of mitochondria from mother to bud in rho^0 cells could be affected. It has been shown that ethanol stimulation can induce H_2O_2 production through the activation of Cdc42, which results in reorganizing actin filaments and increasing cell motility and *in vitro* angiogenesis.[24]

PTK2. PTK2 is a protein kinase involved in polyamine uptake, which regulates ion transport across the plasma membrane of *S. cerevisiae* and is implicated in activation of the yeast plasma membrane H(+)-ATPase (Pma1) in response to glucose metabolism. Natural polyamines are aliphatic cations with multiple functions in cell growth and differentiation. Alterations in the polyamine structure provide a strategy to synthesize analogues that can interfere with the cellular functions of natural polyamines. Analogues of spermine are particularly effective in modifying the synthesis, catabolism, and uptake of natural polyamines. The increased requirement of natural polyamines in cancer cell growth makes it possible to utilize the polyamine pathway as a therapeutic target in cancer cells. We find that *PTK2* is downregulated, suggesting that rho^0 cells may be impaired in polyamine uptake.

TFS1. TFS1 is a Cdc25-dependent nutrient- and ammonia-response cell cycle regulator. It is a suppressor of Cdc25, which encodes the putative GTP exchange factor for Ras1/Ras2 in yeast.[25] Cdc25 acts on Ras and thereby stimulates cAMP production in *S. cerevisiae*. The cAMP-protein kinase A (PKA) pathway in yeast

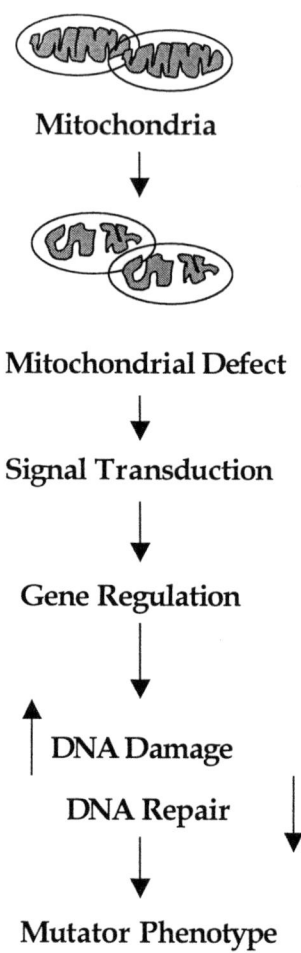

FIGURE 1. Mechanisms depicting how mitochondrial dysfunction signal is transduced, causing changes in gene regulation that may lead to mutator phenotype in the nucleus.

plays a major role in the control of metabolism, stress resistance, and proliferation, particularly in connection with the available nutrient conditions. Extensive information has been obtained on the core section of the pathway, that is, Cdc25, Ras, adenylate cyclase, and PKA and on components interacting directly with this core section, such as the Ira proteins, Cap/Srv2, and the two cAMP phosphodiesterases. We find that *TFS1* is downregulated in rho^0 cells, suggesting that Cdc25 is derepressed and cAMP production may be stimulated in cells with dysfunctional mitochondria.

HSP12. Changing the growth conditions of *S. cerevisiae* by adding glucose to cells growing on a nonfermentable carbon source leads to rapid repression of general stress-responsive genes such as *HSP12*, *SSA3*, and *HSP26*. The major transcription factors governing (stress-induced) transcriptional activation of *HSP12* are Msn2 and Msn4, binding to the general stress-responsive promoter elements (STRE). Glucose repression of *HSP12* is independent of Msn2/4. However, STRE-mediated transcription is the target of repression by low amounts of glucose, suggesting that other factors are involved in STRE-mediated transcriptional regulation of *HSP12*.[26] The *S. cerevisiae* protein kinase Rim15 was identified as a stimulator of meiotic gene expression. Loss of Rim15 causes an additional pleiotropic phenotype in cells grown to stationary phase on rich medium; this phenotype includes defects in trehalose and glycogen accumulation, in transcriptional derepression of *HSP12*, *HSP26*, and *SSA3*, in induction of thermotolerance and starvation resistance, and in proper G1 arrest. These phenotypes are commonly associated with hyperactivity of the Ras/cAMP pathway.[27] The *HSP12* gene encodes one of the two major small heat shock proteins (HSPs) of *S. cerevisiae*. Hsp12 accumulates massively in yeast cells exposed to heat shock, osmotic stress, oxidative stress, and high concentrations of alcohol as well as in early-stationary-phase cells. The *HSP12* promoter region contains five repeats of the STRE element, which are essential to confer wild-type induced levels on osmotic stress, heat shock, and entry into stationary phase. Disruption of the *HOG1* and *PBS2* genes leads to a dramatic decrease in *HSP12* expression in osmotic stressed cells, whereas overproduction of *HOG1* results in a fivefold increase in wild-type induced levels on a shift to a high salt concentration. By contrast, mutations resulting in high PKA activity reduce or abolish the accumulation of *HSP12* mRNA in stressed cells. On the contrary, mutants containing defective PKA catalytic subunits exhibit high basal levels of *HSP12* mRNA. Together, these results suggest that *HSP12* is a target of the HOG-response pathway under negative control of the Ras/cAMP pathway.[28] Furthermore, loss of Gis1 results in a defect in transcriptional derepression on nutrient limitation of various genes that are negatively regulated by the Ras/cAMP pathway (e.g., *SSA3*, *HSP12*, and *HSP26*).[29] Along these lines, we find that *HSP12* is downregulated, predicting that the activity of the Ras/cAMP pathway is likely to be increased in rho^0 cells.

Response of Genes Involved in Transcription due to Mitochondrial Dysfunction (TABLE 2)

BAS1. The gene expression of the Bas1 transcription factor is repressed by adenine and is required for optimal expression of enzymes involved in purine *de novo* synthesis.[30] It has been shown that efficient transcription of yeast AMP biosynthesis genes requires interaction between Bas1 and Bas2, which is promoted in the presence of 5′-phosphoribosyl-4-succinocarboxamide-5-aminoimidazole (SAICAR), a metabolic intermediate whose synthesis is controlled by feedback inhibition of Ade4 acting as the purine nucleotide sensor within the cell.[31] Our results show that *BAS1* expression is upregulated in rho^0 cells, suggesting that adenine levels may be low in these cells.

MSN4. Signal transduction pathways inactivated during periods of starvation are implicated in the regulation of longevity in organisms ranging from yeast to mammals and are mediated by the stress-resistance proteins Msn2/Msn4 and Rim15. De-

TABLE 2. Genes involved in transcription that are upregulated in rho^0 cells

ORF	Gene name	Description	Fold change
YLL011W	SOF1	Involved in 18S pre-rRNA production	2.9
YLL008W	DRS1	RNA helicase of the DEAD box family	3.1
YKR099W	BAS1	Transcription factor	2.4
YKR024C	DBP7	RNA helicase required for 60S ribosomal subunit assembly	2.4
YKL078W	DHR2, JA2	RNA helicase, involved in ribosomal RNA maturation	2.1
YJL033W	HCA4, DBP4	Can suppress the U14 snoRNA rRNA processing function	2.3
YHR169W	DBP8	Strong similarity to Drs1p and other probable ATP-dependent RNA helicases	2.3
YHL020C	OPI1	Negative regulator of phospholipids biosynthesis pathway	2.0
YGL120C	PRP43, JA1	Involved in spliceosome disassembly	2.4
YGL209W	MIG2, MLZ1	C2H2 zinc-finger protein	4.1
YER127W	LCP5	Involved in 18S rRNA maturation	2.2
SNR13	SNR13	Small nucleolar RNA; member of the C/D box family	2.0
YDR123C	INO2, SCS1, DIE1	Basic helix-loop-helix (BHLH) transcription factor	5.6
YDR021W	FAL1	Involved in maturation of 18S rRNA	3.7
YCL054W	SPB1	Required for ribosome synthesis, putative methylase	2.0
YBR238C		Strong similarity to Spt16	2.8
YBR083W	TEC1, ROC1	Ty transcription factor	2.0
YBL054W		Weak similarity to B-myb	3.8
YPR065W	ROX1, REO1	Heme-dependent transcriptional repressor of hypoxic genes	6.1
YOL010W	RCL1 S	Similarity to human RNA 3'-terminal phosphate cyclase	3.6
YNL075W	IMP4	Component of the U3 small nucleolar ribonucleoprotein	2.7
YNL175C	NOP13	Similarity to *S. pombe* Rnp24, Nsr1, and human splicing factor	2.3
YMR239C	RNT1	Double-stranded ribonuclease	3.1
YMR139W	RIM11, GSK3, MDS1	Ser/thr protein kinase	2.1
YML043C	RRN11	RNA polymerase I specific transcription factor	3.8

TABLE 2. Genes involved in transcription that are upregulated in rho^0 cells (*continued*)

ORF	Gene name	Description	Fold change
YKL062W	MSN4	Transcriptional activator	2.8
YKL093W	MBR1	Required for optimal growth on glycerol	2.5
YJR127C	ZMS1	Transcription factor with similarity to regulatory protein Ard1p	2.4
YJL056C	ZAP1	Regulatory protein involved in zinc responsive transcriptional regulation	4.1
YJL103C		Putative regulatory protein	3.4
YAL001C	TFC3, TSV115, FUN24	TFIIIC (subunit, 138 kDa)	3.4
YGR248W	SOL4	Strong similarity to Sol3	5.4
YFL021W	GAT1, NIL1	Transcription factor for nitrogen regulation	2.1
YFL052W		Strong similarity to Mal63, YPR196w and Mal13	2.5
YER028C		Similarity to Mig1	10.4
YEL066W	HPA3	Histone and other protein acetyltransferase	3.0
YDR216W	ADR1	Zinc-finger transcription factor	5.8
YPL031C	PHO85	Cyclin-dependent protein kinase	2.1
YPL230W	USV1	Similarity to Rgm1	4.2
YOR230W	WTM1	Transcriptional modulator	2.1
YOR185C	GSP2, CNR2	GTP-binding protein of the RAS superfamily	2.0
YMR280C	CAT8, DIL1, MSP8	Transcription factor involved in gluconeogenesis	3.0
YML027W	YOX1	Homeodomain protein	2.0
YOR028C	CIN5, SDS15, YAP4, HAL6	Transcriptional activator	2.7
YAR007C	RFA1, BUF2, FUN3, RPA1	DNA replication factor A, 69 kDa subunit	2.1

letion of the *RAS2* gene doubles the mean life span by a mechanism that requires Msn2/4 and Sod2. These findings link mutations that extend chronologic life span in *S. cerevisiae* to superoxide dismutases and suggest that the induction of other stress-resistance genes regulated by Msn2/4 and Rim15 is required for maximum longevity extension.[32] The global regulator of transcription Ccr4/Not complex is believed to contribute to transcriptional regulation by the Ras/cAMP pathway. Msn2/4-dependent transcription, which is known to be under negative control of PKA, is derepressed in all ccr4/Not mutant strains. Mutations in various *NOT* genes result in a synthetic temperature-sensitive growth defect when combined with mutations that compromise cells for PKA activity. The Not3 and Not5 proteins, which are modified and subsequently degraded by stress signals that also lead to increased Msn2/4-dependent activity, show a specific two-hybrid interaction with Tpk2.[33] The Ras/cAMP pathway in *S. cerevisiae* plays a major role in the control of metabolism, stress resistance, and proliferation, particularly in connection with the available nutrient conditions. Extensive information has been obtained on the core section of the pathway, that is, Cdc25, Ras, adenylate cyclase, and PKA, and on components interacting directly with this core section, such as the Ira proteins, Cap/Srv2, and the two cAMP phosphodiesterases. Recent work has now begun to reveal upstream regulatory components and downstream targets of the pathway. A G-protein-coupled receptor system (Gpr1-Gpa2) acts upstream of adenylate cyclase and is required for glucose activation of cAMP synthesis in concert with a glucose phosphorylation-dependent mechanism. Although a genuine signaling role for the Ras proteins remains uncertain, they appear to mediate at least part of the potent stimulation of cAMP synthesis by intracellular acidification. Recently, several new targets of the PKA pathway have been discovered. These include the Msn2 and Msn4 transcription factors mediating part of the induction of STRE-controlled genes by a variety of stress conditions, the Rim15 protein kinase involved in stationary phase induction of a similar set of genes, and the Pde1 low-affinity cAMP phosphodiesterase, which specifically controls agonist-induced cAMP signaling. A major issue that remains to be resolved is the precise connection between the Ras/cAMP pathway and other nutrient-regulated components involved in the control of growth and of phenotypic characteristics correlated with growth, such as the Sch9 and Yak1 protein kinases. Cln3 appears to play a crucial role in the connection between the availability of certain nutrients and Cdc28 kinase activity, but it remains to be clarified which nutrient-controlled pathways control Cln3 levels.[34] We find that Msn4 is downregulated in rho^0 cells, suggesting hyperactivity of the Ras/cAMP pathway in cells with dysfunctional mitochondria. The Msn2/4 transcription factors and the Ras/cAMP pathway control the yeast H_2O_2 response, and strains deleted for MSN2 and MSN4 are hypersensitive to H_2O_2, although they can still adapt to this oxidant. Furthermore, they are unable to induce many proteins of the H_2O_2 stimulon. Strains lacking Pde2, and therefore carrying high intracellular cAMP levels, are hypersensitive to H_2O_2. In the presence of exogenous cAMP, this strain does not induce the entire H_2O_2 Msn2/4 regulon, suggesting that when intracellular cAMP levels are high, the Ras/cAMP pathway negatively affects the H_2O_2 stress response through Msn2/4.[35]

MBR1. The Mbr1 protein is a limiting factor for growth on glycerol medium in late exponential growth phase at low temperature and at high external pH. Deletion of *MBR1* protects cells against stress, while overexpression of this gene has the opposite effect. *MBR1* expression is induced in the late growth phase and is negatively controlled by the PKA. However, Mbr1 is not an essential element of any one of these pathways.[36] Using a genetic screen to select for nuclear suppressors of mitochondrial mutations, which affect the mitochondrial protein synthesis, Valens *et al.*[37] identified Mbr1 as one protein involved in controlling the assembly or the regulation of other genes involved in mitochondrial protein synthesis. We find that *MBR1* is downregulated in rho^0 cells. It is likely due to the increased activity of the Ras/cAMP pathway.

INO2 and OPI1. In *S. cerevisiae*, the phospholipid biosynthetic genes are highly regulated at the transcriptional level in response to the phospholipid precursors inositol and choline. In the absence of inositol and choline, the expression of phospholipid biosynthetic genes is derepressed. The transcription factor complex INO2/INO4, which binds to an upstream activation sequence (UAS) in the promoters, mediates this activation of phospholipid biosynthetic genes. In the presence of inositol and choline, the product of the *OPI1* gene represses transcription controlled by the UAS element (Ashburner et al.[38]). We find that both *OPI1* and *INO2* are upregulated in rho^0 cells, indicating that phospholipid biosynthesis is likely to be affected.

TEC1. The transcription factor Tec1 is involved in pseudohyphal differentiation and agar-invasive growth of *S. cerevisiae* cells. The two highly conserved *RAS* genes, *RAS1* and *RAS2*, of budding yeast are redundant for viability. *RAS2* activates invasive growth using either of two downstream signaling pathways, the filamentation MAPK (Cdc42/Ste20/MAPK) cascade or the PKA-dependent (Cyr1/cAMP/PKA) pathway. Both these *Ras2*-controlled signaling pathways stimulate expression of the filamentation response and depends on the transcription factors Ste12 and Tec1, indicating a cross-talk between the MAPK and the cAMP signaling pathways (Mosch et al.[39]). We find that TEC1 is upregulated in rho^0 cells.

ROX1. The *ROX1* gene encodes a repressor of genes involved in hypoxic functions.[40] Genome-wide gene expression analysis has revealed that the two transcription factors, Upc2 and Rox1, regulate the majority of anaerobically induced genes in *S. cerevisiae*.[41] We find that *ROX1* is upregulated in rho^0 cells, suggesting that the intracellular oxygen level is altered in cells with mitochondrial dysfunction.

PHO85. The Pho85 cyclin-dependent kinase is involved in several signal transduction pathways in *S. cerevisiae*. The cellular responses mediated by Pho85 include cell cycle progression and metabolism of nutrients, and it has been suggested that the kinase activity of Pho85 signals the cell that the current environment is satisfactory.[42] We find that the *PHO85* gene is downregulated in rho^0 cells, suggesting that the intracellular environment is different from that of wild-type cells.

WTM1. The Wtm1 transcriptional repressor (WD repeat-containing transcriptional modulator) was identified as a protein present in a large nuclear complex.[43] Strains deficient in *WTM* expression show increased repression at the silent mating-type locus, HMR, and at telomeres. We find that the *WTM1* gene is downregulated in rho^0 cells and is likely to affect telomeres.

GSP2. The small GTPase Ran encoded by *GSP2* protein contains the ras consensus domains involved in GTP binding and metabolism. Both *RAS2* and *GSP2* expression exhibits carbon source dependency and is affected by the activity of the Ras/cAMP pathway. It has been suggested that Gsp2 plays a role in regulating the activities of the PRP20 complex.[44] The PRP20 protein is required for mRNA export and maintenance of nuclear structure. PRP20 acts as guanine nucleotide exchange factors for the nuclear Ras-like Ran/GSP1 proteins. We find that GSP2 is downregulated in rho^0 cells, confirming that the Ras/cAMP pathway is involved in cells with mitochondrial dysfunction.

CAT8. The transition between the fermentative and the oxidative metabolism (diauxic shift) is associated with major changes in gene expression and protein synthesis. The zinc cluster protein Cat8p affects the expression of many genes during diauxic shift. Many of these gene products are important for the first steps of ethanol utilization, the glyoxylate cycle, and gluconeogenesis. No function involved in the tricarboxylic cycle and the oxidative phosphorylation seems to be controlled by Cat8p.[45] We find that CAT8 is downregulated in rho^0 cells, indicating the difference in metabolic state of rho^0 cells and wild-type cells.

RIM11. The *RIM11* gene is one of four genes encoding glycogen synthase kinase 3 (GSK-3). Recent results suggest that GSK-3 promotes formation of a complex between Msn2 and DNA, which is required for the proper response to different forms of stress.[46] We find that *RIM11* is upregulated in rho^0 cells, suggesting that mitochondrial dysfunction leads to intracellular stress.

CIN5. Cin5 belongs to the family of bZIP proteins. Mutations in this gene affect chromosome stability.[47] It has been shown that overexpression of *CIN5* increases resistance to cisplatin as well as to DNA-alkylating agents. Based on these findings, it has been suggested that Cin5 is involved in drug resistance and might protect yeast against the toxicity of cisplatin and other alkylating agents by regulating the expression of certain genes that confer resistance to DNA-alkylating agents.[48] We found that *CIN5* is downregulated in rho^0 cells.

RFA1. The Rfa protein complex binds to upstream regulatory sequences (URS1), elements that are situated upstream of many genes and mediate negative control of their transcription. Genes regulated through the URS1 site participate in carbon, nitrogen, and inositol metabolism, electron transport, meiosis, sporulation, DNA repair, and mating-type switching.[49] The Rfa protein complex also binds to CAR1 sequences supporting transcriptional activation through the UAS sequences. Furthermore, the Rfa product participates in DNA replication as single-stranded DNA binding proteins as well as in DNA repair, suggesting that this protein complex serves multiple roles in transcription and replication.[50] We find that RFA1 is downregulated in rho^0 cells, which could affect the expression of many genes involved in DNA repair.

Mitochondria-to-Nucleolus Communication in S. cerevisiae *(TABLE 2)*

The nucleolus is the site of ribosome biogenesis and protein synthesis machinery. TABLE 2 shows that 13 of 25 genes that are upregulated in rho^0 cells are involved in ribosome biogenesis and translational control. These genes include SOF1, DRS1, DBP7 and DBP8, DHR2, HCA4, PRP43, LCP5, SNR13, FAL1, SPB1, IMP4, and RRN11. These data suggest that these genes are involved in mitochondria-to-nucleolus communication, and their upregulation is likely to be a feature of human cells lacking proper mitochondrial function.

DISCUSSION

This study reports the identification of genes involved in signal trasnduction and transcription that respond to mitochondrial dysfunction in *S. cerevisiae*. Interestingly, among the genes that are involved in signal transduction, PKA2 was the only gene whose expression was upregulated. Pka2p is a component of the protein kinase PKA in yeast. The PKA in human cells is regulated by the Ras/cAMP pathway and is involved in chromosome instability, neoplastic transformation, apoptosis, as well as angiogenesis.[51,52] Mutations in components of the Ras/cAMP pathway in yeast (*RAS1*, *RAS2*, and *CYR1*) that decrease the activity of PKA result in increased life span, suggesting a role for this protein kinase in aging.[53,54] Decreased PKA activity in yeast results in increased stress resistance through activation of the Msn2/4 stress response pathway, thus establishing a link between stress resistance mediated by the

Ras/cAMP pathway through the Msn2/4 proteins and aging. We find that genes negatively regulated by the Ras/cAMP pathway are differentially expressed in rho^0 cells compared to wild-type yeast cells. Our gene expression analysis suggests that Ras/cAMP pathway in rho^0 cells is activated, which leads us to speculate that this pathway may contribute to neoplastic transformation of cells with mitochondrial dysfunction.

Several tumor suppressors and protooncogenes affect the formation of the mature ribosome, and it has been proposed that this might regulate malignant progression by altering the protein synthesis machinery.[55] For instance, the majority of genes upregulated in N-myc transfected neuroblastomas are ribosomal proteins and genes controlling rRNA maturation.[56] It has been shown that constitutively active PKA leads to doubling of the amount of several ribosomal protein mRNAs.[56] Our study suggests that 15 genes involved in ribosome biogenesis in the nucleolus were upregulated. It is important to note that components of the translation machinery are deregulated in cancer cells and, more importantly, that several protooncogenes and tumor suppressors (such as p53, Rb, NPM, and PTEN) directly regulate ribosome biogenesis.[51] It has also been shown that rRNA synthesis is upregulated in cancer cells.[55]

Recent experiments in our laboratories show that rho^0 cells are hypersensitive to H_2O_2.[57] Further experimental evidence revealed that rho^0 cells either treated or not treated with H_2O_2 are hypermutable. We also find that rho^0 cells have impaired repair of H_2O_2-induced DNA damage.[21a,57] We propose that these phenotypes are a consequence of downregulation of Rfa1 protein in rho^0 cells. Rfa1 is a subunit of the multiprotein complex RFA. RFA is multifunction and is involved in transcription (of DNA repair and other genes, Ref. 49), DNA recombination, DNA replication, and excision repair. It is tempting to propose that downregulation of Rfa1 protein in rho^0 cells may lead to genetic instability of the nuclear genome. Based on these observations, we propose a model as described in FIGURE 1. FIGURE 1 depicts that mitochondrial dysfunction in cells causes changes in expression of genes involved in signal transduction and gene regulation. These alterations may lead to increased damage to nuclear DNA (Delsite *et al.*, manuscript submitted) and/or reduced DNA repair (via downregulation of Rfa1 protein, as just described) that may consequently lead to a mutator phenotype in the nuclear genome. It is likely that an analogous mechanism exists in human cells that may contribute to carcinogenesis.

ACKNOWLEDGMENTS

This research was supported by grants from the National Institutes of Health (RO1-097714) and an American Heart Association Scientist Development Award (9939223N; to K.K.S.).

REFERENCES

1. WARBURG, O. 1930. Metabolism of Tumors. Arnold Constable. London, UK.
2. WARBURG, O. 1956. On the origin of cancer cells. Science **123**: 309–314.
3. SZENT-GYORGYI, A. 1977. Electron biology and cancer. *In* Search and Discovery: A Tribute to Albert Szent-Gyorgyi. B. Kaminer, Ed.: 329–335. Academic Press. New York.

4. WOODS, M.W. & H.G. DUBUY. 1945. Cytoplasmic diseases and cancer. Science **102**: 591–593.
5. SINGH, K.K. 1998. Mitochondrial DNA Mutations in Aging, Disease, and Cancer. Springer. New York.
6. POLYAK, K., Y. LI, H. ZHU, et al. 1998. Somatic mutations of the mitochondrial genome in human colorectal tumours. Nat. Genet. **20**: 291–293.
7. KROEMER, G. & J.C. REED. 2000. Mitochondrial control of cell death. Nat. Med. **6**: 513–519.
8. SINGH, K.K. 2000. Mitochondrial me and the mitochondrial journal. Mitochondrion **1**: 1–2.
9. MODICA-NAPOLITANO, J.S. & K.K. SINGH. 2002. Mitochondrial DNA mutation in cancers. Expert reviews in molecular medicine: http://www.expertreviews.org/. Accession information: (02)00449–Oh.htm (shortcode: tab002ksb).
10. WALLACE, D.C. 1999. Mitochondrial diseases in man and mouse. Science **283**: 1482–1488.
11. TRAVEN, A., J.M. WONG, D. XU, et al. 2001. Interorganellar communication. Altered nuclear gene expression profiles in a yeast mitochondrial dna mutant. J. Biol. Chem. **276**: 4020–4027.
12. EPSTEIN, C.B., J.A. WADDLE, W. HALE IV, et al. 2001. Genome-wide responses to mitochondrial dysfunction. Mol. Biol. Cell **12**: 297–308.
13. DELSITE, R., S. KACHHAP, R. ANBAZHAGAN, et al. 2002. Nuclear genes involved in mitochondria-to-nucleus communication in breast cancer cells. Mol. Cancer **1**: 6.
14. DEVAUX, F., E. CARVAJAL, S. MOYE-ROWLEY, et al. 2002. Genome-wide studies on the nuclear PDR3-controlled respopnse to mitochondrial dysfunction in yeast. FEBS Lett. **515**: 25–28.
15. FLISS, M.S., H. USADEL, O.L. CABALLERO, et al. 2000. Facile detection of mitochondrial DNA mutations in tumors and bodily fluids. Science **287**: 2017–2019.
16. SHERMAN, F., G.R. FINK & J.B. HICKS. 1994. Methods in Yeast Genetics: A Laboratory Manual. Cold Spring Harbor Laboratory Press. Plainview, NY.
17. PAUL, M.F., J. VELOURS, G. ARSELIN DE CHATEAUBODEAU, et al. 1989. The role of subunit 4, a nuclear-encoded protein of the FO sector of yeast miltochondrial ATP synthase, in the assembly of the whole complex. Eur. J. Biochem. **185**: 163–171.
18. TODA, T., S. CAMERON, P. SASS, et al. 1987. Three different genes in *S. cerevisiae* encode the catalytic subunits of the camp-dependent protein kinase. Cell **50**: 277–287.
19. MCENTEE, C.M., R. CANTWELL, M.U. RAHMAN, et al. 1993. Transcription of the yeast mitochondrial genome requires cyclic AMP. Mol. Gen. Genet. **241**: 213–224.
20. GORNER, W., E. DURCHSCHLAG, M.T. MARTINEZ-PASTOR, et al. 1998. Nuclear localization of the C2H2 zinc finger protein Msn2p is regulated by stress and protein kinase A activity. Genes Dev. **12**: 586–597.
21. ROBINSON, L.C. & K. TATCHELL. 1991. TFS1: a suppressor of cdc25 mutations in *Saccharomyces cerevisiae*. Mol. Gen. Genet. **230**: 241–250.
21a. DELSITE, R.O., L.J. RASMUSSEN, A.K. RASMUSSEN, et al. 2003. Mitochondrial impairment is accompanied by oxidative DNA repair in the nucleus. Mutagenesis **18**: 497–503.
22. JOHNSON, S. 2001. The possible crucial role of iron accumulation combined with low tryptophan, zinc and manganese in carcinogenesis. Med. Hypotheses **57**: 539–543.
23. FUKUYAMA, R., A. NAKAYAMA, T. NAKASE, et al. 2002. A newly established neuronal rho-O cell line highly susceptible to oxidative stress accumulates iron and other metals. Relevance to the origin of metal ion deposits in brains with neurodegenerative disorders. J. Biol. Chem. **277**: 41455–41462.
24. QIAN, Y., J. LUO, S.S. LEONARD, et al. 2003. Hydrogen peroxide formation and actin filament reorganization by CDC42 is essential for ethanol-induced *in vitro* angiogenesis. J. Biol. Chem. **278**: 4542–4551.
25. ROBINSON, L.C. & K. TATCHELL. 1991. TFS1: a suppressor of cdc25 mutations in *Saccharomyces cerevisiae*. Mol. Gen. Genet. **230**: 241–250.
26. DE GROOT, E., J.P. BEBELMAN, W.H. MAGER, et al. 2000. Very low amounts of glucose cause repression of the stress-responsive gene HSP12 in *Saccharomyces cerevisiae*. Microbiology **146**: 367–375.

27. REINDERS, A., N. BURCKERT, T. BOLLER, et al. 1998. Saccharomyces cerevisiae campdependent protein kinase controls entry into stationary phase through the Rim15p protein kinase. Genes Dev. **12:** 2943–2955.
28. VARELA, J.C., U.M. PRAEKELET, P.A. MEACOCK, et al. 1995. The Saccharomyces cerevisiae HSP12 gene is activated by the high-osmolarity glycerol pathway and negatively regulated by protein kinase A. Mol. Cell Biol. **15:** 6232–6245.
29. PEDRUZZI, I., N. BURCKERT, P. EGGER, et al. 2000. Saccharomyces cerevisiae Ras/camp pathway controls post-diauxic shift element-dependent transcription through the zinc finger protein Gis1. EMBO J. **19:** 2569–2579.
30. DENIS, V., H. BOUCHERIE, C. MONRIBOT, et al. 1998. Role of the myb-like protein bas1p in Saccharomyces cerevisiae: a proteome analysis. Mol. Microbiol. **30:** 557–566.
31. REBORA, K., C. DESMOUCELLES, F. BORNE, et al. 2001. Yeast AMP pathway genes respond to adenine through regulated synthesis of a metabolic intermediate. Mol. Cell Biol. **21:** 7901–7912.
32. FABRIZIO, P., L.L. LIOU, V.N. MOY, et al. 2003. Functions downstream of Sch9 to extend longevity in yeast genetics. Genetics **163:** 35–46.
33. LENSSEN, E., U. OBERHOLZER, J. LABARRE, et al. 2002. Saccharomyces cerevisiae Ccr4-not complex contributes to the control of Msn2p-dependent transcription by the Ras/camp pathway. Mol. Microbiol. **43:** 1023–1037.
34. THEVELEIN, J.M. & J.H. DE WINDE. 1999. Novel sensing mechanisms and targets for the camp-protein kinase A pathway in the yeast Saccharomyces cerevisiae. Mol. Microbiol. **33:** 904–918.
35. HASAN, R., C. LEROY, A.D. ISNARD, et al. 2002. The control of the yeast H2O2 response by the Msn2/4 transcription factors. Mol. Microbiol. **45:** 233–241.
36. REISDORF, P., E. BOY-MARCOTTE & M. BOLOTIN-FUKUHARA. 1997. The MBR1 gene from Saccharomyces cerevisiae is activated by and required for growth under suboptimal conditions. Mol. Gen. Genet. **255:** 400–409.
37. VALENS, M., T. RINALDI, B. DAIGNAN-FORNIER, et al. 1991. Identification of nuclear genes which participate to mitochondrial translation in Saccharomyces cerevisiae. Biochimie **73:** 1525–1532.
38. ASHBURNER, B.P. & J.M. LOPES. 1995. Autoregulated expression of the yeast INO2 and INO4 helix-loop-helix activator genes effects cooperative regulation on their target genes. Mol. Cell Biol. **15:** 1709–1715.
39. MOSCH, H.U., E. KUBLER, S. KRAPPMANN, et al. 1999. Crosstalk between the Ras2p-controlled mitogen-activated protein kinase and camp pathways during invasive growth of Saccharomyces cerevisiae. Mol. Biol. Cell **10:** 1325–1335.
40. LOWRY, C.V. & R.S. ZITOMER. 1984. Oxygen regulation of anaerobic and aerobic genes mediated by a common factor in yeast. Proc. Natl. Acad. Sci. USA **81:** 6129–6133.
41. KWAST, K.E., L.C. LAI, N. MENDA, et al. 2002. Genomic analyses of anaerobically induced genes in Saccharomyces cerevisiae: functional roles of Rox1 and other factors in mediating the anoxic response. J. Bacteriol. **184:** 250–265.
42. CARROLL, A.S. & E.K. O'SHEA. 2002. Pho85 and signaling environmental conditions. Trends Biochem. Sci. **27:** 87–93.
43. PEMBERTON, L.F. & G. BLOBEL. 1997. Characterization of the Wtm proteins: a novel family of Saccharomyces cerevisiae transcriptional modulators with roles in meiotic regulation and silencing. Mol. Cell Biol. **17:** 4830–4841.
44. BELHUMEUR, P., A. LEE, R. TAM, et al. 1993. GSP1 and GSP2, genetic suppressors of the prp20-1 mutant in Saccharomyces cerevisiae: GTP-binding proteins involved in the maintenance of nuclear organization. Mol. Cell Biol. **13:** 2152–2161.
45. HAURIE, V., M. PERROT, T. MINI, et al. 2001. The transcriptional activator Cat8p provides a major contribution to the reprogramming of carbon metabolism during the diauxic shift in Saccharomyces cerevisiae. J. Biol. Chem. **275:** 76–85.
46. HIRATA, Y., T. ANDOH, T. ASAHARA, et al. 2003. Yeast glycogen synthase kinase-3 activates Msn2p-dependent transcription of stress responsive genes. Mol. Biol. Cell **14:** 302–312.
47. FERNANDES, L., C. RODRIGUES-POUSADA & K.L. STRUHL. 1997. Yap, a novel family of eight bZIP proteins in Saccharomyces cerevisiae with distinct biological functions. Mol. Cell Biol. **17:** 6982–6993.

48. FURUCHI, T., H. ISHIKAWA, N. MIURA, *et al.* 2001. Two nuclear proteins, Cin5 and Ydr259c, confer resistance to cisplatin in *Saccharomyces cerevisiae*. Mol. Pharmacol. **59:** 470–474.
49. SINGH, K.K. & L. SAMSON. 1995. Replication protein A binds to regulatory elements in yeast DNA repair and DNA metabolism genes. Proc. Natl. Acad. Sci. USA **92:** 4907–4911.
50. LUCHE, R.M., W.C. SMART, T. MARION, *et al.* 1993. *Saccharomyces cerevisiae* BUF protein binds to sequences participating in DNA replication in addition to those mediating transcriptional repression (URS1) and activation. Mol. Cell Biol. **13:** 5749–5761.
51. MATYAKHINA, L., S.M. LENHERR & C.A. STRATAKIS. 2002. Protein kinase A and chromosomal stability. Ann. N.Y. Acad. Sci. **968:** 148–157.
52. TORTORA, G. & F. CIARDIELLO. 2002. Protein kinase A as target for novel integrated strategies of cancer therapy. Ann. N.Y. Acad. Sci. **968:** 139–147.
53. SUN, J., S.P. KALE, A.M. CHILDRESS, *et al.* 1994. Divergent roles of RAS1 and RAS2 in yeast longevity. J. Biol. Chem. **269:** 18638–18645.
54. RUGGERO, D. & P.P. PANDOLFI. 2003. Does the ribosome translate cancer? Nat. Rev. Cancer **3:** 179–192.
55. BOON, K., H.N. CARON, R. VAN ASPEREN, *et al.* 2001. N-myc enhances the expression of a large set of genes functioning in ribosome biopgenesis and protein synthesis. EMBO J. **20:** 1383–1393.
56. KLEIN, C. & K. STRUHL. 1994. Protein kinase A mediates growth-regulated expression of yeast ribosomal protein genes by modulating RAP1 transcriptional activity. Mol. Cell Biol. **14:** 1920–1928.
57. RASMUSSEN, A.K., A. CHATTERJEE, L.J. RASMUSSEN, *et al.* 2003. Mitochondria-mediated nuclear mutator phenotype in *Saccharomyces cerevisiae*. Nucleic Acids Res. **31:** 3909–3917.

Enhanced Detection of Deleterious Mutations by TTGE Analysis of Mother and Child's DNA Side by Side

HAEYOUNG KWON, DUAN JUN TAN, REN-KUI BAI, AND LEE-JUN C. WONG

Institute for Molecular and Human Genetics, Georgetown University Medical Center, Washington, DC 20007, USA

ABSTRACT: Mitochondrial DNA (mtDNA) disorders represent a group of heterogeneous diseases that are caused by mutations in mtDNA. We examined 45 pairs of mother and the affected child, by screening the entire mitochondrial genome with temporal temperature gradient gel electrophoresis (TTGE), using 32 pairs of overlapping primers. TTGE is an effective method of mutation detection. It detects and distinguishes heteroplasmic mutations from homoplasmic mutations. By running the mother and child's DNA samples side by side and sequencing only the DNA fragments showing different TTGE patterns, excessive sequencing can be avoided, particularly because most sequence variations represent benign polymorphisms. Mutations identified by sequencing were further confirmed by PCR/ASO (allele-specific oligonucleotide) dot blot analysis or PCR/RFLP (restriction fragment length polymorphism). A total of seven differences in sequence between mother and child pairs were identified: A189G, T5580C, G5821A, C5840T, A8326G, G12207A, and G15995A. All but two mutations were novel. The most significant are the A8326G, G12207A, and G15595A mutations. The A8326G is located at the anticodon region of tRNA^{Lys}, right next to the first nucleotide of the triplet codon, and it is highly conserved throughout evolution. The G12207A mutation is located at the first base of tRNA^{ser} (AGY). The G15995A mutation occurs at a stem region that results in the disruption of the first base pair at the anticodon loop of tRNA^{Pro} and is highly conserved throughout evolution from sea urchins to mammals. Running TTGE side by side with DNAs from mother and the affected child is a novel method to detect deleterious mutations.

KEYWORDS: mitochondrial DNA mutation; temporal temperature gradient gel electrophoresis (TTGE) mutation detection

INTRODUCTION

The human mitochondrial genome is a circular, double-stranded DNA with 16,569 base pairs. It encodes 13 of more than 80 polypeptide subunits of the respiratory chain complexes, 2 ribosomal RNAs, and 22 transfer RNAs.[1] All other enzymes and factors that are essential for mitochondrial DNA transcription, trans-

---

Address for correspondence: Lee-Jun C. Wong, Ph.D., Institute for Molecular and Human Genetics, Georgetown University Medical Center, M4000, 3800 Reservoir Road, NW, Washington, DC 20007. Voice: 202-444-0760; fax: 202-444-1770.
wonglj@georgetown.edu

Ann. N.Y. Acad. Sci. 1011: 299–303 (2004). © 2004 New York Academy of Sciences.
doi: 10.1196/annals.1293.028

lation, and replication are encoded by the nuclear genome, translated in the cytoplasm and then imported into the mitochondrion.[2] Hence, a mitochondrial disorder can be caused by mutations in either the nuclear or the mitochondrial genome, or both. Mitochondrial DNA disorders are also characterized by maternal inheritance, heteroplasmy, and the threshold effect. Thus, the clinical manifestations of a disease depend on the type of mutation, the proportion of mutant mtDNA, and the tissues involved. Diagnosis may be achieved by mutational analysis of mtDNA.

In this study, we screened for mutations in mtDNA samples from 45 mother–child pairs using an effective screening method, temporal temperature gradient gel electrophoresis (TTGE).[3] DNA samples from the asymptomatic mother and affected child were analyzed side by side. The fragments that showed differences in TTGE banding patterns between mother and child's samples were sequenced. Here the results of such mutational analysis are reported.

MATERIAL AND METHODS

Samples used in this study were from patients referred to the Molecular Diagnostics Laboratory at the Institute for Molecular and Human Genetics at Georgetown University Medical Center for molecular diagnosis of mitochondrial disorders. Total DNA was extracted from the patients' peripheral blood lymphocytes and muscles specimens according to published procedures.[4] The polymerase chain reaction (PCR) conditions, the primers used in PCR amplification, the size of the fragment, the temperature ranges, and the ramp rates for TTGE analysis of the whole mitochondrial genome have been described.[5] DNA fragments showing differences in banding patterns between mother and child's DNA on TTGE analysis were followed by DNA sequencing using a Big Dye terminator cycle sequencing kit (Perkin-Elmer, Applied Biosystems), according to the manufacturer's protocols. To confirm the heteroplasmic state of the mutation, allele-specific oligonucleotide (ASO) dot blot analysis and PCR/RFLP (restriction fragment length polymorphism) were performed.[6]

RESULTS AND DISCUSSION

Forty-five affected children and their asymptomatic mothers were screened for DNA mutations, spanning the entire genome of mitochondrial DNA using TTGE. In total, 136 homoplasmic and 11 heteroplasmic banding patterns were detected (data not shown). Although homoplasmic alterations were usually not followed by sequencing, some of the unusual ones were sequenced. Seven differences were detected in the sequences of the child compared to that of the respective mother (TABLE 1). Except for A189G and G5821A, they were all novel mutations. Mutations T5580C, G5821A, C5840T, A8326G, and G15995A were detected from a patient with cystic fibrosis who was also suspected of having a mitochondrial cytopathy.[7] The result of the TTGE analysis was that the proband was homoplasmic for the A8326G and heteroplasmic for G15995A, whereas the asymtomatic mother was heteroplasmic for A8326G and had no nucleotide substitution at nucleotide position (np) 15995 (FIG. 1A). These results were also confirmed by PCR/ASO (FIG. 1B). The A8326G mutation is located in the anticodon region, tRNA^{Lys} (np 8295–8364),

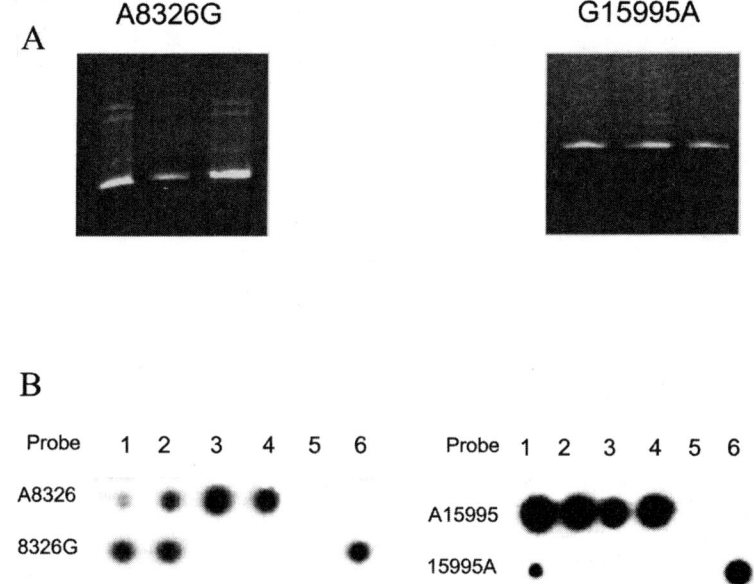

FIGURE 1. mtDNA mutations found in mother and child. (**A**) The mutation of A8326G and G15995A by TTGE. *Lane 1*, the asymtomatic mother; *lane 2*, affected child; *lane 3*, child/mother mix. (**B**) The confirmation of mutations by ASO dot blot analysis. The *top strip* was hybridized with normal ASO probe, and the *bottom strip* was hybridized with mutant ASO probe. *Lane 1*, affected child; *lane 2*, mother; *lanes 3 and 4*, normal controls; *lane 5*, no DNA template control; and *lane 6*, synthetic positive control. From Wong et al.[7] Reproduced with permission from the *American Journal of Medical Genetics*.

TABLE 1. Mutations found to be different in mother and affected child

Mutation	Gene	Heteroplasmy/Homoplasmy	Novel/Reported
T5580C	Noncoding	Heteroplasmy	Novel
G5821A	tRNA Cysteine	Heteroplasmy	Reported
C5840T	tRNA Tyrosine	Heteroplasmy	Novel
A8326G	tRNA Lysine	Homoplasmy	Novel
G12207A	tRNA Serine	Homoplasmy	Novel
G15995A	tRNA Proline	Heteroplasmy	Novel
G189A	D–loop	Heteroplasmy	Reported

right next to the first nucleotide of the triplet codon (FIG. 2A). The G15995A disrupts the first base pair of the stem region at the anticodon loop of tRNA^{Pro} (np 15955–16023) (FIG. 2B). Both nucleotides are highly conserved throughout evolution and are not present in 130 normal controls. The G12207A mutation is located at the first base of tRNA^{Ser}(AGY). The mutation disrupts the base pairing with the 3' end of the tRNA where the amino acid is charged. This mutation was detected both in the

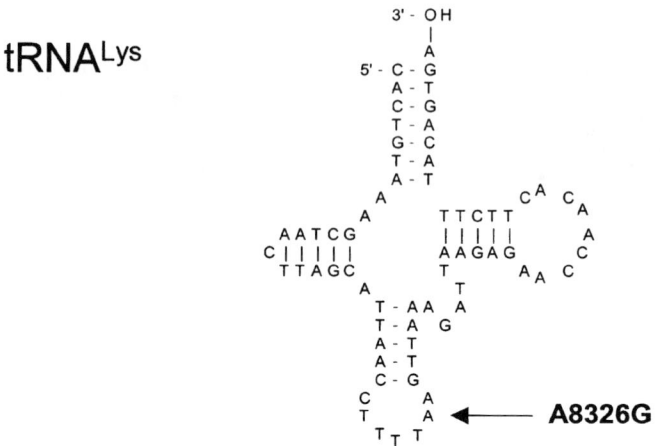

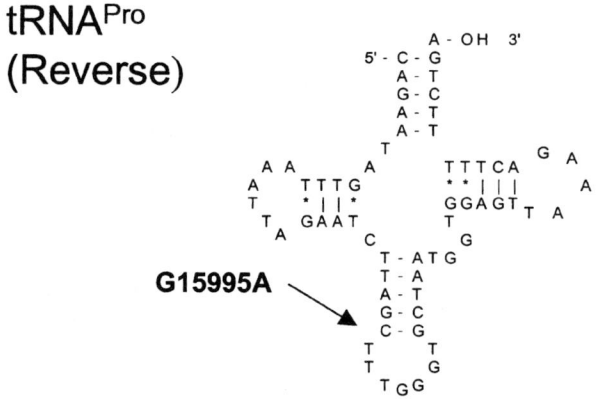

FIGURE 2. Structure of tRNA [Lys] (*top*) and tRNA [Pro] (*bottom*). The anticodon and the mutated nucleotide positions are indicated. From Wong *et al.*[7] Reproduced with permission from the *American Journal of Medical Genetics*.

proband's blood and in muscle. The G12207A mutation is not present in 130 normal controls.

TABLE 2 shows novel missense mutations, which were detected both in the proband and in the proband's mother. Some of the missense mutations result in drastic alterations such as changes from a hydrophilic to a hydrophobic structure (S45F). Most sequencing results show no differences between mother and proband.

Not many differences in sequences were noted between mothers and affected children, although these patients had clinical features characteristic of mitochondrial disorders. This may be due to the fact that about 85% of the polypeptide subunits of

TABLE 2. Novel missense variations present in both mother and affected child

Mutation	Gene	Amino acid change	Heteroplasmy/ Homoplasmy	Novel/Reported
A6663G	COI	I254V	Homoplasmy	Novel
C10192T	ND3	S45F	Homoplasmy	Novel
C11177T	ND4	P140S	Homoplasmy	Novel
A13907G	ND3	N524S	Homoplasmy	Novel
G15596A	Cytb	V284I	Homoplasmy	Novel

the respiratory chain complexes are encoded by the nuclear genes. It is possible that the molecular defect in these patients lies in the nuclear genes. Nevertheless, analyzing mother and child side by side by TTGE eliminated excessive sequencing and is a novel method to detect potential pathogenic mutations.

REFERENCES

1. ANDERSON, S. 1981. Sequence and organization of the human mitochondrial genome. Nature **290:** 457–465.
2. TZAGOLOFF, A. 1986. Genetics of mitochondrial biogenesis. Annu. Rev. Biochem. **55:** 249–285.
3. CHEN, T.J. 1999. Detection of mitochondrial DNA mutations by temporal temperature gradient gel electrophoresis. Clin. Chem. **45:** 1162–1167.
4. LAHIRI, D. 1991 A rapid non-enzymatic method for the preparation of HMW DNA from blood for RFLP studies. Nucleic Acids Res. **19:** 5444.
5. WONG, L.-J.C. 2002. Comprehensive scanning of the whole mitochondrial genome for mutations. Clin. Chem. **48:** 1901–1912.
6. WONG, L.-J.C. 1997. Direct detection of multiple point mutations in mitochondrial DNA. Clin. Chem. **43:** 1857–1861.
7. WONG, L.-J.C., et al. 2002. A cystic fibrosis patient with two novel mutations in mitochondrial DNA: mild disease led to delayed diagnosis of both disorders. Am. J. Med. Genet. **113:** 59–64.

Quantitative PCR Analysis of Mitochondrial DNA Content in Patients with Mitochondrial Disease

REN-KUI BAI, CHERNG-LIH PERNG, CHANG-HUNG HSU, AND LEE-JUN C. WONG

Institute for Molecular and Human Genetics, Georgetown University Medical Center, Washington DC 20007, USA

ABSTRACT: Molecular diagnosis of mitochondrial DNA disorder is usually focused on point mutations and large deletions. In the absence of detectable mtDNA mutations, abnormal amounts of mtDNA, either depletion or elevation, can be indicative of mitochondrial dysfunction. The amount of mitochondrial DNA (mtDNA), however, varies among individuals of different ages and among different tissues within the same individual. To establish a range of mtDNA levels, we analyzed 300 muscle and 200 blood specimens from patients suspected of having a mitochondrial disorder by real-time quantitative polymerase chain reaction (PCR) method. Copy numbers were calculated from the standard curve and threshold cycle number using TaqMan probes; 6FAM 5′TTACCGGGCTCTGCCATCT3′-TAMRA and VIC-5′AGCAATAACAGGTCTGTGATG3′-TAMRA for mtDNA and 18S rRNA gene (nDNA), respectively. The copy number ratio of mtDNA to nDNA was used as a measure of mtDNA content in each specimen. The mtDNA content in muscle increases steadily from birth to about 5 years of age; thereafter, it stays about the same. On the contrary, the mtDNA content in blood decreases with age. The amount of mtDNA in skeletal muscle is about 5–20 times higher than that in blood. About 7% of patients had mtDNA levels in muscle below 20% of the mean of the age-matched group, and about 10% of patients had muscle mtDNA levels 2- to 16-fold higher than the mean of the age-matched group. Patients with abnormal levels of mtDNA, either depletion or proliferation, had significant clinical manifestations characteristic of mitochondrial disease in addition to abnormal respiratory enzymes and mitochondrial cytopathies. Cardiomyopathy, lactic acidosis, abnormal brain MRI findings, hypotonia, developmental delay, seizures, and failure to thrive are general clinical pictures of patients with mtDNA depletion. The average age of patients with mtDNA depletion is 4.1 years, compared to 23.6 years in patients with mtDNA proliferation. Mutations in nuclear genes involved in mtDNA synthesis and deoxynucleotide pools are probably the cause of mtDNA depletion. Our results demonstrate that real time quantitative PCR is a valuable tool for molecular screening of mitochondrial diseases.

KEYWORDS: mtDNA depletion; quantitative analysis of mtDNA; real-time PCR analysis; mitochondrial DNA copy number

Address for correspondence: Lee-Jun C. Wong, Ph.D., Institute for Molecular and Human Genetics, Georgetown University Medical Center, M4000, 3800 Reservoir Rd., NW, Washington, DC 20007. Voice: 202-784-0760; fax: 202-784-1770.
wonglj@georgetown.edu

Ann. N.Y. Acad. Sci. 1011: 304–309 (2004). © 2004 New York Academy of Sciences.
doi: 10.1196/annals.1293.029

INTRODUCTION

Point mutations and large deletions in mtDNA account for the molecular defects in a small portion of patients with mitochondrial respiratory deficiency.[1–3] Possibly the respiratory defects in some of these patients are caused by a quantitative deficiency in mtDNA content rather than specific mutations. Previous studies indicate that mtDNA depletion due to mutations in the thymidine phosphorylase (TP) gene are responsible for the mitochondrial neurogastrointestinal encephalomyopathy (MNGIE) syndrome.[4–6] Mutations in nuclear genes that are involved in mtDNA synthesis or maintenance of deoxynucleotide pools may affect the biogenesis of mitochondria and therefore affect the mtDNA content. On the other hand, defective mitochondria are often proliferated. Thus, mtDNA amplification could be a compensatory mechanism in response to inefficient mitochondrial respiratory function. Here, we report the use of a real-time quantitative PCR assay to determine the mtDNA content in muscle and blood. Quantitative alterations in mtDNA may have implications in molecular defects of nuclear or mitochondrial genes.

MATERIAL AND METHODS

Specimens and DNA Isolation

Patients were referred to the Molecular Genetics Laboratory, Institute for Molecular and Human Genetics at Georgetown University Medical Center for molecular diagnosis of mitochondrial disorders. Total DNA was isolated from 300 muscle specimens using proteinase K digestion followed by standard phenol/chloroform extraction and ethanol precipitation.[7] Blood DNA was extracted by a salting-out method[8] from peripheral blood lymphocytes from 200 patients.

Primers and Probes

The primers for RT Q-PCR analysis of mtDNA are mtF3212 (5′CACCCAAGAACAGGGTTTGT3′) and mtR3319 (5′TGGCCATGGGTATGTT-GTTAA3′), those for the nuclear DNA (nDNA), 18S rRNA gene are 18S1546F (5′TAGAGGGACAAGTGGCGTTC3′) and 18S1650R (5′CGCTGAGCCAGTCA-GTGT3′).[9] The TaqMan probes; 6FAM-5′TTACCGGGCTCTGCCATCT3′-TAMRA and VIC-5′AGCAATAACAGGTCTGTGATG3′-TAMRA, were labeled at the 5′ end with a fluorescent reporter, 6FAM and VIC, for the mtDNA and the nDNA 18S rRNA gene, respectively, whereas the 3′ ends were labeled with a quencher TAMRA. The 10 µL PCR reaction contains 1× TaqMan Universal PCR Master Mix (ABI P/N 4304437), 500 nM of each primer, 200 nM of TaqMan probe, and 0.2–2 ng of total genomic DNA extract. PCR conditions are 2 min at 50°C, 10 min at 95°C, followed by 40 cycles of 15 s of denaturation at 95°C and 60 s of annealing/extension at 60°C. Real-time quantitative analysis was performed on the Sequence Detector System ABI-Prism 7700.[9]

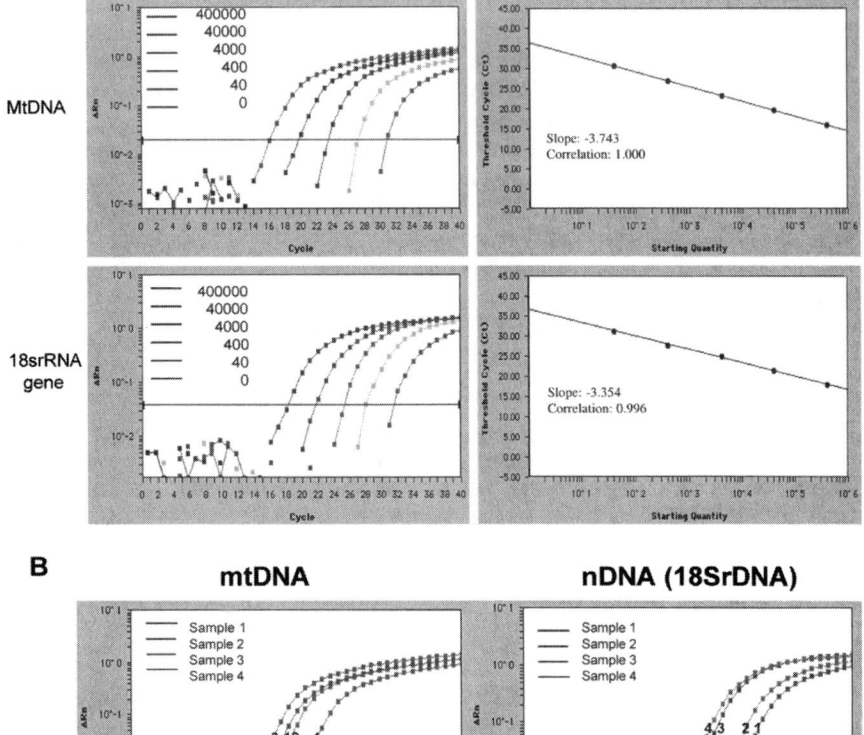

Sample 1: mtDNA/nDNA = Copy number of mtDNA/Copy number of 18S rRNA gene = 154683/648 = 238.7
Sample 2: mtDNA/nDNA = 89354/1613 = 55.4
Sample 3: mtDNA/nDNA = 320945/20261 = 15.84
Sample 4: mtDNA/nDNA = 11975/37312 = 0.32

FIGURE 1. Real-time PCR analysis. Standard curve of the mtDNA and the 18S rRNA gene. The cloned PCR products from both the mtDNA and the 18S rDNA were serially diluted for real-time PCR. The copy number in each dilution was calculated from the actual DNA concentration. The threshold cycle number (C_T) was plotted against the copy number of the DNA template at the start of PCR. DNA samples of unknown copy numbers were analyzed using the same conditions. The copy number of mtDNA and nDNA was calculated from the C_T number and by use of the standard curve.

TABLE 1. Comparison between the two groups of patients with mtDNA depletion and mtDNA proliferation

	mtDNA proliferation (>twofold of mean)	mtDNA depletion (<30% of mean)
Total number	29/300	27/300
Average age	23.6 years	4.1 years
Male/female	12/17	11/16
<10-year-old	15 (9/15 >3.6-fold)	24
Major clinical manifestation	Muscle weakness Exercise intolerance Fatigability Ophthalmoplegia Lactic acidosis Seizure Hypotonia Developmental delay Abnormal histochemistry Abnormal respiratory enzyme	Cardiomyopathy Low plasma carnitine Failure to thrive Abnormal MRI Lactic acidosis Seizure Hypotonia Developmental delay Abnormal histochemistry Abnormal respiratory enzyme Ophthalmoplegia Elevated CSF protein

Standard Curve

Standard DNA solutions for the mitochondrial genome and the nuclear 18S rRNA gene (nDNA) were generated from PCR products cloned in a vector of pCR2.1-TOPO. Serial dilutions were made and RT Q-PCR reactions were performed as just described to construct the standard curve from the C_T values and the number of copies of the standard plasmid DNA (FIG. 1A).

Determination of mtDNA/nDNA Ratio as a Measure of mtDNA Content

FIGURE 1B illustrates a sample run. The copy number of the mtDNA and the nDNA is calculated using the threshold cycle number (C_T) and intrapolating from the standard curve. The ratio of the copy number of mtDNA to the copy number of nDNA is the measurement of mtDNA content. FIGURE 1B shows four specimens with mtDNA/nDNA ratio from 0.32 to 239.

RESULTS

The ratio of mtDNA/nDNA in muscle varies from 0.1 to 1700, and in blood varies from 0.05 to 23. Evidently, some patients had markedly elevated mtDNA and some had severely depleted mtDNA, particularly in muscle. The average mtDNA content was calculated for each age group by excluding the samples in the highest and lowest 10%. The results show that the mtDNA content increases from birth to about 5 years of age and remains at about the same level after that (FIG. 2A). On the contrary, the mtDNA content in blood decreases with age (FIG. 2B). The amount of mtDNA in skeletal muscle is about 5–20 times higher than that in blood. About 7% of patients

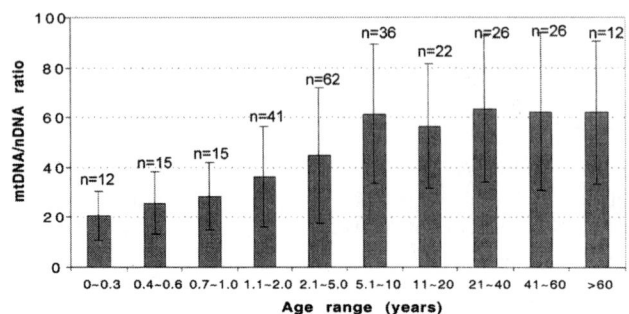
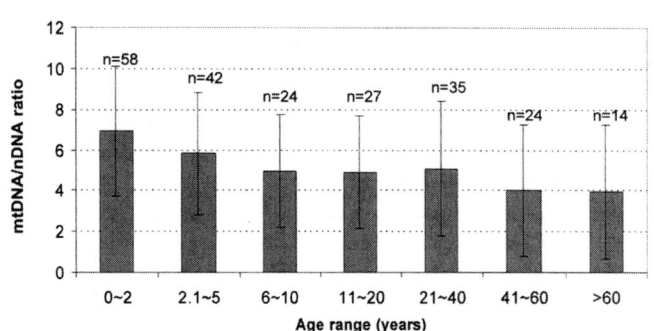

FIGURE 2. mtDNA content in muscle and blood of various age groups of patients measured by mtDNA/18Sr DNA copy number ratios. (**A**) Muscle specimens; (**B**) blood specimens.

had mtDNA levels in muscle below 20% of the age-matched mean, and about 10% of patients had muscle mtDNA levels 2- to 16-fold higher than that of the age-matched mean. Patients with an abnormally high or low level of mtDNA shared some significant clinical features of mitochondrial disease as well as abnormal respiratory enzymes and mitochondrial cytopathies (TABLE 1). Cardiomyopathy, lactic acidosis, abnormal brain MRI, hypotonia, developmental delay, seizure, and failure to thrive are general clinical manifestations that seem to be associated with young patients (<10) with mtDNA depletion (<30% of age-matched mean). However, muscle weakness, exercise intolerance, ophthalmoplegia, and lactic acidosis seem to be the predominant clinical symptoms of patients with elevated mtDNA content. The average age, 4.1 and 23.6 years for patients with mtDNA depletion and mtDNA proliferation, respectively, is significantly different (TABLE 1).

DISCUSSION AND CONCLUSION

Most patients with severe mtDNA depletion are young children, with an average age of 4.1 years. We sequenced the TP gene of these patients and found mutations in only 1 patient (data not shown) who was 20 years old. Although mutations in the TP gene are the main cause of mtDNA depletion in adult patients with MNGIE, TP gene mutations do not seem to be the reason for mtDNA depletion in children <10 based on these results. It is very likely that mutations in other nuclear genes involved in mtDNA synthesis or deoxynucleotide pools, such as polymerase gamma, DNA helicase, nucleotide translocase, or thymidine kinase, are responsible for the mtDNA depletion in these patients. The patients who had mtDNA proliferation were much older, and their muscle biopsies showed evidence of ragged-red fibers. The increase in mtDNA content in these cases correlates with mitochondrial proliferation and this is probably a compensatory mechanism for defective mitochondria. Knowing the mtDNA content in muscle tissue will facilitate the search for the molecular defects responsible for the mitochondrial disease.

REFERENCES

1. LIANG, M.H. & L.-J.C. WONG. 1998. Yield of mtDNA mutation analysis in 2000 patients. Am. J. Med. Genet. **77:** 385–400.
2. WONG, L.-J.C. 2001. Recognition of mitochondrial DNA deletion syndrome with non-neuromuscular multisystemic manifestation. Genet. Med. **3:** 399–404.
3. WONG, L.-J.C., M.-H. LIANG, H. KWON, et al. 2002. Comprehensive scanning of the whole mitochondrial genome for mutations. Clin. Chem. **48:** 1901–1912.
4. NISHINO, I., A. SPINAZZOLA & M. HIRANO. 1999. Thymidine phosphorylase gene mutations in MNGIE, a human mitochondrial disorder. Science **283:** 689–692.
5. NISHINO, I., A. SPINAZZOLA, A. PAPADIMITRIOU, et al. 2000. Mitochondrial neurogastrointestinal encephalomyopathy: an autosomal recessive disorder due to thymidine phosphorylase mutations. Ann. Neurol. **47:** 792–800.
6. HIRANO, M., R. MARTI, C. FERREIRO-BARROS, et al. 2001. Defects of intergenomic communication: autosomal disorders that cause multiple deletions and depletion of mitochondrial DNA. Semin. Cell Dev. Biol. **12:** 417–427.
7. WONG, L.-J.C. & C. LAM. 1997. Alternative, noninvasive tissues for quantitative screening of mutant mitochondrial DNA. Clin. Chem. **43:** 1241–1243.
8. LAHIRI, D. & J. NURNBERGER, JR. 1991. A rapid non-enzymatic method for the preparation of HMW DNA from blood for RFLP studies. Nucleic Acids Res. **19:** 5444.
9. WONG, L.-J.C. & R. BAI. 2002. Real-time quantitative PCR analysis of mitochodnrial DNA in patients with mitochondrial disease. Am. J. Hum. Genet. Suppl. **71:** 501.

Somatic Mitochondrial DNA Mutations in Oral Cancer of Betel Quid Chewers

DUAN-JUN TAN,[a] JULIA CHANG,[b] WOAN-LING CHEN,[b] LESLEY J. AGRESS,[a] KUN-TU YEH,[b] BAOTYAN WANG,[b] AND LEE-JUN C. WONG[a]

[a]*Institute for Molecular and Human Genetics, Georgetown University Medical Center, Washington, DC, USA*

[b]*Changhua Christian Hospital, Chang-Hua, Taiwan*

ABSTRACT: Somatic mitochondrial DNA alteration is a general phenomenon that occurs in cancerous cells. Although numerous mtDNA mutations have been identified in various tumors, the pathogenic significance of these mutations remains unclear. In order to better understand the role of mtDNA mutations in the neoplastic process of oral cancer, the occurrence of mtDNA mutations in oral squamous cell carcinomas was screened by temporal temperature gradient gel electrophoresis (TTGE). The entire mitochondrial genome was amplified with 32 pairs of overlapping primers. The DNA fragments showing different banding patterns between normal and tumor mtDNA were sequenced for the identification of the mutations. Fourteen of 18 (77.8%) tumors had somatic mtDNA mutations with a total of 26 mutations. Among them, 6 were in mRNA coding region. Three were missense mutations (C14F, H186R, T173P) in NADH dehydrogenase subunit 2 (ND2). One frameshift mutation, 9485delC, was in cytochrome *c* oxidase subunit III. Eight (44%) tumors had insertion or deletion mutations in the np303-309 poly C region of the D-loop. Our results demonstrate that somatic mtDNA mutations occur in oral cancer. The missense and frameshift mutations in the evolutionary conserved regions of the mitochondrial genome may have functional significance in the pathogenesis of oral cancer.

KEYWORDS: oral cancer; mitochondrial DNA mutation

INTRODUCTION

Oral cancer is the fourth leading cause of cancer in Taiwanese males. People who are addicted to chewing betel nut are at increased risk of developing oral cancer. Betal quid contains tender areca nuts with husk. Cell cultures treated with areca nut extract were found to have increased oxidative DNA damage as evidenced by the generation of 8-hydroxy-2′-deoxy guanosine.[1–3] The masticatory stimulant lime generates an alkaline environment and accelerates the production of reactive oxygen species (ROS).[3]

Address for correspondence: Lee-Jun C. Wong, Ph.D., Institute for Molecular and Human Genetics, Georgetown University Medical Center, M4000, 3800 Reservoir Road, NW, Washington, DC 20007. Voice: 202-444-0760; fax: 202-444-1770.
wonglj@georgetown.edu

The human mitochondrial genome is a circular double-stranded DNA of 16.6 kb encoding 13 respiratory chain protein subunits, 22 tRNAs, and 2 rRNAs.[4] Each cell contains hundreds to thousands of mitochondria, and each mitochondrion contains 2–10 copies of mitochondrial DNA (mtDNA).[5] The major function of mitochondria is to produce energy to support cellular activities through the oxidative phosphorylation pathway. During this process ROS are generated. Due to the lack of protective histone proteins, mtDNA is an easy target for oxidative DNA damage. In addition, the limited DNA repair mechanisms allow mtDNA mutations to accumulate. Thus, the mutation rate of mtDNA is at least 10 times higher than that of nuclear DNA. The roles of mitochondria in energy metabolism, the generation of ROS, aging, and the initiation of apoptosis suggest that the mitochondria play an important role in tumorigenesis.[6–8]

Recently, somatic mtDNA mutations have been reported in various types of tumors.[6,8–14] Most of the somatic mtDNA mutations are in the noncoding D-loop region.[10–12,14] Significant missense mutations, although rare, usually have functional effects. In this paper, we report on the occurrence of mtDNA mutations in oral tumors and speculate on the possible significance of these mutations.

MATERIALS AND METHODS

Tissue Specimens and DNA Extraction

Seventeen pairs of primary oral squamous cell carcinomas and the surrounding tissues, as well as one oral adenoid cystic carcinoma and the surrounding tissue, were obtained from patients with a history of betel quid chewing at Changhua Christian Hospital, Changhua, Taiwan. All tissues were snap frozen in liquid nitrogen and stored at −80°C until they were analyzed. DNA was isolated from frozen tissues by proteinase K digestion and phenol/chloroform extraction. Total DNA was quantified using fluorescent Hoechst dye H33258 with DyNA Quant 200 (Amersham Biosciences, Uppsala, Sweden) according to the manufacturer's protocol. DNA was diluted to 5 ng/µL for use in PCR reactions.

Mutational Analysis of the Entire Mitochondrial Genome

The temporal temperature gradient gel electrophoresis (TTGE) mutation detection method was used for screening mtDNA mutations. The 32 pairs of primers for PCR and the conditions for TTGE have been described in detail in previous publications.[15,16] In TTGE analysis, a single band shift represents a homoplasmic DNA alteration, and a multiple banding pattern represents a heteroplasmic mutation. The DNA fragments from normal and tumor tissues of the same patient were analyzed side-by-side. Any DNA fragments showing differences in banding patterns between normal and tumor samples were sequenced for the identification of the exact mutations. The DNA sequencing was performed by use of a big dye terminator cycle sequencing kit (Perkin Elmer, Wellesley, MA) and analyzed using an ABI 377 (Applied Biosystem, Foster City, CA) automated sequencer. The results of DNA sequence analysis were compared with the complete human mitochondrial sequence deposited in GenBank (accession no. J01415)[17] using MacVector 7.0 (Oxford Molecular Ltd., Oxford, England) software. Any DNA sequence differences between the tumor and the matched normal mtDNA were scored as somatic mutations. Sequence variations found in both tumor and matched normal mtDNA, but different from that recorded in the GenBank, were scored as germline variations. Each was then checked against the Mitomap database.[4] Those not recorded in the database were categorized as novel mtDNA variations, and those that appeared in the database were categorized as either reported polymorphisms or known mutations. The amino acid sequence alignment for evolutionarily conserved status was analyzed by use of Vector NTI Advance 6.0 (InforMax Inc., Bethesda, MD) software.

RESULTS

TTGE analysis showed a number of different banding patterns between tumor and surrounding tissues (FIG. 1A and C). Sequencing revealed a homoplasmic G4510T mutation (lane tu in FIG. 1A) in the NADH dehydrogenase subunit 2 (ND2) region of tumor 3 from a heteroplasmic G4510T in the surrounding tissue (lane nl in FIG. 1A). A heteroplasmic A4986C mutation was found in tumor 18 (FIG. 1C). TABLE 1 summarizes the somatic mtDNA mutations found in the mitochondrial genome of 18 oral tumors. Fourteen of 18 (77.8%) tumors harbored somatic mtDNA mutations. Among the total of 26 somatic mutations, 20 were in the hypervariable D-loop region (76.9%), five in ND regions (19.2%), and one in cytochrome c oxidase subunit (COX) III (3.8%). The mutant mitochondria may be present in either the homoplasmic or heteroplasmic state. Three of the mutations in the mRNA region were potentially harmful missense/nonsense mutations because they occurred at the evolutionarily highly conserved amino acid residues with a nonconserved amino acid

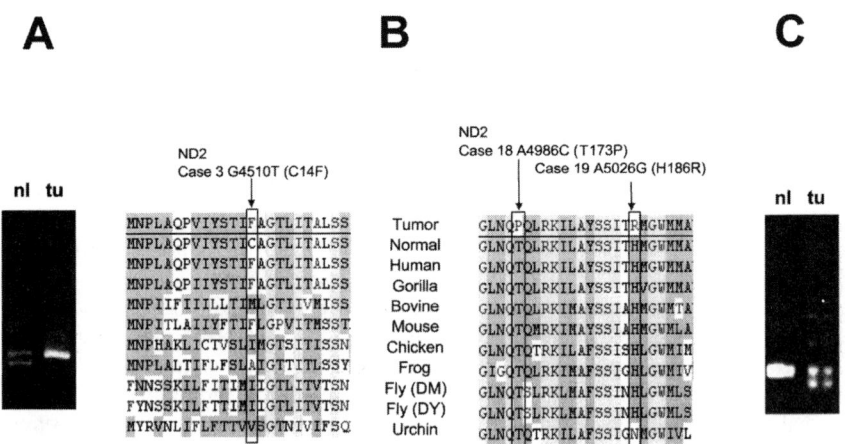

FIGURE 1. TTGE analysis of mtDNA mutations. (**A**) Heteroplasmic to homoplasmic G4510T (C14F) mutation in ND2 region of case 3. (**B**) Amino acid sequence alignment of ND2 region of mtDNA. DM, *Drosophila melanogaster*; DY, *Drosophila yakuba*. (**C**) Homoplasmic to heteroplasmic A4986C (T173P) mutation in ND2 region of case 18.

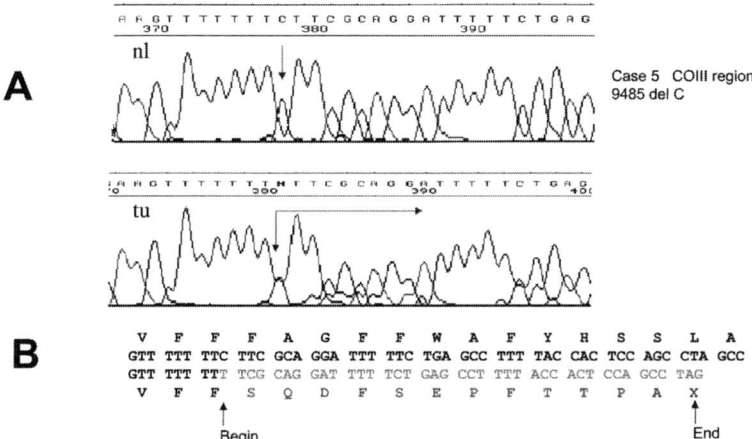

FIGURE 2. A frame-shift mutation, 9485delC in COXIII region of case 5. (**A**) Sequence results in the region of normal (*upper panel*) and tumor (*lower panel*) showing a deletion of 9485C. (**B**) The truncated protein results from 9485delC.

change (FIG. 1B). The A4986C changes threonine at position 173 to proline, and A5026G changes a histidine residue at 186 to arginine in ND2. These amino acids are highly conserved throughout evolution, from sea urchin and fruit flies to mice and humans (FIG. 1C). One somatic mutation is a frameshift mutation in cytochrome *c* oxidase subunit III (9485delC), which results in a truncated polypeptide of 104 amino acid residues (FIG. 2). The G4510T mutation causes the substitution of cysteine with phenylalanine at amino-acid 14 of the ND2 subunit. This mutation was heteroplasmic in the surrounding tissue, but homoplasmic wild type in the tumor, a change back from variant to wild type. This amino acid is not conserved throughout evolution. The effect of the missense and frameshift mutations on the pathologic mechanism of tumorigenesis is currently not clear.

DISCUSSION

In this report, we present a comprehensive study of somatic mtDNA mutations in human oral cancer. A survey of the reported somatic mtDNA mutations in various primary carcinomas revealed that mutations in the mRNA region are infrequent compared to mutations in the noncoding D-loop region. Most of these mutations are silent mutations. A high frequency of missense and nonsense mutations characterizes the mtDNA mutation spectrum in oral cancer tissues (TABLE 1). In addition, there is a high frequency (58%, 15/26) of homoplasmy to heteroplasmy changes that occur in oral cancer (TABLE 1). Most of the reported somatic mtDNA mutations are in the homoplasmic state.[18] Another characteristic feature of oral cancer is the large-scale mtDNA deletion (data not shown). Large deletions were not detected in 39 breast and 20 hepatic tumors (unpublished observation). These unique characteristics may reflect the unique chemical carcinogenesis in oral cancer.

TABLE 1. Somatic mtDNA mutations in oral cancer

Case	Gene	Somatic mutations	Sequence pattern[a]	Codon	AA Change[b]	Function
3	D-loop	C318T	ht→hm			Replication primer
3	ND2	G4510T	ht→hm	TGT→TTT	C14F[c]	NADH dehydrogenase
4	D-loop	T204C	hm→ht			H-strand origin
4	D-loop	G207A	hm→ht			H-strand origin
4	D-loop	C246T	hm→ht			mtTF1 binding site
4	D-loop	C489T	hm→ht			
5	COX III	9485 del C frameshift	hm→ht	frameshift		Cytochrome c oxidase
6	D-loop	303-309 ins C($C_8/C_9 \to C_9/C_8$)	ht→ht (low→high%)			Conserved sequence block II
7	D-loop	303-309 ins C($C_8 \to C_8/C_9$)	hm→ht			Conserved sequence block II
8	D-loop	303-309 ins C($C_8 \to C_8/C_9$)	hm→ht			Conserved sequence block II
9	D-loop	303-309 ins CC($C_8 \to C_9/C_{10}$)	hm→ht			Conserved sequence block II
10	D-loop	C204T	hm→hm			H-strand origin
10	D-loop	303-309 del CCCC($C_8 \to C_4$)	hm→hm			Conserved sequence block II
10	D-loop	C313A	hm→hm			Conserved sequence block II
10	D-loop	A207G	ht→hm			H-strand origin
10	ND3	C10245T	ht→hm	CTA→TTA	L63L	NADH dehydrogenase
11	D-loop	66-71 del G	hm→ht			
12	D-loop	303-309 ins C($C_8 \to C_8/C_9$)	hm→ht			Conserved sequence block II
13	D-loop	303-309 ins C($C_9/C_{10} \to C_{10}/C_9$)	ht→ht (low→high%)			Conserved sequence block II
14	D-loop	C222T	hm→ht			Conserved sequence block I
14	ND4	T11794C	hm→ht	TCT→TCC	S345S	NADH dehydrogenase
14	D-loop	T16320C	ht→hm			Hypervariable segment 1
14	D-loop	C16419A	ht→ht (low→high%)			
18	D-loop	303-309 ins CC($C_8 \to C_8/C_{10}$)	hm→ht			Conserved sequence block II
18	ND2	A4986C	hm→ht	ACC→CCC	T173P	NADH dehydrogenase
19	ND2	A5026G	hm→ht	CAC→CGC	H186R	NADH dehydrogenase

[a] ht, heteroplasmic; hm, homoplasmic.
[b] AA, amino acid.
[c] Missense mutations are in bold.
Total mutation: 26; hm→ht: 15; hm→ht: 5; hm→hm: 3; ht→ht: 3.

In Taiwan, betel quid chewing is the main cause of oral squamous-cell carcinoma and oral submucous fibrosis. The areca nuts and lime can induce oxidative DNA damage to buccal mucosa cells which leads to carcinogenesis.[1–3] These missense mutations in evolutionarily conserved amino acids are predicted to have a structural/functional impact on oxidative energy metabolism.

The mononucleotide repeats between np 303 and 309 are among the most polymorphic of mitochondrial microsatellites. This homopolymeric C-stretch is part of the conserved sequence block II located within the regulatory D-loop region and is involved in the formation of a persistent RNA-DNA hybrid that leads to the initiation of mtDNA heavy-strand replication.[19] In this study, we found that 44% (8/18) of tumors had insertion or deletion in the np 303–309 region, consistent with the previous report that np 303–309 is a mutation hot spot.[20]

Many germline variations were identified. Although the biochemical consequence of homoplasmic polymorphisms are considered too subtle to cause any effect on oxidative phosphorylation, long-term accumulation of the subtle differences in oxidative phosphorylation activity may eventually result in oxidative stress. Thus, in the late onset of a disease such as cancer, germline mtDNA variations can potentially play a role in modifying the risk of developing cancer.

In conclusion, the incidence of somatic mtDNA mutations is high in oral tumors. Missense and frameshift mutations that result in significant structural/functional alterations may play an important role in the tumorigenesis of oral cancer. More extensive biochemical and molecular studies will be necessary for determination of the pathologic effect of these somatic mutations.

REFERENCES

1. LIU, T.Y., C.L. CHEN & C.W. CHI. 1996. Oxidative damage to DNA induced by areca nut extract. Mutat. Res. **367:** 25–31.
2. FRIESEN, M., G. MARU, V. BUSSACHINI, et al. 1988. Formation of reactive oxygen species and of 8-hydroxy-2′-deoxyguanosine in DNA in vitro with betel-quid ingredients. IARC Sci. Publ. **89:** 417–421.
3. NAIR, U.J., M. FRIESEN, I. RICHARD, et al. 1990. Effect of lime composition on the formation of reactive oxygen species from areca nut extract in vitro. Carcinogenesis **11:** 2145–2148.
4. MITOMAP. 2001. A human mitochondrial genome database. http://www.mitomap.org.
5. ATTARDI, G. & G. SCHATZ. 1988. Biogenesis of mitochondria. Annu. Rev. Cell Biol. **4:** 289–333.
6. FLISS, M.S., H. USADEL, O.L. CABALLERO, et al. 2000. Facile detection of mitochondrial DNA mutations in tumors and bodily fluids. Science **287:** 2017–2019.
7. COPELAND, W.C., J.T. WACHSMAN, F.M. JOHNSON & J.S. PENTA. 2002. Mitochondrial DNA alterations in cancer. Cancer Invest. **20:** 557–569.
8. POLYAK, K., Y. LI, H. ZHU, et al. 1998. Somatic mutations of the mitochondrial genome in human colorectal tumours. Nat. Genet. **20:** 291–293.
9. HIBI, K., H. NAKAYAMA, T. YAMAZAKI, et al. 2001. Mitochondrial DNA alteration in esophageal cancer. Int. J. Cancer **92:** 319–321.
10. LIU, V.W.S., H.H. SHI, A.N.Y. CHEUNG, et al. 2001. High incidence of somatic mitochondrial DNA mutations in human ovarian carcinomas. Cancer Res. **61:** 5998–6001.
11. KIRCHES, E., G. KRAUSE, M. WARICH-KIRCHES. et al. 2001. High frequency of mitochondrial DNA mutations in glioblastoma multiforme identified by direct sequence comparison to blood samples. Int. J. Cancer **93:** 534–538.
12. TAN, D-J., R. BAI & L-J.C. WONG. 2002. Comprehensive scanning of somatic mitochondrial DNA mutations in breast cancer. Cancer Res. **62:** 972–976.

13. TAMURA, G., S. NISHIZUKA, C. MAESAWA, *et al.* 1999. Mutations in mitochondrial control region DNA in gastric tumours of Japanese patients. Eur. J. Cancer **35:** 316–319.
14. PARRELLA, P., Y. XIAO, M.S. FLISS, *et al.* 2001. Detection of mitochondrial DNA mutations in primary breast cancer and fine-needle aspirates. Cancer Res. **61:** 7623–7626.
15. WONG, L-J.C., M-H. LIANG, H. KWON, *et al.* 2002. Comprehensive scanning of the whole mitochondrial genome for mutations. Clin. Chem. **48:** 1901–1912.
16. CHEN, TJ., R. BOLES & L-J.C. WONG. 1999. Detection of mitochondrial DNA mutations by temporal temperature gradient gel electrophoresis. Clin. Chem. **45:** 1162–1167.
17. ANDERSON, S., A.T. BANKIER, B.G. BARRELL, *et al.* 1981. Sequence and organization of the human mitochondrial genome. Nature **290:** 457–465.
18. PENTA, J.S., F.M. JOHNSON, T. WACHSMAN & W.C. COPELAND. 2001. Mitochondrial DNA in human malignancy. Mutat. Res. **488:**119–133.
19. LEE, D.Y. & D.A. CLAYTON. 1998. Initiation of mitchondrial DNA replication by transcription and R-loop processing. J. Biol. Chem. **273:** 30614–30621.
20. SANCHEZ-CESPEDES, M., P. PARRELLA, S. NOMOTO, *et al.* 2001. Identification of a mononucleotide repeat as a major target for mitochondrial DNA alterations in human tumors. Cancer Res. **61:**7015–7019.

Association of the Mitochondrial DNA 16189 T to C Variant with Lacunar Cerebral Infarction

Evidence from a Hospital-Based Case-Control Study

CHIA-WEI LIOU,[a] TSU-KUNG LIN,[a] FENG-MEI HUANG,[a] TZU-LING CHEN,[b] CHENG-FENG LEE,[b] YAO-CHUNG CHUANG,[a] TENG-YEOW TAN,[a] KU-CHOU CHANG,[a] AND YAU-HUEI WEI[b]

[a]*Department of Neurology, Chang Gung Memorial Hospital, Kaohsiung, Taiwan*

[b]*Department of Biochemistry and Center for Cellular and Molecular Biology, National Yang-Ming University, Taipei, Taiwan*

ABSTRACT: A transition of T to C at nucleotide position 16189 in the hypervariable D-loop region of mitochondrial DNA (mtDNA) has attracted research interest for its probable correlation with increasing insulin resistance and development of diabetes mellitus (DM) in adult life. In this article, we present our observations of the positive relationship between this variant and cerebral infarction. Six hundred and one subjects in two groups—one with cerebral infarction (307 cases), the other with no cerebral infarction (294 cases)—were recruited. Their clinical features, fasting blood sugar and insulin levels, and insulin resistance index, were recorded. Patients with cerebral infarction were further categorized into four different subgroups according to the TOAST criteria for stroke classification. The results showed the occurrence of the mtDNA 16189 variant in 34.2% of patients with cerebral infarction and in 26.5% of normal controls. The difference in the occurrence rates between the two groups was statistically significant ($P = 0.041$). Further studies of the occurrence rate in each stroke subgroup revealed that the variant occurred at the highest frequency in the small vessel subgroup (41.5%). The difference in occurrence rate between this subgroup and the normal controls is highly significant ($P = 0.006$). These results correlated well with the findings of significantly increased levels of average fasting blood insulin and a higher index of average insulin resistance in the small vessel subgroup of patients harboring this mtDNA variant. Taken together, we suggest that the mtDNA 16189 variant is a predisposing genetic factor for the development of insulin resistance and may be related to various phenotypic expressions in adult life such as development of DM and vascular pathologies involved in stroke and cardiovascular diseases.

KEYWORDS: mitochondrial DNA; T16189C polymorphism; insulin resistance; lacunar cerebral infarction

---

Address for correspondence: Chia-Wei Liou, M.D., Department of Neurology, Chang Gung Memorial Hospital, 123, Ta-Pei Road, Niao-Sung Hsiang, Kaohsiung 833, Taiwan. Voice: 886-7-7317123, ext. 2283; fax: 886-7-7318762.
cwliou@ms22.hinet.net

Ann. N.Y. Acad. Sci. 1011: 317–324 (2004). © 2004 New York Academy of Sciences.
doi: 10.1196/annals.1293.031

INTRODUCTION

A proper recognition of risk factors for stroke has provided clear targets for primary and secondary prevention. This has allowed many countries with high stroke rates to lessen the incidence of stroke in the past few years.[1–3] However, there are still some unrecognized risk factors responsible for the persistently high rate of stroke in certain at-risk populations.[4,5] Of these factors, insulin resistance has received increasing attention.[6] Insulin resistance commonly exists in the populations of most developed countries and has recently attracted clinical attention for its characteristic effects on vascular pathophysiology.[7] The presence of insulin resistance and changes associated with certain clinical conditions does not necessarily mean the presence of the diabetic syndrome. However, the consequences of hyperinsulinemia resulting from a persistently hyperglycemic state may interfere with the biological functions of the vascular endothelium, which ultimately results in progressive vascular arteriosclerosis.[8–10] Although the detailed pathology of insulin resistance in vessels is only partially understood, the importance of its impact on the onset of stroke cannot be underestimated.

Stroke rates, especially presentations of lacunar cerebral infarctions and intracerebral hemorrhages in small vessels, are more prevalent in east Asians of Mongolian origin.[11–13] Although Asian countries share similar known risk factors with most Western countries, the incidence rates of stroke subtypes are different, with a greater predominance of small vessel disease. This propensity toward certain subtypes of stroke suggests that other factors, including ethnic or cultural differences, may play a role. In our recent studies on the relationship between the nucleotide position (np) 16189 polymorphism of mitochondrial DNA (mtDNA) and insulin resistance, we found a positive correlation between the presence of the mtDNA 16189 variant and increased insulin resistance and the subsequent development of diabetes mellitus (DM).[14] A high prevalence rate of this variant in the general population was found not only in our study but also in other studies conducted in other east Asian countries.[15–17] In this paper, we report that the increased insulin resistance associated with the occurrence of the mtDNA 16189 variant is also related to certain subtypes of cerebral infarction. The implications of this sequence variation in the D-loop of mtDNA in the pathogenesis of stroke and cardiovascular diseases are discussed.

MATERIALS AND METHODS

A total of 601 unrelated Taiwanese subjects over age 40 were enrolled in this study. The subjects were recruited and divided into two groups according to the following criteria: (1) Patients proven to have cerebral infarction based on both clinical presentation and positive brain computed tomography (CT) or magnetic resonance imaging (MRI) findings and (2) subjects with no previous history of cerebral infarction yet who also received a brain CT or MRI examination to rule out any undetected silent cerebral infarction. Group 1 consisted of 307 patients with cerebral infarction who had been admitted into Kaohsiung Chang Gung Memorial Hospital for therapy during the period from July 2001 to December 2001. Group 2 comprised 294 healthy subjects who were randomly selected after regular health examinations to participate in this study. There were no significant differences in age, sex, and body-mass index of the subjects in these two groups. Informed prior consents were acquired from all subjects. Different subtypes of stroke in Group 1 patients were classified according to the TOAST criteria based on clinical features, brain imaging, cardiac imaging, ultrasonography of extracranial and intracranial large arteries, magnetic resonance angiography (MRA), and other remarkable laboratory findings.[18]

Any patient with a history of DM or whose blood sample study showed possible glucose intolerance (fasting sugar ≥126 nmol/L) was excluded from this study.

Mitochondrial DNA was extracted from peripheral leukocytes and the region of interest amplified using polymerase chain reaction (PCR) techniques as described previously.[19] Two primer pairs were used—the forward primer consisted of np 15971-15990 of mtDNA and the reverse primer was an oligonucleotide spanning np 16471-16452 of mtDNA. The presence of the mtDNA 16189 variant was determined by using a combination of PCR and restriction fragment length polymorphism (RFLP) analysis with the restriction enzyme *Mnl* I. PCR products were digested with 1 U of the enzyme for at least 1 h at 37°C and subjected to electrophoresis with both positive and negative controls on a 2% agarose gel for 45 min at 80 V. DNA restriction fragments were visualized by UV transillumination of the gel stained with ethidium bromide.

For quantitative assessment of pancreatic β-cell function and peripheral insulin resistance, we also checked blood levels of glucose and insulin in the fasting state. Insulin resistance was determined using homeostasis model assessment (HOMA): fasting blood glucose concentration (mg/dL) × fasting blood insulin concentration (mU/L)/22.5. Pancreatic β-cell function was assessed using the formula: 20 × serum insulin concentration/(serum glucose concentration −3.5).[20]

Statistical analysis was performed using Student's *t*-test, Chi-square test, and multiple regression analysis. Data are expressed as mean ± SD. A difference between groups with $P < 0.05$ is considered statistically significant.

RESULTS

RFLP analysis demonstrated the presence of the 16189 variant in both study groups. It was present in 34.2% (105 out of 307) of Group 1 patients with cerebral infarction and 26.5% (78 out of 294) of Group 2 subjects with no cerebral infarction. Although a higher occurrence rate of the 16189 variant was noted in Group 1 patients, its statistical significance was borderline, compared to the rate of subjects in Group 2 ($P = 0.041$). After further comparison between the 16189 variant rates in different subgroups of Group 1 patients, a higher variant rate (41.5%) in the small vessel subgroup was noted. The difference was statistically significant in comparison with the variant rate in the normal control group after adjustment with various risk factors for stroke ($P = 0.006$, see TABLE 1).

The results of the evaluation of average insulin resistance (HOMA-IR) and pancreatic β-cell function (HOMA-beta cell) for each subject are listed in TABLE 2. These data were calculated separately for the two groups of subjects with and without cerebral infarction and/or various stroke subgroups of patients within the cerebral infarction group. Subjects with and without the 16189 variant in each group and stroke subgroups were recorded. A comparison between the infarction and non-infarction group revealed a higher level of average fasting blood insulin in the

TABLE 1. Clinical characteristics of the cases and controls

	Patients with cerebral infarction (CI)					Normal controls
	CIGp1	CIGp2	CIGp3	CIGp4	CI Total	
Number	68	51	94	94	307	294
Age (years)	67.7 ± 9.9	67.8 ± 9.2	65.7 ± 9.4	60.6 ± 14.5	66.2 ± 10.5	65.8 ± 9.6
Percentage male	63	60	57	63	61	49
Hypertension status (%)	85.9	70.8	87.1	69.2	84.6	45
Hypercholesteroemia status (%)	56	40.9	61	50	56.6	44.2
Hypertriglycedemia status (%)	44	33.3	31.8	28.6	35.6	25.5
Smoking status (%)	34.5	27.4	30.6	28.2	30.2	25
16189 Wild-type (n)	44	37	55	66	202	216
Variant (n)	24	14	39	28	105	78
Percentage of variant (%)	35.3	27.5	41.5 *	29.8	34.2 #	26.5

Stroke subtypes were determined according to the TOAST Classification for CVD. CI, cerebral infarction; CIGp1, large-artery atherosclerosis; CIGp2, cardioembolism; CIGp3, small vessel occlusion (lacune); CIGp4, stroke of other determined and undetermined etiology; CI Total, all cases with cerebral infarction.

* $P < 0.006$, in comparison between CIGp3 and normal controls; # $P < 0.041$, in comparison between CI total and normal controls.

TABLE 2. Characteristic data of the cases and controls

	Patients with cerebral infarction					Normal controls
	CIGp1	CIGp2	CIGp3	CIGp4	CI Total	
Fasting plasma glucose (mmol/L)						
16189 wild type	5.53 ± 0.67	5.7 ± 0.79	5.66 ± 0.62	5.47 ± 0.67	5.58 ± 0.68	5.36 ± 0.52
16189 variant	5.41 ± 0.53	5.31 ± 0.43	5.59 ± 0.5	5.25 ± 0.47	5.42 ± 0.49	5.48 ± 0.51
Fasting plasma insulin (pmol/L)						
16189 wild type	9.05 ± 4.52	9.29 ± 5.21	8.61 ± 4.31	9.45 ± 5.93	9.11 ± 5.05	8.63 ± 4.21
16189 variant	10.25 ± 4.52	9.07 ± 4.36	11.17 ± 4.59 *	9.9 ± 5.17	10.34 ± 4.7	8.87 ± 5.84
HOMA IR						
16189 wild type	2.22 ± 1.34	2.35 ± 1.22	2.16 ± 1.19	2.30 ± 1.76	2.26 ± 1.52	2.06 ± 0.97
16189 variant	2.46 ± 1.06	2.14 ± 0.83	2.77 ± 1.02 *	2.31 ± 1.07	2.49 ± 1.02#	2.16 ± 1.32
HOMA beta-cell function						
16189 wild type	89.2 ± 31.9	84.5 ± 38.5	79.7 ± 29.9	95.9 ± 41.9	87.6 ± 35.8	92.8 ± 28.3
16189 variant	107.3 ± 30.4	100.2 ± 28.4	106.9 ± 30.6 *	113.1 ± 34.1	107.8 ± 31.2	89.6 ± 39.1

HOMA, homeostasis modal assessment; IR, insulin resistance.

* $P < 0.01$, in comparison between 16189 wild type and 16189 variant; # $P < 0.05$, in comparison between 16189 wild type and 16189 variant.

former. There was a higher level of insulin resistance in the infarction group (see TABLE 2). Moreover, higher levels of fasting blood insulin and insulin resistance were also found in subjects harboring the 16189 variant in both the infarction and non-infarction groups. However, only the differences within the infarction groups were statistically significant ($P < 0.05$). Higher levels of fasting blood insulin and insulin resistance were also found in patients of the partial infarction subgroup harboring the 16189 variant. The difference was particularly significant for the patients in the lacunar infarction subgroup ($P < 0.01$).

DISCUSSION

In the present study, we observed that the incidence of the mtDNA 16189 variant in patients with cerebral infarction is higher than that in the subjects without cerebral infarction. However, the difference showed only borderline statistical significance. Upon further analysis of the occurrence rate of the mtDNA 16189 variant in the various subgroups of stroke, we found a higher incidence in the subgroup of patients with small vessel disease than in the other subgroups of stroke patients. The difference in the occurrence rate of the mtDNA variant is significant compared to the occurrence rate in the small vessel subgroup and the normal control group. This result suggests that the increased incidence of 16189 variant in patients with cerebral infarction is confined to the small vessel subgroup. It is concluded from the data in TABLE 2 that the incidence of 16189 variant is significantly higher than that of the small vessel group and less significantly for the other groups. The result also suggests that the mtDNA 16189 variant exerts its influence on vascular pathology more predominantly in small vessel disease. The mtDNA 16189 variant has been reported as an additional candidate gene, other than nuclear genes, for the determination of insulin resistance in humans and for probable development of DM in adult life.[21] This suggestion was based on an observation of a higher occurrence rate of 16189 variant in diabetic patients, an increased fasting blood insulin level as well as insulin resistance in normal adults harboring the variant.[22] The higher occurrence rate of mtDNA 16189 variant in the Taiwanese people as well as other east Asian peoples was shown in previous studies.[15–17] We think the 16189 T to C incidence among Asians and other ethnic groups could have some relation to the relatively higher rate of small vessel disease in this area. Further study is required to clarify this issue in the future.

Increased insulin resistance has been observed in patients with cerebral infarction but without DM.[23,24] This observation, combined with an estimation of increased odds ratio between 1.6 and 2.6 in many epidemiological studies, suggests that the degree of insulin resistance is a risk factor for cerebral infarction.[25–28] The pathogenesis linking increased insulin resistance and risk for stroke remains unknown, but increasing evidence of the pathological effects of insulin on blood vessels offers a plausible explanation. It was also suggested that compensatory hyperinsulinemia due to increased insulin resistance is an important initiator.[24–26] However, the relation between insulin resistance and different subtypes of stroke has rarely been described.[29] Pathogenic differences between different subtypes of stroke may suggest different molecular mechanisms are involved. In the present study, we observed a significant relation between the presence of the mtDNA 16189 variant and lacunar cerebral infarction. The fasting blood insulin levels are significantly higher in the patients harboring the mtDNA 16189 np variant in small vessel subgroup. These results suggest that the mtDNA 16189 variant influences the pathophysiology of human cerebral vascular disease involved with small vessels.

The first hypervariable segment in the control region of the human mtDNA contains a homopolymeric tract of cytosines between 16184 and 16193 and is interrupted at np 16189 by a thymine. This variant is commonly found in the general population, resulting in an uninterrupted homopolymeric C tract.[30] The consequence of the T to C transition remains unknown, but it is thought that this change may affect the transcription and replication of the mitochondrial genome.[31] From this study, it seems this consequence may also be involved in the development of cerebral infarction with a similar underlying pathogenesis of increased insulin resistance. The pathology is more pronounced in stroke patients of the small vessel subtype. Data collected from this and other research groups suggest that the mtDNA 16189 variant is a common genotype for the development of insulin resistance. It may lead to various phenotypic expressions in adult life such as the development of DM and different vascular pathologies involved in cerebral infarction and cardiovascular disease. However, the 16189 variant may simply be a functionless marker of the mitochondrial genome that co-segregates with the DM pathologies. It is also possible that there are some other genetic markers in the nucleus that co-segregate with the 16189 variant in DM patients. Therefore, elucidating the mechanism underlying the pathological implication of mtDNA sequence variation at np 16189 warrants further investigations.

ACKNOWLEDGMENTS

This work was supported by research grants from the National Science Council, Executive Yuan, Republic of China (NSC-90-2314-B-182A-062; NSC-90-2314-B-182A-079; NSC-90-2320-B010-079; and NSC-91-2314-B-182A-040).

REFERENCES

1. UEDA, K., T. OMAE, Y. HIROTA, et al. 1981. Decreasing trend in incidence and mortality from stroke in Hisayama residents, Japan. Stroke **12:** 154–160.
2. TUOMILEHTO, J., R. BONITA, A. STEWART, et al. 1991. Hypertension, cigarette smoking, and the decline in stroke incidence in eastern Finland. Stroke **22:** 7–11.
3. BROWN, R.D., J.P. WHISNANT, J.D. SICKS, et al. 1996. Stroke incidence, prevalence and survival: secular trends in Rochester, Minnesota, through 1989. Stroke **27:** 373–380.
4. WOLF, P.A., R.B. D'AGOSTINO, M.A. O'NEAL, et al. 1992. Secular trends in stroke incidence and mortality. The Framingham Study. Stroke **23:** 1551–1555.
5. GILLUM, R.F. & C.T. SEMPOS. 1997. The end of the long-term decline in stroke mortality in the United States? Stroke **28:** 1527–1529.
6. KERNAN, W.N., S.E. INZUCCHI, C.M. VISCOLI, et al. 2002. Insulin resistance and risk for stroke. Neurology **59:** 809–815.
7. MATHER, K., T.J. ANDERSON & S. VERMA. 2001. Insulin action in the vasculature: physiology and pathophysiology. J. Vasc. Res. **38:** 415–422.
8. HU, R.M., E.R. LEVIN, A. PEDRAM & H.J.L. FRANK. 1993. Insulin stimulates production and secretion of endothelin from bovine endothelial cells. Diabetes **42:** 351–358.
9. SCHERRER, U., D. RANDIN, P. VOLLENWEIDER, et al. 1994. Nitric oxide release accounts for insulin's vascular effects in humans. J. Clin. Invest. **94:** 2511–2515.

10. BARON, A.D. & M.J. QUON. 1999. Insulin action and endothelial function. *In* Insulin Resistance—The Metabolic Syndrome X. 1st edit. G.M. Reaven & A. Lews, Eds.: 247–260. Humana Press. Totowa, NJ.
11. KAY, R., J. WOO, L. KREEL, et al. 1992. Stroke subtypes among Chinese living in Hong Kong: the Shatin Stroke Registry. Neurology **42:** 985–987.
12. YIP, P.K., J.S. JENG, T.K. LEE, et al. 1997. Subtypes of ischemic stroke: a hospital-based stroke registry in Taiwan (SCAN-IV). Stroke **28:** 2507–2512.
13. TANIZAKI, Y., Y. KIYOHARA, I. KATO, et al. 2000. Incidence and risk factors for subtypes of cerebral infarction in a general population: the Hisayama study. Stroke **31:** 2616–2622.
14. LIOU, C.W., P.W. WANG, T.L. CHEN, et al. 2002. The relationship of mitochondrial DNA 16189 polymorphism to the pathogenesis of type 2 diabetes mellitus complicated by cerebral infarction. 4th Symposium on Molecular Diabetology in Asia, Abstr. A42. Shanghai.
15. HORAI, S. & K. HAYASAKA. 1990. Intraspecific nucleotide sequence differences in the major noncoding region of human mitochondrial DNA. Am. J. Hum. Genet. **46:** 828–842.
16. JI, L., L. GAO & X. HAN. 2001. Association of 16189 variant (T→C transition) of mitochondrial DNA with genetic predisposition to type 2 diabetes in Chinese populations. Zhonghua Yi Xue Za Zhi **81:** 711–714.
17. KIM, J.H., K.S. PARK, Y.M. CHO, et al. 2002. The prevalence of the mitochondrial DNA 16189 variant in non-diabetic Korean adults and its association with higher fasting glucose and body mass index. Diabet. Med. **19:** 681–684.
18. ADAMS, H.P., B.H. BENDIXEN, L.J. KAPPELLE, et al. 1993. Classification of subtype of acute ischemic stroke. Definitions for use in a multicenter clinical trial. Stroke **24:** 35–41.
19. SHIH, K.D., T.C. YEN, C.Y. PANG & Y.H. WEI. 1991. Mitochondrial DNA mutation in a Chinese family with myoclonic epilepsy and ragged-red fiber disease (MERRF). Biochem. Biophys. Res. Commun. **174:** 1109–1116.
20. MATTHEWS, D.R, J.P. HOSKER, A.S. RUDENSKI, et al. 1985. Homeostasis model assessment: insulin resistance and beta-cell function from fasting plasma glucose and insulin concentrations in man. Diabetologia **28:** 412–419.
21. POULTON, J. 1998. Does a common mitochondrial DNA polymorphism underlie susceptibility to diabetes and the thrifty genotype? Trends Genet. **14:** 387–389.
22. POULTON, J., M. SCOTT BROWN, A. COOPER, et al. 1998. A common mitochondrial DNA variant is associated with insulin resistance in adult life. Diabetologia **41:** 54–58.
23. GERTLER, M.M., H.E. LEETMA, R.J. KOUTROUBY & E.D. JOHNSON. 1975. The assessment of insulin, glucose and lipids in ischemic thrombotic cerebrovascular disease. Stroke **6:** 77–84.
24. PYORALA, M., H. MIETTINEN, M. LAAKSO & K. PYORALA. 1998. Hyperinsulinemia and risk of stroke in healthy middle-aged men. The 22 year follow-up results of the Helsinki Policeman Study. Stroke **29:** 1860–1866.
25. FOLSOM, A.R., M.L. RASMUSSEN, L.E. CHAMBLESS, et al. 1999. Prospective associations of fasting insulin, body fat distribution, and diabetes with risk of ischemic stroke. Diabetes Care **22:** 1077–1083.
26. WANNAMETHEE, S.G., I.J. PERRY & A.G. SHAPER. 1999. Nonfasting serum glucose and insulin concentrations and the risk of stroke. Stroke **30:** 1780–1786.
27. LINDAHL, B., B. DINESEN, M. ELIASSON, et al. 2000. High proinsulin levels precede first-ever stroke in a nondiabetic population. Stroke **31:** 2936–2941.
28. LAKKA, H.M., T.A. LAKKA, J. TUOMILEHTO, et al. 2000. Hyperinsulinemia and the risk of cardiovascular death and acute coronary and cerebrovascular events in men. Arch. Intern. Med. **160:** 1160–1168.
29. SHINOZAKI, K., H. NARITOMI, T. SHIMIZU, et al. 1996. Role of insulin resistance associated with compensatory hyperinsulinemia in ischemic stroke. Stroke **27:** 37–43.
30. BENDALL, K.E. & B.C. SYKES. 1995. Length heteroplasmy in the first hypervariable segment of the human mtDNA control region. Am. J. Hum. Genet. **57:** 248–256.

31. MARCHINGTON, D.R., J. POULTON, A. SELLAR & I.J. HOLT. 1996. Do sequence variants in the major non-coding region of the mitochondrial genome influence mitochondrial mutations associated with disease. Hum. Mol. Genet. **5:** 473–479.
32. POULTON, J., J. LUAN, V. MACAULAY, *et al.* 2002. Type 2 diabetes is associated with a common mitochondrial variant: evidence from a population-based case-control study. Hum. Mol. Genet. **11:**1581–1583.

Mechanisms of Cell Death Induced by Cadmium and Arsenic

SHIRO JIMI, MASANOBU UCHIYAMA, AYA TAKAKI, JYUNJI SUZUMIYA, AND SYUJI HARA

Central Laboratory for Pathology and Morphology, Medical Informatics and Research Unit, First Department of Internal Medicine, Fukuoka University, 7-45-1 Nanakuma, Jonanku, Fukuoka 814-0180, Japan

ABSTRACT: Cadmium (Cd) and arsenic (As) are known toxic metals in humans. As trioxide (As_2O_3) has been recently used as a mitochondria-targeting drug in acute promyelocytic leukemia. In the present study, we examined the intracellular action of these metals using rat kidney tubular cells and cells tolerant to the metals. The cells were cultured with $CdCl_2$ (1–10 μM) or As_2O_3 (1–2.5 μM). Cells tolerant to Cd and As (Cd-T and As-T, respectively) were defined as cells that survived at toxic concentrations of each metal. Both Cd and As induced cell toxicity in a dose-dependent fashion, which was accompanied by fragmented DNA and decreased mitochondrial membrane potential. Intracellular glutathione (GSH) increased with the increase of Cd and As concentration. In Cd-T and As-T cells, GSH levels were twice those observed in normal cells. When each metal-tolerant culture was exposed to the other different metal, i.e., As or Cd, the protective property was maintained. However, when buthionine sulfoximine (BSO) was added to the metal-tolerant cultures, apoptosis was restored in both Cd-T and As-T. Our results indicate that (1) although GSH is increased in NRK52E by the addition of Cd and As, mitochondria-mediated apoptosis can be still induced, (2) the protective property against metal-induced cytotoxicity is identical in Cd-T and As-T cultures, and (3) although GSH was higher in the metal-tolerant cell lines, depression of GSH by BSO induced apoptosis. We conclude that Cd- and As-induced apoptosis is mediated by an identical mechanism involving intracellular GSH reactive oxidation.

KEYWORDS: cadmium; arsenic; apoptosis; oxidation

INTRODUCTION

Various metals are known to induce toxic effects in humans. Chronic poisoning by cadmium (Cd) is known as Itai-Itai disease in Japan, and we previously found that its target organelles are the mitochondria in the kidney.[1,2] In some Asian countries, arsenic (As) is also known to induce severe toxicity and to exhibit carcinogenicity.[3] Conversely, As trioxide has recently been utilized as a mitochondria-targeting drug

Address for correspondence: Shiro Jimi, Ph.D., Central Laboratory for Pathology and Morphology, School of Medicine, Fukuoka University, 7-45-1 Nanakuma, Jonanku, Fukuoka 814-0180, Japan. Voice: +81-92-801-1011 ext. 3562; fax: +81-92-874-9029.
sjimi@fukuoka-u.ac.jp

for acute promyelocytic leukemia, and beneficial effects have been reported in the United States[4] and China.[5] Therefore, the main intracellular action of Cd and As is presumed to involve a unique cell death pathway, which may involve non-membrane-mediated signal transduction or a novel pathway.

Our group at the Fukuoka University Hospital in Japan is currently conducting a clinical trial using As. Patients with leukemia who acquired tolerance to antineoplastic agents have been treated with As. However, it remains unclear as to why As can selectively induce cell death in neoplastic cells but in not normal cells *in vivo*. The mechanism of action of As in cells is also unclear. The present study was designed to clarify the intracellular actions of As and Cd. For this purpose, we compared the effects of these metals on normal kidney tubular epithelial cells and on cells tolerant to high doses of these metals.

MATERIALS AND METHODS

Normal kidney tubular epithelial cells (TEC: NRK52E Rat, ATCC, US) and their metal-tolerant counterparts were cultured in 5% fetal bovine serum (FBS)-containing Dulbecco's modified Eagle's medium (DMEM) with Cd ($CdCl_2$) at concentrations of 1, 5, and 1.0 µM, or As (As_2O_3) at concentrations of 1 and 2.5 µM. Metal-tolerant cells were prepared by incubation of the NRK52E cells with increasing concentrations of Cd or As for more than 1 and 2 months, respectively. Cd-tolerant cells could survive in a concentration of 10 µM Cd (Cd-10T), and As-tolerant cells could survive in a concentration of 2.5 µM As (As-2.5T). The number of apoptotic cells was assayed by nuclear staining with DAPI (Sigma Chemical Co., St. Louis, MO). Intracellular glutathione (GSH) was measured using a total glutathione quantification kit (Dojin Molecular Technologies Inc., Japan). Mitochondrial mass and membrane potential were determined by MitoTracker Green FM (Molecular Probes, Eugene, OR) and Rhodamine 123 (Molecular Probes), respectively. Immunohistochemistry on monolayer cultures was performed using anti-oxidized phosphatidylcholine (oxPC) antibody (DLH3) for oxidized lipid[6] and anti-metallothionein antibody against MT-1 and MT-2 (Dako, Denmark).

RESULTS

Cytotoxic Effects of Cadmium and Arsenic

When Cd and As were added to cultured cells, a cytotoxic effect was observed after 48 h of incubation at concentrations of more than 10 µM Cd and 2.5 µM As, respectively (FIG. 1A). Morphologically, condensation of core chromatin and nuclear fragmentation were noted in the cultures. DNA extracted from the cultures was fragmented into various sizes and showed a ladder pattern on agarose gel electrophoresis (data not shown). The cytotoxic effects of both Cd and As were therefore mediated via the apoptosis cell death pathway.

When Cd or As was added to normal NRK52E cells, the number of apoptotic cells increased in a dose-dependent fashion after 48 h of incubation (FIG. 1A). However, Cd-10T cells and As-2.5T cells acquired an apoptosis-resistant phenotype for each

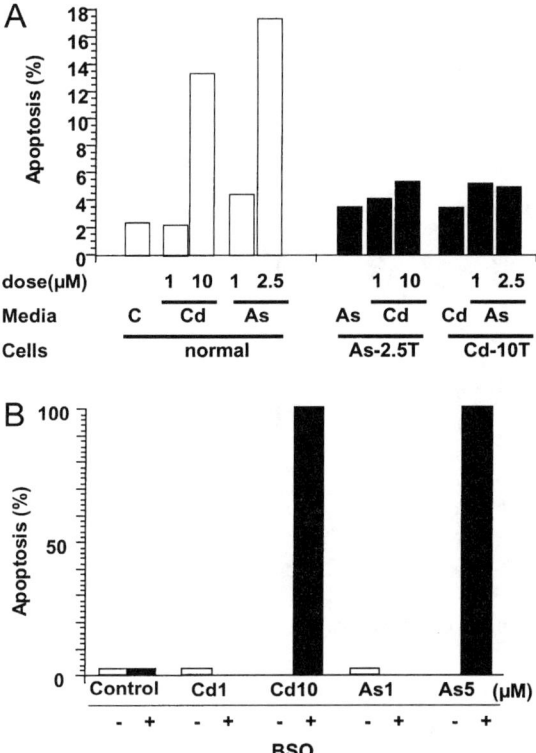

FIGURE 1. Induction of apoptosis by Cd and As in normal NRK52E cells and metal-tolerant cells (**A**), and the effect of GSH inhibition on apoptosis by buthionine sulfoximine (BSO) (**B**). Data are mean of triplicated data.

respective metal (FIG. 1A). When each metal-tolerant culture was exposed to the other metal, i.e., As or Cd, the apoptosis-resistant property was maintained (FIG. 1A).

Mitochondrial Mass and Membrane Potential

Physiological changes in the mitochondria of cells treated with Cd and As for 48 h were examined by flow cytometric analysis after loading with MitoTracker Green FM and Rhodamine 123. The mitochondrial membrane potential decreased dose-dependently for Cd and As, but the mitochondrial mass did not change (FIG. 2).

Metallothionein Expression

Metallothionein was immunohistochemically analyzed in normal NRK52E cells exposed to Cd and in Cd-10T cells, and its expression was found to be higher in the latter cells (Cd-tolerant) than in normal cells exposed to 10 µM Cd (FIG. 3a and b, respectively). However, no immunoreactivity for metallothionein was detected in normal cells exposed to As or in As-2.5T cells (FIG. 3c and d, respectively).

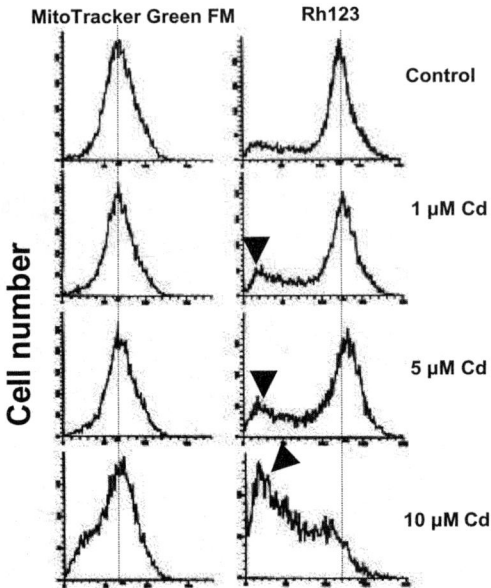

FIGURE 2. Changes of mitochondrial membrane potential and mitochondrial mass induced by Cd.

GSH Levels in Cells Exposed to Cadmium and Arsenic

Enzyme-linked immunosorbent assay (ELISA) was used to measure intracellular total GSH concentration before and after the addition of Cd or As to cultures. The concentration of GSH per cell increased significantly after 48 h of incubation with Cd or As (FIG. 4A), in a dose-dependent fashion. Comparison of intracellular GSH levels between normal and metal-tolerant cells (FIG. 4B) showed that the level was significantly higher in Cd-10T (~2-fold, $P<0.05$) and As-2.5T (~4.5-fold, $P<0.01$) than in normal cells.

Intracellular Oxidation in Cells Exposed to Cadmium and Arsenic

A lipid peroxidation product (oxPC) was formed intracellularly after the addition of Cd or As (FIG. 3e: control; f: 10 µM Cd; g: 2.5 µM As). There was no significant difference in the extent of oxPC formation between Cd- and As-exposed cells. oxPC was localized in the mitochondria after exposure to either metal. The control cell cultures exhibited only trace levels of oxPC.

Effect of GSH Inhibition on Apoptosis in Normal Cells

The effect of GSH inhibition on apoptosis of NRK52E cells exposed to Cd or As was examined. Buthionine sulfoximine (BSO), a GSH inhibitor, had no effect on apoptosis *per se* (FIG. 1B). However, when BSO was added to the normal cells together with Cd or As, the level of apoptosis was significantly enhanced compared with that

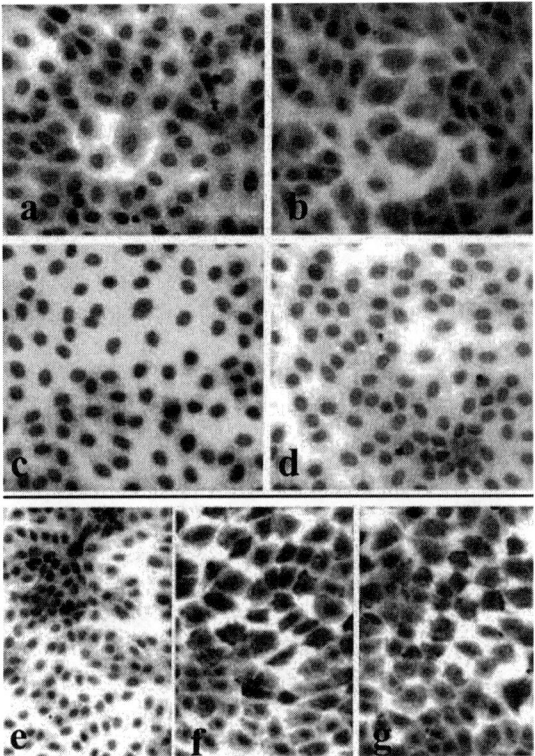

FIGURE 3. Immunohistochemical expression of metallothionein and oxPC. Metallothionein expression in normal NRK52E cells exposed to 10 μM Cd (**a**), and 2.5 μM As (**c**), and 10 μM Cd-tolerant cells (**b**) and 2.5 μM As-tolerant cells (**d**). Intracellular peroxidation in normal NRK52E cells (**e**) and cells exposed to 10 μM Cd (**f**) and 2.5 μM As (**g**).

in cells exposed to each metal alone. GSH inhibition was also analyzed in Cd-10T and As-2.5T cells (FIG. 1B). Addition of BSO strongly restored apoptosis in Cd-10T cells exposed to 1 and 10 μM Cd and in As-2.5T cells exposed to 1 and 2.5 μM As.

DISCUSSION

We have previously shown that chronic Cd exposure induces kidney mitochondria dysfunction in rats[1] and mitochondrial DNA is deleted in the tubular epithelial cells,[2,7] which are already in a post-mitotic state.[7] In the present study, we examined the direct mechanisms of action of metals, including Cd and As, on cell death over a short time period using cultured cells. Exposure of cultured cells to Cd and As resulted in the formation of peroxidation product (oxPC) and a decrease in mitochondrial membrane potential, culminating in apoptotic cell death. Accumulation of oxidized products was also noted in the proximal tubular epithelial cells of Cd-exposed

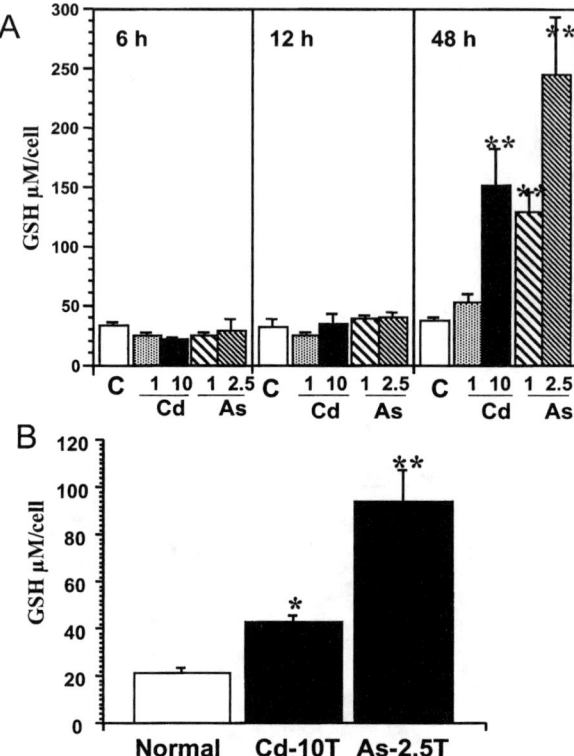

FIGURE 4. Intracellular glutathione (GSH) concentrations in NRK52E cells incubated with Cd and As (**panel A**), and in metal-tolerant cells (**panel B**). Data are mean ± SD. *$P<0.05$, **$P<0.01$, compared with normal (C) cells, by Student's t-test.

rats, especially in the mitochondria.[7] These results suggest that the cytotoxic effects of Cd and As are mediated through induction of intramitochondrial membrane oxidation.

On the other hand, GSH, a powerful scavenger of intracellular reactive oxygen species (ROS),[8] also accumulated in response to Cd and As, probably as a response to critical intracellular oxidation. Interestingly, the intracellular concentrations of GSH in metal-tolerant and apoptosis-resistant cells were higher than in normal cells. However, when BSO, a GSH inhibitor, was added to the metal-tolerant cells, most of the cells died via apoptosis. These results indicate that GSH is important for cell survival by scavenging intracellular oxidation products induced by Cd and As.

To investigate whether the mechanisms of cytotoxicity induced by Cd and As are identical, Cd- and As-tolerant cultures were each exposed to the other metal, and the results showed that they maintained the protective property. Therefore, both Cd- and As-induced cytotoxicity appear to be regulated by a similar mechanism.

Taken together with previous findings, our present results indicate that apoptosis induced by Cd and As involves disruption of mitochondrial functions due to GSH-reactive intracellular oxidization.

ACKNOWLEDGMENT

This study was supported by a Grant-in-Aid for Scientific Research from the Ministry of Education, Science and Culture of Japan (No. 13670234).

REFERENCES

1. TAKEBAYASHI, S., S. JIMI, M. SEGAWA, et al. 2000. Cadmium induces osteomalacia mediated by proximal tubular atrophy and disturbances of phosphate reabsorption. A study of 11 autopsies. Pathol. Res. Pract. **196:** 653–663.
2. TAKEBAYASHI, S., S. JIMI, M. SEGAWA, et al. 2003. Mitochondrial DNA deletion in tubules is the result of Itai-Itai disease. Clin. Exp. Nephrol. **7:** 18–26.
3. BAUDOUIN, C., M. CHARVERON, R. TARROUX, et al. 2002. Environmental pollutants and skin cancer. Cell. Biol. Toxicol. **18:** 341–348.
4. SOIGNET, S.L., P. MASLAK, Z.G. WANG, et al. 1998. Complete remission after treatment of acute promyelocytic leukemia with arsenic trioxide. N. Engl. J. Med. **339:** 1341–1348.
5. CHEN, G.Q., X.G. SHI, W. TANG, et al. 1997. Use of arsenic trioxide (As_2O_3) in the treatment of acute promyelocytic leukemia (APL): I. As_2O_3 exerts dose-dependent dual effects on APL cells. Blood **89:** 3345–3353.
6. JIMI, S., N. UESUGI, K. SAKU, et al. 1999. Possible induction of renal dysfunction in patients with lecithin:cholesterol acyltransferase deficiency by oxidized phosphatidylcholine in glomeruli. Arterioscler. Thromb. Vasc. Biol. **19:** 794–801.
7. TAKAKI, A., S. JIMI & H. IWASAKI. 2004. Chronic cadmium exposure-induced expression of senescence-associated enzyme and accumulation of mitochondrial DNA deletion in rats. Ann. N.Y. Acad. Sci. **1011:** this volume.
8. KITO, M., Y. AKAO, N. OHISHI, et al. 2002. Arsenic trioxide-induced apoptosis and its enhancement by buthionine sulfoximine in hepatocellular carcinoma cell lines. Biochem. Biophys. Res. Commun. **291:** 861–867.

Cadmium-Induced Nephropathy in Rats Is Mediated by Expression of Senescence-Associated Beta-Galactosidase and Accumulation of Mitochondrial DNA Deletion

AYA TAKAKI,[a] SHIRO JIMI,[b] MASARU SEGAWA,[b] AND HIROSHI IWASAKI[a]

[a]*Department of Pathology and* [b]*Central Laboratory for Morphology and Pathology, Fukuoka University School of Medicine, Jonanku, Fukuoka 814-0180, Japan*

ABSTRACT: Long-term exposure to cadmium (Cd) induces perturbation of kidney proximal tubular epithelial cells. Mitochondrial dysfunction in renal cortical cells may contribute to the pathogenesis of Cd-induced nephropathy. In this study, we examined the accumulation of mitochondrial DNA (mtDNA) with a large deletion and cellular senescence in the renal cortex. Wistar rats at 8 weeks of age were intraperitoneally injected with 1 mL of 1 mM $CdCl_2$ or saline, 3 times/week for 5, 20, 40, or 80 weeks. Mitochondrial Cd content in the renal cortex was quantified by atomic absorption analysis. Cytochrome *c* oxidase (CCO) and senescence-associated beta-galactosidase (SA-β-gal) activity were determined in renal cortex by enzyme-histochemistry. mtDNA in total DNA extracted from the renal cortex was amplified by PCR, and mtDNA deletions, including 4,834-bp (nt8118-nt12937) deletion, were determined and semiquantified. After 40 weeks of Cd injection, Cd levels in the renal cortex reached a saturation level, and 30% of the level of the whole-cell fraction was found in the mitochondria. CCO activity in the renal cortex, which was predominantly found in proximal tubular cells, decreased after 40 weeks of Cd exposure. Expression of SA-β-gal was detected primarily in the proximal tubular cells and significantly increased after 80 weeks of Cd exposure. After 40 weeks of study, accumulation of 4,834-bp deletion in mtDNA was evident in both groups of rats; however, the amount of the deletion was significantly greater in Cd-treated rats than in control rats. Our results indicate that long-term Cd exposure induced a post-regenerative state of proximal tubular cells, which accelerated accumulation of 4,834-bp mtDNA deletions in the renal cortex, suggesting that Cd may be a senescence acceleration factor for kidney proximal tubular epithelial cells, which results in Cd-induced nephropathy.

KEYWORDS: chronic cadmium intoxication; mitochondrial DNA deletion; senescence

Address for correspondence: Shiro Jimi, Ph.D., Central Laboratory for Pathology and Morphology, School of Medicine, Fukuoka University, 7-45-1 Nanakuma, Jonanku, Fukuoka 814-0180, Japan. Voice: +81-92-801-1011 ext. 3562; fax: +81-92-874-9029.
sjimi@fukuoka-u.ac.jp

INTRODUCTION

Long-term exposure to low-dose cadmium (Cd) induces chronic Cd intoxication. The kidney is one of the main organs that retains Cd at a high level, especially in the renal cortex.[1] When Cd accumulates in the renal cortex, it induces proximal tubular damage, and the mitochondria in proximal tubular cells show morphological abnormalities and dysfunction.[2] However, most previous studies that examined Cd intoxication were performed in acute or subacute settings. Therefore, the exact mechanisms of chronic Cd intoxication are still unknown.

Recently, human mitochondrial DNA (mtDNA) with large-scale deletions has been detected in mitochondrial diseases and various age-associated disorders.[3,4] In general, genetic mtDNA diseases exhibit distinct symptoms during the aging process. Even in normal individuals, mtDNA deletion has been identified in various tissues with aging, especially in post-mitotic tissue.[5–7] The same abnormality has been also described in rats.[8–10] The development of senescence-associated disorders may involve accumulation of mtDNA damage and/or less energy production as a background.

In the present study, we used a rat model of chronic Cd intoxication to investigate mtDNA alteration during long-term exposure to Cd for more than one year and a half. The aim of our study was to test our hypothesis that long-term exposure to Cd is associated with accumulation of mtDNA deletion, which results in Cd-induced nephropathy. Our results showed that long-term Cd exposure resulted in a post-regenerative state of proximal tubular epithelial cells, associated with substantial accumulation of 4,834-bp mtDNA deletion in the renal cortex. This may be the direct cause of proximal tubular dysfunction in Cd-treated rats.

MATERIALS AND METHODS

Experimental Animals

We used 32 female Wistar rats (*Rattus norvgcus*) at 8 weeks of age. They were intraperitoneally injected 3 times/week with 1 mL of 1 mM $CdCl_2$ (Sigma Chemical Co., St. Louis, MO) or saline for 5, 20, 40, or 80 weeks. They had free access to tap water and normal chow (CE2, Nihon Clea Japan Inc., Tokyo) and had a normal night–day rhythm. After the end of study, 8 animals at each time point were anesthetized, and blood was drawn from the left ventricle. Renal tissue samples were excised, and the renal cortex was isolated. For mtDNA analysis, the renal cortex was frozen in liquid nitrogen. For morphological study, 1-mm thick kidney tissue slices were fixed in 5% buffered formalin.

Quantification of Cadmium Content in Cellular Fractions in Renal Cortex

For determination of Cd contents, whole or mitochondrial fractions of renal cortical tissue (500 mg) in 5 mL of 0.25 M sucrose solution were homogenized by a Teflon homogenizer in ice, and washed three times with sucrose solution at $50 \times g$ for 10 min at 4°C to remove tissue fragments. The sediment was used as a whole-cell fraction. Tissue was subsequently suspended in 0.25 M sucrose solution, and centrifuged at 600 × g for 10 min at 4°C to remove the nuclear fraction. The supernatant was then centrifuged at 8,500 × g for 10 min at 4°C, resuspended in 0.25 M sucrose solution, and then centrifuged again at the same gravity. The pellet was obtained and used as the mitochondrial fraction. Electron microscopy showed that more than 99% of this fraction consisted of mitochondria. Cd contents of the whole-cell fraction and the mitochondrial fraction were determined by an atomic absorbance spectrometer (Hitachi, Tokyo) using a Cd standard solution (Wako, Osaka).

Enzyme-Histochemistry of Cytochrome c Oxidase

Light and electron microscopic detection of cytochrome c oxidase (CCO) activity was conducted by the method of Moraczewski and Anderson[11] with some modification. In brief, frozen sections were placed on glass or in 0.1 M phosphate buffer (pH 7.4) with 5% glucose, for light and electron microscopy, respectively. Sections were reacted for 60 min at 37°C with 0.1 mM phosphate buffer (pH 7.4) containing 1.4 mM 3.3′-diaminobentidine, 0.8 mM cytochrome c, 248 mM sucrose, and 0.01% catalase. After reaction, sections were washed three times in phosphate buffer.

Enzyme-Histochemistry of Senescence-Associated β-Galactosidase

The kidney was removed and fixed for 30 min with 2% formaldehyde and frozen in optimal cutting temperature (OCT) compound. Frozen sections were washed, and incubated at 37°C with freshly prepared senescence-associated β-gal (SA-β-gal) stain solution: 1 mg of 5-bromo-4-chloro-3-indolyl β-D-galactoside (X-gal) per mL (stock solution = 20 mg/mL of dimethylformamide), 40 mM citric acid, sodium phosphate, pH 6.0, 5 mM potassium ferrocyanide, 5 mM potassium ferricyanide, 150 mM NaCl, and 2 mM $MgCl_2$. Staining was evident in 2–4 h.[12]

Mitochondrial DNA Preparation and PCR

Renal cortex was isolated and frozen until used. The cortical tissue (100 mg) was homogenized in 3 mL of buffer containing 0.1 M NaCl, 25 mM Tris-HCl (pH 8.0) and 25 mM EDTA. The homogenate was centrifuged at 800 × g for 10 min to precipitate the nuclear fraction. The supernatant was again centrifuged at 7,000 × g for 10 min to yield the mitochondrial fraction. This mitochondrial fraction was digested after addition of 0.5% sodium dodecyl sulfate (SDS) and 20 μg/mL RNase A at 37°C for 2 h. Digests were incubated with 100 μg/mL proteinase K at 60°C for 12 h. Mitochondrial DNA was extracted with phenol and chloroform, then precipitated with 0.3M CH_3COONa plus ethanol and suspended in TE (pH 7.5). mtDNA fragments were amplified in a 50-μL reaction mixture containing 100 ng of mtDNA, 0.2 mM of each of the dNTPs, 1.0 μL (5 units) of AmpliTaq Gold DNA polymerase (Applied Biosystems, Foster City, CA), 0.5 μM of each primer (L4834, nt7825-7844: 5′-TTTCTTCCCAAACCTTTCCT-3′; H4834 nt13117-13099: 5′-AAGCCTGCTAGGATGCTTC-3′), 15 mM Tris-HCl (pH 8.0), 50 mM KCl, 3 mM $MgCl_2$, and 6 μg/mL of albumin. The reactions were carried out with a PCR thermal cycler (Takara Shuzo Co., Tokyo). Following a hot start with 95°C for 12 min, amplification was performed for 30 cycles with 10 s of denaturation at 94°C, 30 s of annealing at 60°C, and 2 min of extension at 72°C; and a final extension was performed for 5 min at

72°C. PCR products were separated in 2% EtBr/agarose gel and visualized by UV transillumination.

Statistical Analysis

All data were expressed as mean ± SEM. Differences between groups were examined for statistical significance using the Student's t-test and one-way analysis of variance (ANOVA). A P value less than 0.05 denoted the presence of a statistically significant difference.

RESULTS AND DISCUSSION

Accumulation of Cadmium in Cellular Fractions in Renal Cortex

Cd accumulation in the nuclear fraction was evident in renal cortical tissues of Cd-treated rats, but the amount was lower during the experimental periods (data not shown). Cd content in the mitochondrial fraction of the renal cortex continued to rise until 40 weeks of treatment, and then it reached a plateau. After 40 weeks, 30% of the Cd content in the whole-cell fraction was distributed in the mitochondria (FIG. 1). However, after 80 weeks of treatment, Cd contents decreased. These results indicate that the mitochondrion is a main organelle that traps Cd in renal tubular epithelial cells.

Enzyme-Histochemistry of CCO Activity

CCO activity, which is a key enzyme for mitochondrial function, was examined by the enzyme-histochemistry technique. Electron microscopy showed localization of the enzyme activity in the cristae of the mitochondria of proximal tubular epithelial cells (data not shown). The cortical activity of CCO in Cd-treated rats was gradually suppressed after 20 weeks of Cd-exposure, and significantly decreased after

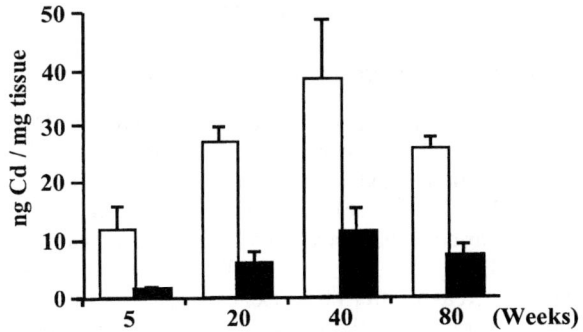

FIGURE 1. Cd contents in whole-cell fractions (*open bars*) and mitochondrial fractions (*closed bars*) in the renal cortex of Cd-treated rats. Data are mean ± SEM in ng/mg tissue.

TABLE 1. Cytochrome *c* oxidase (CCO) activity in the mitochondria of proximal tubular epithelial cells in the control and Cd-treated rats

	Experimental period (weeks)			
	5	20	40	80
Control (% area)	62.3 ± 0.81	59.0 ± 11.1	62.2 ± 3.3	61.9 ± 2.7
Cd (% area)	60.2 ± 6.1	56.8 ± 3.5	35.8 ± 2.9*	41.3 ± 7.5*

Data are mean ± SEM. *$P < 0.05$ compared with the control.

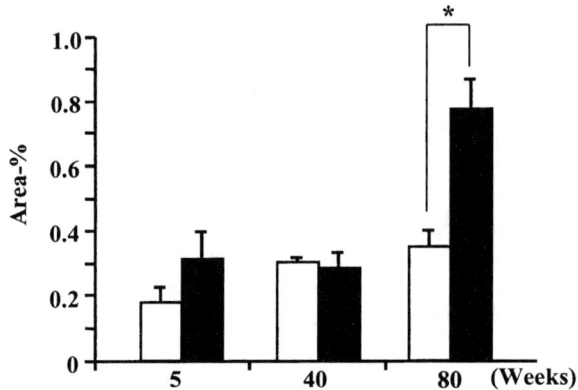

FIGURE 2. Senescence-associated β-galactosidase (SA-β-gal) activity expressed as % area of the renal cortex of control rats (*open bars*) and Cd-treated rats (*closed bars*). Data are means ± SEM; *$P <0.05$.

40 weeks of Cd-exposure ($P <0.05$), to about 60% of the control level (TABLE 1). The parallel decrease of CCO activity and accumulation Cd in the mitochondria (FIG. 1) suggests that Cd may directly affect mitochondrial functions.

Enzyme-Histochemistry of SA-β-Gal Activity

Failure of proliferation by mitotic cells could result in intracellular accumulation of undesirable materials, such as harmful proteins or genes, resulting in cell stress. Accordingly, we focused in the next series of experiments on changes in senescence-associated β-gal (SA-β-gal), a maker of cell senescence.[13] The level of SA-β-gal activity in proximal tubular cells was lower in the control rats than in Cd-treated rats, but increased slightly during the aging process (FIG. 2). On the other hand, SA-β-gal activity significantly increased after 80 weeks of study in Cd-treated rats ($P <0.05$, FIG. 2). These findings imply that the mechanisms responsible for the regeneration of proximal epithelial cells may be damaged in rats treated with Cd for 80 weeks, as well as those involved in other cell functions such as apoptosis, allowing survival of cells with intracellular defects.

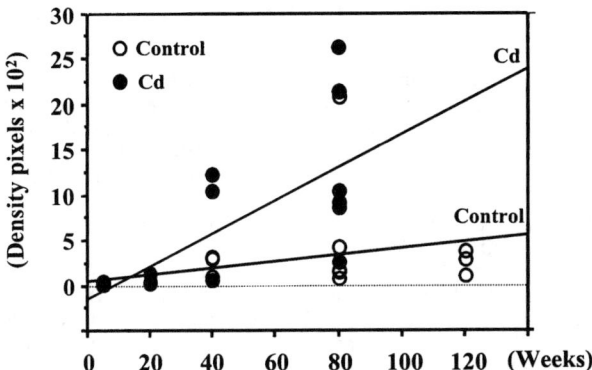

FIGURE 3. Relation between the density of the 4,834-bp deletion band in control rats (*open circles*) and Cd-treated rats (*closed circles*). The amount of the deletion significantly increased ($P < 0.05$) with time in Cd-treated rats, but not in the control rats.

MtDNA Deletion in the Renal Cortex

Finally, we examined the rat renal cortex for age-associated mtDNA deletion, which is 4,834-bp long (located at nt8118-12937), having a 16-bp direct repeat at the edges. In this study, one region of mtDNA, which contained the 4,834-bp deletion, was amplified by the PCR technique. The 4,834-bp deletion shown by a 459-bp product was confirmed by *Ban*II, which cleaved the product into two by digesting at one 16-bp direct repeat. A trace level of deletion was found in both the control and Cd-treated rats until 20 weeks of study. However, after 40 weeks of study, the age-associated mtDNA deletion was detected in the control and Cd-exposed rats (50% of cases were positive in both groups). However, all rats possessed age-associated mtDNA deletion at the end of the 80-week study. When the amount of 4,834-bp mtDNA deletion was quantified by a semiquantitative PCR technique, the mtDNA deletion was increased with long-term Cd treatment (FIG. 3). A significant difference ($P < 0.05$) was found between the control and Cd-treated rats at 80 weeks. Although the deletion was detectable after 40 weeks (48 weeks of age) in control rats, the amount of the deletion slowly increased during the normal aging process. These results indicate that Cd is an accelerator, but not a producer, of the accumulation of age-associated mtDNA deletion.

Considered together, our results indicate that long-term exposure to Cd intoxication could result in renal proximal tubular dysfunction by phenotypic alteration of epithelial cells into a postmitotic and nondegenerative state, which, in turn, provides a chance for accelerated accumulation of age-associated mtDNA deletion.

ACKNOWLEDGMENT

This study was supported by a Grant-in-Aid for Scientific Research from the Japanese Ministry of Education, Science and Culture (No. 13670234).

REFERENCES

1. SVARTENGREN, M., C.G. ELINDER, L. FRIBERG & B. LIND. 1986. Distribution and concentration of cadmium in human kidney. Environ. Res. **39:** 1–7.
2. TAKEBAYASHI, S., S. JIMI, M. SEGAWA & Y. KIYOSHI. 2000. Cadmium induces osteomalacia mediated by proximal tubular atrophy and disturbances of phosphate reabsorption. A study of 11 autopsies. Pathol. Res. Pract. **196:** 653–663.
3. WALLANCE, D.C. 1999. Mitochondrial diseases in man and mouse. Science **283:** 1482–1488.
4. LARSSON, N.G. & D.A. CLAYTON. 1995. Molecular genetic aspect of human mitochondria disorders. Annu. Rev. Genet. **29:** 151–178.
5. CORTOPASSI, G.A. & N. ARNHEIM. 1990. Detection of a specific mitochondrial DNA deletion in tissues of older individuals. Nucleic Acids Res. **18:** 6927–6933.
6. CORRAL-DEBRINSKI, M., T. HORTON, M.T. LOTT, *et al.* 1992. Mitochondrial DNA deletions in human brain: regional variability and increase with advanced age. Nature Genet. **2:** 324–329.
7. ZHANG, C., A. BAUMER, R.J. MAXWELL, *et al.* 1992. Multiple mitochondrial DNA deletions in an elderly human individual. FEBS Lett. **297:** 34–38.
8. GADALETA, M.N., G. RAINALDI, A.M. LEZZA, *et al.* 1992. Mitochondrial DNA copy number and mitochondrial DNA deletion in adult and senescent rats. Mutat. Res. **275:** 181–193.
9. FILSER, N., C. MARGUE & C. RICHTER. 1997. Quantification of wild-type mitochondrial DNA and its 4.8-kb deletion in rat organs. Biochem. Biophys. Res. Commun. **233:** 102–107.
10. YOWE, D.L. & B.N. AMES. 1998. Quantitation of age-related mitochondrial DNA deletions in rat tissues shows that their pattern of accumulation differs from that of humans. Gene **209:** 23–30.
11. MORACZEWSKI, A. & R.C. ANDERSON. 1966. The determination by quantitative histochemistry of the effect of phenothiazines on brain cytochrome c oxidase activity. J. Histochem. Cytochem. **14:** 64–76.
12. CHOI, J., I. SHENDRIK, M. PEACOCKE, *et al.* 2000. Expression of senescence-associated beta-galactosidase in enlarged prostates from men with benign prostatic hyperplasia. Urology **56:** 160–166.
13. DIMRI, G.P., X. LEE, G. BASILE, *et al.* 1995. A biomarker that identifies senescent human cells in culture and in aging skin in vivo. Proc. Natl. Acad. Sci. USA **92:** 9363–9367.

Investigation of Common Mitochondrial Point Mutations in Korea

SEON-JOO KWON,[a] SUNG-SUP PARK,[b] JONG-MIN KIM,[a] TAE-BEOM AHN,[a] SEUNG HYUN KIM,[c] JUHAN KIM,[c] SUNG-HYUN LEE,[d] CHOONG-KUN HA,[e] MOO-YOUNG AHN,[f] AND BEOM S. JEON[a]

[a]*Department of Neurology, Seoul National University Medical Research Center, Clinical Research Institute, Seoul, Korea*

[b]*Departments of Neurology and Clinical Pathology, Seoul National University College of Medicine, Seoul, Korea*

[c]*Department of Neurology, Hanyang University, College of Medicine, Seoul, Korea*

[d]*Department of Neurology, Chungbuk National University, College of Medicine, Cheongju, Korea*

[e]*Department of Neurology, Inha University, College of Medicine, Incheon, Korea*

[f]*Department of Neurology, Sooncheonhyang University Hospital, Seoul, Korea*

ABSTRACT: Between 1997 and 2002, 65 patients with suspected mitochondrial diseases were screened for the mitochondrial point mutations A3243G, T3271C, A8344G, and T8356C. Among these patients, 15 were found to have one of these mutations: 12 with A3243G and 3 with A8344G. The phenotypes of A3243G and A8344G mutations were MELAS and MERRF, respectively. Many asymptomatic family members had the same mutations. In this report, detailed clinical and laboratory findings are presented.

KEYWORDS: mitochondrial disease; point mutation; MELAS; MERRF

INTRODUCTION

Mitochondrial disorders may present various neurologic features including encephalopathy, myopathy, and hearing loss.[1] Neurologic diseases such as mitochondrial myopathy, encephalopathy, lactic acidosis, and stroke-like episodes (MELAS), and myoclonus epilepsy and ragged-red fibers (MERRFs) are good examples. However, mitochondrial disorders show clinical, biochemical, and molecular heterogeneity. The lack of standardized diagnostic criteria poses difficult challenges in making a diagnosis. Walker *et al.*[2] suggested a diagnostic classification for respiratory chain encephalomyopathies, whereas others[3,4] tried to establish a

Address for correspondence: Beom S. Jeon, M.D., Department of Neurology, Seoul National University College of Medicine, 28 Yeongondong, Jongno-gu, Seoul, Korea. Voice: 82-2-760-2876; fax: 83-2-3672-4949.
brain@snu.ac.kr

consensus of diagnostic criteria for mitochondrial disorders by evaluating a comprehensive set of current diagnostic criteria.

Still, clinical heterogeneity and the lack of standardized diagnostic measures pose serious obstacles to clinicians' confirmation or exclusion of these diseases.[5] Laboratory studies include cell redox status and respiratory chain function (e.g., lactate and lactate/pyruvate ratio), numerical or structural abnormalities of mitochondria in tissue biopsies, gene studies, and enzyme histochemistry with its own limitations and pitfalls. Only research laboratories devoted to mitochondrial disorders could complete the studies listed. Looking for mitochondrial DNA point mutation is one of the easier approaches and an effective diagnostic method.[6,7]

As of March 2000, however, 114 mitochondrial mutations had already been described.[8] Therefore, it is not easy to cover all known mutations in a diagnostic laboratory. Among the reported mutations, A3243G and A8344G are the most common pathogenic mutations of mtDNA. They are genetically and phenotypically distinct.[9,10] The A3243G mutation was originally described in a patient with MELAS, which is clinically characterized by recurrent stroke-like episodes, migraine, nausea, and vomiting. However, chronic progressive external ophthalmoplegia, diabetes mellitus, myopathy, cardiomyopathy, and gastrointestinal dysfunction have also been associated with this mutation.[11,12] Approximately 80% of the reported MELAS cases are known to be caused by the A3243G transition and 10% by the T3271C transition. The A8344G mutation was originally reported in MERRF, which is mainly characterized by progressive myoclonus epilepsy, ataxia, and myopathy, as well as optic atrophy, peripheral neuropathy, hearing loss, and dementia.[13] Over 90% of the MERRF cases are due to the A8344G (80%) or the T8356C (15%) mutation. We started looking into the aforementioned common mitochondrial point mutations (A3243G, T3271C, A8344G, and T8356C) to establish a referral diagnostic laboratory for mitochondrial disorders in Korea.

METHODS

Patients and Clinical Phenotypes

Sixty-three blood samples were referred from 1997 to 2002. When a patient was reported to have features of MELAS, we examined the A3243G and the T3271C mutation. If the patient had features of MERRF, mutations at 8344 and 8356 were examined. The medical records of patients found to have the mutations were reviewed. When possible, family members were also studied.

Enzyme Histochemistry

Results of muscle biopsy specimens obtained by open biopsy were available in eight patients. The collected samples were immediately frozen in liquid nitrogen-cooled isopentane. Fresh-frozen blocks of biopsied muscle samples were sectioned at a thickness of 10 μm, processed with modified Gomori-trichrome and succinyl dehydrogenase (SDH), and examined under a light microscope.

TABLE 1. Primers, PCR reactions

	Site	Primers
A3243G	Forward	5 -AGGACAAGAGAAATAAAGGC-3
	Backward	3 -ATGCGTTTCCGGGGTTGCAC-5
T3271C	Forward	5 -AGGACAAGAGAAATAAAGGC-3
	Backward	3 -AATTCCAGTCTCCAAGTTAAGGAGAAGAAT-5
A8344G	Forward	5 -AGCCCACTGTAAAGCTAACT-3
	Backward	3-GGTTGTGCCGAAATGTCACTTTACGGGGTT-5
T8356C	Forward	5-AGCCCACTGTAAAGCTAACT-3
	Backward	3-ATCTCACTTTACGGGGTTGATTTATGATGGCA-5

TABLE 2. Number of requested tests for diagnosis of mitochondrial diseases

	Point mutation	Total tested (n)	Positive results (n)
MELAS	A3243G	52	12
	T3256C	40	0
MERRF	A8344G	27	3
	T8356C	26	0

PCR-RFLP Analysis

DNA was extracted from whole blood. An Eppendorf Mastercycler 5330 machine was used for PCR. The total volume of the PCR reaction was 25.0 L, consisting of 2.5 L of ×10 reaction buffer, 2.0 L of 10 mM dNTPs, 2.5 L of each primer, 0.2 mL of Ex Taq polymerase, 2 L of DNA template, and 14.3 L of distilled water. The primers used are summarized in TABLE 1. Electrophoresis was performed on 4% 1:3 NuSieve agarose gel with the addition of 10 µL of ethidium bromide (EtBr) in 0.5× TAE buffer solution. The 3 mL of PCR-RFLP product was loaded with the 2 µL of loading buffer and run at 100 V for 30 min.

DNA Sequencing

Sequence analyses were performed with a Thermo Sequenase radiolabeled terminator cycle sequencing kit (USB, Amersham, UK). The suspected site of point mutation was confirmed by visual inspection of the electropherogram.

RESULTS

Sixty-three patients were studied. The number of patients tested and the number of positive results are listed in TABLE 2.

Molecular Genetic Study

Fifteen patients were confirmed to have the mtDNA point mutations. Twelve patients had the A3243G mutations and three had the A8344G mutations. A3256G

TABLE 3. Summary of clinical and laboratory findings

	Patients tested (n)	A3243G	Patients tested (n)	A8344G
Number of patients	52	12[a]	27	3
Age of onset (range)	12	20 ± 13 yr (3–51)	3	50 ± 11 yr (37–60)
Sex ratio (M:F)	12	5:1	3	2:1
Clinical phenotypes	12	All MELAS	3	All MERRF
Main initial symptom	12	Stroke	3	Myoclonus
Neurological manifestation (frequency)	12	Stroke (10), seizure (9), dementia (3), myopathy (3), deafness (3)	3	Myoclonus (3), seizure (2), deafness (1), ataxia (1)
Extraneurological manifestation (frequency)	12	Nephropathy (3), cardiomyopathy (2), endocrinopathy (2)[b]	3	None
Histology (frequency)	6	RRF (4), SDH+ (1),[c] nonspecific myopathy (4)	2	RRF(2), SDH+(2)
Mean lactate level (serum:CSF) mmol/L	9	8.6 ± 3.2 : 20.5 ± 10.3	0	No one tested
Molecular study-family	6 members from 4 families	5[d]	6 members from 3 families	4[e]

[a] Four patients (26%) died at the time of the study.
[b] One hypogonadotropic hypogonadism; one hypoparathyroidism.
[c] Succinyl dehydrogenase-positive muscle.
[d] A brother and a mother from one MELAS family and a brother from another MELAS family were symptomatic. Two mothers from two different families were asymptomatic.
[e] Four asymptomatic patients (sister, mother, daughter, and son) from one MERRF family.

and T8356C mutations were not found. DNA sequencing for the point mutations was performed in the available family members. Of the twelve families with A3243G mutation, 6 members from 4 families were studied and 5 from 4 families were identified to have the same mutation. Out of 6 members from the 3 families with A8344G mutations, 4 asymptomatic members from one family showed the same mutation.

Clinical Features

Seven of 15 patients (46%) who were confirmed to have the mutations met major clinical criteria[4]: six of them had MELAS and one had MERRF. The others met minor clinical criteria.[4] The mean age of onset was 20 ± 13 years (range: 3–51 years) for the A3243G mutation group and 50 ± 11 years (range: 37–60 years) for the A8344G mutation group. The male/female ratio was 6.5:1, and four patients (26%) had died. Common neurologic features in the A3243G group were as follows: stroke (10/12); seizure (9/12), dementia (3/12), myopathy (3/12), and deafness (3/12); in the A8344G group: myoclonus (3/3), seizure (2/3), deafness (1/3), and ataxia (1/3). Of these, the most common clinical features were stroke and seizure (8/15 [53%] and

7/15 [46%], respectively). Five patients showed extraneurologic manifestations: two had cardiomyopathy and three had nephropathy. Two of three nephropathic patients had endocrinopathy; two had hypogonadotrophic hypogonadism and the other hypoparathyroidism. All nine patients who were checked for lactate had elevated levels in the serum (mean ± SD: 8.6 ± 3.2 mmol/L) and cerebrospinal fluid (20.5 ± 10.3 mmol/L). Electrophysiologic studies were performed in six patients. Only one showed myopathy, whereas two had mild sensory polyneuropathy. Three patients had clinically affected family members (TABLE 3).

Muscle Biopsy

The results of muscle biopsy were available in eight patients. Four (50%) had ragged-red fibers (RRFs) on modified Gomori-trichrome staining; three met the major histologic criteria (>2% RRF in skeletal muscle) and one satisfied minor criteria (any RRFs if <30 years of age). One patient showed vacuolar myopathy with no RRFs.

DISCUSSION

We found A3243G and A8344G mutations in patients with clinically suspected mitochondrial disorders in Korea. The clinical features of our patients with these mutations were consistent with those of previous reports.

Our data show a low positive rate in confirming the presence of mitochondrial diseases. The main cause of the low positive rate may be the imprecise clinical diagnosis—our data were collected from patients who had been identified by inconsistent clinical criteria of the referring neurologist. Some of the patients referred to us may have had mitochondrial diseases other than MELAS and MERRF. More importantly, we examined only four mutations sites and did not study all four mutations in all samples. The selected sample could be another factor affecting the positive rate. It is well known that patients harboring the A3243G mutation usually have higher levels of mutant mtDNA in muscle than in rapidly dividing tissues such as blood. Therefore, selection of the affected organ or tissues with heavy mutation loads is also important in increasing the detection sensitivity. Southern blotting is another method that can increase the sensitivity in low mutant-loaded individuals.[9,10] Despite all of these limitations, we found that the A3243G and A8344G mutations are the most common mtDNA mutations in Korea, which can be used as a guide to efficiently screen mitochondrial diseases. Currently, we are expanding the list of mutations to be tested. In our study, maternal family members showed a high positive rate (90%), even though they were completely asymptomatic. Considering the heteroplasmic characteristics of mitochondrial disease and threshold effects, the high positive rate is not extraordinary. The high positive rate of asymptomatic family members suggests that the molecular screening of affected family members could be helpful in genetic counseling and prenatal diagnosis.[14,15]

REFERENCES

1. GOTO, Y. 1995. Clinical features of MELAS and mitochondrial DNA mutations. Muscle Nerve **3:** S107–S112.
2. WOLF, N.L. & J.A. SMEITINK. 2002. A mitochondrial disorders: a proposal for consensus diagnostic criteria for infants and children. Neurology **59:** 1402–1405.
3. WALKER, U.A., S. COLLINS, E. BYRNE, et al. 1996. Respiratory chain encephalomyopathies: a diagnostic classification. Eur. Neurol. **36:** 260–267.
4. BERNIER, F.P., A. BONEH, X. DENNETT, et al. 2002. Diagnostic criteria for respiratory chain disorders in adults and children. Neurology **59:** 1406–1411.
5. DIMAURO, S., E. BONILLA, D.C. DE VIVO, et al. 1999. Does the patient have a mitochondrial encephalopathy? J. Child Neurol. **14:** S23–S35.
6. PARRA, D., A. GONZALEZ, C. MUGUETA, et al. 2001. Laboratory approach to mitochondrial diseases. J. Physiol. Biochem. **57:** 267–284.
7. GILLIS, L. & E. KAYE. 2002. Diagnosis and management of mitochondrial diseases. Pediatr. Clin. North Am. **49:** 203–219.
8. WHITE, S. 1998. Molecular mechanisms of mitochondrial disorders. Thesis, Department of Paediatrics. University of Melbourne. Australia.
9. CHINNERY, P.F., N. HOWELL, R.N. LIGHTOWLERS, et al. 1997. Molecular pathology of MELAS and MERRF: the relationship between mutation load and clinical phenotypes. Brain **120:** 1713–1721.
10. KIM, D.S., D.S. JUNG, K.H. PARK, et al. 2001. Histochemical and molecular genetic study of MELAS and MERRF in Korean patients. J. Korean Med. Sci. **17:** 103–112.
11. MORAES, C.T., S. DIMAURO, M. ZEVIANI, et al. 1989. Mitochondrial DNA deletions in progressive external ophthalmoplegia and Kearns-Sayre syndrome. N. Engl. J. Med. **320:** 1293–1299.
12. LAFORET, P., F. ZIEGLER, D. STERNBERG, et al. 2000. MELAS (A3243G) mutation of mitochondrial DNA: a study of the relationships between the clinical phenotype in 19 patients and morphological and molecular data. [In French] Rev. Neurol. (Paris) **156:** 1136–1147.
13. HIRANO, M., G. SILVESTRI, D.M. BLAKE, et al. 1994. Mitochondrial neurogastrointestinal encephalomyopathy (MNGIE): clinical, biochemical and genetic features of an autosomal recessive mitochondrial disorder. Neurology **44:** 721–727.
14. THORBURN, R. 2001. Mitochondrial disorders: genetics, counseling, prenatal diagnosis and reproductive options. Am. J. Med. Genet. **106:** 102–114.
15. DIMAURO, S., K. TANJI, E. BONILLA, et al. 2002. Mitochondrial abnormalities in muscle and other aging cells: classification, causes, and effects. Muscle Nerve **26:** 597–697.

Leber's Hereditary Optic Neuropathy: The Spectrum of Mitochondrial DNA Mutations in Iranian Patients

M. HOUSHMAND,[a] F. SHARIFPANAH,[a] A. TABASI,[b] M.-H. SANATI,[a] M. VAKILIAN,[a] SH. LAVASANI,[a] AND S. JOUGHEHDOUST[a]

[a]*National Research Center for Genetic Engineering and Biotechnology, Tehran, Iran*

[b]*Farabi Eye Hospital, Tehran, Iran*

ABSTRACT: We studied 14 patients with Leber's hereditary optic neuropathy (LHON) to investigate the mtDNA haplotypes associated with the primary mutation(s). Eleven patients carried the mitochondrial DNA (mtDNA) G11778A mutation, while one had the T14484C mutation; one patient had the G3460A mutation and one the G14459A mutation. The Iranian G11778A LHON mutation was not associated with two mtDNA haplogroups—M (0.0% compared with 3.2% in healthy controls) and J (7.7% compared with 10% in healthy controls). Our results showed a similarity in the pattern of LHON primary point mutations between Iranian families with LHON and those of Russian, European, and North American origin. Our results also do not support an association between mtDNA haplogroups J and M with LHON primary point mutations.

KEYWORDS: mitochondria; primary point mutation; Leber's hereditary optic neuropathy; mutation and families

INTRODUCTION

Leber's hereditary optic neuropathy (LHON) is a maternally inherited form of central visual loss that occurs subacutely in young adult men.[1] Experimental support for mitochondrial inheritance was obtained when LHON was first associated conclusively with an inherited mutation in the mitochondrial genome (mtDNA).[2] Wallace *et al.*[4] were the first group to identify a LHON mutation when they showed that a high proportion of LHON families carried a mutation at nucleotide G11778A that results in the substitution of histidine instead of the highly conserved arginine at position of 340 in the ND4 subunit of complex I (NADH-ubiquinone oxidoreductase).[4] The G11778A mutation is found in 50–70% of all LHON pedigrees.[5] Since the study by Wallace *et al.*,[4] hundreds of LHON patients worldwide have been analyzed to identify other LHON mtDNA mutations. There is now a consensus that transitions at nucleotides G3460A, T14484C, and G14459A (ND1: A52T; ND6: M64V; and

---

Address for correspondence: Massoud Houshmand, Ph.D., National Research Center for Genetic Engineering and Biotechnology, 19# Abbass Shafie Alley, Ghods Str. Enghlab Ave., P.O. Box 14155-6343, Tehran, Iran. Voice: +98 21 6415143; fax: +98 21 6419834.

massoudh@nrcgeb.ac.ir

ND6: A72V, respectively) are also pathogenic LHON mutations.[2,6] These four LHON mutations account for >95% of multigenerational LHON pedigrees of northern European descent,[5] and each of these mutations has arisen multiple times within the human population.[7]

Beyond this broad agreement about the G11778A, G3460A, T14484C, and G14459A mutations, it is not yet clear how many other mtDNA mutations may have an etiological or pathogenic role in LHON. Some investigators maintain that there are numerous mutations associated with LHON and that these can have primary, secondary, or intermediate levels of pathogenicity.[8] Thus, more than eight point mutations in the ND6 gene were associated with LHON; these observations suggest that this gene is a hot spot for LHON.[18]

Recently, several reports showed a relationship between diseases and mitochondrial haplogroups. Hofmann et al.[19] concluded that certain European mtDNA haplogroups determine a genetic susceptibility to various disorders. Sudoyo et al.[23] found association between the G11778A LHON mutation and two mtDNA haplogroups—M (47%) and a novel lineage, BM (37%). They suggested that the combination of A10398G and other SNPs, specific for the haplogroups J, M, or BM, might act synergistically to increase the penetrance of the LHON mutations, thus allowing their detection.[24]

The aim of this study was to define the prevalence of a panel of mitochondrial DNA (mtDNA) mutations associated with LHON in Iranian patients with LHON and to find whether there is an association between these mutations and mitochondrial haplogroups J and M.

MATERIALS AND METHODS

Samples were gathered from 14 patients and 22 relatives of six families. All the patients were defined clinically as having LHON on the basis of painless, subacute, bilateral optic neuropathy. All patients were above 20 years of age, except one (13 years of age). mtDNA was extracted from whole blood and assayed for a panel of primary LHON-associated mtDNA mutations (at nucleotide positions G11778A, G3460A, T14484C, and G14459A) by polymerase chain reaction (PCR)-based and RFLP methods as described previously.[17,20–22] The presence of mitochondrial haplogroups J and M were investigated by the PCR-RFLP method in patients with primary point mutations as described previously.[23] These haplogroups were investigated in 169 healthy individuals for haplogroup J and in 218 individuals for haplogroup M.

RESULTS AND DISCUSSION

On the basis of our study, 14 patients were positive for one of four primary LHON mutations, 13 males (93%) and one female (7%). The G11778A mutation was found in all the females (100%) of our patient's' family and no one showed the LHON phenotypes. There is a large excess of affected males, but not all males at risk develop the disease. These observations could be explained by the existence of an X-linked visual loss susceptibility gene. The mechanism of incomplete penetrance and male predominance in LHON remains unclear.[9–12]

One of our patients who had the G11778A mutation showed the LHON phenotype earlier than his brother with the same level of the mutation because of consumption of tobacco and alcohol. Several clinical features of the disease imply that nuclear genes might also be involved in its expression. Cock and his colleagues suggested that the nuclear environment can influence the expression of the biochemical defect in LHON patients with the G3460A mutation.[13]

Brown et al. found the four primary point mutations in most but not all families with LHON in Russia, and suggested that the spectrum of mtDNA mutations associated with LHON in Russia is similar to that in Europe and North America.[14]

Dogulu and his colleagues studied three LHON primary point mutations (G11778A, G3460A, and T14484C) in 32 Turkish probands. Of the 32 probands, 3 carried the T14484C mutation, one carried the G11778A mutation, and one carried the G3460A mutation.[15]

In our study, we detected 78% of patients with the G11778A mutation, 7% with the G3460A mutation, 7% with the T14484C mutation, and 7% with the G14459A mutation. Our results showed the similarity between Iranian families with LHON and Russian, European, and North American families.

There is a significant association between levels of mutant mtDNA and manifestation of the disease phenotype. Because a high proportion of families with the G3460A mutation demonstrate heteroplasmy, this is likely to be a significant factor in disease expression.[16] Thresholds of more than 90% for this mutation were reported earlier by Jun et al.[17] The mother of their patient had more than 70% of the mutated mtDNAs found in the patient's blood.[17] We have found 36% heteroplasmic G14459A in one of our patients. Jun et al. suggested that the G14459A mutation may alter the coenzyme Q–binding site of complex I.[24]

We suggest that there must be an additional pathogenic factor in our patient who showed the LHON phenotype with lower percentage of G14459A heteroplasmy at the earlier of age.

Torroni et al.[25] suggested that one ancient combination of haplogroup J-specific mutations increases both the penetrance of the two primary mutations, G11778A and T14484C, and the risk of disease expression. The results of Brown et al. strongly support a role for haplogroup J in the expression of certain LHON mutations.[26] However, Yen et al. found a high frequency of the G11778A mtDNA mutation in Chinese patients with LHON. No specific multimutation pattern such as the European mtDNA haplogroup J was found by their study.[27]

We investigated 14 patients with LHON and 169 DNA samples from normal healthy controls for haplogroup J and 218 individuals for haplogroup M as normal controls. We found one patient with LHON who carried haplogroup J (7.7%) while 17 of 169 persons in the normal group were positive for haplogroup J (10%) and 7 of 218 persons were positive for haplogroup M (3.2%). Haplogoup M was not detected in any LHON patients. Although haplogroup M is reportedly common in the Asian population, we could find only 7 individuals in our samples, suggesting that this haplogroup is not common in the Iranian population. Different explanations have been proposed for the association between haplogroup J and the LHON mutations G11778A/ND4 and T14484C/ND6, including the unconventional proposal that mtDNA haplogroup J may exert a protective rather than a detrimental effect.[28]

We conclude that there is no association between Iranian LHON primary point mutations and haplogroups J or M.

REFERENCES

1. NIKOSKELAINEN, E., M.-L. SAVONTAUS, O.P. WANNE, *et al.* 1987. Leber's hereditary optic neuroretinopathy, a maternally inherited disease: a geneologic study in four pedigrees. Arch. Ophthalmol. **105:** 665–671.
2. HUOPONEN, K., J. VILKKI, P. AULA, *et al.* 1991. A new mtDNA mutation associated with Leber hereditary optic neuroretinopathy. Am. J. Hum. Genet. **48:** 1147–1153.
3. RIORDAN-EVA, P. 2000. Neuro-ophthalmology of mitochondrial diseases. Curr. Opin. Ophthalmol. **11:** 408–412.
4. WALLACE, D.C., G. SINGH, M.T. LOTT, *et al.* 1988. Mitochondrial DNA mutation associated with Leber's hereditary optic neuropathy. Science **242:** 1427–1430.
5. MACKEY, D.A., R.J. OOSTRA, T. ROSENBERG, *et al.* 1996. Primary pathogenic mtDNA mutations in multigenerational pedigrees with Leber hereditary optic neuropathy. Am. J. Hum. Genet. **59:** 481–485.
6. HOWELL, N. 1997. Leber hereditary optic neuropathy: how do mitochondrial DNA mutations cause degeneration of the optic nerve? J. Bioenerg. Biomembr. **29:** 165–173.
7. BROWN, M.D., A. TORRONI, C.L. RECKORD, *et al.* 1995. Phylogenetic analysis of Leber's hereditary optic neuropathy mitochondrial DNAs indicates multiple independent occurrences of the common mutations. Hum. Mutat. **6:** 311–325.
8. BROWN, M.D. & D.C. WALLACE. 1994. Molecular basis of mitochondrial DNA disease. J. Bioenerg. Biomembr. **26:** 273–289.
9. BU, X.D. & J.I. ROTTER. 1991. X chromosome-linked and mitochondrial gene control of Leber hereditary optic neuropathy: evidence from segregation analysis for dependence on X chromosome inactivation. Proc. Natl. Acad. Sci. USA **15:** 8198–8202.
10. CHALMERS, R.M., M.B. DAVIS, M.G. SWEENEY, *et al.* 1996. Evidence against an X-linked visual loss susceptibility locus in Leber hereditary optic neuropathy. Am. J. Hum. Genet. **59:** 103–108.
11. SWEENEY, M.G., M.B. DAVIS, A. LASHWOOD, *et al.* 1992. Evidence against an X-linked locus close to DXS7 determining visual loss susceptibility in British and Italian families with Leber hereditary optic neuropathy. Am. J. Hum. Genet. **51:** 741–748.
12. OOSTRA, R.J., S. KEMP, P.A. BOLHUIS, *et al.* 1996. No evidence for "skewed" inactivation of the X-chromosome as cause of Leber's hereditary optic neuropathy in female carriers. Hum. Genet. **97:** 500–505.
13. COCK, H.R., S.J. TABRIZI, J.M. COOPER, *et al.* 1998. The influence of nuclear background on the biochemical expression of 3460 Leber's hereditary optic neuropathy. Ann. Neurol. **44:** 187–193.
14. BROWN, M.D., S. ZHADANOV, J.C. ALLEN, *et al.* 2001. Novel mtDNA mutations and oxidative phosphorylation dysfunction in Russian LHON families. Hum. Genet. **109:** 33–39.
15. DOGULU, C.F., H. TOPALOGLU, V. SEYRANTEPE, *et al.* 2001. Mitochondrial DNA analysis in the Turkish Leber's hereditary optic neuropathy population. Eye (Lond.) **15:** 183–188.
16. BLACK, G.C., K. MORTEN, A. LABORDE, *et al.* 1996. Leber's hereditary optic neuropathy: heteroplasmy is likely to be significant in the expression of LHON in families with the 3460 ND1 mutation. Br. J. Ophthalmol. **80:** 915–917.
17. JUN, A.S., M.D. BROWN & D.C. WALLACE. 1994. A mitochondrial DNA mutation at nucleotide pair 14459 of the NADH dehydrogenase subunit 6 genes associated with maternally inherited Leber hereditary optic neuropathy and dystonia. Proc. Natl. Acad. Sci. USA **91:** 6206–6210.
18. LUBERICHS, J., B. LEO-KOTTLER, D. BESCH, *et al.* 2002. A mutational hot spot in the mitochondrial ND6 gene in patients with Leber's hereditary optic neuropathy. Graefes Arch. Clin. Exp. Ophthalmol. **240:** 96–100.

19. HOFMANN, S., M. JAKSCH, R. BEZOLD, et al. 1997. Population genetics and disease susceptibility: characterization of central European haplogroups by mtDNA gene mutations, correlation with D loop variants and association with disease. Hum. Mol. Genet. **6:** 1835–1846.
20. MOSHIMA, Y. 1995. Risk of false-positive molecular genetic diagnosis of Leber hereditary optic neuropathy. Am. J. Ophthalmol. **119:** 245–246.
21. HWANG, Y.M. & H.W. PARK. 1996. Carbon monoxide poisoning an epigenetic factor for Leber hereditary optic neuropathy. Korean J. Ophthalmol. **10:** 122–123.
22. KOBAYASHI, Y., H. SHARPE & N. BROWN. 1994. Single cell analysis of intercellular heteroplasmy of mtDNA in Leber hereditary optic neuropathy. Am. J. Hum. Genet. **55:** 206–209.
23. SUDOYO, H., H. SURYADI, P. LERTRIT, et al. 2002. Asian-specific mtDNA backgrounds associated with the primary G11778A mutation of Leber's hereditary optic neuropathy. J. Hum. Genet. **47:** 594–604.
24. JUN, A.S., I.A. TROUNCE, M.D. BROWN, et al. 1996. Use of transmitochondrial cybrids to assign a complex I defect to the mitochondrial DNA-encoded NADH dehydrogenase subunit 6 gene mutation at nucleotide pair 14459 that causes Leber hereditary optic neuropathy and dystonia. Mol. Cell. Biol. **16:** 771–777.
25. TORRONI, A., M. PETROZZI, L. D'URBANO, et al. 1997. Haplotype and phylogenetic analyses suggest that one European-specific mtDNA background plays a role in the expression of Leber hereditary optic neuropathy by increasing the penetrance of the primary mutations 11778 and 14484. Am. J. Hum. Genet. **60:** 1107–1121.
26. BROWN, M.D., E. STARIKOVSKAYA, O. DERBENEVA, et al. 2002. The role of mtDNA background in disease expression: a new primary LHON mutation associated with Western Eurasian haplogroup J. Hum. Genet. **110:** 130–138.
27. YEN, M.Y., A.G. WANG, W.L. CHANG, et al. 2002. Leber's hereditary optic neuropathy—the spectrum of mitochondrial DNA mutations in Chinese patients. Jpn. J. Ophthalmol. **46:** 45–51.
28. CARELLI, V., L. VERGANI, B. BERNAZZI, et al. 2002. Respiratory function in cybrid cell lines carrying European mtDNA haplogroups: implications for Leber's hereditary optic neuropathy. Biochim. Biophys. Acta **1588:** 7–14.

Index of Contributors

Agress, L.J., 310–316
Ahn, M.-Y., 339–344
Ahn, T.-B., 339–344
Alam, T.I., 61–68
Ando, F., 36–44
Attardi, G., 272–283

Bai, R.-K., 299–303, 304–309
Bai, Y., 272–283
Bernacchia, A., 86–100
Bianchi, C., 86–100
Biondi, A., 86–100
Bovina, C., 86–100

Castelli, G.P., 86–100
Chang, J., 310–316
Chang, K.-C., 317–324
Chen, T.-L., 317–324
Chen, W.-L., 310–316
Cho, B.Y., 205–216
Cho, H., 123–132
Cho, Y.M., 205–216
Choi, Y.S., 69–77
Chomyn, A., 272–283
Chou, W.-P., 154–167
Chuang, Y.-C., 317–324

Dambueva, I.K., 21–35
Deng, J.-H., 272–283
Derenko, M.V., 21–35
DiMauro, S., 217–231, 232–245
Dorzhu, C.M., 21–35

Falasca, A.I., 86–100
Formiggini, G., 86–100
Fuku, N., 7–20
Furuichi, M., 101–111

Genova, M.L., 86–100
Guan, M.-X., 259–271
Guo, L.-J., 7–20
Guo, M.-J., 45–56

Ha, C.-K., 339–344
Hamasaki, N., 61–68
Hara, S., 325–331
Hattori, N., 193–204
Hirano, S., 101–111
Hong, C.-H., 133–145
Houshmand, M., 345–349
Hsu, C.-H., 304–309
Hu, P., 272–283
Huang, F.-M., 317–324

Ichinoe, A., 101–111
Ide, Y., 101–111
Inagaki, H., 78–85
Iwasaki, H., 332–338

Jeon, B.S., 339–344
Jimi, S., 325–331, 332–338
Jin, C.J., 205–216
Jin, Y.-T., 154–167
Jou, M.-J., 45–56, 112–122
Jou, S.-B., 45–56
Joughehdoust, S., 345–349

Kajimoto, Y., 168–176
Kamino, K., 36–44
Kanazawa, A., 78–85
Kaneto, H., 168–176
Kang, D., 61–68
Kanki, T., 61–68
Kashiwagi, A., 78–85
Kawamori, R., 193–204
Kim, J., 339–344
Kim, J.-M., 339–344
Kim, J.-Y., 146–153
Kim, S.H., 339–344
Kim, S.Y., 205–216
Kurata, M., 7–20
Kwon, H., 299–303
Kwon, S.-J., 339–344

Lavasani, Sh., 345–349
Lee, C.-F., 133–145, 317–324
Lee, H.K., ix, 1–6, 205–216

Lee, J.-H., 123–132
Lee, K.-U., 69–77
Lee, M.-S., 146–153
Lee, S.-H., 339–344
Lenaz, G., 86–100
Lin, T.-K., 317–324
Liou, C.-W., 317–324
Liu, C.-Y., 133–145

Maliarchuk, B.A., 21–35
Mancuso, M., 217–231, 232–245

Nagai, Y., 78–85
Naini, A., 232–245
Nakabeppu, Y., 101–111
Nakagawa, Y., 177–184
Nakayama, H., 61–68
Nishio, Y., 78–85
Nomiyama, T., 193–204

Ohno, M., 101–111
Ohsawa, I., 36–44
Ohta, S., 36–44, 193–204

Pak, Y.K., 69–77
Park, D.J., 205–216
Park, H.K., 205–216
Park, J.S., 272–283
Park, K.S., 205–216
Park, S.-S., 339–344
Park, S.Y., 146–153
Peng, T.-I., 45–56, 112–122
Perng, C.-L., 304–309
Piao, L., 193–204
Pich, M.M., 86–100

Rasmussen, A.K., 284–298
Rasmussen, L.J., 284–298
Rychkov, S.Y., 21–35

Sakumi, K., 101–111
Sanati, M.-H., 345–349
Sasaki, N., 61–68
Segawa, M., 332–338
Sharifpanah, F., 345–349

Shieh, D.-B., 154–167
Shimokata, H., 36–44
Shin, C.S., 205–216
Singh, K.K., 284–298
Song, X., 272–283
Suzuki, S., 185–192
Suzumiya, J., 325–331

Tabasi, A., 345–349
Takaki, A., 325–331, 332–338
Takeyasu, T., 7–20
Takio, K., 61–68
Tan, D.-J., 299–303, 310–316
Tan, T.-Y., 317–324
Tanaka, M., 7–20
Tanaka, Y., 193–204
Tay, S., 217–231
Tominaga, Y., 101–111
Tsuchimoto, D., 101–111

Uchino, H., 193–204
Uchiyama, M., 325–331

Vakilian, M., 345–349

Wang, B., 310–316
Watada, H., 193–204
Wei, Y.-H., 133–145, 154–167, 317–324
Wong, L.-J.C., 246–258, 299–303, 304–309, 310–316
Wong, T.-Y., 154–167
Wu, H.-Y., 45–56

Xu, J.-X., 57–60

Yagi, T., 272–283
Yeh, K.-T., 310–316
Yoon, G., 123–132
Yoon, Y.-S., 123–132
Yoshimura, D., 101–111

Zakharov, I.A., 21–35

OHIO UNIVERSITY LIBRARY

Please return this book as soon as you have finished with it. In order to avoid a fine it must be returned by the latest date stamped below. All books are subject to recall after two weeks or immediately if needed for reserve.

DEC 0 2 2004

NOV 1 0 2004

DEC 0 2 2005

CF